VOLUME ONE HUNDRED AND FIVE

SEMICONDUCTORS AND SEMIMETALS

Semiconductor Quantum Science and Technology

SERIES EDITORS

CHENNUPATI JAGADISH

Distinguished Professor
Department of Electronic Materials Engineering
Research School of Physics and Engineering
Australian National University
Canberra, ACT2601, Australia

ZETIAN MI

Professor
Department of Electrical Engineering and Computer Science
University of Michigan
1310 Beal Avenue
Ann Arbor, MI 48109
United States of America

VOLUME ONE HUNDRED AND FIVE

SEMICONDUCTORS AND SEMIMETALS

Semiconductor Quantum Science and Technology

Edited by

STEVEN T. CUNDIFF

Department of Physics; Department of Electrical Engineering and Computer Science, University of Michigan, Ann Arbor, MI, United States

MACKILLO KIRA

Department of Electrical Engineering and Computer Science; Department of Physics, University of Michigan, Ann Arbor, MI, United States

ELSEVIER

ACADEMIC PRESS

An imprint of Elsevier

Academic Press is an imprint of Elsevier
50 Hampshire Street, 5th Floor, Cambridge, MA 02139, United States
525 B Street, Suite 1650, San Diego, CA 92101, United States
The Boulevard, Langford Lane, Kidlington, Oxford OX5 1GB, United Kingdom
125 London Wall, London, EC2Y 5AS, United Kingdom

First edition 2020

Notices
Knowledge and best practice in this field are constantly changing. As new research and experience broaden our understanding, changes in research methods, professional practices, or medical treatment may become necessary.

Practitioners and researchers must always rely on their own experience and knowledge in evaluating and using any information, methods, compounds, or experiments described herein. In using such information or methods they should be mindful of their own safety and the safety of others, including parties for whom they have a professional responsibility.

To the fullest extent of the law, neither the Publisher nor the authors, contributors, or editors, assume any liability for any injury and/or damage to persons or property as a matter of products liability, negligence or otherwise, or from any use or operation of any methods, products, instructions, or ideas contained in the material herein.

ISBN: 978-0-12-823773-1
ISSN: 0080-8784

Publisher: Zoe Kruze
Acquisitions Editor: Jason Mitchell
Editorial Project Manager: Chris Hockaday
Production Project Manager: Abdulla Sait
Cover Designer: Alan Studholme

Typeset by SPi Global, India

Contents

Contributors

Marc Aßmann
Experimentelle Physik 2 Technische Universität Dortmund, Dortmund, Germany

Motoaki Bamba
Department of Physics, Kyoto University, Kyoto; PRESTO, Japan Science and Technology Agency, Saitama, Japan

Andrey Baydin
Department of Electrical and Computer Engineering, Rice University, Houston, TX, United States

Manfred Bayer
Experimentelle Physik 2 Technische Universität Dortmund, Dortmund, Germany

Samir Bounouar
Technische Universität Berlin, Institute of Solid State Physics, Berlin, Germany

Steven T. Cundiff
Department of Physics; Department of Electrical Engineering and Computer Science, University of Michigan, Ann Arbor, MI; JILA, University of Colorado and National Institute of Standards and Technology, Boulder, CO, United States

Marcelo Davanco
National Institute of Standards and Technology, Gaithersburg, MD, United States

Brandon Demory
Department of Electrical Engineering and Computer Science, University of Michigan, Ann Arbor, MI, United States

Hui Deng
Physics Department, University of Michigan, Ann Arbor, MI, United States

Edward Flagg
Department of Physics and Astronomy, West Virginia University, Morgantown, WV, United States

Weilu Gao
Department of Electrical and Computer Engineering, University of Utah, Salt Lake City, UT, United States

Jiaqi Hu
Physics Department, University of Michigan, Ann Arbor, MI, United States

Mackillo Kira
Department of Electrical Engineering and Computer Science; Department of Physics, University of Michigan, Ann Arbor, MI, United States

Junichiro Kono
Department of Physics and Astronomy; Department of Electrical and Computer Engineering; Department of Materials Science and NanoEngineering, Rice University, Houston, TX, United States

Pei-Cheng Ku
Department of Electrical Engineering and Computer Science, University of Michigan, Ann Arbor, MI, United States

Andreas Muller
Department of Physics, University of South Florida, Tampa, FL, United States

Kai Müller
Walter Schottky Institut and Department of Electrical and Computer Engineering, Technische Universität München, Garching, Germany

Nicolas Marquez Peraca
Department of Physics and Astronomy, Rice University, Houston, TX, United States

Stephan Reitzenstein
Technische Universität Berlin, Institute of Solid State Physics, Berlin, Germany

Georgios Roumpos
JILA, University of Colorado and National Institute of Standards and Technology, Boulder, CO, United States

Tuba Sarwar
Department of Electrical Engineering and Computer Science, University of Michigan, Ann Arbor, MI, United States

Rohan Singh
Department of Physics, Indian Institute of Science Education and Research Bhopal, Bhopal, India

Glenn S. Solomon
Joint Quantum Institute, National Institute of Standards and Technology & University of Maryland, College Park, MD, United States

Shuo Sun
JILA and Department of Physics, University of Colorado, Boulder, CO, United States

Takeshi Suzuki
ISSP, The University of Tokyo, Kashiwa, Japan

Chu-Hsiang Teng
Department of Electrical Engineering and Computer Science, University of Michigan, Ann Arbor, MI, United States

Xiaodong Xu
Department of Physics; Department of Materials Science and Engineering, University of Washington, Seattle, WA, United States

Wang Yao
Department of Physics, The University of Hong Kong; HKU-UCAS Joint Institute of Theoretical and Computational Physics at Hong Kong, Pok Fu Lam, Hong Kong, China

Hongyi Yu
Guangdong Provincial Key Laboratory of Quantum Metrology and Sensing & School of Physics and Astronomy, Sun Yat-Sen University (Zhuhai Campus), Zhuhai, China

Long Zhang
Physics Department, University of Michigan, Ann Arbor, MI, United States; Department of Physics, Xiamen University, Xiamen, People's Republic of China

CHAPTER ONE

Toward scalable III-nitride quantum dot structures for quantum photonics

Pei-Cheng Ku*, Tuba Sarwar, Brandon Demory, and Chu-Hsiang Teng
Department of Electrical Engineering and Computer Science, University of Michigan, Ann Arbor, MI, United States
*Corresponding author: e-mail address: peicheng@umich.edu

Contents

1. Introduction

Semiconductor heterostructures are the foundation for modern optoelectronic devices which power the internet, displays, lighting, and many other crucial technologies today (Alferov, 1996; Kroemer, 1996). Semiconductor quantum dots are heterostructures with confinement potential sufficiently strong in all three dimensions that energy quantization becomes non-negligible in every direction (Ledentsov et al., 1998). When an electric charge is spatially confined in such a tiny volume, its electronic structure becomes similar to that of an atom and generates unique optoelectronic properties that are not observable even in other nanoscale heterostructures such as quantum wells. Semiconductor quantum dots have been incorporated into a wide variety of devices including lasers (Arakawa and Sakaki, 1982), displays, light-emitting diodes, solar cells, and sensors. In addition, because of the atomic-like energy structure, quantum dots have

Semiconductors and Semimetals, Volume 105
ISSN 0080-8784
https://doi.org/10.1016/bs.semsem.2020.09.002

been considered an ideal candidate for a future integrated platform of quantum information processing. In this chapter, we will discuss the promises and challenges of using quantum dots to realize an integrated quantum photonic chip, which is likely a critical component for future quantum network with applications ranging from secure communication to interconnecting quantum processors (Kimble, 2008) and from atomic clocks (Kómár et al., 2014) to photonic quantum computing (Brod et al., 2019).

In a quantum photonic chip, quantum dots can be used to generate, detect, and process photonic quantum information. In direct bandgap semiconductors, the photon can strongly interact with the exciton, an electron-hole pair bound by the electrostatic force. In the following, we will focus our discussions on digital quantum information in which the information is encoded in the form of qubits. Unlike in an atom, electrons and holes in semiconductor quantum dots can be independently addressed. Moreover, quantum dots possess a rich electronic structure comprising of multiple valence bands. These unique properties give rise to both opportunities and challenges of using quantum dots to process quantum information. For example, one can prepare a quantum dot with an electron and encode the quantum information using the robust spin degree of freedom of the electron, and then interface it with a photonic qubit via the excitation of a trion in which the spin of the second electron must be opposite to the initial spin qubit due to the Pauli's exclusion principle (Atature, 2006).

The interaction between a quantum dot and a photonic qubit needs to remain coherent during the entire process. Any incoherence degrades the fidelity of the qubit. As a result, quantum dots for a quantum photonic circuit will need to satisfy more stringent requirements than quantum dots used for other classical applications. These requirements include the stability, the radiative efficiency, and the coherence of excitons as well as the determinism of the exciton–photon interaction.

In semiconductors, the stability of excitons is governed by the exciton binding energy which characterizes the Coulomb interaction between the electron and hole. The exciton binding energy in bulk semiconductors is given by

$$E_B = -\frac{\mu}{m_0 \varepsilon_r^2 n^2} R_y$$

where $R_y = 13.6$ eV is the Rydberg energy, μ is the reduced effective mass of the electron and hole, ε_r is the static dielectric constant, and n is the exciton quantum number. For most cases, we operate at the lowest exciton state, i.e., $n=1$. Table 1 compares the exciton binding energies in a few common semiconductors. Wide bandgap semiconductors tend to have more stable excitons. Generally, the operating temperature T needs to be sufficiently low such that kT is only a fraction of the exciton binding energy.

Another important parameter concerning the exciton is its Bohr radius which is given by

$$a_B = \frac{m_0 \varepsilon_r}{\mu} n^2 \times a_H$$

where a_H is the hydrogen Bohr radius (0.53 Å). The Bohr radii of a few common semiconductors are summarized in Table 1 too. Intuitively, weakly bound excitons tend to have smaller radii. When the dimension of a heterostructure is much larger than the exciton Bohr radius, two excitons with the same energies but spatially separated can coexist in the same heterostructure. Hence for unique addressability of a single exciton for quantum information processing, the quantum dot needs to have a dimension comparable or smaller than the exciton Bohr radius. When the quantum dot dimension is smaller than twice the bulk Bohr radius value, the actual Bohr radius of the exciton is compressed due to the spatial confinement. This can increase the exciton binding energy by more than an order of magnitude (Xiao, 1999). Although smaller quantum dots tend to increase the stability of excitons, the interaction strength with the photon can also become weaker. Moreover, there will be a higher demand on the materials quality at the interface between the quantum dot and the matrix. A larger exciton, on the other hand, is more susceptible to decoherence from the environment. For example, a charge nearby can cause the exciton energy

Table 1 Exciton binding energies (E_B) and Bohr radii (a_B) of a few common direct bandgap semiconductors.

	E_B (meV)	a_B (nm)	E_G (eV)
InAs	1.3	37	0.35
InN	6.0	7.9	0.68
InP	6.1	9.4	1.34
GaAs	4.2	12	1.42
ZnO	60	2.3	3.37
GaN	26	2.8	3.4

The bandgap E_G (at 300 K) is included for comparison.

to fluctuate over time, a phenomenon called the spectral wandering or spectral diffusion.

Decoherence is much harder to control. The major exciton decoherence mechanism in quantum dots is from exciton–phonon interactions. The decoherence time scale can be estimated from the linewidth of the zero-phonon line using single-dot spectroscopy. At low temperature when phonon interactions are more suppressed, the exciton decoherence time ranges from a few to hundreds of picoseconds depending on the quantum dot material, geometry (size/shape), and fabrication method (Borri et al., 2001; Holmes et al., 2015). Because the radiative lifetime of excitons is typically on the order of a nanosecond, the fast decoherence can completely destroy the quantum coherence by the time the exciton–photon interaction is complete. To overcome this challenge, one can speed up the radiative process using the Purcell's effect by embedding the quantum dot in an optical cavity such that exciton–photon interaction is completed before the decoherence kicks in. In situations when qubit needs to be stored in a quantum dot beyond the interaction with one photonic qubit, e.g., creating a cluster state comprising of a network of entanglement between photons (Schwartz et al., 2016), the exciton coherence is usually insufficiently long and it becomes necessary to store the qubit in the spin degree of freedom. Over 1 s spin coherence time has been shown in InAs quantum dots using the nuclear spin fluctuation freezing technique (Chow et al., 2016; Sun et al., 2012; Wagner et al., 1999; Xu et al., 2009). The ability to controllably prepare the quantum dot in a trion state is therefore of great importance for many quantum photonic functionalities.

The last but perhaps the most challenging problem is the ability to deterministically couple the exciton to a specific photonic mode for deterministic single photon emission and quantum gates. This requires the engineering of the optical local density of states in the space containing the quantum dot, e.g., with a three-dimensional photonic crystal (Aoki et al., 2008; Barth et al., 2006; Noda, 2000). Significant advances have been made in recent years with >60% photon extraction efficiency reported from quantum dots in micropillar cavities (Somaschi et al., 2016; Wang et al., 2019). The current fabrication process is, however, difficult to scale to a more complex quantum photonic circuit due to the random locations of the dots. In this chapter, we will discuss scalable quantum dot structures in which the properties and their interactions with photons can be tailored by engineering means. We will focus our discussions on epitaxial quantum dot structures. Compared to colloidal or gate-defined quantum dots, epitaxial quantum dots can be more easily integrated with optical cavities, which is essential to mitigate decoherence and enable deterministic coupling with photons. We will briefly review research advances in scaling InAs quantum dot structures but will dedicate most of this chapter on III-nitride based quantum dot structures.

2. Site-controlled quantum dot heterostructures

2.1 Brief review of site-controlled InAs quantum dots

High-quality InAs quantum dots with properties suitable for quantum information processing have been predominantly fabricated using molecular beam epitaxy. With optimizations of the growth process and operating at liquid helium temperature, InAs quantum dots exhibit excellent spin and exciton coherence, near unity radiative efficiency, and have been extensively studied in the context of quantum photonics including single and entangled photon sources as well as quantum gates for photonic qubits (Shields, 2007). InAs dots have been typically grown using the Stranski–Krastanov growth mode by which island-like structures are randomly nucleated when the thickness of the wetting layer exceeds the critical thickness due to lattice mismatch (Ledentsov et al., 1998). The size, shape, and location of the dots are often random. Without proper control of these properties, it will be extremely difficult to realize even a simple quantum photonic circuit incorporating more than a few InAs quantum dots.

There have been many approaches to control the properties of InAs quantum dots such as patterning the substrate with nanoscale pits to force the nucleation of quantum dots at the locations of these pits during the Stranski–Krastanov growth. Although the radiative efficiency of these dots can be very poor due to the nonradiative surface recombinations at the etched interface, the subsequent growth of additional layers of quantum dots can have high quality while still vertically aligned to the first layer of "dark" dots. In addition to the Stranski–Krastanov growth, InAs quantum dots can also been formed by the capillary effect using metal-organic chemical vapor deposition when a thin quantum well is deposited on a nonflat surface such as an inverted pyramid. The thickness (composition) of quantum well near the apex or valley of the nonflat surface is increased, resulting in a three-dimensional confinement potential. InAs quantum dots using capillary effects have shown remarkable homogeneity (e.g., 1 meV inhomogeneous linewidth; Mohan et al., 2010) and integration of these structures in photonic crystal cavities have been reported (Delgoffe et al., 2019; Rigal et al., 2015). More recently, droplet epitaxy emerges as another viable growth method for site-controlled quantum dots (Gurioli et al., 2019). Metallic nanoparticles (droplets) are first deposited on the substrate. These droplets are then exposed to arsenic-rich environment to be converted into III-As quantum dot nanostructures.

Site-controlled InAs quantum dots can also be realized by postgrowth selection. The properties of quantum dots are measured after the growth using photo- or cathodo-luminescence. Dots with desired properties are then selected for subsequent device fabrication. Methods include measurement and selection of dots followed by in situ lithography of photonic structures (Gschrey et al., 2015; He et al., 2017; Nowak et al., 2014) or pick-and-place using micromanipulator (Davanco et al., 2017; Elshaari et al., 2017; Kim et al., 2017; Zadeh et al., 2016).

2.2 Site-controlled III-nitride quantum dots

2.2.1 Strain induced III-nitride quantum dots

Although InAs quantum dots have been the de facto choice for proof-of-concept demonstrations of a variety of quantum photonic functionalities on a semiconductor platform, the small exciton binding energy (~a few meV) demands an extremely low operating temperature. The limited wavelength range (0.9–1.1 μm) also limits certain applications such as free-space or fiber-based quantum networks and interfaces with other qubits such as trap ions or nitrogen vacancy centers in diamonds. Recently, group-III nitride (AlN, GaN, InN) quantum dots have attracted attention from research community due to a much large exciton binding energy (e.g., 26 meV in bulk GaN) and equally mature material growth and device fabrication knowledge base thanks to the developments of short-wavelength laser and light-emitting diodes. As a result, excitons can remain stable at a much higher operating temperature. In quantum dots, the exciton binding energy can be further enhanced. Indeed, single photon emission has been experimentally observed up to 350 K (Deshpande et al., 2014; Holmes et al., 2014, 2016b; Tamariz et al., 2020; Zhang et al., 2013). In the extreme case, the theory has predicted an exciton binding energy up to 1.4 eV in an InN nanowire with a diameter ~1 nm (Bayerl and Kioupakis, 2014).

Unlike InAs quantum dots for which the Stranski–Krastanov growth is the most commonly used fabrication process, the small exciton Bohr radius in III-nitride semiconductors requires the quantum dot dimension to be an order of magnitude smaller. A smaller dot requires a larger lattice mismatch in the wetting layer to nucleate during the Stranski–Krastanov growth. This presents a unique challenge for the growth of III-nitride quantum dots. As a result, much attention has been focused on alternate growth modes including droplet epitaxy (Jarjour et al., 2007), capillary effect (Arita et al., 2017; Cho et al., 2018; Holmes et al., 2013, 2014, 2016b), and dot-in-wire structures (Bhattacharya et al., 2017). These methods in combination with lithography can directly enable site-controlled III-nitride quantum dots with smaller dimensions. However, the control over the quantum dot size and shape is still difficult. The inhomogeneous broadening linewidth of III-nitride quantum dot array is generally much larger than that of InAs quantum dot array.

Another way of creating quantum dots is to generate a lateral confinement potential in a quantum-well structure without physical interfaces. This concept is not new and in fact predated the Stranski–Krastanov approach (Deng et al., 2020). For example, one can create a lateral potential well in a two-dimensional electron gas to form a quantum dot (Fulton and Dolan, 1987). In III-nitride, one can generate the lateral confinement potential by using a nonuniform strain profile. The piezoelectric effect in III-nitride semiconductors then translates the strain profile into the modulation of the energy bandgap. The strain induced confinement potential is not unique to III-nitride and can be observed in general in semiconductors and monolayer transition metal dichalcogenides. But the large piezoelectric effect in III-nitride can effectively amplify the confinement potential. In Fig. 1, the radial confinement of an electron in an InGaN nanodisk embedded in a GaN nanopillar is considered. The nanopillar geometry allows the strain in the InGaN quantum well to be laterally relaxed around the perimeter and results in a lateral confinement potential which confines both electrons and holes toward the center of the nanodisk (Zhang et al., 2014). We also compare the electron wavefunctions (at the top of the nanodisk) with and without the piezoelectric field by artificially turning off the piezoelectric effect (by assuming zinc-blende III-nitride). The results are shown in Fig. 1. It can be observed that the piezoelectric effect can significantly enhance the quantum confinement.

2.2.2 Fabrication of strain induced quantum dots

Strain induced III-nitride quantum dots can be fabricated by either the top-down or bottom-up process. Illustrated in Fig. 2 is a top-down process (Lee and Ku, 2012; Zhang et al., 2016b). A single (or multiple if vertically stacked quantum dots is desired) quantum-well structure is first grown using molecular beam epitaxy or metal-organic chemical vapor deposition. An optional

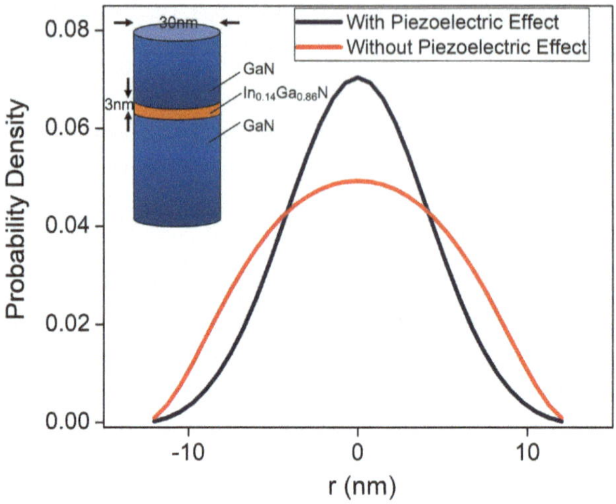

Fig. 1 The effect of the piezoelectric field on the lateral confinement of an electron in an $In_{0.14}Ga_{0.86}N/GaN$ nanodisk with a thickness of 3 nm and a diameter of 30 nm as illustrated in the inset. The nanodisk is embedded in a nanopillar along the crystal [0001] direction. The strain profile was first calculated using the continuum elastic theory to minimize the total strain energy. The strain profile is then used to obtain the eigenenergy and wavefunction (envelope function) of the electron. The process is repeated by assuming zinc-blende InGaN and GaN in which the piezoelectric field along the [0001] is zero. The electron is now confined radially by the nanopillar/air interface. It can be observed that the piezoelectric effect enhances the electron confinement toward the center of the nanodisk.

p–n junction structure can be added for electrical control of the quantum dot. The substrate can be either sapphire, GaN, or silicon. After growth, the sample is patterned by electron-beam lithography to define the size and cross-sectional geometry of the quantum dot. The dimension of the lithographic pattern will need to account for the etch bias. A two-stage dry/wet etching process has been found to produce high-quality quantum dots. First, the quantum well(s) is etched by inductively coupled plasma reactive ion etching (ICP-RIE). The sloped sidewall is then straightened by an anisotropic wet etching process using diluted KOH. Once the sidewall becomes vertical (parallel to the [0001] direction), the etching rate is considerably slowed due to the crystal dependence of the etching. This allows the final quantum-dot dimension to be precisely controlled. If electrical control of dots is not required, one can finish the fabrication by passivating the quantum dot sidewall. If electrical control is needed, the sample is then planarized, e.g., by spin-on-glass (SOG) and etched back before contact electrodes are defined.

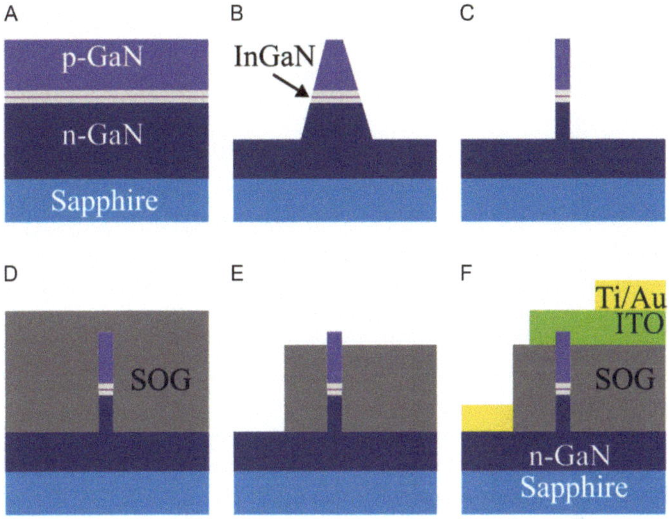

Fig. 2 The illustration of a top-down fabrication process for the strain induced III-nitride quantum dot. (A) The quantum-well structure identical to the vertical stack of the quantum dot(s) is grown by epitaxy. (B) The nanopillar is formed by reactive ion etching. (C) The sidewall is straightened by anisotropic wet etching. (D) The sample is then planarized and (E) etched back to expose the tip of the nanopillar for the subsequent metallization step (F). *Reprinted from Zhang, L., Teng, C., Ku, P., Deng, H., 2016. Site-controlled InGaN/GaN single-photon-emitting diode. Appl. Phys. Lett. 108, 153102. https://doi.org/10.1063/1.4945984, with the permission of AIP Publishing.*

A bottom-up process (Deshpande et al., 2013) forms the nanopillar using a catalyst process and the nanodisk is grown directly during the nanopillar growth. The bottom-up process requires molecular beam epitaxy as metal-organic chemical vapor deposition tends to form core-shell structure when heterostructures are grown in a nanowire. The bottom-up process can be more easily extended to quantum dots with a large lattice mismatch. However, the thickness and compositional uniformity in a bottom-up strain induced quantum dot is more difficult to achieve.

2.2.3 Effective quantum dot dimension

The strain-induce lateral confinement potential can enable the formation of quantum dots with an effective dimension much smaller than the physical diameter of the nanopillar. Fig. 3 shows the effective quantum dot dimension versus the actual nanopillar diameter. The effective quantum dot is assumed to be a nanodisk of the same thickness and composition bounded by an infinite potential barrier (Robinett, 1996). We look for the diameter of

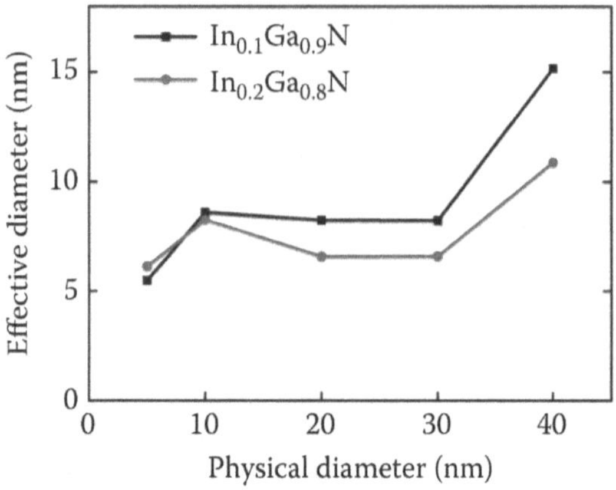

Fig. 3 The effective quantum-dot dimension versus the physical diameter of an InGaN nanodisk embedded in a GaN nanopillar as illustrated in the inset of Fig. 1. The strain induced confinement potential spatially confines the electron and hole toward the center of the nanodisk, enabling the quantum dot to be formed at a dimension much larger than the size of the exciton. *Reproduced with permission of The Licensor through PLSclear. Ku, P.-C., Teng, C.-H., Deng, H., 2017. III-nitride semiconductor single photon sources. in: Bi, W., Kuo, H., Ku, P.-C., Shen, B. (Eds.), Handbook of GaN Semiconductor Materials and Devices. Taylor & Francis, 661–669.*

the effective quantum disk at which the quantum confinement for the electron and hole becomes the same as that in the actual InGaN nanodisk. For an $In_{0.2}Ga_{0.8}N$/GaN nanodisk with a diameter up to 30 nm, the effective quantum-dot dimension 8 nm is comparable to the diameter of the exciton, ~6 nm. Fig. 3 suggests that the strain induced confinement potential effectively creates a strong quantum confinement to spatially trap the exciton at the center of the nanodisk. Single photon emission has been observed up to 90 K from such a structure (Zhang et al., 2013).

2.2.4 Charge-tunable III-nitride quantum dots

When quantum dots are used to process quantum information, the robust spin coherence is far more superior to exciton coherence. Coherence of electron spin in III-nitride semiconductors has been shown to be robust and can persist beyond room temperature (Buß et al., 2010, 2014). To interface with a photonic qubit, we can create a trion. Due to the Pauli's exclusion principle, the two electrons in a trion must have opposite spins.

To create a trion, i.e., an exciton plus an additional charge, a p–n junction can be added to the nanopillar structure. As shown in Fig. 4, the application of a voltage across the InGaN quantum dot can change the band

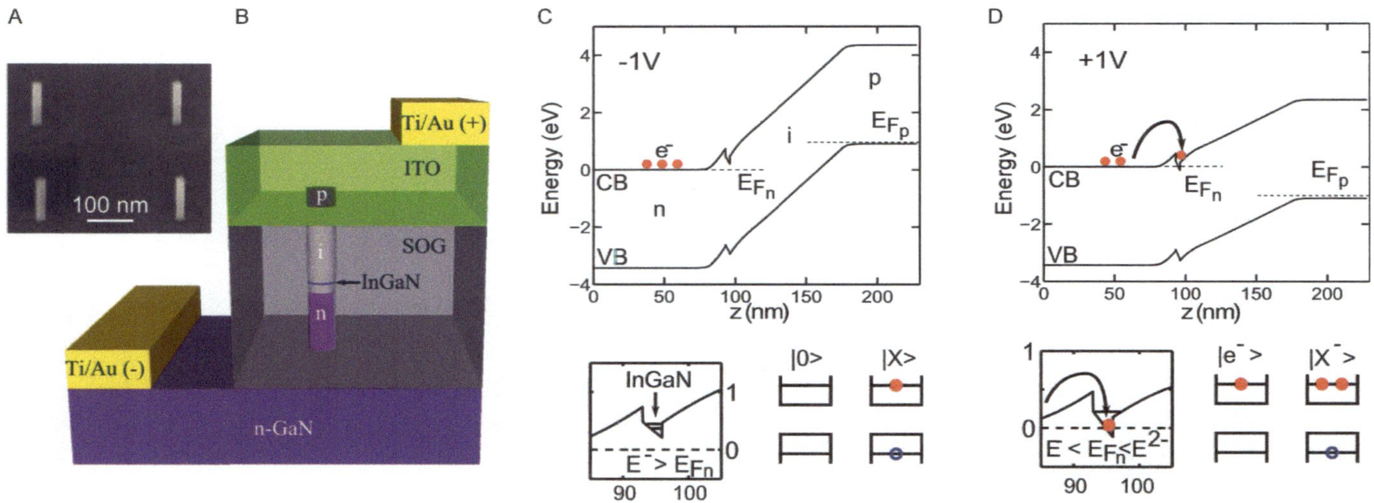

Fig. 4 (A) The scanning electron micrograph and (B) the schematic of a strain induced $In_{0.15}Ga_{0.85}N$/GaN quantum dot biased with a voltage source. The thickness and diameter of the nanodisk are 3 and 30 nm, respectively. (C) and (D) The band diagram and the illustration of the injection of an electron into the quantum dot and the subsequent excitation into the trion state. When the voltage is further increased, a second electron can be injected into the quantum dot and a doubly negatively charged trion can be formed. *Reprinted with permission from Zhang, L., Teng, C.-H., Ku, P.-C., Deng, H., 2016. Charge-tunable indium gallium nitride quantum dots. Phys. Rev. B 93 (8), 085301. Copyright (2016) by the American Physical Society.*

diagram and inject one or more electrons in the dot. The number of electrons injected is determined by the voltage. Fig. 5 shows the emission from a neutral exciton and a negatively charged trion from the same InGaN quantum dot biased at different voltages (Zhang et al., 2016a), demonstrating the ability to prepare the III-nitride quantum dot in a trion state.

2.2.5 Quantum dot emission wavelength range

Group-III nitride semiconductors exhibit direct bandgaps across a wide spectral range from 190 nm through 1.9 μm. Fig. 6 shows the wavelength range of various III-nitride nanopillar structures fabricated along the [0001] direction (Chung et al., 2020). Upon the formation of quantum dots with a nanopillar diameter up to a few tens of nanometers, III-nitride quantum dots can cover a wavelength range from 300 nm through 900 nm. Hence, III-nitride quantum dots can potentially provide an ideal platform interfacing with various quantum nodes including neutral atoms, trap ions, and nitrogen vacancies in diamonds.

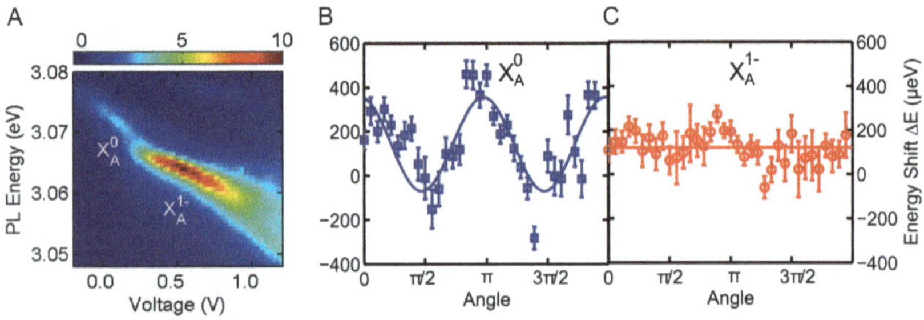

Fig. 5 (A) Voltage-dependent photoluminescence spectra from a single InGaN quantum dot shown in Fig. 3. The two branches correspond to emission from neutral exciton X^0 and negatively charged trion X^{1-}. The subscript A denotes A exciton. The voltage-dependence of the emission energy is due to the quantum-confined Stark effect. The presence of an extra electron also screens the piezoelectric field and leads to a much brighter emission of the trion. (B) Polarization-resolved photoluminescence measurement. Due to unintentional processing variations, the circular InGaN nanodisk can exhibit a slight asymmetry, resulting in a fine-structure splitting, manifested by the slight energy shift between the *x*- and *y*-polarized light. (C) The extra electron in the trion suppresses the fine-structure splitting. The two electrons in the conduction band with opposite spins have a total spin of 0. As a result, the electron–hole Coulomb exchange interaction and thus the fine-structure splitting vanish. *Reprinted with permission from Zhang, L., Teng, C.-H., Ku, P.-C., Deng, H., 2016. Charge-tunable indium gallium nitride quantum dots. Phys. Rev. B 93 (8), 085301. Copyright (2016) by the American Physical Society.*

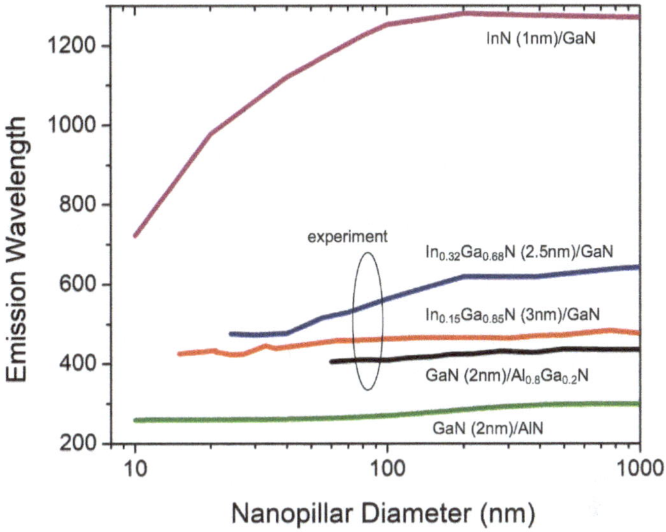

Fig. 6 Strain induced III-nitride quantum dots can cover a wavelength from 300 nm through 900 nm by properly choosing the composition, thickness, diameter, and barrier material. As shown here is a compilation of five different III-nitride nanopillar materials. The thickness is shown in the parenthesis. The data for $In_{0.32}Ga_{0.68}N/GaN$, $In_{0.15}Ga_{0.85}N/GaN$, and $GaN/Al_{0.8}Ga_{0.2}N$ were from experiments (Chung et al., 2020; Zhang et al., 2014).

2.3 Challenges of III-nitride quantum dots

2.3.1 Exciton energy fluctuation and dephasing

To date, the emission linewidth of both III-nitride quantum dot array and indivudal dot has been much broader than that of InAs quantum dots. The two major mechanisms that broadens the linewidth are the strong interactions between the exciton and acoustic phonons, and the exciton energy fluctuation due to nearby charges. The piezoelectric field in III-nitride separates the electrons and holes and creates a permanent dipole moment which then interacts with environmental charges in the vicinity of the dot to shift the exciton energy. The fluctuation of charges then broadens the exciton emission linewidth (Holmes et al., 2015). The spectral diffusion can be directly observed by monitoring the exciton emission wavelength as a function of time (Bardoux et al., 2006; Gao et al., 2019; Ostapenko et al., 2010). The spectral diffusion can be suppressed by applying an external electric field to suppress charge fluctuation (Somaschi et al., 2016) or using the Purcell's effect (Lyasota et al., 2019).

In III-nitride, one can further reduce spectral diffusion by suppressing the permanent dipole moment, e.g., by using an ultrathin quantum dot active region which extends the electron and hole wavefunctions into the barrier and increases their spatial overlap. For example, Fig. 7 shows the estimated permanent dipole moment in an InN/GaN nanopillar structure of different InN thicknesses. By reducing the thickness of an InN/GaN quantum dot from 8 to 2 ML, the permanent dipole moment can be reduced by more than two orders of magnitude. Ultrathin III-nitride may also further enhance the exciton binding energy. For example, it has been shown that monolayer GaN possesses an exciton binding energy exceeding 200 meV (Bayerl and Kioupakis, 2019; Wu et al., 2020).

Once spectral diffusion is mostly suppressed, the transform-limited linewidth of a III-nitride quantum dot can be much narrower. A 27 μeV homogeneous linewidth has been reported at 4 K from a GaN/AlGaN quantum dot formed by well-width fluctuations in a quantum well (Arita et al., 2017). This level of coherence can likely be preserved by embedding the quantum dot in an optical cavity with a moderate quality factor ~10–100 (Demory et al., 2015).

2.3.2 Radiative efficiency

The strong piezoelectric field in III-nitride along the [0001] direction separates the electrons from holes and significantly increases the radiative lifetime. A long radiative lifetime not only reduces the radiative efficiency but also degrades coherence, allowing nonradiative recombination and dephasing to prevail. Radiative lifetime can be modified via the Purcell's effect by embedding the quantum dot inside an optical cavity. High-quality

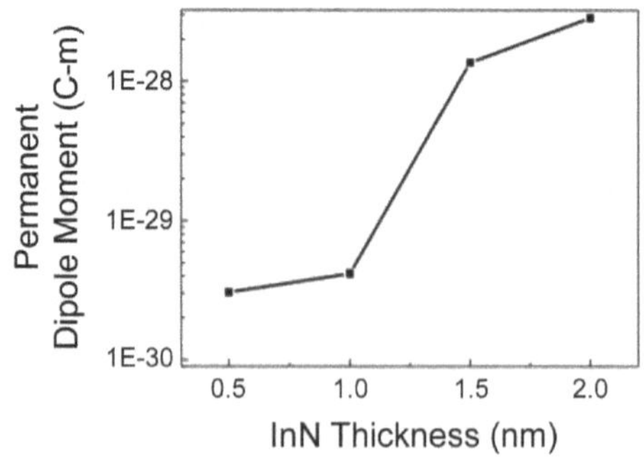

Fig. 7 The permanent dipole moment of the (ground-state) exciton in an InN/GaN nanopillar ($D \gg 30$ nm). The effective-mass approximation was used which may become invalid when the InN thickness is only a few monolayers (ML).

factor III-nitride based optical cavities have been shown (Sergent et al., 2012; Vico Triviño et al., 2012) although not with a deterministic placement of quantum dot. An alternative approach is to use a self-aligned metallic cavity with a site-controlled III-nitride quantum dot embedded (Demory et al., 2015, 2017). Despite a lower Purcell's factor due to metal absorption, a metallic cavity has a broad resonance. Hence the quantum-dot cavity resonance alignment can be greatly simplified.

Fig. 8 shows two different metallic cavities. The metallic/dielectric layers can be directly deposited without the need to align to the quantum dot. The plasmonic resonance of the silver was used to generate the Purcell's effect. The Al_2O_3 spacer was included to reduce the loss although further analysis suggested (Fig. 8C and D) the swapping of silver and Al_2O_3 and the addition of another silver layer can result in a much higher Purcell's factor without degrading the quantum efficiency. Fig. 8B shows the experimental measurements which confirmed that a simple metallic cavity can reduce the lifetime of a strain induced III-nitride quantum dot by an average of 15 times and enhance the intensity by an average of 10.9 times. The quantum dot was confirmed to be a single photon emitter with and without the silver optical cavity. The average Purcell's factor measured was 46.

2.3.3 Fine control of exciton wavelength

When more than one quantum dots are integrated in the same quantum photonic circuit, the exciton energies need to be within the cavity linewidth which needs to be smaller than the homogeneous linewidth. Hence exciton energies from the two quantum dots need to be fine-tuned to a high precision. In strain induced III-nitride quantum dots, the lateral and transverse confinement potentials are determined by the nanodisk diameter and thickness, respectively. Hence any microscopic variation of these parameters can lead to exciton energy fluctuation. We have experimentally measured 30 nominally identical InGaN/GaN quantum dots from the same sample undergoing the same fabrication process (Teng et al., 2019). The results are summarized in Table 2. Both the exciton energies (E) and lifetimes (γ) are shown together with their statistics. The experimental results were also compared to numerical simulations by considering the three possible sources of fluctuations: the random alloy which is inherent to the ternary nature of the InGaN compound, the well-width fluctuation due to growth imperfection, and the inaccurate diameter due to patterning error. In the case of random alloys and well-width fluctuations, the electron and hole wavefunctions are distorted by microscopic fluctuations of the confinement

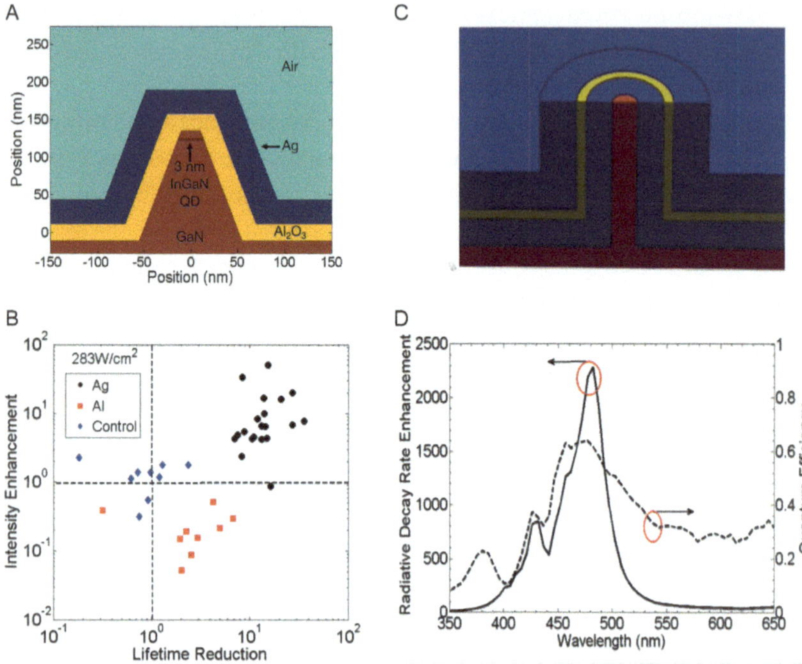

Fig. 8 (A) and (B) Two different metallic optical cavities that can be fabricated around a strain induced III-nitride quantum dot without alignment. (B) Experimentally measured emission intensity and lifetime from a single quantum dot. The measurement was repeated over multiple different quantum dots. The quantum dots are divided into three sections. The control section does not receive any metal or dielectric coating. The Ag (silver) section has a structure shown in (A). The Al (aluminum) section has the same structure as the Ag section except the silver layer is replaced by aluminum. Silver and aluminum have similar reflectivity in the wavelength range of interests. The comparison of the Ag to the Al sections can rule out the change of emission intensity due to the enhanced reflection from the laser excitation. The average Purcell's factor measured was 46 with the maximum over 120. (C) and (D) Another metallic cavity with two silver layers separated by an Al_2O_3 spacer. The calculation shows a Purcell's factor > 2000 and radiative efficiency >0.6 can be achieved. *Panels A and B: Reprinted with permission from Demory, B., Hill, T.A., Teng, C.-H., Zhang, L., Deng, H., Ku, P.-C., 2015. Plasmonic enhancement of single photon emission from a site-controlled quantum dot. ACS Photonics 2 (8), 1065–1070. https://doi.org/10.1021/acsphotonics.5b00086. Copyright (2015) American Chemical Society.*

potential as shown in Fig. 9. The result is a slight change of the exciton Bohr radius and electron-hole wavefunction overlap, which results in change in exciton energy and lifetime.

The results in Table 2 suggests that all three sources contribute significantly to the variation of both the static (exciton energy) and dynamic

Table 2 Statistics of exciton energies (E) and decay rates (γ) obtained from experiments and numerical simulations of 30 individual InGaN quantum dots.

	E (eV)	ΔE (meV)	γ (GHz)	$\Delta\gamma$ (GHz)
Experiment (all data)	2.8893	30.4	0.1462	0.1085
Experiment (three outliers removed)	2.8834	20.8	0.1241	0.036
Theory (diameter variation)	2.8682	10.7	0.0627	0.0123
Theory (RA only)	2.8542	10.8	0.1002	0.0162
Theory (WWF only)	2.8690	8.8	0.0810	0.0072
Theory (total IB)		17.6		0.0216

Both the averages and standard deviations are shown. Numerical calculations separately considered three different inhomogeneous broadening (IB) mechanisms: random alloys (RA), well-width fluctuations (WWF) of 2 ML and diameter variation (1.7 nm from SEM measurements). The total IB was determined using $\Delta = \sqrt{\Delta_{Dia}^2 + \Delta_{RA}^2 + \Delta_{WWF}^2}$.

Reprinted from Teng, C.-H., Zhang, L., Deng, H., Ku, P.-C., 2019. Mechanisms of inhomogeneous broadening in InGaN dot-in-wire structures. J. Appl. Phys. 126 (8), 083104. https://doi.org/10.1063/1.5111343, with the permission of AIP Publishing.

(lifetime) properties of excitons. Hence, additional strategies are needed to control and fine-tune the quantum dot resonance.

Using photoelectrochemical etching, Xiao et al. showed that the inhomogeneous broadening can be reduced by factor of three (Xiao et al., 2014). Because the photoelectrochemical etching can only be triggered when holes are generated by optical excitation, only quantum dots with emission wavelengths longer than the excitation laser will be etched until the emission is blue shifted to be shorter than the laser. Another strategy of fine-tuning the wavelength is by applying a small voltage across the quantum dot. As shown in Fig. 4A, both the exciton and trion energies can be tuned over 10 meV by varying the voltage bias by ~0.5 V. Increasing the voltage range can further increase the tuning range. For example, in Zhang et al. (2016b), 30 meV tuning was reported over a 5 V bias range.

3. Tailoring exciton–photon interactions

There are numerous ways to encode a photonic qubit including photon number, polarization, dual rail, and time bin. Regardless of the encoding scheme, the design of a chip-scale quantum photonic circuit is often sensitive to the polarization of light. In the majority of epitaxial quantum dot structures, the lowest energy exciton consists of a hole from the heavy hole

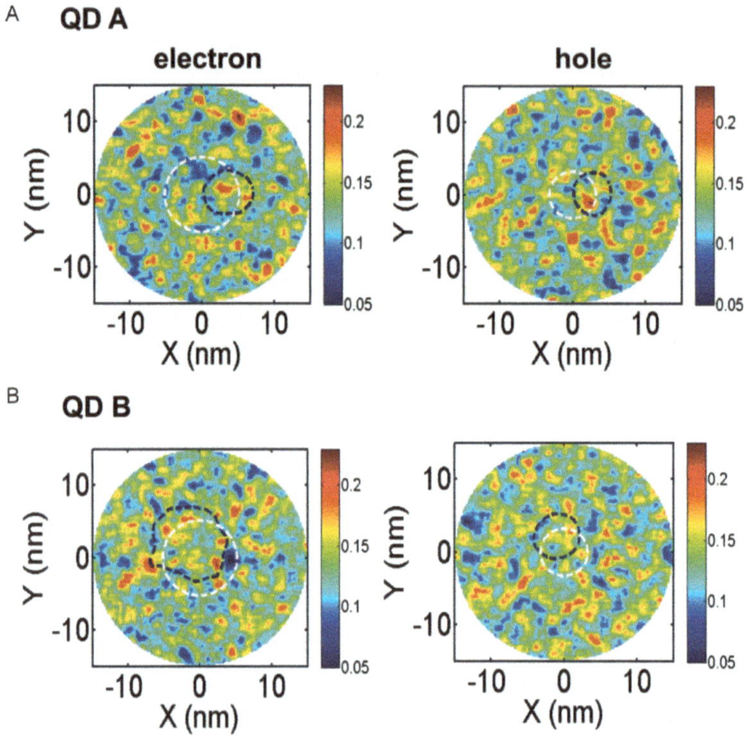

Fig. 9 The illustration of the effect of random alloy on the microscopic electron and hole wave functions in an $In_{0.14}Ga_{0.86}N/GaN$ quantum dot. The $In_{0.14}Ga_{0.86}N$ diameter and thickness are 30 and 3 nm, respectively. The electron and hole wave functions are shown, with excitonic effect taken into account, as the white and black dashed lines which represent the full-width-half-maximum of the $|\text{wavefunction}|^2$. The white and black dashed lines are results without and with considerations of the random-alloy effect, respectively. The background heat map is the indium composition randomly generated when the random-alloy effect is taken into account. Two quantum dots are simulated with identical parameters but with different indium compositional maps as the random-alloy effect is stochastic. In both cases, the wavefunctions are distorted from the original without the random-alloy effect. *Reprinted from Teng, C.-H., Zhang, L., Deng, H., Ku, P.-C., 2019. Mechanisms of inhomogeneous broadening in InGaN dot-in-wire structures. J. Appl. Phys. 126 (8), 083104. https://doi.org/10.1063/1.5111343, with the permission of AIP Publishing.*

band. For an ideal, perfectly symmetric quantum dot grown on the (001) substrate, the two degenerate exciton states with opposite spins result in unpolarized light emission, which can be exploited to generate polarization-entangled photon pairs (Stevenson et al., 2006). But in reality, the two states are often non-degenerate due to unintentional processing variations (Krizhanovskii et al., 2005; Ohno et al., 2011). An external polarization filter will greatly reduce the quantum efficiency. Many applications require certain minimum efficiencies. For example, boson sampling requires >50% efficiency in order to demonstrate superiority to classical algorithms (Neville et al., 2017). Hence it is desirable to be able to intentionally introduce anisotropy.

To control the polarization direction, one can introduce intentional anisotropy to the quantum dot (Lundskog et al., 2014; Mrowiński et al., 2018; Teng et al., 2015; Wang et al., 2017) or to the photon local density of states (Coles et al., 2016; Lodahl et al., 2017; Söllner et al., 2015; Wang et al., 2019). Linearly polarized emission has often been reported from III-nitride quantum dots (Bayerl and Kioupakis, 2014; Holmes et al., 2016a; Jemsson et al., 2014; Lundskog et al., 2014; Wang et al., 2017; Zhang et al., 2013). In strain induced III-nitride quantum dots, polarization control can be realized postgrowth. This is an advantage when quantum dots with different polarizations are needed in the same quantum photonic circuit. One example is a single photon source with an externally tunable linear polarization state (Ku et al., 2017). As shown in Fig. 10, III-nitride strain induced quantum dots in elliptically shaped nanopillars can generate linearly polarized photons with a degree of linear polarization as high as 100% and

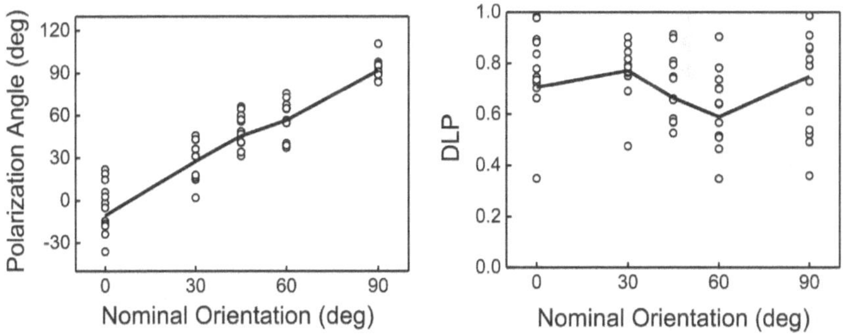

Fig. 10 Quantum dot emission is analyzed by its polarization direction and the degree of linear polarization (DLP). 60 strain induced InGaN/GaN quantum dots were measured. The cross section of all the quantum dots is elliptical with a nominal aspect ratio of 1.64 although the actual cross-sectional shape was found to have variations from dot to dot. Photon antibunching was confirmed for at least one quantum dot randomly selected from each orientation. *Reprinted from Teng, C.-H., Zhang, L., Hill, T.A., Demory, B., Deng, H., Ku, P.-C., 2015. Elliptical quantum dots as on-demand single photons sources with deterministic polarization states. Appl. Phys. Lett. 107 (19), 191105. https://doi.org/10.1063/1.4935463, with the permission of AIP Publishing.*

polarization direction continuously tunable via the orientation of the ellipse (Teng et al., 2015).

The highly pure polarization state from a strain induced quantum dot can be attributed to the small energy spacings between the valence bands. The quantum-dot anisotropy mixes the heavy hole, light hole, and split-off bands. The mixing leads to the anisotropy of exciton wave function and linearly polarized emission. Fig. 11 shows the calculated as well as measured degree of polarization (DLP) from an elliptically shaped quantum dot. Generally speaking, the DLP increases with the degree of asymmetry of the ellipse, characterized by the ratio between the long and short axes. What is interesting is the rate of increase for DLP is different for different materials. The rate of increase is faster for materials with smaller split-off energy. In InGaN, the split-off energy is 58 times smaller than that in CdTe. For a large split-off energy, the valence mixing is dominated by the heavy hole and light hole. The heavy hole wavefunction has only the X and Y components (with Z along the [001] or [0001] direction) while the light hole wavefunction has an additional Z component. The quantum-dot anisotropy in the X and Y directions fixes the emission polarization while the extra Z component randomizes it. The split-off band also

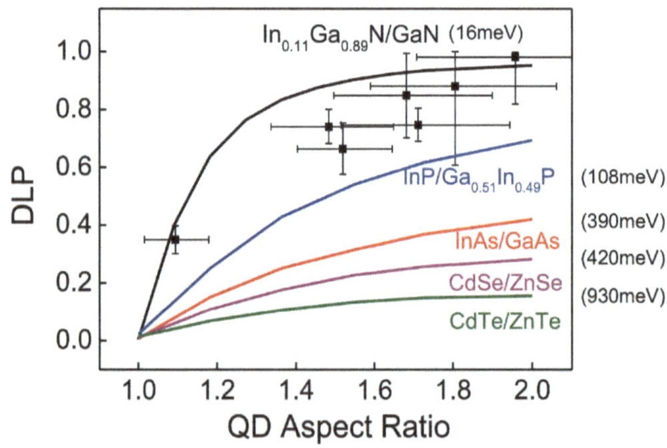

Fig. 11 Experimentally measured DLP's of InGaN/GaN quantum dots with elliptical cross sections and different aspect ratios. The result is compared with theoretically calculated DLP for different quantum dot materials. The rate of increase for DLP is faster for materials with a smaller split-off energy (given by the numbers in the parentheses). *Reprinted from Teng, C.-H., Zhang, L., Hill, T.A., Demory, B., Deng, H., Ku, P.-C., 2015. Elliptical quantum dots as on-demand single photons sources with deterministic polarization states. Appl. Phys. Lett. 107 (19), 191105. https://doi.org/10.1063/1.4935463, with the permission of AIP Publishing.*

has a Z component which can cancel out the randomization caused by the light hole. In materials with a smaller split-off energy, such as III-N, the mixing of the split-off band can be significantly stronger and results in a highly pure linearly polarized emission as shown in Fig. 11.

4. Conclusion

Semiconductor quantum dots are a promising candidate toward future quantum information processing on an integrated platform. Tremendous progress has been made in the past two decades for the fabrication, control, and integration of quantum dots with waveguides and optical cavities. The majority of results reported to date have been based on InAs self-assembled quantum dots fabricated using the Stranski–Krastanov growth mode. Yet, the random nature of the growth complicates the potential applicability of these materials in a large-scale quantum photonic circuit.

Scalable quantum dots represent an important technological step toward practical applications. The properties of these quantum dots and their interactions with photons need to be tailored by engineering means. Most efforts have been focused on the control of quantum-dot locations. The ability to deterministically place a quantum dot also enables the integration with an optical cavity to mitigate decoherence, suppress spectral diffusion, and allow further control of exciton–photon interactions.

The exciton binding energy in group-III nitride semiconductors is 20 times larger than that of InAs. Large exciton binding energy allows excitons to be stable even at room temperature. Single photon emission has been reported in various III-nitride quantum dots above room temperature. III-nitride quantum dots also cover a wavelength range from 300 to 900 nm that can be directly interfaced with qubits in neutral atoms, trap ions, and diamonds. These unique properties made III-nitride quantum dots a topic of interests in recent years.

Strain induced III-nitride quantum dots are introduced to enable quantum dot location control. First, a quantum well is grown. The lateral confinement potential is then induced via strain relaxation using the nanopillar structure. The piezoelectric effect in combination with strain relaxation generates a lateral confinement potential for both electrons and holes at the center of the nanopillar. A strain induced III-nitride quantum dot can have a physical diameter much larger than the size of the confinement potential. Hence, a strong quantum confinement can be achieved even with a much larger structure which is much easier to control. It also minimizes the negative impact of the surface.

With the addition of a p–n junction and electrical contacts, a single electron can be selectively injected into a strain induced III-nitride quantum dot. When excited, the negatively charged trion can interface with a photonic qubit while allowing the electron spin to hold the quantum information. Spin coherence can be very robust in III-nitride semiconductors, even at room temperature. Spin based photonic quantum information processing has been extensively studied in InAs quantum dots and can likely be applied to III-nitride materials in the near future.

The strong piezoelectric field in III-nitride results in a long radiative lifetime which not only makes the exciton more susceptible to dephasing but lowers the radiative efficiency. Metallic optical cavities self-aligned to strain induced III-nitride quantum dots have been shown to result in an average Purcell's factor of 46 without compromising the ability to emit a single photon. Further increase of the Purcell's factor to >2000 while maintaining a quantum efficiency $>60\%$ is possible using dielectric cavities or metallic cavities with alternating metal/dielectric layers. A Purcell's factor of 100 or more is likely to be sufficient to enable coherent exciton–photon interactions.

Microscopic compositional fluctuations or the random-alloy effect inherent to any ternary compound semiconductors can distort the electron and hole wavefunctions and introduce a large inhomogeneous broadening in an array of apparently uniform strain induced III-nitride quantum dots, even when the quantum dot dimensions are perfectly uniform. Using a ternary alloy as the quantum-dot matrix will likely have a similar effect as well-width fluctuations. Hence, binary quantum dot material and matrix, such as GaN/AlN and InN/GaN, are the best way to minimize inhomogeneous broadening caused by the microscopic compositional and thickness fluctuations. Fine tuning quantum dot resonance can also be achieved by a voltage bias. It has been shown that the exciton resonance can be shifted by about 5 meV/V in a strain induced III-nitride quantum dot.

The last but perhaps the most challenging need for quantum-dot based photonic quantum information processing is the ability to precisely control the exciton–photon interaction, allowing the exciton to interact with only one photonic mode. This is especially crucial for the implementation of a deterministic quantum gate. The control of the exciton–photon interaction requires both controlling the exciton wavefunction and the photonic local density of states. The latter is often accomplished by an optical cavity which will greatly benefit from scalable quantum dot structures. Controlling the exciton wave function can be achieved by intentionally introducing anisotropy to the quantum dot. Using an elliptical cross section for the strain induced III-nitride quantum dot, highly pure linearly polarized single photon emission has been observed with the polarization direction externally tunable via the orientation of the quantum dot. The near unity degree of linear polarization is attributed to a small split-off energy in III-nitride materials. The result can be applied to build a BB84 single photon source for quantum key distribution by combining four different elliptical quantum dots with orientations of 0°, 45°, 90°, and 135° (Bennett and Brassard, 2014).

The future quantum photonic chips will likely require multiple materials to be integrated on the same platform. III-nitride quantum dots can play an crucial role in interfacing with qubits in various modalities including photons, quantum dot spins, atoms, trap ions, and color centers in diamonds. The ability to operate beyond the cryogenic temperature is appealing. Many challenges are still yet to be overcome. But the future is bright as the III-nitride LED!

References

Alferov, Z.I., 1996. The history and future of semiconductor heterostructures from the point of view of a Russian scientist. Phys. Scr. T 68, 32–45. https://doi.org/10.1088/0031-8949/1996/t68/005.

Aoki, K., et al., 2008. Coupling of quantum-dot light emission with a three-dimensional photonic-crystal nanocavity. Nat. Photonics 2 (11), 688–692. https://doi.org/10.1038/nphoton.2008.202.

Arakawa, Y., Sakaki, H., 1982. Multidimensional quantum well laser and temperature dependence of its threshold current. Appl. Phys. Lett. 40 (11), 939–941. https://doi.org/10.1063/1.92959.

Arita, M., et al., 2017. Ultraclean single photon emission from a GaN quantum dot. Nano Lett. 17 (5), 2902–2907. https://doi.org/10.1021/acs.nanolett.7b00109.

Atature, M., 2006. Quantum-dot spin-state preparation with near-unity fidelity. Science 312 (5773), 551–553. https://doi.org/10.1126/science.1126074.

Bardoux, R., et al., 2006. Photoluminescence of single GaN/AlN hexagonal quantum dots on Si(111): spectral diffusion effects. Phys. Rev. B Am. Phys. Soc. 74 (19), 195319. https://doi.org/10.1103/PhysRevB.74.195319.

Barth, M., et al., 2006. Imaging single quantum dots in three-dimensional photonic crystals. Phys. Rev. Lett. 96 (24), 23–26. https://doi.org/10.1103/PhysRevLett.96.243902.

Bayerl, D., Kioupakis, E., 2014. Visible-wavelength polarized-light emission with small-diameter InN nanowires. Nano Lett. 14 (7), 3709–3714. https://doi.org/10.1021/nl404414r.

Bayerl, D., Kioupakis, E., 2019. Room-temperature stability of excitons and transverse-electric polarized deep-ultraviolet luminescence in atomically thin GaN quantum wells. Appl. Phys. Lett. 115 (13). https://doi.org/10.1063/1.5111546. AIP Publishing LLC.

Bennett, C.H., Brassard, G., 2014. Quantum cryptography: public key distribution and coin tossing. Theoret. Comput. Sci. 560 (P1), 7–11. https://doi.org/10.1016/j.tcs.2014.05.025. Elsevier B.V.

Bhattacharya, P., et al., 2017. III-nitride electrically pumped visible and near-infrared nanowire lasers on (001) silicon. In: Semiconductors and Semimetals, first ed. Elsevier Inc., https://doi.org/10.1016/bs.semsem.2016.07.002.

Borri, P., et al., 2001. Ultralong dephasing time in InGaAs quantum dots. Phys. Rev. Lett. 87 (15), 157401–157404. https://doi.org/10.1103/PhysRevLett.87.157401.

Brod, D.J., et al., 2019. Photonic implementation of boson sampling: a review. Adv. Photonics 1 (03), 1. https://doi.org/10.1117/1.ap.1.3.034001.

Buß, J.H., et al., 2010. Temperature dependence of electron spin relaxation in bulk GaN. Phys. Rev. B 81, 155216. https://doi.org/10.1103/PhysRevB.81.155216.

Buß, J.H., et al., 2014. High temperature electron spin dynamics in bulk cubic GaN: nanosecond spin lifetimes far above room-temperature. Appl. Phys. Lett. 105 (18). https://doi.org/10.1063/1.4901108.

Cho, J.H., et al., 2018. Strongly coherent single-photon emission from site-controlled InGaN quantum dots embedded in GaN nanopyramids. ACS Photonics 5 (2), 439–444. https://doi.org/10.1021/acsphotonics.7b00922.

Chow, C.M., et al., 2016. Non-local optical nuclear spin locking in an InAs quantum dot molecule: extended two-electron spin coherence. Phys. Rev. Lett. 117, 077403.

Chung, K., et al., 2020. Wavelength tuning in the purple wavelengths using strain-controlled Al x Ga 1–x N/GaN disk-in-wire structures. Appl. Phys. Lett. 116 (4), 041102. https://doi.org/10.1063/1.5140996. AIP Publishing LLC.

Coles, R.J., et al., 2016. Chirality of nanophotonic waveguide with embedded quantum emitter for unidirectional spin transfer. Nat. Commun. 7 (1), 11183. https://doi.org/10.1038/ncomms11183. Nature Publishing Group.

Davanco, M., et al., 2017. Heterogeneous integration for on-chip quantum photonic circuits with single quantum dot devices. Nat. Commun. 8 (1), 1–12. https://doi.org/10.1038/s41467-017-00987-6. Springer US.

Delgoffe, A., et al., 2019. Integrated quantum photonics using site-controlled quantum dots and tailored-potential photonic crystals. In: *2019 Conference on Lasers and Electro-Optics, CLEO 2019 - Proceedings*, pp. 7–8, https://doi.org/10.23919/CLEO.2019.8749658.

Demory, B., et al., 2015. Plasmonic enhancement of single photon emission from a site-controlled quantum dot. ACS Photonics 2 (8), 1065–1070. https://doi.org/10.1021/acsphotonics.5b00086.

Demory, B., et al., 2017. Improving the radiative efficiency of InGaN quantum dots via an open top cavity. ACS Photonics 4, 795.

Deng, G.-W., Xu, N., Li, W.-J., 2020. Gate-defined quantum dots: fundamentals and applications BT—quantum dot optoelectronic devices. In: Yu, P., Wang, Z.M. (Eds.), Quantum Dot Optoelectronic Devices. Springer International Publishing, Cham, pp. 107–133.

Deshpande, S., et al., 2013. Electrically driven polarized single-photon emission from an InGaN quantum dot in a GaN nanowire. Nat. Commun. 4 (1), 1675. https://doi.org/10.1038/ncomms2691. Nature publishing group.

Deshpande, S., et al., 2014. Electrically pumped single-photon emission at room temperature from a single InGaN/GaN quantum dot. Appl. Phys. Lett. 105 (14). https://doi.org/10.1063/1.4897640.

Elshaari, A.W., et al., 2017. On-chip single photon filtering and multiplexing in hybrid quantum photonic circuits. Nat. Commun. 8 (1), 379. https://doi.org/10.1038/s41467-017-00486-8. Springer US.

Fulton, T.A., Dolan, G.J., 1987. Observation of single-electron charging effects in small tunnel junctions. Phys. Rev. Lett. 59 (1), 109–112. https://doi.org/10.1103/PhysRevLett.59.109.

Gao, K., et al., 2019. Spectral diffusion time scales in InGaN/GaN quantum dots. Appl. Phys. Lett. 114 (11). https://doi.org/10.1063/1.5088205.

Gschrey, M., et al., 2015. Highly indistinguishable photons from deterministic quantum-dot microlenses utilizing three-dimensional in situ electron-beam lithography. Nat. Commun. 6 (May), 7662. https://doi.org/10.1038/ncomms8662.

Gurioli, M., et al., 2019. Droplet epitaxy of semiconductor nanostructures for quantum photonic devices. Nat. Mater. 18 (8), 799–810. https://doi.org/10.1038/s41563-019-0355-y. Springer US.

He, Y.-M., et al., 2017. Deterministic implementation of a bright, on-demand single-photon source with near-unity indistinguishability via quantum dot imaging. Optica 4 (7), 802. https://doi.org/10.1364/OPTICA.4.000802.

Holmes, M., et al., 2013. Measurement of an exciton Rabi rotation in a single GaN/Al xGa1-xN nanowire-quantum dot using photoluminescence spectroscopy: evidence for coherent control. Phys. Rev. Lett. 111 (5), 1–4. https://doi.org/10.1103/PhysRevLett.111.057401.

Holmes, M.J., et al., 2014. Room-temperature triggered single photon emission from a III-nitride site-controlled nanowire quantum dot. Nano Lett. 14 (2), 982–986. https://doi.org/10.1021/nl404400d.

Holmes, M., et al., 2015. Spectral diffusion and its influence on the emission linewidths of site-controlled GaN nanowire quantum dots. Phys. Rev. B 92 (11). https://doi.org/10.1103/PhysRevB.92.115447.

Holmes, M.J., et al., 2016a. In: Chyi, J.-I., et al. (Eds.), Linearly polarized single photons from small site-controlled GaN nanowire quantum dots. Proc. SPIE, p. 97481E, https://doi.org/10.1117/12.2212455.

Holmes, M.J., et al., 2016b. Single photons from a hot solid-state emitter at 350 K. ACS Photonics 3 (4), 543–546. https://doi.org/10.1021/acsphotonics.6b00112.

Jarjour, A.F., et al., 2007. Cavity-enhanced blue single-photon emission from a single InGaN/GaN quantum dot. Appl. Phys. Lett. 91 (5), 1–4. https://doi.org/10.1063/1.2767217.

Jemsson, T., et al., 2014. Linearly polarized single photon antibunching from a site-controlled InGaN quantum dot. Appl. Phys. Lett. 105 (8). https://doi.org/10.1063/1.4893476.

Kim, J.H., et al., 2017. Hybrid integration of solid-state quantum emitters on a silicon photonic chip. Nano Lett. 17 (12), 7394–7400. https://doi.org/10.1021/acs.nanolett.7b03220.

Kimble, H.J., 2008. The quantum internet. Nature 453 (7198), 1023–1030. https://doi.org/10.1038/nature07127.

Kómár, P., et al., 2014. A quantum network of clocks. Nat. Phys. 10 (8), 582–587. https://doi.org/10.1038/nphys3000.

Krizhanovskii, D.N., et al., 2005. Individual neutral and charged InxGa1-xAs-GaAs quantum dots with strong in-plane optical anisotropy. Phys. Rev. B Condens. Matter Mater. Phys. 72 (16), 8–11. https://doi.org/10.1103/PhysRevB.72.161312.

Kroemer, H., 1996. Band offsets and chemical bonding: the basis for heterostructure applications. Phys. Scr. T 68, 10–16. https://doi.org/10.1088/0031-8949/1996/t68/001.

Ku, P.-C., Teng, C.-H., Deng, H., 2017. III-nitride semiconductor single photon sources. In: Bi, W., et al. (Eds.), Handbook of GaN Semiconductor Materials and Devices. Taylor & Francis, pp. 661–669.

Ledentsov, N.N., et al., 1998. Quantum dot heterostructures: fabrication, properties, lasers (review). Sem. Ther. 32 (4), 343–365. https://doi.org/10.1134/1.1187396.

Lee, L.K., Ku, P.C., 2012. Fabrication of site-controlled InGaN quantum dots using reactive-ion etching. Phys. Stat. Solidi (c) 9, 609.

Lodahl, P., et al., 2017. Chiral quantum optics. Nature 541 (7638), 473–480. https://doi.org/10.1038/nature21037. Nature Publishing Group.

Lundskog, A., et al., 2014. Direct generation of linearly polarized photon emission with designated orientations from site-controlled InGaN quantum dots. Light Sci. Appl. 3 (1). https://doi.org/10.1038/lsa.2014.20.

Lyasota, A., et al., 2019. Limiting the spectral diffusion of nano-scale light emitters using the purcell effect in a photonic-confined environment. Sci. Rep. 9 (1), 1–9. https://doi.org/10.1038/s41598-018-37677-2. Springer US.

Mohan, A., et al., 2010. Record-low inhomogeneous broadening of site-controlled quantum dots for nanophotonics. Small 6 (12), 1268–1272. https://doi.org/10.1002/smll.201000341.

Mrowiński, P., et al., 2018. Photonic engineering of highly linearly polarized quantum dot emission at telecommunication wavelengths. Phys. Rev. B 97 (16), 1–11. https://doi.org/10.1103/PhysRevB.97.165427.

Neville, A., et al., 2017. Classical boson sampling algorithms with superior performance to near-term experiments. Nat. Phys. 13 (12), 1153–1157. https://doi.org/10.1038/nphys4270.

Noda, S., 2000. Full three-dimensional photonic bandgap crystals at near-infrared wavelengths. Science 289 (5479), 604–606. https://doi.org/10.1126/science.289.5479.604.

Nowak, A.K., et al., 2014. Deterministic and electrically tunable bright single-photon source. Nat. Commun. 5 (1), 3240. https://doi.org/10.1038/ncomms4240.

Ohno, S., et al., 2011. Optical anisotropy and photoluminescence polarization in single InAlAs quantum dots. Appl. Phys. Lett. 98 (16), 1–4. https://doi.org/10.1063/1.3583453.

Ostapenko, I.A., et al., 2010. Large internal dipole moment in InGaN/GaN quantum dots. Appl. Phys. Lett. 97 (6), 2010–2013. https://doi.org/10.1063/1.3477952.

Rigal, B., et al., 2015. Site-controlled quantum dots coupled to a photonic crystal molecule. Appl. Phys. Lett. 107 (14). https://doi.org/10.1063/1.4932228.

Robinett, R.W., 1996. Visualizing the solutions for the circular infinite well in quantum and classical mechanics. Am. J. Phys. 64 (4), 440–446. https://doi.org/10.1119/1.18188.

Schwartz, I., et al., 2016. Deterministic generation of a cluster state of entangled photons. Science 354 (6311), 434–437. https://doi.org/10.1126/science.aah4758.

Sergent, S., et al., 2012. High-Q (>5000) AlN nanobeam photonic crystal cavity embedding GaN quantum dots. Appl. Phys. Lett. 100 (12). https://doi.org/10.1063/1.3695331.

Shields, A.J., 2007. Semiconductor quantum light sources. Nat. Photonics 1 (4), 215–223. https://doi.org/10.1038/nphoton.2007.46.

Söllner, I., et al., 2015. Deterministic photon-emitter coupling in chiral photonic circuits. Nat. Nanotechnol. 10 (9), 775–778. https://doi.org/10.1038/nnano.2015.159. Nature Publishing Group.

Somaschi, N., et al., 2016. Near optimal single photon sources in the solid state. Nat. Photonics 10 (2), 1–6. https://doi.org/10.1038/nphoton.2016.23.

Stevenson, R.M., et al., 2006. A semiconductor source of triggered entangled photon pairs. Nature 439 (7073), 178–182. https://doi.org/10.1038/nature04446.

Sun, B., et al., 2012. Persistent narrowing of nuclear-spin fluctuations in InAs quantum dots using laser excitation. Phys. Rev. Lett. 108 (May), 1–4. https://doi.org/10.1103/PhysRevLett.108.187401.

Tamariz, S., et al., 2020. Toward bright and pure single photon emitters at 300 K based on GaN quantum dots on silicon. ACS Photonics 7 (6), 1515–1522. https://doi.org/10.1021/acsphotonics.0c00310.

Teng, C.-H., et al., 2015. Elliptical quantum dots as on-demand single photons sources with deterministic polarization states. Appl. Phys. Lett. 107 (19). https://doi.org/10.1063/1.4935463.

Teng, C.-H., et al., 2019. Mechanisms of inhomogeneous broadening in InGaN dot-in-wire structures. J. Appl. Phys. 126 (8), 083104. https://doi.org/10.1063/1.5111343. AIP Publishing LLC.

Vico Triviño, N., et al., 2012. High quality factor two dimensional GaN photonic crystal cavity membranes grown on silicon substrate. Appl. Phys. Lett. 100 (7). https://doi.org/10.1063/1.3684630.

Wagner, H.P., et al., 1999. Interaction-induced effects in the nonlinear coherent response of quantum-well excitons. Phys. Rev. B 60 (7), 4454–4457. American Physical Society. http://link.aps.org/doi/10.1103/PhysRevB.60.4454.

Wang, T., et al., 2017. Deterministic optical polarisation in nitride quantum dots at thermoelectrically cooled temperatures. Sci. Rep. 7 (1), 12067. https://doi.org/10.1038/s41598-017-12233-6. Springer US.

Wang, H., et al., 2019. Towards optimal single-photon sources from polarized microcavities. Nat. Photonics 13 (11), 770–775. https://doi.org/10.1038/s41566-019-0494-3.

Wu, Y., et al., 2020. Monolayer GaN excitonic deep ultraviolet light emitting diodes. Appl. Phys. Lett. 116 (1). https://doi.org/10.1063/1.5124828.

Xiao, Z., 1999. Exciton binding energy in spherical quantum dots in a magnetic field. J. Appl. Phys. 86 (8), 4509–4514. https://doi.org/10.1063/1.371394.

Xiao, X., et al., 2014. Quantum-size-controlled photoelectrochemical fabrication of epitaxial InGaN quantum dots. Nano Lett. 14 (10), 5616–5620. https://doi.org/10.1021/nl502151k. American Chemical Society.

Xu, X., et al., 2009. Optically controlled locking of the nuclear field via coherent dark-state spectroscopy. Nature 459 (7250), 1105–1109. https://doi.org/10.1038/nature08120. Macmillan Publishers Limited. All rights reserved.

Zadeh, I.E., et al., 2016. Deterministic integration of single photon sources in silicon based photonic circuits. Nano Lett. 16 (4), 2289–2294. https://doi.org/10.1021/acs.nanolett.5b04709.

Zhang, L., et al., 2013. Single photon emission from site-controlled InGaN/GaN quantum dots. Appl. Phys. Lett. 103 (19), 192114.

Zhang, L., et al., 2014. How much better are InGaN/GaN nanodisks than quantum wells—oscillator strength enhancement and changes in optical properties. Appl. Phys. Lett. 104 (5), 51116. https://doi.org/10.1063/1.4864083.

Zhang, L., Teng, C.-H., et al., 2016a. Charge-tunable indium gallium nitride quantum dots. Phys. Rev. B 93 (8), 085301.

Zhang, L., Teng, C., et al., 2016b. Site-controlled InGaN/GaN single-photon-emitting diode. Appl. Phys. Lett. 108, 153102. https://doi.org/10.1063/1.4945984.

CHAPTER TWO

Microcavity exciton polaritons

Long Zhang[a,b,*], Jiaqi Hu[a], and Hui Deng[a]
[a]Physics Department, University of Michigan, Ann Arbor, MI, United States
[b]Department of Physics, Xiamen University, Xiamen, People's Republic of China
*Corresponding author: e-mail address: zhanglong@xmu.edu.cn

Contents

1. Introduction

Exciton polaritons are quasiparticles formed by strong coupling between cavity photons and semiconductor excitons. The photon component makes their effective mass very small, and therefore enables quantum degeneracy at high temperature. The exciton component results in strong nonlinearity, and therefore provides a semiconductor platform to explore nonlinear optics and many-body physics. As dissipative and interacting bosons, exciton polaritons exhibit distinct properties compared to semiconductor lasers or cold atomic systems while also share many similarities with both. They form a system of deep fundamental interest and unique prospects for technological applications.

Semiconductors and Semimetals, Volume 105
ISSN 0080-8784
https://doi.org/10.1016/bs.semsem.2020.10.001

This chapter is organized as follows: First, we will give a brief overview of the linear properties of microcavity polaritons and a variety of nonlinear macroscopic quantum phenomena, including polariton condensates, superfluidity, vortices, Berezinskii–Kosterlitz–Thouless phase and Bardeen–Cooper–Schrieffer phase. Then we will review recent progress on manipulating the quantum states of polaritons by controlling their photon or exciton components, aimed at creating new quantum states and implementing quantum simulations. Lastly, we review progress on new materials with larger binding energies and oscillator strength, which may enable room temperature exciton polariton devices.

2. Exciton polaritons in semiconductor microcavity

2.1 Strong coupling and exciton polaritons

2.1.1 Exciton

The quasi-vacuum of a semiconductor is the state with a filled valence band and an empty conduction band. When a semiconductor is pumped optically or electrically, electrons from the valence band can be excited to the conduction band, leaving positively charged quasiparticles (holes) behind in the valence band. The electron and hole are bound together via Coulomb interaction, forming a bound pair: an exciton, which is analogous to a hydrogen atom where an electron is bound to a proton. Compared to a hydrogen atom in vacuum, an exciton has a much smaller effective mass and the Coulomb interaction is screened. Therefore an exciton has a larger Bohr radius and smaller binding energy. Depending on the reduced exciton mass and dielectric constant, two types of excitons are formed, the Frenkel and Wannier–Mott excitons. Frenkel excitons are more tightly bound, having a radius comparable to the interatomic distance, and localized to a molecule or atom (Böer and Pohl, 2018). They are commonly found in organic semiconductors. Wannier–Mott excitons extend over many lattice constants and are free to move through the lattice. They are typically found in inorganic semiconductors. We compare the typical values of the reduced mass, Bohr radius and binding energy of the two types of excitons and atoms in Table 1. Excitons with binding energies larger than ∼25 meV (room temperature thermal energy), will be stable at room temperature (RT). (See more discussions in Section 3).

Table 1 Parameters comparison.

Systems	Atomic gases	Frenkel exciton	Wannier–Mott exciton
Effective mass m^*/m_e	10^3	10^{-1}	10^{-1}
Bohr radius a_B	10^{-1}Å	10 Å	10–100 Å
Binding energy	10 eV	1 eV	0.01–1 eV

To characterize how strongly excitons can interact with photons, the oscillator strength f of an exciton is defined in analogy to the atomic oscillator strength:

$$f = \frac{2m^*\omega}{\hbar} |\langle u_v|\boldsymbol{r}\cdot\boldsymbol{e}|u_c\rangle|^2 \frac{V}{\pi a_B^3}. \tag{1}$$

Here, $m^* = \left(m_e^{-1} + m_h^{-1}\right)^{-1}$ is the reduced mass of the exciton, $|u_c\rangle$ and $|u_v\rangle$ represent the Bloch wavefunctions of the electron and hole, V is the quantization volume, and a_B^3 is the Bohr radius of the exciton (Deng et al., 2010). Strong exciton-photon coupling is more readily established in semiconductors with strong exciton oscillator strengths. To enhance the exciton-photon coupling, cavities can be used to enhance the local field.

2.1.2 Semiconductor microcavity

A microcavity is an optical resonator with size comparable to the wavelength of light in at least one dimension (Alexey, 2007). The quality factor and the mode volume are two key parameters of a microcavity. The quality factor (Q-factor) is defined as the ratio of the frequency ω_c to the full linewidth at half maximum (FWHM) $\delta\omega_c$ of the cavity resonance:

$$\mathrm{Q} = \frac{\omega_c}{\delta\omega_c} \tag{2}$$

The Q-factor is a measure of the rate at which optical energy decays from within the cavity (from absorption, scattering or leakage through the imperfect mirrors) and Q^{-1} is the fraction of energy lost in a single round-trip around the cavity. Equivalently, the exponentially decaying photon number has a lifetime given by $\tau = \frac{1}{\delta\omega_c} = \frac{\mathrm{Q}}{\omega_c}$. A microcavity with a higher Q-factor can provide a longer photon lifetime, which is often important for building up coherence.

The mode volume is a measure of the spatial confinement of resonant mode inside the cavity. For an optical resonator, the mode volume can be defined as the volume integral of the electric field intensity in and around the cavity, normalized to the maximum field intensity:

$$V_{eff} = \frac{\int_V \epsilon(\boldsymbol{r})|\boldsymbol{E}(\boldsymbol{r})|^2 d^3\boldsymbol{r}}{\max\left[\epsilon(\boldsymbol{r})|\boldsymbol{E}(\boldsymbol{r})|^2\right]}, \tag{3}$$

where $\epsilon(\boldsymbol{r})$ is the dielectric constant, $|E(\boldsymbol{r})|$ is the electric field strength, and V is a quantization volume encompassing the resonator and with a boundary in the radiation zone of the cavity (Srinivasan et al., 2006). A smaller mode volume means a stronger local field and greater enhancement of the interaction between electric dipoles and the optical field.

Different types of microcavities have been explored to engineer light matter interactions, such as Fabry–Pérot (FP) cavities, photonic crystals, whispering gallery cavities, and plasmonic cavities. Fig. 1 lists a variety of cavities with their typical quality factor and mode volume labeled. The cavity can be designed based on the gain material and the requirements on field confinement (mode volume) and Q-factor (Wang et al., 2017a).

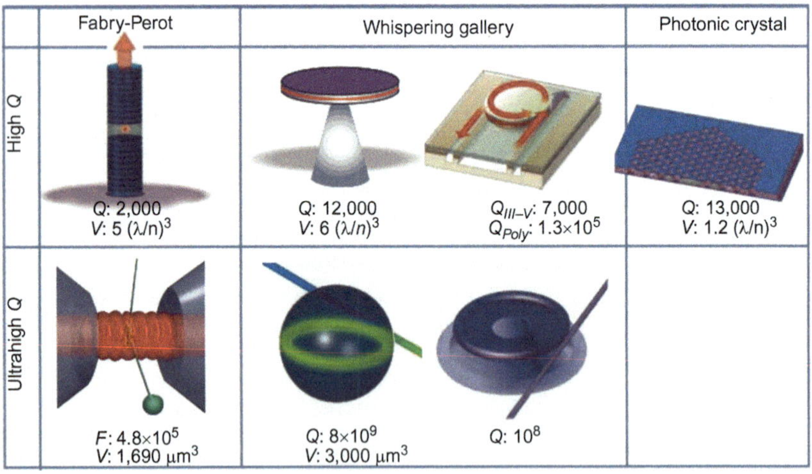

Fig. 1 Different types of microcavities with quality factors (Q) and mode volume (V) labeled. *From Vahala, K.J., 2003. Optical microcavities. Nature 424 (6950), 839–846. https://doi.org/10.1038/nature01939.*

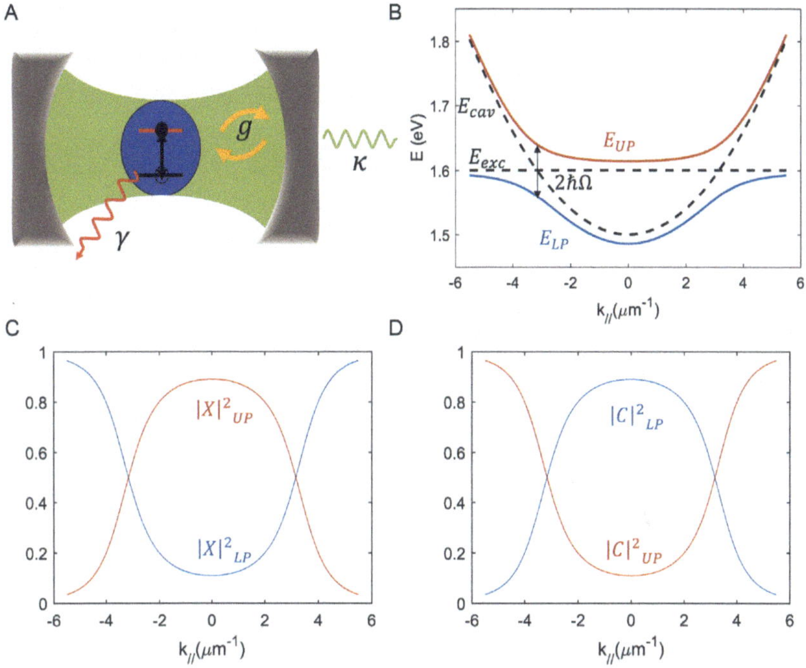

Fig. 2 Exciton polaritons (A) Schematic shows a two level system interacts with an optical cavity. γ and κ represent the decay rate of uncoupled cavity photon and the two level system respectively. g represents the interaction strength. (B)The dispersion of polaritons. (C) and (D) The Hopfield coefficients of polaritons.

2.1.3 Strong coupling and exciton polaritons

First we consider a two-level atom, placed in a resonant high-Q cavity, as schematically shown in Fig. 2A. The atom can be excited by absorbing a cavity photon, and then de-excited by emitting the photon. If the photon can be coherently re-absorbed and re-emitted in a reversible way, this system is in the strong coupling regime. Quantitatively, the energy exchange rate between atom and cavity, also known as the Rabi-frequency, needs to be faster than the decay rate of cavity photon and atom. Spectroscopically, the strongly mixed modes anticross in energy, with an energy gap called the Rabi splitting. Analogously, an electron-hole pair in a semiconductor can be created by absorbing a photon and can recombine by emitting a photon. If the exciton and cavity photon exchange energy in a reversible way, the system enters the strong coupling regime, where the exciton and photon strongly couple to form two new eigenstates, called exciton polaritons. Here we consider a system with quantum wells embedded in the planar cavity, where a two-dimensional (2D) exciton resonance couples to the planar microcavity modes. Using the rotating wave approximation, the linear Hamiltonian of the system is written in the second quantization form as:

$$\hat{H}_{\text{pol}} = \hat{H}_{\text{cav}} + \hat{H}_{\text{exc}} + \hat{H}_{\text{I}} = \sum E_{\text{cav}}(k_{\|}, k_c)\hat{a}^{\dagger}_{k_{\|}}\hat{a}_{k_{\|}} + \sum E_{\text{exc}}(k_{\|})\hat{b}^{\dagger}_{k_{\|}}\hat{b}_{k_{\|}} + \sum g(\hat{a}^{\dagger}_{k_{\|}}\hat{b}_{k_{\|}} + \hat{a}_{k_{\|}}\hat{b}^{\dagger}_{k_{\|}}) \tag{4}$$

Here, $\hat{a}^{\dagger}_{k_{\|}}$ is the photon creation operator with an in-plane wavenumber $k_{\|}$ and longitudinal wavenumber $k_c = \mathbf{k} \cdot \hat{\mathbf{z}}$ determined by the cavity resonance. $\hat{b}^{\dagger}_{k_{\|}}$ is the exciton creation operator with an in-plane wave number $k_{\|}$. g is the exciton-photon dipole interaction strength. The above Hamiltonian can be diagonalized by the transformations.

$$\hat{P}_{k_{\|}} = X_{k_{\|}}\hat{b}_{k_{\|}} + C_{k_{\|}}\hat{a}_{k_{\|}} \tag{5}$$

$$\hat{Q}_{k_{\|}} = -C_{k_{\|}}\hat{b}_{k_{\|}} + X_{k_{\|}}\hat{a}_{k_{\|}}. \tag{6}$$

Then $\hat{H}_{\text{pol}}$ becomes

$$\hat{H}_{\text{pol}} = \sum E_{\text{LP}}(k_{\|})\hat{P}^{\dagger}_{k_{\|}}\hat{P}_{k_{\|}} + \sum E_{\text{UP}}(k_{\|})\hat{Q}^{\dagger}_{k_{\|}}\hat{Q}_{k_{\|}}. \tag{7}$$

Here $(\hat{P}_{k_{\|}}, \hat{P}^{\dagger}_{k_{\|}})$ and $(\hat{Q}_{k_{\|}}, \hat{Q}^{\dagger}_{k_{\|}})$ are the creation and annihilation operators of the new $E_C(\boldsymbol{k}_{\|})$ and of the new quasiparticles or eigenmodes of the system. They are called the lower polariton (LP) and upper polariton (UP), corresponding to the two branches of lower and higher eigenenergies, respectively. Thus a polariton is a linear superposition of an exciton and a photon with the same in-plane wave number. Since both excitons and photons are bosons, so are the polaritons. The exciton and photon fractions in each LP and UP are given by the amplitude squared of $X_{k_{\|}}$ and $C_{k_{\|}}$, which are referred to as the Hopfield coefficients (Hopfield, 1958). They satisfy:

$$X^2_{k_{\|}} + C^2_{k_{\|}} = 1. \tag{8}$$

We define detuning as the energy difference between exciton and photon $\Delta E(k_{\|}) = E_{\text{cav}}(k_{\|}) - E_{\text{exc}}(k_{\|}, k_c)$. $X_{k_{\|}}$ and $C_{k_{\|}}$ are given by:

$$\left|X_{k_{\|}}\right|^2 = \frac{1}{2}\left(1 + \frac{\Delta E(k_{\|})}{\sqrt{\Delta E(k_{\|})^2 + 4g^2}}\right)$$
$$\left|C_{k_{\|}}\right|^2 = \frac{1}{2}\left(1 - \frac{\Delta E(k_{\|})}{\sqrt{\Delta E(k_{\|})^2 + 4g^2}}\right) \tag{9}$$

At zero detuning, LP and UP are exactly half photon and half exciton, as shown in Fig. 2C and D. The energies of the polaritons, which are the eigenenergies of the Hamiltonian, are deduced from the diagonalization procedure as

$$E_{LP,UP}(k_{\|}) = \frac{1}{2}\left[E_{\text{exc}} + E_{\text{cav}} \pm \sqrt{4g^2 + (E_{\text{exc}} - E_{\text{cav}})^2}\right] \tag{10}$$

When the uncoupled exciton and cavity photon are at the same energy, $E_X = E_C$, LP and UP energies have the minimum separation $2\hbar\Omega = E_{UP} - E_{LP} = 2\sqrt{g^2 - \frac{1}{4}(\gamma - \kappa)^2}$, which is the Rabi splitting. Due to the coupling between the exciton and photon modes, the new polariton energies anticross when the cavity and exciton energies are tuned across each other, as shown in Fig. 2B.

When taking into account the finite lifetime of the cavity photon and exciton, the eigenenergy is modified as

$$E_{LP,UP}(k_{\|}) = \frac{1}{2}[E_{\text{exc}} + E_{\text{cav}} + i(\kappa + \gamma)] \pm \sqrt{g^2 + \frac{1}{4}[E_{\text{exc}} - E_{\text{cav}} + i(\kappa - \gamma)]^2}. \tag{11}$$

where γ is the out-coupling rate of a cavity photon, κ is the nonradiative decay rate of an exciton.

Note that, a nonvanishing Rabi splitting $2\hbar\Omega$, obtained for $g > 1/2|\gamma - \kappa|$, is an insufficient condition for strong coupling. Reversible energy exchange between the exciton and photon modes is equivalent to requiring the two coupled modes to be spectrally separable. The minimum mode splitting needs to be greater than the sum of the half linewidths of the modes (Savona et al., 1995):

$$2\hbar\Omega > \gamma + \kappa, \text{ or, } g > \sqrt{(\gamma^2 + \kappa^2)/2}. \tag{12}$$

Since a polariton is the mixed state of an exciton and a photon, its properties, including effective mass, lifetime, nonlinearity, and density of states, will depend on the fraction of exciton and photon, or, the Hopfield coefficients X and C. These properties will in turn affect the dynamics and coherence of polaritons. The effective masses of polaritons are given by:

$$\frac{1}{m_{LP}} = \frac{|X|^2}{m_{\text{exc}}} + \frac{|C|^2}{m_{\text{cav}}}, \tag{13}$$

$$\frac{1}{m_{UP}} = \frac{|C|^2}{m_{\text{exc}}} + \frac{|X|^2}{m_{\text{cav}}}, \tag{14}$$

where m_{exc} is the effective exciton mass of its center-of-mass motion and m_{cav} is the effective cavity photon mass, which is typically on the order of $10^{-4} m_{\text{exc}}$ (Deng et al., 2010). Since $m_{\text{cav}} \ll m_{\text{exc}}$,

$$m_{LP,UP}(k_{\|}) \simeq m_{\text{cav}}/|C|^2, \tag{15}$$

$$m_{UP}(k_{\|}) \simeq m_{\text{cav}}/|X|^2. \tag{16}$$

For a microcavity with $C(k_{\|} = 0)$ of the order of one, $m_{LP} \sim 10^{-4} m_{\text{exc}}$. This very small effective mass makes possible high critical temperatures of phase transitions for the system.

The lifetime of the polaritons is directly determined by κ and γ as

$$\begin{aligned} \gamma_{LP} &= |X|^2\kappa + |C|^2\gamma, \\ \gamma_{UP} &= |C|^2\kappa + |X|^2\gamma. \end{aligned} \tag{17}$$

The lifetimes of LPs is critical for achieving quantum degeneracy. In typical high quality GaAs quantum well microcavity samples available today, $\gamma = 1 - 100\ \text{ps}^{-1}$ and $\kappa \sim \text{ns}^{-1}$, hence the lifetime of LP is mainly determined by the cavity photon lifetime and proportional to the $|C|^2$. In the new emerging 2D semiconductors, the γ and κ are both on the order of 1 ps, so the polariton lifetime remains similar at different detunings. A polariton decays in the form of emitting a photon with the same $k_{\|}$ and energy. The one-to-one correspondence between the internal polariton mode and the external out-coupled photon mode lends convenience to experimental access to the system.

2.2 Phase transitions of exciton polaritons

As a composite electron, hole and photon system with strong interactions among the constituents, exciton polaritons exhibit a rich variety of macroscopic quantum phenomena, such as Bose–Einstein condensation (BEC), superfluidity, Berezinskii–Kosterlitz–Thouless (BKT) phase, and Bardeen–Cooper–Schrieffer (BCS) phase. As an open-dissipative system, exciton polaritons bring new opportunities to re-examine these concepts in particle-hole systems and nonequilibrium regimes. At the same time, thanks to the very low effective mass of exciton polaritons, these phenomena can take place at high critical temperatures, promising a bridge to technological applications, as we will review in the next section.

2.2.1 Polariton condensate

In the density regime where excitons are well bound, exciton polaritons can be described as weakly interacting bosons. Analogous to bosonic atomic gasses, a large number of polaritons may occupy the same quantum state, forming a condensate. Such a polariton condensate provides a source of coherent light, called a polariton laser, with a threshold carrier density far below the band inversion density. After the theoretical prediction of a bulk polariton laser in 1996 (Imamogˇlu et al., 1996) and a series of experimental efforts (Huang et al., 2002; Kozuma et al., 1999; Savvidis et al., 2000), microcavity polariton condensation and lasing have been established and demonstrated in many labs in many material systems (Bajoni et al., 2008; Balili et al., 2007; Deng et al., 2006; Kasprzak et al., 2006). We discuss below a couple common roadblocks to achieving condensation and common characters of polariton condensates.

2.2.1.1 Energy relaxation bottleneck

By definition of a phase transition, coherence in a polariton condensate should occur spontaneously instead of inherit from a pump laser. This requires incoherent pumping, such as electrical injection or nonresonant optical pumping at a higher energy than that of the polariton condensate. The injected carriers need to be cooled down through incoherent phonon scattering. The energy relaxation from hot carriers to excitons and exciton-like polaritons at large $k_{||}$ typically takes place over picosecond timescales, through inelastic exciton-phonon interaction, as shown in Fig. 3A. However, phonon scattering is significantly reduced toward LPs at small $k_{||}$ where mixing with photons leads to a much lower energy density of states and often a much shorter lifetime. Consequently, LPs may accumulate around the deflection point of the dispersion, forming a relaxation bottleneck and preventing condensation (Tassone et al., 1997).

Fortunately the bottleneck effect may be overcome by nonlinear polariton–polariton scattering and stimulated scattering. Polariton–polariton scattering results from mainly the exciton exchange interactions. Energy and

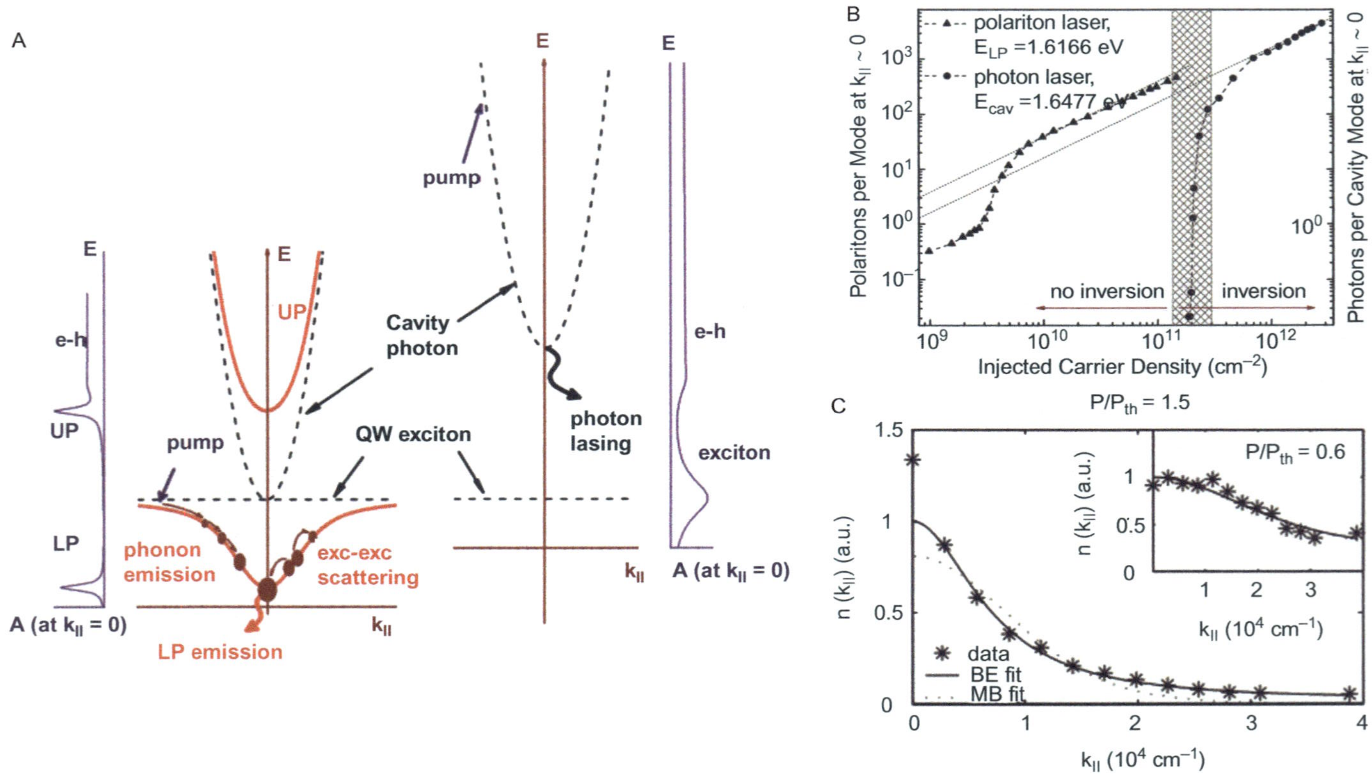

Fig. 3 Polariton condensate (A) Schematics of the experimental principles for a photon laser (*right panel*) compared to LP condensation (*left panel*). (C) Number of LPs and cavity photons per mode vs injected carrier density for a polariton condensate triangles, the scheme in the *left panel* of (A) and a photon laser circles, the scheme in the *right panel* of (A), respectively. The *gray zone* marks the population inversion densities from band edge to 15 meV above the band edge. (B) The measured LP population per state vs $k_{||}$ stars, compared with BE *solid line* and MB *dotted line* distribution functions at pump rates $P/P_{th} = 1.5$ and 0.6 inset. At $P/P_{th} = 0.6$, the fitted BE and MB distribution curves almost overlap (Deng et al., 2003). *Panel (A): From Deng, H., Haug, H., Yamamoto, Y., 2010. Exciton-polariton Bose-Einstein condensation. Rev. Mod. Phys. 82 (2), 1489–1537. https://doi.org/10.1103/RevModPhys.82.1489. ISBN: 0034-6861; Panel (B): From Deng, H., Weihs, G., Snoke, D., Bloch, J., Yamamoto, Y., 2003. Polariton lasing vs. photon lasing in a semiconductor microcavity. Proc. Natl. Acad. Sci. 100 (26), 15318–15323. https://doi.org/10.1073/pnas.2634328100.*

in-plane momentum of the polaritons are conserved in this process. The scattering is negligible at very low densities but is enhanced by bosonic final state stimulation when the occupancy in the low-$k_{\|}$ states approaches degeneracy (Deng et al., 2010).

2.2.1.2 Exciton saturation and transition to photon lasing

When the carrier density in a QW approaches the Mott density, $1/a_B^2$ in 2D, excitons dissociate into electron and hole plasma. Above the Mott density, optical gain accompanied with conduction valence band population inversion appears. The system may form a BCS state (see Section 2.2.4) or conventional photon lasing, where the photon emission is stimulated into cavity mode (Fig. 3B). Using multiple QWs can increase the saturation density by diluting the exciton density per QW, so that polariton lasing can be achieved before saturation (Deng et al., 2002). Both polariton lasing and photon lasing have been observed in the same GaAs microcavity (Bajoni et al., 2008; Deng et al., 2003), where the threshold of polariton lasing is typically one to two orders of magnitude lower than that of photon lasing.

2.2.1.3 Thermodynamics of polariton condensate

As a driven-dissipative system, the thermal and quantum dynamics of exciton polaritons are determined by the interplay between polariton lifetime τ_{LP}, LP-LP scattering rate $\gamma_{LP\text{-}LP}$, and LP-phonon scattering rate $\gamma_{LP\text{-}phonon}$. With long τ_{LP} and strong $\gamma_{LP\text{-}LP}$, quantum degeneracy at the LP ground state ($k_{\|}=0$) and a Bose–Einstein (BE) distribution of polariton condensates with quasiequilibrium was observed (Fig. 3C) (Deng et al., 2006; Kasprzak et al., 2006). However, the extracted LP temperature is higher than the lattice temperature, indicating the nonthermal equilibrium of the polariton condensates. To reach polariton condensation with thermal equilibrium, longer τ_{LP} and/or stronger $\gamma_{LP\text{-}phonon}$ are required, which can be implemented in two ways. One way is to increase the exciton fraction of LP by increasing the detuning $\delta = E_C - E_X$, so that τ_{LP} and $\gamma_{LP\text{-}phonon}$ are increased (Deng et al., 2006). The other way is to increase the cavity quality factor, which leads to a longer τ_{LP} (Sun et al., 2017b).

2.2.1.4 Coherence property

In polariton condensation, a macroscopic number of particles occupy a single-quantum state, leading to quantum correlations on the macroscopic scale, which can be quantified by coherence functions (Glauber, 1963). The wavefunction of the condensate serves as the order parameter.

A finite condensate wavefunction manifests as a finite off-diagonal elements of the density matrix of the system, often referred to as off-diagonal long-range order. It can be measured through the first-order spatial coherence functions by Billet interferometers (Richard et al., 2005) and Michelson interferometers (Deng et al., 2007; Kasprzak et al., 2006; Roumpos et al., 2012), from which increased spatial coherence above condensation threshold was observed (Deng et al., 2007; Kasprzak et al., 2006; Roumpos et al., 2012). Second- and higher order coherence functions of the state further characterize the statistical nature of the condensate and distinguish it from a thermal mixture. The second order temporal coherence function of polariton condensates was first measured in 2D planar microcavities (Deng et al., 2002), where the coherence is limited by intensity fluctuations from mode competition, reservoir-induced intensity fluctuations, and condensation depletion. Achieving intensity fluctuation at the Poisson limit of $g^2(0) \sim 1$ requires stable single-mode polariton lasing (Kim et al., 2016; Zhang et al., 2014), which was demonstrated in three-dimensional confined polariton systems with strong polarization anisotropy, implemented by a hybrid grating-DBR cavity (Kim et al., 2016) and by an etched pillar cavity (Klaas et al., 2018).

2.2.2 Polariton superfluidity

Superfluidity, the ability of a quantum fluid to flow without friction, is one of the most spectacular phenomena occurring in degenerate gases of interacting bosons. It has been intensively studied in exciton polariton condensates, especially the unique properties stemming from their intrinsically nonequilibrium nature (Keeling and Berloff, 2009).

A quantum-field-theoretical formulation for a weakly interacting BEC system was first developed by Bogoliubov, which predicted the phonon-like linear excitation spectrum at the low-momentum regime (Bogolyubov, 1959). The Bogoliubov excitation spectrum of polaritons has been studied both theoretically and experimentally with reasonable agreement (Fig. 4A) (Utsunomiya et al., 2008; Wouters and Carusotto, 2007). The superfluid critical velocity is eight orders of magnitude higher than that for atomic BEC. In addition to the normal Bogoliubov excitation, a ghost branch with a negative dispersion has been predicted (Wouters and Carusotto, 2007) and recently observed in a polariton condensate close to thermal equilibrium (Ballarini et al., 2020; Pieczarka et al., 2020).

In addition to the Bogliubov excitaion, quantized vortices may form in a superfluid. Quantized vortices pinned at disorders were first observed in 2008

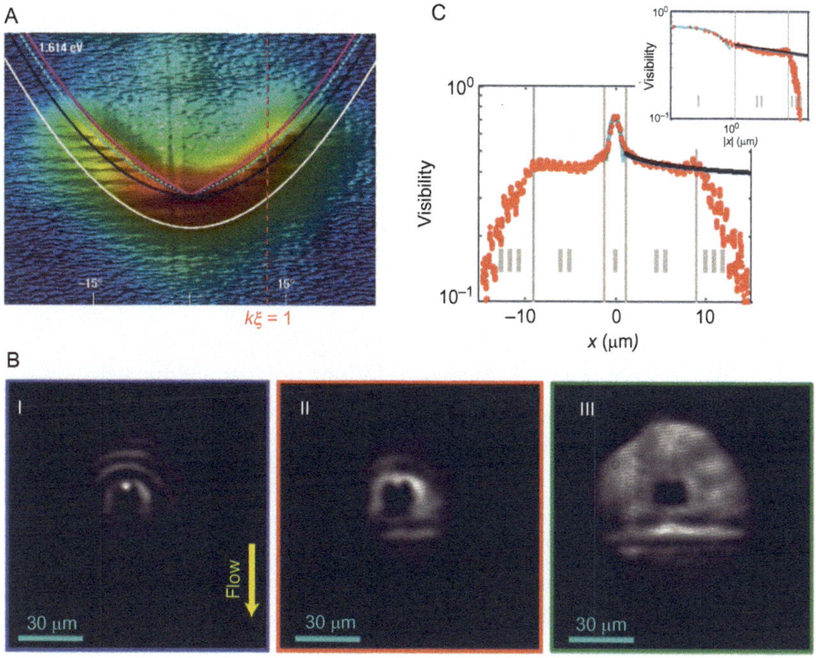

Fig. 4 Various phase transitions of polaritons. (A) Time integrated dispersion of lower polariton branch at the pumping rate of $4P_{th}$ (P_{th} is the polariton lasing threshold). Four theoretical curves represent the Bogoliubov excitation energy on the basis of the homogeneous model (*pink line*), and the Bogoliubov excitation curve based on the local-density approximation (*light-blue dotted line*), the quadratic dispersion curve starting from the condensate energy (*black line*), and the noninteracting free-polariton quadratic dispersion curve(*white line*). (B) The experimental near-field images of the excitation spot around a defect for the three excitation densities: I below threshold, II around threshold, III above threshold. At low density (B-I), the polariton fluid scatters on the defect, giving rise to interference fringes. At high powers the fringes are significantly affected by polariton–polariton interactions (B-II) and eventually show the onset of a superfluid regime (B-III). (C) The visibility of interference fringes measured by Michelson interferometer as a function of spatial distance. The regions I–III are selected manually. In region II, a power-law fit with exponent $a_p = 0.082$ is shown. (*Insert*: The same in a log-log plot, where the power-law decay appears as a *straight line*). *Panel (A): From Utsunomiya, S., Tian, L., Roumpos, G., Lai, C.W., Kumada, N., Fujisawa, T., Kuwata-Gonokami, M., Löffler, A., Höfling, S., Forchel, A., Yamamoto, Y., 2008. Observation of Bogoliubov excitations in exciton-polariton condensates. Nat. Phys. 4 (9), 700–705. https://doi.org/10.1038/nphys1034; Panel (B): From Amo, A., Lefrère, J.A., Pigeon, S., Adrados, C., Ciuti, C., Carusotto, I., Houdré, R., Giacobino, E., Bramati, A., 2009. Superfluidity of polaritons in semiconductor microcavities. Nat. Phys. 5 (11), 805–810. https://doi.org/10.1038/nphys1364; Panel (C): From Nitsche, W.H., Kim, N.Y., Roumpos, G., Schneider, C., Kamp, M., Höfling, S., Forchel, A., Yamamoto, Y., 2014. Algebraic order and the Berezinskii-Kosterlitz-Thouless transition in an exciton-polariton gas. Phys. Rev. B 90(20), 205430. https://doi.org/10.1103/PhysRevB.90.205430.*

(Lagoudakis et al., 2008). Bound, counter-circulating vortex–antivortex pairs were also observed in more homogeneous samples (Roumpos et al., 2011).

A striking feature of superfluidity is the absence of density excitations when the fluid hits a localized static obstacle at flow speeds below some critical velocity. Such evidence was presented by the measurement of Rayleigh scattering of a polariton condensate in the presence of an impurity (Amo et al., 2009). A polariton condensate was created by resonant pumping at certain angle with a continuous wave laser, and centered around natural defects in the sample. At low densities, Rayleigh scattering by the defect was observed. With increasing density n, the speed of sound $c_s = \sqrt{\hbar n/m}$ of the polariton condensates increases. When polariton flow velocity v_p is lower than c_s, Landau criterion for superfluidity is satisfied and suppressed polariton scattering can be observed (Fig. 4B).

2.2.3 Berezinskii–Kosterlitz–Thouless phase

Microcavity exciton polaritons are 2D gasses, which allows investigation of interesting physics peculiar to 2D, such as the Berezinskii–Kosterlitz–Thouless (BKT) transition, driven by the interplay of long-range order and thermally excited vortices (Berezinskiĭ, 1972). At low temperatures, local orders may form in a 2D Bose gas accompanied by thermally excited vortices. With further cooling of the system below a critical temperature T_{BKT}, bound vortex–antivortex pairs form, which stabilize the phase to a power-law decay versus distance. The critical temperature of the BKT transition is

$$k_B T_{BKT} = n_s \frac{\pi \hbar^2}{2m^2} \tag{18}$$

where n_s is the superfluid density. While the observation of vortices and vortex–antivortex pairs may suggest a connection to a BKT phase (Lagoudakis et al., 2008; Roumpos et al., 2011), such pairs could be a result of disorders. In a true BKT phase, the pairs associated with phase fluctuations are thermally activated. One way to identify them is to conduct single-short measurements, which is often hindered by the low number of photons from a short-lived condensate (Byrnes et al., 2014). Single-short polariton condensate has been demonstrated recently in a high quality GaAs microcavity with polariton lifetime of ~200 ps (Estrecho et al., 2018), which may enable studies of spontaneous vortices in the future.

Another experimental evidence for the BKT transition is the power-law decay of the spatial coherence function $g^{(1)}$ and the dependence of the power-law exponent on the density:

$$g^{(1)}(x, -x) = \frac{n_s}{n}\left(\frac{|x|}{\lambda}\right)^{-a_p}, \tag{19}$$

where n_s and n are the superfluid and total densities, λ is the characteristic length on the order of the healing length, and the exponent is $a_p = \left(n_s\lambda_T^2\right)^{-1}$. The critical density from Eq. (18) implies that $a_p = \frac{1}{4}$ at the BKT threshold and $a_p < \frac{1}{4}$ above the threshold. This has been demonstrated experimentally (Nitsche et al., 2014) (Fig. 4C).

2.2.4 Bardeen–Cooper–Schrieffer phase

At high carrier densities beyond the Mott transition density, the exciton binding energy is screened by the surrounding charge carriers, and thus an exciton polariton laser is expected to continuously transform to a standard photon laser based on electron-hole plasma (Szymanska et al., 2003). Interestingly, theoretical studies predicted the existence of another many-body phase with BCS-like correlation at the intermediate densities. Such a phase has been predicted first by models where excitons are approximated by localized two-level systems (Keeling et al., 2005), which was later extended to include exciton–exciton interactions through a contact potential (Yamaguchi et al., 2013). To include realistic internal electron-hole structure and their Coulomb interaction, models are developed, first in the Hartree–Fock approximation for a closed system at $T = 0$ (Byrnes et al., 2010; Kamide and Ogawa, 2010), and most recently extended to beyond Hartree–Fock approximation for a finite temperature, non-equilibrium system (Hu et al., 2019). The BCS-like phase consists of a condensate made of Coulomb or photo-induced bound electron-hole pairs, with a gap opening in both conduction and valence bands. Early experiments showed a photoluminescence sideband at high densities that could be possibly explained by the polariton BCS theory (Horikiri et al., 2016). This BCS-like phase has been recently reported in an unconventional polarization-selective cavity, wherein with increasing density, a smooth crossover from a polariton BEC sate to a BCS-like sate was identified by measurement of fermionic gain in its electronic reservoir due to population inversion (Hu et al., 2019).

2.3 Polariton devices

With the exciton polaritons resided on the semiconductor chip platform, the study of the rich many-body phenomena exhibited by polaritons is driven not only by scientific curiosity and discovery, but also by the promise of turning those phenomena to novel optoelectronic devices, such as low threshold lasers, ultrafast switches, and quantum simulators.

2.3.1 Polariton lasers

Polariton lasers are formed via stimulated scattering into the condensate below the band inversion density. Therefore they offer the advantage of ultralow threshold and potentially faster response time compared to conventional lasers driven by band inversion. Polariton lasers accompanied with spontaneous coherence have been shown to exhibit thresholds two or more orders of magnitude lower than that of conventional photon lasers. A practical polariton laser needs to be electrically pumped, operate at room temperature, and maintain strong coherence at the presence of room temperature thermal fluctuation. The demonstration of polariton lasing with electrical injection in GaAs microcavities at cryogenic temperatures was reported (Fig. 5A) (Bhattacharya et al., 2013; Schneider et al., 2013), however, the operation temperature is limited by the small binding energy of GaAs. Room temperature polariton lasing under optical pumping has been demonstrated in new material systems (see Section 3). Most recently, RT polariton light emission diodes (LED) and polariton laser have been demonstrated in several systems (Bhattacharya et al., 2014; Graf et al., 2017; Gu et al., 2019; Jayaprakash et al., 2019). The coherence property of room temperature polariton lasers has only been characterized by measuring the first-order coherence function (Christopoulos et al., 2007; Daskalakis et al., 2014; Zhang et al., 2015b). More thorough study of the coherence properties including intensity stability of these room temperature lasers is yet to be done (Kim et al., 2016).

2.3.2 Polariton switches

Photonic information processing has shown many advantages over electronics, such as low heating and high speed operation. However, small nonlinearity in typical photonic systems and limitations in integration with electronic materials hamper its development. Microcavity exciton polaritons, as mixed states of photon and exciton, shed light on the next generation devices combing both strong nonlinearity of excitons and coherence and

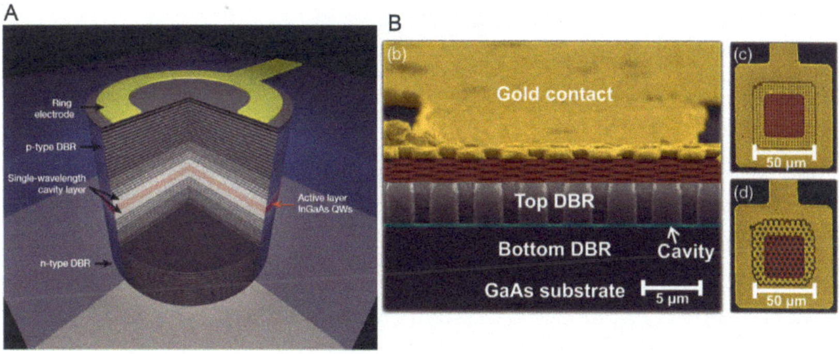

Fig. 5 Polariton devices. (A) Schematics of an electrically contacted 20-mm-diameter micropillar with four quantum wells (QWs) in the cavity, sandwiched by gradually doped distributed Bragg reflectors (DBRs). (B) Scanning electron microscope (SEM) image of half-etched polariton lattice with top and bottom electrical contacts. *Panel (A) From Schneider, C., Rahimi-Iman, A., Kim, N.Y., Fischer, J., Savenko, I.G., Amthor, M., Lermer, M., Wolf, A., Worschech, L., Kulakovskii, V.D., Shelykh, I.A., Kamp, M., Reitzenstein, S., Forchel, A., Yamamoto, Y., Höfling, S., 2013. An electrically pumped polariton laser. Nature 497 (7449), 348–352. https://doi.org/10.1038/nature12036; https://doi.org/10.1103/PhysRevA.69.023809 Panel (B) From Suchomel, H., Klembt, S., Harder, T.H., Klaas, M., Egorov, O.A., Winkler, K., Emmerling, M., Thomale, R., Höfling, S., Schneider, C., 2018. Platform for electrically pumped polariton simulators and topological lasers. Phys. Rev. Lett. 121 (25), 257402. https://doi.org/10.1103/PhysRevLett.121.257402.*

low dephasing of light. The ultrafast dynamics of polaritons will enable fast responses to external triggers, and the rapid motion of polaritons will allow the switches to be nonlocal (Amo et al., 2010).

Most of the polariton switches are designed based on polariton bistability. If the lower polariton branch is driven near resonantly but blue detuned by at least the polariton linewidth, with increasing pumping power, the polariton state will blue shift. When it is on resonance with the laser energy, the polariton population sharply increases, giving rise to an S-shaped dependence of the polariton population as a function of power. Inversely when reducing the power, however, the transition back to the low-population regime occurs at a lower power (Baas et al., 2004). Such a polariton bistable nonlinearity is typically employed in proposed (Liew et al., 2008) and demonstrated polariton logic devices (Amo et al., 2010; Ballarini et al., 2013). By using a novel bistability named as spin bifurcation (Ohadi et al., 2015), which does not rely on coherence from outside of the system, an electrical spin switch was demonstrated (Dreismann et al., 2016). Most recently, an all optical spin switch based on organic polaritons was demonstrated at room temperature (Zasedatelev et al., 2019).

2.3.3 Quantum states and quantum simulators

Other than the analogues of classical devices such as polariton lasers and polariton switches, quantum states of polaritons have also been studied and proposed for future applications. Squeezed polariton states (Boulier et al., 2014) and polariton antibunching (Delteil et al., 2019; Muñoz Matutano et al., 2019) have been reported.

The realization of polariton lattices (see Section 3) propelled the field of quantum simulations based on microcavity polaritons (Berloff et al., 2017; Lai et al., 2007). The polariton system can be seen as a simulator for interacting bosons with a Hamiltonian determined by the polariton landscape. Furthermore, by incorporating spin dynamics controllable by light polarization, exciton polaritons have the potential to study spin Hamiltonian and explore magnetic order. Recently, an electronically driven square and honeycomb lattice of exciton polaritons has been reported, paving the way toward real world devices based on polariton lattices for on-chip applications (Fig. 5B) (Suchomel et al., 2018).

Polariton systems also provide a versatile platform to simulate quantum fluids and study quantum hydrodynamics (Boulier et al., 2020). Quantum turbulence (Boulier et al., 2016), phase bistability in one-dimensional dark solitons (Goblot et al., 2016), and multistable dissipative gap solitons in a 1D flat band (Goblot et al., 2019) have been studied in microcavity polariton systems. A complete theoretical scheme of quantum computing with exciton polariton condensates formed in semiconductor micropillars has also been proposed (Ghosh and Liew, 2020).

The effects of dissipation and dephasing, which are absent in purely photonic systems, have to be considered in realizing polariton quantum devices. These can be suppressed by symmetry-protected topological states, which have been proposed (Karzig et al., 2015; Nalitov et al., 2015) and observed in polariton system (Klembt et al., 2018).

3. Control of polariton modes

Exciton polaritons have become a versatile platform for a wide range of research including many-body physics, quantum optics, nonequilibrium dynamics, polariton devices. Many of these studies involve engineering of the polariton modes. For example, by reducing the spatial size of the polariton mode, the mode properties, coherence, and interactions of a polariton laser can be modified (Bajoni et al., 2008; Kim et al., 2016; Klaas et al., 2018; Zhang et al., 2014), and the enhanced interaction is desirable for quantum photonic studies (Delteil et al., 2019; Muñoz Matutano et al., 2019). By constructing a periodic potential, energy band structures can be realized in polaritons, enabling novel fundamental phenomena, such as polariton topological insulators (Klembt et al., 2018) and Hamiltonian simulators (Berloff et al., 2017). Several reviews have already covered many important engineered polariton systems (Byrnes et al., 2014; Fraser, 2017; Gibbs et al., 2011; Sanvitto and Kéna-Cohen, 2016), as well as techniques specific to creating a potential landscape (Schneider et al., 2017). Being composite quasiparticles, exciton-polaritons benefit from many degrees of freedom for tuning and control. In this section, we give a brief review on the engineering and control of polariton via its photonic and excitonic modes.

3.1 Control of photonic modes

The photon mode is determined by and confined in a cavity. Modifications to the cavity can introduce changes to the photon mode. Properties of the polariton mode, such as spatial distribution, energy-wavevector dispersion, and decay rate, crucially depend on the photonic mode and therefore can be tailored to a great extent. The most widely used cavities have been Fabry–Pérot (FP) cavities formed by distributed Bragg reflectors (DBRs). More recently, many unconventional structures, such as photonic crystals, have emerged, which allow greater flexibility in mode engineering and control.

3.1.1 Planar DBR-DBR cavity

The Fabry–Pérot microcavity made of two DBR mirrors has been an archetypal structure for exciton polaritons (Weisbuch et al., 1992). Formed by two parallel DBR mirrors, the photonic resonant mode has a quasi-2D geometry, featuring a parabolic in-plane dispersion at small wavevectors, with tight confinement in the other direction. Two properties are readily modified in such planar cavities: the quality factor and resonance energy of the cavity mode. For III–V semiconductors, monolithic DBR-DBR cavities, including the QWs embedded inside, can be grown epitaxially, with minimal defects or impurities. By increasing the number of the quarter-wavelength layers in the DBR, quality factors higher than 3×10^5 has been achieved (Sun et al., 2017a). The cavity layer thickness typically decreases from the center of the wafer toward the edge, leading to a gradual change of the cavity mode resonance. The polarization dependence of the reflection phase of the mirrors can give rise to a TE-TM energy splitting of the cavity mode (Panzarini et al., 1999) which contributes to the optical spin Hall effect of polaritons (Kavokin et al., 2005; Leyder et al., 2007). By filling the cavity with liquid crystal medium, stronger anisotropy can be introduced, tunable by applying a voltage that changes the orientation of the molecules, and used to realize different forms of spin-orbit coupling (Lekenta et al., 2018; Rechcińska et al., 2019).

3.1.2 Deposition

To modulate the planar cavity mode in a nondestructive manner, metal films can be deposited on the top surface of the sample. These metal films will locally shorten the effective cavity length and increase the resonance energy due to suppression of the electromagnetic field inside them (see Fig. 6). The modulation strength is limited by the thickness of the upper DBR and is typically on the order of 0.1 meV. As such, this method is best suited for creating a weak periodic modulation for polaritons. It was used to demonstrate the first 1D periodic potential for microcavity potentials, leading to the observation of phase-locked condensate arrays (Lai et al., 2007). Later, many different types of 2D lattices were created and corresponding polariton

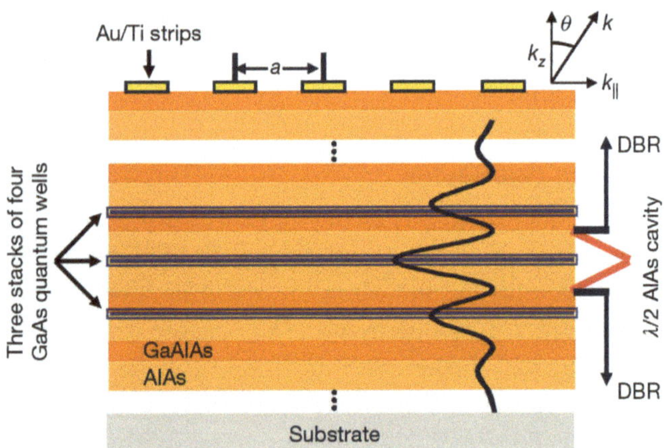

Fig. 6 A schematic of a cavity polariton array formed by depositing periodic thin metallic strips (Au/Ti) on top of a microcavity structure. The 1.4 μm-wide strips are equally spaced with a pitch distance $a = 2.8$ μm. *From Lai, C.W., Kim, N.Y., Utsunomiya, S., Roumpos, G., Deng, H., Fraser, M.D., Byrnes, T., Recher, P., Kumada, N., Fujisawa, T., Yamamoto, Y., 2007. Coherent zero-state and π-state in an exciton-polariton condensate array. Nature 450 (7169), 529–532. https://doi.org/10.1038/nature06334.*

bands and band-polariton phases have been studied, with an emphasize on dynamic condensation in high-orbital states (Kim et al., 2011, 2013, 2014; Kusudo et al., 2013; Masumoto et al., 2012).

3.1.3 Etch and overgrowth

To create a stronger modulation of the cavity mode, apertures are embedded close to the center of the cavity (El Daïf et al., 2006). To achieve this, the epitaxial chamber is connected to lithography and etching systems. After the growth of the bottom DBR and cavity layer with QWs, the cladding of the cavity is patterned through lithography and etching. The growth of the top DBR continues afterwards, resulting in a patterned cavity. The pattern embedded at the bottom of the top DBR modulates the cavity length and the cavity resonance up to a few meV. Exciton-polariton luminescence and condensate have been studied in single potential traps (Kaitouni et al., 2006; Nardin et al., 2010; Winkler et al., 2013) as well as lattices (Harder et al., 2020; Winkler et al., 2015, 2016) (Fig. 7).

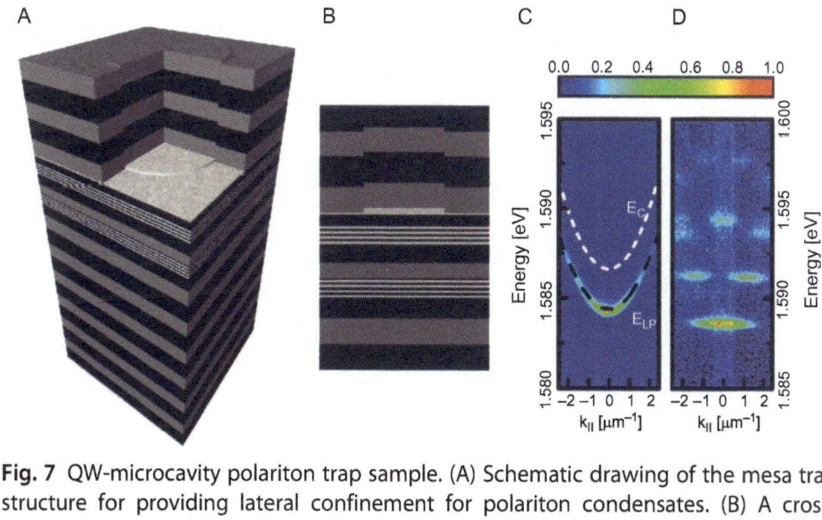

Fig. 7 QW-microcavity polariton trap sample. (A) Schematic drawing of the mesa trap structure for providing lateral confinement for polariton condensates. (B) A cross-sectional drawing. (C) and (D) Energy-momentum dispersion of a trap with a diameter of 30 μm and 4 μm respectively. In (C) a fit to the lower polariton dispersion (*dashed black line*) and the corresponding bare photonic dispersion (*dashed white line*) is shown. *From Winkler, K., Fischer, J., Schade, A., Amthor, M., Dall, R., Geßler, J., Emmerling, M., Ostrovskaya, E.A., Kamp, M., Schneider, C., Höfling, S., 2015. A polariton condensate in a photonic crystal potential landscape. New J. Phys. 17 (2), 023001. https://doi.org/10.1088/1367-2630/17/2/023001.*

3.1.4 Etching

As an alternative to modulating the local cavity length, lateral confinement can be imposed by modifying the cavity's physical dimension, through means of lithography and etching (Tartakovskii et al., 1998). Polariton lasing in such lower-dimensional structures was first demonstrated in 0D pillar cavities around 3μm in diameter (see Fig. 8) (Bajoni et al., 2008). The strong confinement leads to stronger nonlinearity and reduced intensity fluctuations (Kim et al., 2016; Klaas et al., 2018). In reflection measurement, with the discrete spectrum and the spatial mode matching the pump laser, strong intensity squeezing was observed (Boulier et al., 2014). Similarly, quasi-1D devices have been created, which can be used to guide polariton flow (Fischer et al., 2014; Goblot et al., 2016; Sturm et al., 2014; Wertz et al., 2010, 2012).

When the micropillars overlap, localized modes at neighboring sites can couple and form new eigenmodes (Galbiati et al., 2012), exhibiting effects such as Josephson oscillation and self-trapping (Abbarchi et al., 2013). The polarization dependent coupling strength can be harnessed to simulate spin-orbit coupling effects in a molecule-like device (Sala et al., 2015). Extending to a larger array, a lattice pattern can be imposed on the polaritons.

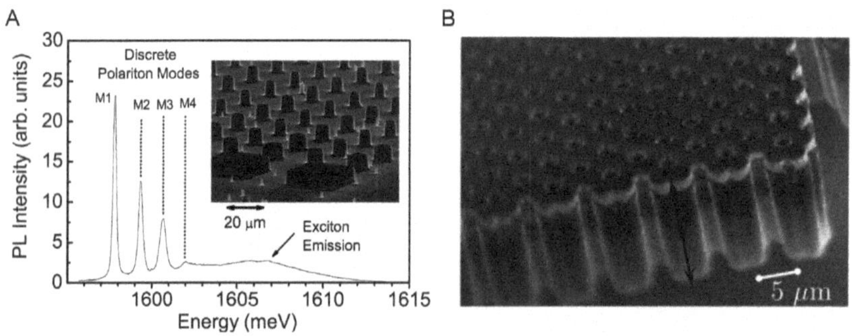

Fig. 8 Etched micropillars. (A) PL spectrum measured on a single 4 μm diameter micropillar; *inset*: scanning electron micrography of a micropillar array. (B) Scanning electron microscope image of a corner of the microstructure. One hexagon of pillars is *underlined* with *blue disks*. The *dark arrows* show the growth axis of the cavity. *Panel (A): From Bajoni, D., Senellart, P., Wertz, E., Sagnes, I., Miard, A., Lemaître, A., Bloch, J., 2008. Polariton laser using single micropillar GaAs–GaAlAs semiconductor cavities. Phys. Rev. Lett. 100 (4), 047401. https://doi.org/10.1103/PhysRevLett.100.047401; Panel (B): From Jacqmin, T., Carusotto, I., Sagnes, I., Abbarchi, M., Solnyshkov, D.D., Malpuech, G., Galopin, E., Lemaître, A., Bloch, J., Amo, A., 2014. Direct observation of dirac cones and a flatband in a honeycomb lattice for polaritons. Phys. Rev. Lett. 112 (11), 116402. https://doi.org/10.1103/PhysRevLett.116.116402.*

In 1D modulated wire and 1D Lieb lattice, effects related to interaction and disorder such as gap soliton (Goblot et al., 2019; Tanese et al., 2013) and localization (Baboux et al., 2016) have been observed. 2D honeycomb (Jacqmin et al., 2014; Suchomel et al., 2018) and Lieb (Whittaker et al., 2018) lattices have also been studied. With an external magnetic field that breaks the time reversal symmetry, topological edge modes have been found in a honeycomb lattice (Klembt et al., 2018).

Deep etching into the bottom DBR provides effective lateral confinement of the cavity mode and can be used to create a deeper periodic potential. But it also leads to etched surfaces in the QWs, which may lead to undesirable effects such as surface potentials and increased nonradiative recombination. To avoid these, sometimes only the top DBR is etched while the QWs are left intact (Klembt et al., 2018; Suchomel et al., 2018).

3.1.5 Concave mirrors

Concave DBR mirrors give rise to a focusing photon mode, which also leads to laterally confined polaritons. Cavities have been constructed using a concave mirror as the top mirror, either fabricated as a standalone mirror or integrated into the end of a fiber (see Fig. 9). Such concave mirrors are placed

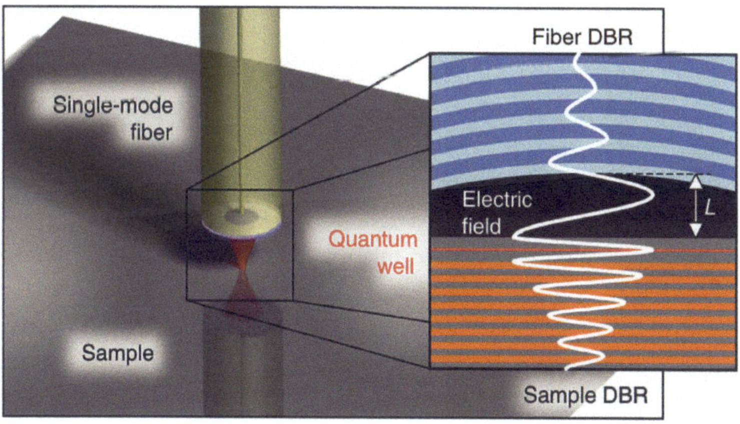

Fig. 9 Semiintegrated microcavity consisting of a curved fiber-end mirror and an integrated planar DBR mirror, with a quantum well grown on top. The *inset* shows the electric field inside the cavity. The quantum well is sandwiched within the surface layer of the semiconductor DBR in a way that the electric field at its position is maximal. The cavity length L is defined as the distance between the two mirror surfaces and is controlled by a piezo actuator. *From Besga, B., Vaneph, C., Reichel, J., Estève, J.A., Reinhard, A., Miguel-Sánchez, J., Imamoğlu, A.A., Volz, T., 2015. Polariton boxes in a tunable fiber cavity. Phys. Rev. Appl. 3 (1), 014008. https://doi.org/10.1103/PhysRevApplied.3.014008.*

above the bottom cavity using piezo-controlled stages to achieve good lateral and angular alignment. Adjusting the distance between the external concave mirror and the bottom half of the cavity enables in-situ tuning of the cavity resonance (Besga et al., 2015; Dufferwiel et al., 2014, 2015c). Polariton lasing has been observed in such a system (Besga et al., 2015; Scafirimuto et al., 2018), and the TE-TM anisotropy can be strong enough to induce energy splittings between different vortex states (Dufferwiel et al., 2015a). The lateral confinement leads to enhanced interactions and photon antibunching has been measured (Delteil et al., 2019; Muñoz Matutano et al., 2019).

3.1.6 Subwavelength grating mirror

The DBR can also be replaced by slab photonic crystals, which enables not only potential landscape engineering, but also modification of other fundamental properties of the cavity mode, such as its polarization state and energy-momentum dispersion. While the design flexibility of slab photonic crystals have been intensely researched since the late 1990s with many remarkable demonstrations (Istrate and Sargent, 2006), a challenge in using them for polaritons is the requirement of high reflectance. It was demonstrate in 2007 that broad-band high reflectance may be obtained in subwavelength high index contrast gratings (Chang-Hasnain and Yang, 2012; Huang et al., 2007; Qiao et al., 2018). The period is below the wavelength to prohibits diffraction orders. High reflectivity is achieved by destructive interference of internal modes at the interface. Using such a subwavelength grating mirror to replacing the DBR, a polariton microcavity was constructed, supporting single-mode polariton lasing (Zhang et al., 2014) (see Fig. 10). The gratings are usually suspended to suppress nonzeroth order diffraction. Recently, monolithic variants have been demonstrated for polariton lasers, allowing higher mechanical strength and stability (Kim et al., 2019).

The subwavelength grating reflectors provide many degrees of freedom to control the polariton modes. Strong lateral confinement and potential landscapes of the polaritons can be achieved by controlling the sizes and patterns of the grating, without the need to etch through the QWs or to embed structures inside the cavity structure (Zhang et al., 2014). For suspended gratings, stress-induced bending of the grating can also be utilized to modulate the cavity length; therefore a potential landscape can be engineered via the tethering patterns around the gratings (Kim et al., 2020; Zhang et al., 2015a). These quasi-0D devices exhibit discrete states and shot-noise-limited intensity stability owing to suppressed mode competition and condensate depletion (Kim et al., 2016).

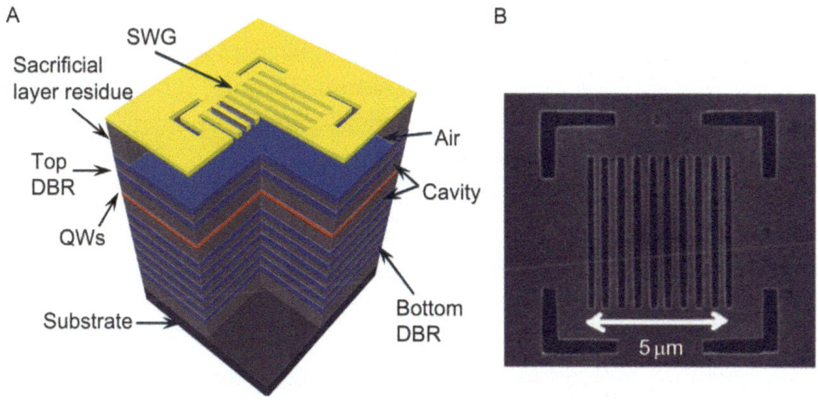

Fig. 10 Examples of the hybrid cavity. (A) A schematic of a 0D hybrid cavity with a SWG mirror. (B) Top-view SEM image of a fabricated 0D cavity with a SWG of 5 × 5 μm in size. *From Zhang, B., Wang, Z., Brodbeck, S., Schneider, C., Kamp, M., Höfling, S., Deng, H., 2014. Zero-dimensional polariton laser in a subwavelength grating-based vertical microcavity. Light Sci. Appl. 3 (1), e135. https://doi.org/10.1038/lsa.2014.16.*

A grating also has distinct features unavailable with DBRs, such as strong polarization dependence and controllable reflectivity and reflection phase. By designing the reflection phase of grating, the energy-momentum dispersion of the polaritons can be engineered, such as to produce a lighter or heavier effective mass, flat band polaritons, and Mexican-hat dispersions (Wang et al., 2015). The grating generally provides high reflectivity in one linear polarization and low reflectivity in the orthogonal one, correspondingly the cavity has high and low quality factors for the two polarizations, respectively. Hence strong coupling and weak coupling coexist at the same spatial location when grating mirrors are used (Zhang et al., 2014). This provides access to the polariton condensate and the exciton reservoir simultaneously, and has enabled the identification of the BCS regime of polariton lasing (Hu et al., 2019).

3.1.7 Photonic crystal

Instead of using a slab photonic crystal as a mirror, the photonic crystal can function as a resonator by itself, with even greater freedom of mode engineering. The photonic band structure can be tailored directly via the lattice structure instead of relying on the modulation of a Fabry–Pérot resonance. A defect in the photonic crystal can form a resonance in the band gap that has high quality factor and diffraction limited mode volume (Akahane et al., 2003).

However, integration to a 2D exciton medium in the photonic crystal is challenging for conventional semiconductors. The QW will be patterned together with the photonic crystal. The exposed surfaces in the QW, with a very large surface to volume ratio, may lead to unwanted surface states, surface charging, and increase of nonradiative recombination. Surface passivation becomes crucial. With III-P-based materials, strong coupling between a QW embedded inside a photonic crystal has been achieved and low threshold polariton lasing has been observed (Azzini et al., 2011). With III-As materials, strong coupling was reported when the QW is placed above the photonic crystal (Bajoni et al., 2009). Here the QW couples to the evanescent field of the photonic crystal mode with a reduced coupling strength.

With the discovery of monolayer van der Waals semiconductors, it was recognized that, without out-of-plane dangling bonds, these materials are naturally resistant to surface effects and form stable, atomically thin crystals, which can be placed directly on top of a photonic crystal slab, where the field enhancement remains large. Strong coupling between monolayer WSe_2 or WS_2 and a 1D photonic crystal has been demonstrated, which also features strongly anisotropic dispersions and tunable Fano lineshapes (Zhang et al., 2018).

3.2 Control of excitons

The polariton modes can also be controlled via the exciton components. A versatile method is using an optical pump to create a mean-field potential for excitons, which leads to interesting yet complicated dynamics. Strain-induced piezo-electrical fields have been effective in creating confinement potentials on a relatively large scales of many tens of microns. Over a similar length scale, external electric and magnetic fields are also used to manipulate the excitons and polaritons.

3.2.1 Optical pumping

Excitons, as approximate, composite bosons, feature both a nonlinearity due to Coulomb interactions and saturation of the oscillator strength due to phase space filling of the constituent fermionic particles. Both effects lead to a shift of the polariton resonance dependent on the carrier density. Therefore a spatially nonuniform carrier density distribution produced through optical pumping can be utilized to create an effective potential landscape for polaritons.

Optical-pump created potentials have been demonstrated to confine and quantize 1D exciton polariton states (Wertz et al., 2010) (Fig. 11A), where the potential size and height can be tuned in-situ by the pumping position and power. Following this work, optical pumping has also been used to create and control polariton vortices (Ma et al., 2020; Sanvitto et al., 2011) and

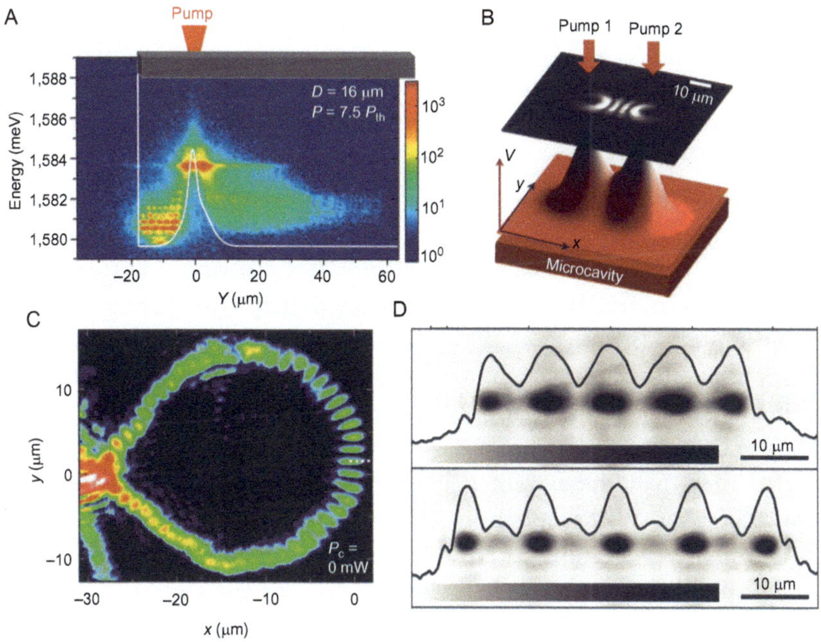

Fig. 11 Exciton polaritons engineered by optical pumping. (A) Discrete polariton states in a trapped potential induced by optical pumping. (B) Optical trapping induced by two pump spots. (C) Real-space image of a polariton Mach–Zehnder interferometer. (D) Experimental realization of a linear chain of five equidistant polariton nodes with lattice constants of 9 μm. *Panel (A): From Wertz, E., Ferrier, L., Solnyshkov, D.D., Johne, R., Sanvitto, D., Lemaître, A., Sagnes, I., Grousson, R., Kavokin, A.V., Senellart, P., Malpuech, G., Bloch, J., 2010. Spontaneous formation and optical manipulation of extended polariton condensates. Nat. Phys. 6 (11), 860–864. https://doi.org/10.1038/nphys1750; Panel (B): From Tosi, G., Christmann, G., Berloff, N.G., Tsotsis, P., Gao, T., Hatzopoulos, Z., Savvidis, P.G., Baumberg, J.J., 2012. Sculpting oscillators with light within a nonlinear quantum fluid. Nat. Phys. 8 (3), 190–194. https://doi.org/10.1038/nphys2182; Panel (C): From Sturm, C., Tanese, D., Nguyen, H., Flayac, H., Galopin, E., Lemaître, A., Sagnes, I., Solnyshkov, D., Amo, A., Malpuech, G., Bloch, J., 2014. Alloptical phase modulation in a cavity-polariton Mach-Zehnder interferometer. Nat. Commun. 5 (1), 3278. https://doi.org/10.1038/ncomms4278; Panel (D): From Berloff, N.G., Silva, M., Kalinin, K., Askitopoulos, A., Töpfer, J.D., Cilibrizzi, P., Langbein, W., Lagoudakis, P.G., 2017. Realizing the classical XY Hamiltonian in polariton simulators. Nat. Mater. 16 (11), 1120–1126. https://doi.org/10.1038/nmat4971.*

an oscillating quantum fluid (Tosi et al., 2012) (Fig. 11B). Toward applications, it has been used to demonstrate polariton transistors (Gao et al., 2012), interferometers (Sturm et al., 2014) (see Fig. 11C), and Hamiltonian simulators (Berloff et al., 2017) (Fig. 11D).

Not only the real part of the potential, the imaginary part of the potential can also be controlled to precisely shape the pump beam to study non-Hermitian physics and create chiral modes in the polariton system (Gao et al., 2015, 2018). Moreover, in contrast to population-induced potentials, the dynamic stark shift induced by an optical pump has also been demonstrated to shift the polariton energy (Hayat et al., 2012).

3.2.2 Strain-induced piezo-electric field

By applying external strain, the band gap of the semiconductor and exciton energy can be tuned. Trapping excitons in a harmonic potential was implemented in quantum wells by pressing a sharp tip onto the backside of the substrate (Vörös et al., 2006). When applied to a microcavity, the energy shift of exciton can be transferred to polariton via the Hopfield coefficients, leading to a harmonic potentials for polaritons (Balili et al., 2007). The depth of the potential can be several tens of meV, which is much larger than the potential induced by optical pumping (typically below 5 meV). With the strain potential, polariton condensation in a GaAs system was demonstrated via nonresonant and nonlocal pumping, where optically injected carriers diffused toward the center of the trap to form the condensate.

3.2.3 Electrical field

Applying electric field to the microcavity will result in a static quantum-confined stark effect that tunes the energy of the excitons and thereby the energy of the polaritons. This effect has been demonstrated both below (Gessler et al., 2014) and above condensation threshold (Tsotsis et al., 2014). Optical bistability (Amthor et al., 2015) and ultralow power spin switches have been observed (Dreismann et al., 2016) in electrical driven polariton condensates (Fig. 12A and B).

Excitons can also be polarized by an electric field. Strong coupling between polarized excitons and photons will result in polaritons with oriented dipole momentum, called dipolaritons, which have enhanced nonlinearity due to the long-range dipolar interactions. Dipolaritons have been implemented in two types of structures. The first is coupled asymmetric QWs, as shown in Fig. 12C. By applying electric field, the Fermi levels of two QWs can be aligned, and resonant tunneling of electron between the two wells will result in strong hybridization between spatially direct exciton and indirect excitons Fig. 12D (Christmann et al., 2011; Cristofolini et al., 2012). Such hybrid excitons possess both large oscillator strength and dipole momentum, which are also tunable by electric

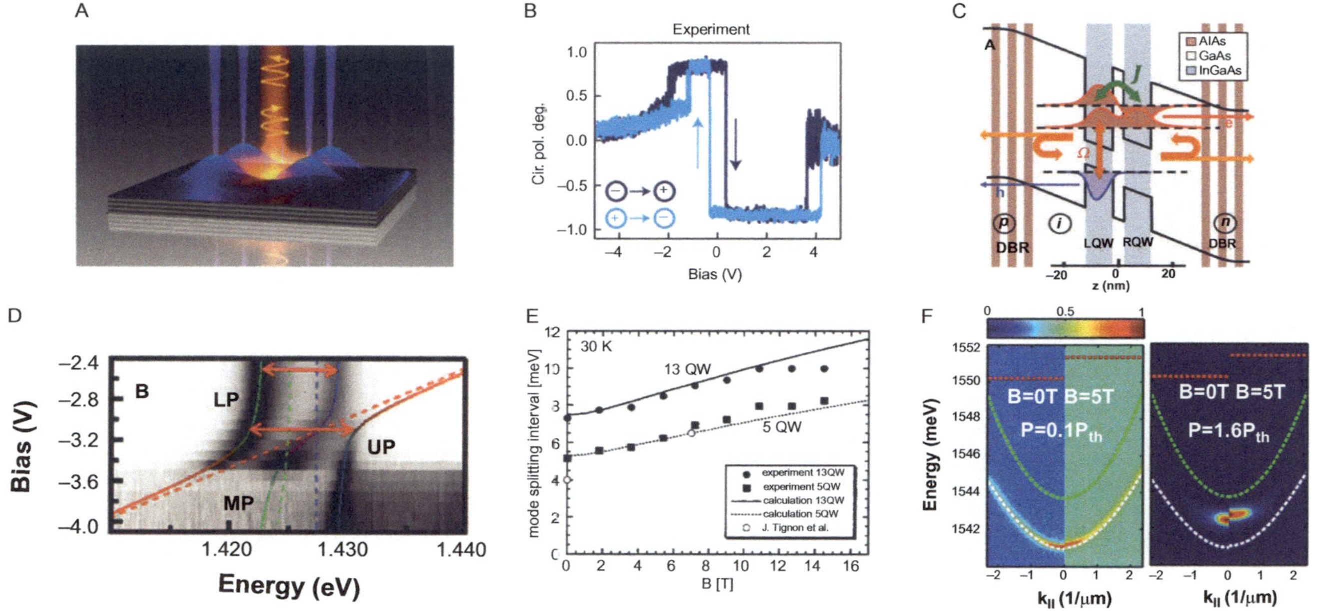

Fig. 12 Exciton polaritons controlled by electric and magnetic fields. (A) Schematic of a trapped polariton condensate created by four laser beams. The resulting condensate is confined to the center and shows circular polarization. (B) Circular polarization of emission versus applied bias under continuous excitation. Bias swept from − 5V to +5V (*dark blue*) and back (*light blue*). Positive values represent right-circular polarization. (C) Schematic band structure of coupled quantum wells in a microcavity at tunneling resonance. As a result of tunneling coupling J, electrons (*red wave functions*) extend over both wells while holes (*blue*) are confined in the LQW. (D) Normalized PL spectra with bias voltage. *Green*, *blue*, and *red dashed lines* are the uncoupled direct exciton, cavity photon, and indirect excitons. *Solid lines* are fits to the three polariton branches. (E) Increased Rabi splitting in microcavity by magnetic field. (F) Magnetic response of microcavity polariton in the linear (*left*) and polariton condensate regime (*right*). *Panel (A) and (B): From Dreismann, A., Ohadi, H., Del Valle-Inclan Redondo, Y., Balili, R., Rubo, Y. G., Tsintzos, S.I., Deligeorgis, G., Hatzopoulos, Z., Savvidis, P.G., Baumberg, J.J., 2016. A sub-femtojoule electrical spin-switch based on optically trapped polariton condensates. Nat. Mater. 15 (10), 1074–1078. https://doi.org/10.1038/nmat4722; Panel (C) and (D): From Cristofolini, P., Christmann, G., Tsintzos, S.I., Deligeorgis, G., Konstantinidis, G., Hatzopoulos, Z., Savvidis, P.G., Baumberg, J.J., 2012. Coupling quantum tunneling with cavity photons. Science 336 (6082), 704–707. https://doi.org/10.1126/science.1219010; Panel (E): From Tanaka, T., Zhang, Z., Nishioka, M., Arakawa, Y., 1996. Magnetic field dependence of exciton oscillator strength by measurements of magnetoexciton-polariton mode splitting in quantum wells with a microcavity. Appl. Phys. Lett. 69 (7), 887–889. https://doi.org/10.1063/1.117977; Panel (F): From Fischer, J., Savenko, I.G., Fraser, M.D., Holzinger, S., Brodbeck, S., Kamp, M., Shelykh, I.A., Schneider, C., Höfling, S., 2014. Spatial coherence properties of one dimensional exciton-polariton condensates. Phys. Rev. Lett. 113 (20), 203902. https://doi.org/10.1103/PhysRevLett.113.203902.*

field. The enhanced nonlinearity may also leads to more efficient scattering of polaritons to the branch bottom, thereby lowering the polariton lasing threshold (Togan et al., 2018; Tsintzos et al., 2018). The second structure to realize dipolariton is a wide QW, where an electric field perpendicular to the QW plane induces charge separation of the electron and hole in the exciton, and thus a net electric dipole moment. It has been demonstrated in a waveguide polariton system (Rosenberg et al., 2016), where an abnormal enhancement up to 200-fold enhancement of the nonlinearity was observed (Rosenberg et al., 2018).

3.2.4 Magnetic field

The magnetic field response of excitons includes three parts: diamagnetic response, Zeeman splitting, and oscillator strength, all of which lead to changes in the polariton energy.

Exciton diamagnetic response under a Faraday configuration in the low-field limit can be described as $\delta E_{dia,X} = \kappa_X B^2$, where, κ_X is diamagnetic coefficient used for estimating the exciton binding energies (Walck and Reinecke, 1998).

In addition to the diamagnetic effect, the total angular momentum of an exciton interacts with a magnetic field and gives rise to Zeeman splitting. The degenerate exciton states with opposite spins will split with an energy difference given by $\Delta E = g_X \mu_B B$ (Rahimi-Iman et al., 2011), where g_X is the Lande g-factor for an exciton, and μ_B is the Bohr magneton.

A magnetic field will also reduce the exciton Bohr radius and as a result enhance the exciton oscillator strength and normal mode splitting of polaritons (Tanaka et al., 1996) (Fig. 12E). Large polariton splitting helps to keep the system in the strong coupling regime when the microcavity is pumped at high densities in studies of macroscopic quantum phenomena (Bhattacharya et al., 2013; Schneider et al., 2013).

Above the polariton lasing threshold, a spinor polariton condensate will also be influenced by a magnetic field. Nonequilibrium spin Meissner effect has been reported (Fischer et al., 2014) (Fig. 12F). Without a magnetic field, the polariton condensate sate is linearly polarized. The magnetic field tends to orient the condensate and make the system circularly polarized through the Zeeman effect. However, the polariton–polariton interaction favors a linear polarization. At weak magnetic fields, this competition will generate elliptically polarized condensate and suppress Zeeman splitting (Fischer et al., 2014; Larionov et al., 2010).

4. Exciton polaritons in new materials

Polariton condensation provides a novel mechanism for energy-efficient lasers and promise many device applications. Reliable room temperature operation is important for a practical device, which may not been possible with a GaAs or CdTe system where the exciton binding energy is smaller than the room temperature thermal energy (~25 meV). Excitons with larger binding energies can be found in wider band gap inorganic semiconductors, such as GaN and ZnO, organic semiconductors with Frenkel excitons, Perovskite, and two-dimensional semiconductors. In this section, we will give a brief review of exciton polaritons realized in these materials.

4.1 Wide band gap inorganic semiconductors

Excitons in wide band gap semiconductors, such as GaN and ZnO, have larger binding energies and correspondingly larger exciton-photon coupling strengths. Therefore, excitons and polaritons in these materials are stable at the room temperature. However, compared to III-As semiconductors, fabricating high quality microcavities for GaN and ZnO is more challenging as a result of the larger lattice mismatches between different alloys.

4.1.1 GaN polariton

The first demonstration of strong exciton-photon coupling at room temperature was in GaN in 2015, between bulk GaN excitons in the cavity layer and the FP mode (Semond et al., 2005). The FP cavity was made of an AlN/AlGaN bottom DBR and a 10 nm thick aluminum layer as the top mirror. The Rabi splitting was about 60 meV. However, due to the very low cavity quality factor (~40), polariton lasing was not achieved. By optimizing the lattice matching condition between the mirror and bulk GaN and increasing the number of DBR pairs, the cavity Q was improved to 3000, and room temperature polarion lasing in bulk GaN was reported (Christopoulos et al., 2007). In parallel, GaN quantum wells, which allow better field overlap and electrical injection, were also successfully integrated with high quality cavities, and polariton lasing was demonstrated (Christmann et al., 2008a) (Fig. 13A). To further lower the lasing threshold, defect-free GaN nanowires with high quantum efficiencies were integrated inside DBR-DBR cavities. A polariton laser with a threshold three orders of magnitude lower than the corresponding photon laser was demonstrated (Das et al., 2011),

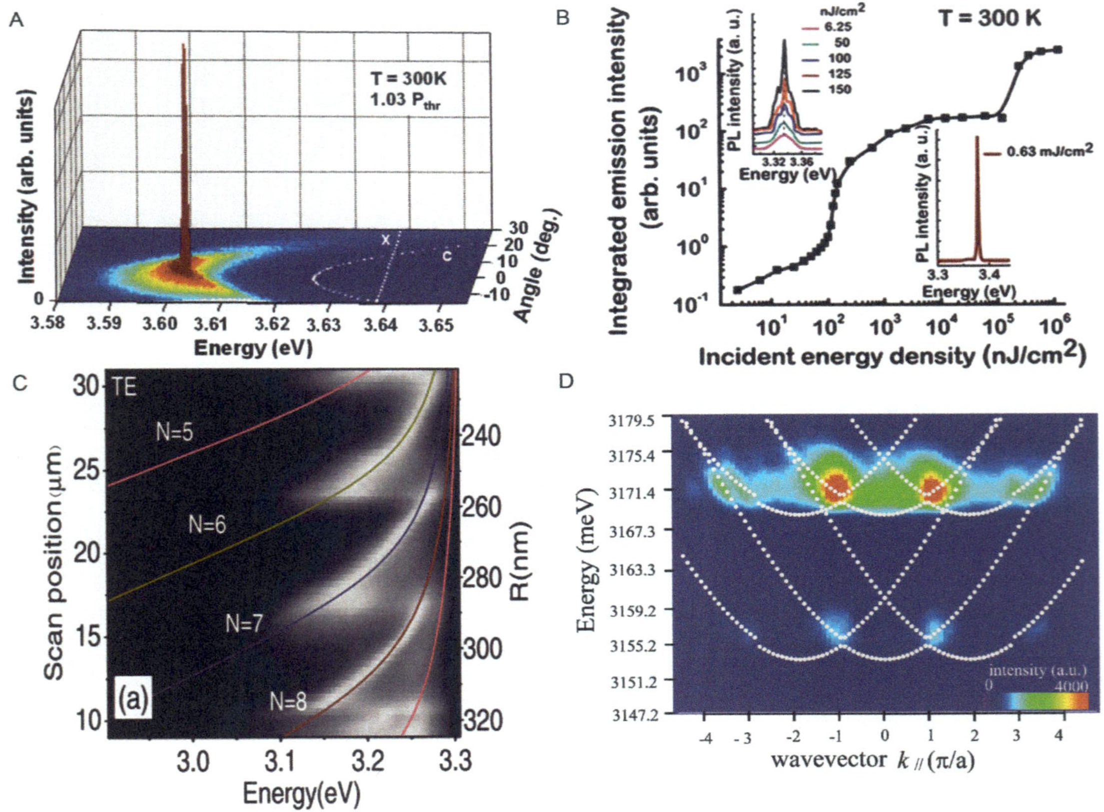

Fig. 13 Strong coupling and polariton lasing in GaN and ZnO. (A) Three-dimensional representation of the far-field emission with emission intensity displayed on the vertical axis above lasing threshold. C and X represents cavity and exciton energies (Christmann et al., 2008b). (B) Variation of the emission intensity with excitation energy at 300 K clearly showing the two thresholds for polariton and photon lasing. The insets show the corresponding emission spectra (Das et al., 2011). (C) PL mapping along the ZnO tapered arm with detection polarization of TE Sun et al. (2008). (D) Polariton condensate in a ZnO superlattice (Zhang et al., 2015b).

as shown by Fig. 13B. An electrically pumped GaN polariton laser was also reported in 2014 (Bhattacharya et al., 2014).

4.1.2 ZnO

ZnO is another material to realize polariton laser at room temperature. Its excitons feature an even larger exciton binding energy (60 meV) than that of GaN excitons and an oscillator strength three orders of magnitude larger than that of GaAs. The first observation of exciton polaritons in ZnO was also reported in a FP cavity, where a bulk ZnO layer was sandwiched between two DBRs (Shimada et al., 2008). However lasing could not be sustained at temperatures higher than 120 K due to the relatively low quality factor of the cavity. With improved cavity Q-factor, room temperature strong coupling and polariton lasing was demonstrated independently by two groups (Li et al., 2013; Lu et al., 2012), where the cavity quality factor was improved to ~3000 (Li et al., 2013), and the Rabi splitting was as high as 230 meV.

One special feature of ZnO is that the crystal can form high quality microrod resonators during the growth via chemical vapor deposition, without the need for heterostructures or complicated fabrication processes. The hexagonal cross section of a microrod can host multiple whispering gallery modes with Q factors above 1000. Strong exciton-photon coupling has been confirmed through angle-resolved PL measurement of the lower polariton dispersion (Trichet et al., 2011). By tuning the growth condition, the microrods can be synthesized as tetrapods or as rods with uniform radius along the growth axis. The tetrapod shape allows continuously variation of the cavity resonance, and thus the detuning between exciton and photon, similar to variation of the cavity thickness in FP cavities. This allows a clear demonstration of the anticrossing of the polariton modes across zero detuning (Sun et al., 2008) (Fig. 13C). In such a high-Q 1D polariton system, a variety of coherent polariton phenomena have been observed at room temperature, including polariton lasing and parametric scattering (Xie et al., 2012), and weak lasing in a polariton supperlattice (Zhang et al., 2015b) (Fig. 13D).

4.2 Organic semiconductors

In inorganic semiconductors, Wannier–Mott excitons are formed and delocalized among many lattice sites in the relatively homogeneous crystals. In contrast, organic semiconductors host Frenkel excitons localized at individual molecules (Karl, 1974). A Frenkel exciton has a much smaller Bohr radius of ~1 nm, a much larger binding energy ~1 eV, and a large oscillator strength. Room temperature exciton-photon strong coupling is readily achieved in cavities with moderate quality factors of ~ 100. The first organic exciton polariton was demonstrated in tetra-(2,6-t-butyl)phenol-porphyrin zinc (4TBPPZn) embedded in a FP cavity(Lidzey et al., 1998). The cavity was composed of nine pairs of SiO_2/SiN_x as bottom mirror and a silver film as the top mirror. The anticrossing dispersion was observed in the reflectivity spectrum with a Rabi splitting of 160 meV, as shown by Fig. 14A. However, it has been difficult to reach polariton lasing in such organic semiconductors for a few reasons: surface roughness of the organic material lowers the quality of the cavity; nonradiative loss especially through Auger recombination is strong; and the damage threshold under laser pumping is low.

To mitigate these problems, anthracene, a molecular crystal of high quantum yield (~0.99) and high damage threshold, was identified. To preserve the pristine optical quality of anthracene while embedding it in a cavity, two DBRs were pre-fabricated and prepatterned with gold stripes. The two DBRs were bonded by aligning the gold stripes and applying pressure. This formed an empty cavity with cavity length defined by the thickness of gold stripes. The cavity was then placed in an environment of liquid anthracene, which could penetrate the channels and form single crystals (Kéna-Cohen et al., 2008). This fabrication technique avoids having any deposition onto the anthracene (Fig. 14A). The first room-temperature polariton lasing in organic semicondcutors was demonstrated using such as a device (Kéna-Cohen and Forrest, 2010) (Fig. 14B).

The carrier relaxation in organic semiconductors is dominated by exciton-phonon interactions, and the reservoir excitons relax to the polariton ground state mainly via nonradiative emission of optical phonons (Kéna-Cohen and Forrest, 2010). This is in sharp contrast to inorganic semiconductors, wherein polariton–polariton scattering plays a crucial rule in populating the polariton ground state. Due to the relatively inefficient energy relaxation compared to the nonradiative loss, the polariton lasing threshold is not significantly lower than that of photon lasing.

Later it was found that carrier dynamics vary strongly among different organic semiconductors with different molecular and crystal structures. Polariton lasing was observed also in organic materials directly grown by thermal evaporation and spin coating (Daskalakis et al., 2014; Torres and Kaner, 2014), indicating that noncrystalline organic semiconductors may also allow polariton lasing.

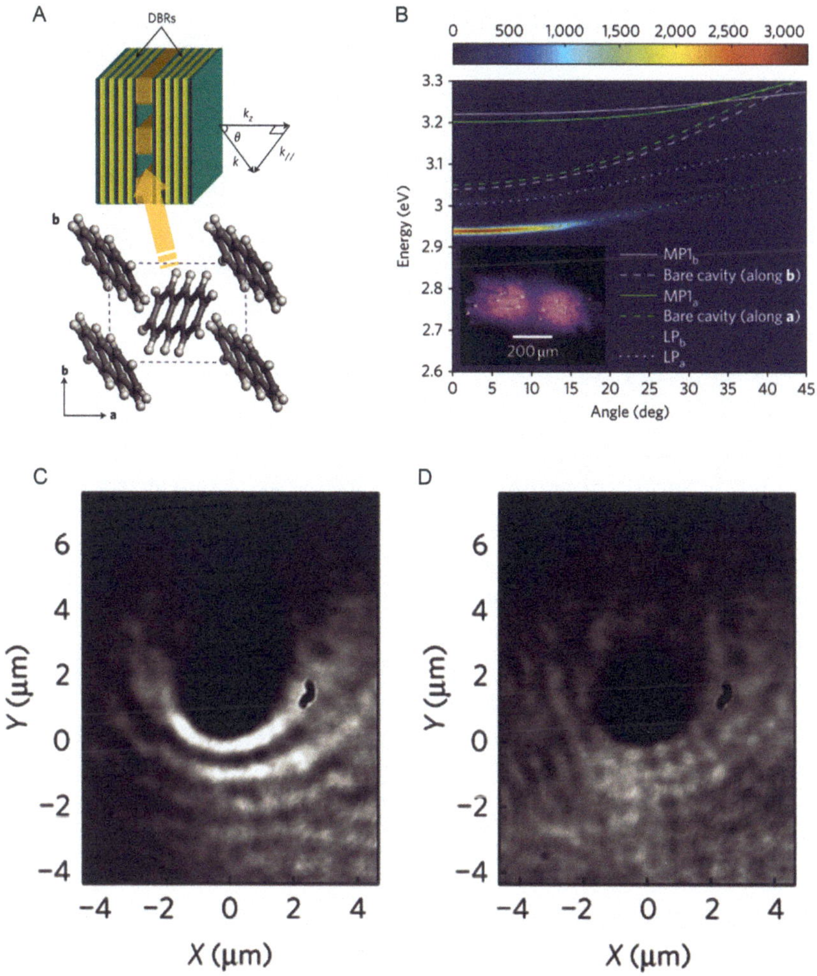

Fig. 14 Strong coupling and polariton lasing in organic semiconductors. (A) *Top panel*: Schematic showing the fabricated structure and empty channels before filling with anthracene. *Bottom panel*: Anthracene crystal structure viewed along the [001] stacking direction. (B) Angle-resolved photoluminescence measured above threshold. The *insets* show the emission spot imaged using a color charge coupled device camera. (C) Real-space polariton distribution for flow with group velocity of 19 μm ps^{-1} from the bottom to the top of the image across an artificial defect. The polariton density is $0.5 \times 10^6 \mu m^{-2}$. (D) Real-space intensity distribution at a polariton density of $10^7 \mu m^{-2}$. The interference fringes and the shadow cone beyond the defect vanish almost perfectly, revealing the physics of a fluid with zero viscosity. *Panel (A): From Kéna-Cohen, S., Forrest, S.R., 2010. Room-temperature polariton lasing in an organic single-crystal microcavity. Nat. Photon. 4 (6), 371–375. https://doi.org/10.1038/nphoton.2010.86; Panel (B): From Kéna-Cohen, S., Forrest, S.R., 2010. Room-temperature polariton lasing in an organic single-crystal microcavity. Nat. Photon. 4 (6), 371–375. https://doi.org/10.1038/nphoton.2010.86; Panel (C and D): From Lerario, G., Fieramosca, A., Barachati, F., Ballarini, D., Daskalakis, K.S., Dominici, L., De Giorgi, M., Maier, S.A., Gigli, G., Kéna-Cohen, S., Sanvitto, D., 2017. Room-temperature superfluidity in a polariton condensate. Nat. Phys. 13 (9), 837–841. https://doi.org/10.1038/nphys4147. ArXiv: 1609.03153.*

In 2017, signatures of superfluidity were observed in the organic polariton systems (Lerario et al., 2017) (Fig. 14C and D). Propagating polaritons were first created by resonant optical pumping. When the polaritons hit an embedded defect, polaritons in the supersonic regime exhibited Rayleigh scattering, while polaritons in the superfluidity regime showed scattering-free propagation and vortex–antivortex nucleation.

Although polariton lasing has been observed in various organic semiconductors, the relaxation dynamics and mechanisms are still elusive, requiring further experimental and theoretical study (Qu et al., 2020).

4.3 Perovskites

Lead halide perovskites have become a rapidly growing research field due to their recent success in photovoltaics, photonics, and optoelectronics. Lead halide perovskites exhibit large light absorption coefficient, large exciton binding energy (30–250 eV), and high quantum yield, which are all favorable for realizing room temperature polariton lasers.

The typical chemical formula of perovskite compounds can be written as ABX_3 or A_2BX_4, where A is a metallic or organic cation, B is a metallic cation and X is an anion that bonds to both. If A is an organic cation, the crystal is called hybrid organic-inorganic perovskites (HOIPs), which can be treated as multiple quantum wells made of inorganic layers with organic cations barriers. If A is a metallic cation, the crystal is an all-inorganic material, called all-inorganic perovskites (AIPs). Compared to HOIPs, AIPs show narrower exciton linewidth, larger optical gain, and much better stability when exposed to moisture and light (Du et al., 2019).

Similar to organic semiconductors, perovskites can integrate with any substrate without lattice matching issues. Exciton polaritons have been observed in perovskites integrated with various photonic architectures. The first demonstration was in a 1D grating with perovskites filling the grooves. The detuning was tuned by the period of gratings, and the Rabi splitting was found to be 100 meV (Fujita et al., 1998). Larger Rabi splitting around 150 meV was observed in a FP cavity composed of a dielectric bottom mirror and silver top mirror (Brehier et al., 2006). By controlling the growth condition, perovskites can form nanowires, where the whispering gallery mode or FP mode of the cross section can provide even better overlap between the field and the excitons, resulting in a Rabi splitting of up to 656 meV in $CsPbBr_3$ (Du et al., 2018). However, due to the very long cavity length, multiple cavity modes couple with the exciton simultaneously, making it difficult to identify the lasing modes. A more controllable and practical optical structure for perovskite systems is the DBR-DBR FP cavity, which has higher Q with a small mode volume and a single longitudinal cavity mode. Room temperature exciton polaritons have been demonstrated in both HOIPs (Fieramosca et al., 2019) and AIPs (Su et al., 2017) integrated with DBR-BR FP cavity (see Fig. 15A and B), with Rabi splitting of 170 meV and 265 meV respectively.

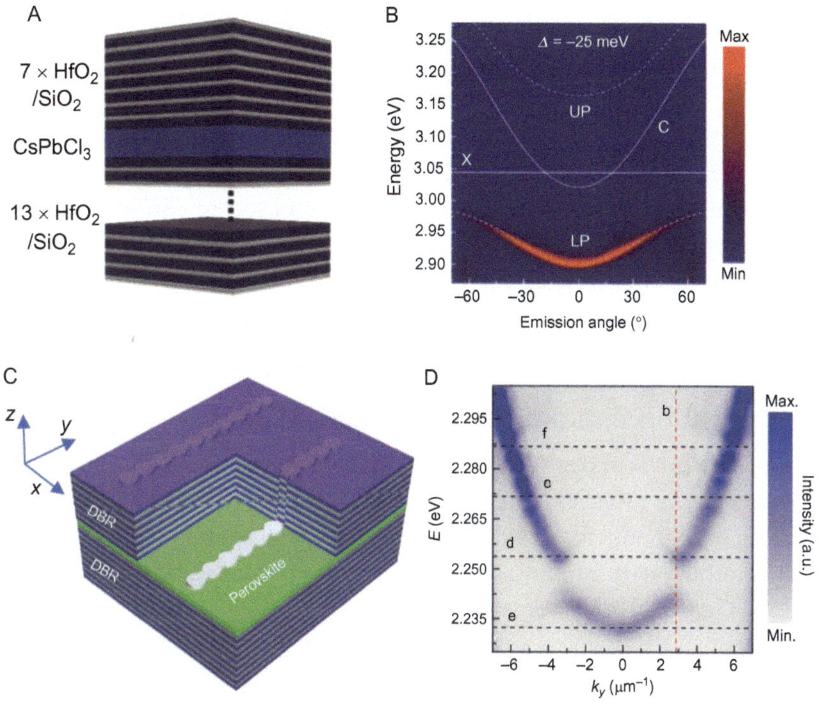

Fig. 15 (A) Schematic of the perovskite microcavity, which consists of a $CsPbCl_3$ crystalline nanoplatelet embedded in a planar microcavity formed by a bottom and a top HfO_2/SiO_2 DBR. (B) Angle-resolved photoluminescence spectrum of perovskite microcavity. (C) Schematic representation of the one-dimensional perovskite lattice structure, where the lattice pattern is created by patterning the polymethyl methacrylate (PMMA) spacer layer. (D) Momentum space polariton dispersion showing large bandgap of 13.3 meV. The *red dashed line* represents the edge of the first Brillouin zone. *Panels (A) and (B) From Su, R., Diederichs, C., Wang, J., Liew, T.C.H., Zhao, J., Liu, S., Xu, W., Chen, Z., Xiong, Q., 2017. Room-temperature polariton lasing in all-inorganic perovskite nanoplatelets. Nano Lett. 17 (6), 3982–3988. https://doi.org/10.1021/acs.nanolett.7b01956; Panels (C) and (D): From Su, R., Ghosh, S., Wang, J., Liu, S., Diederichs, C., Liew, T.C.H., Xiong, Q., 2020. Observation of exciton polariton condensation in a perovskite lattice at room temperature. Nat. Phys. 16 (3), 301–306. https://doi.org/10.1038/s41567-019-0764-5.*

Interestingly, strong nonlinearity of polaritons was demonstrated in HOIP polaritons (Fieramosca et al., 2019). The nonlinear coefficient $g \sim 3\ \mu eV \cdot \mu m^{-2}$ is comparable with the highest value reported in GaAs waveguide cavity and 3D confined structures, but are observed at room temperature. To demonstrate a polariton condensate, the AIPs are more promising due to their excellent optical gain properties and better optical stability under high-power laser pumping. Room temperature polariton lasing was demonstrated in an all inorganic $CsPbCl_3$ nanoplatelet embedded in a DBR-DBR cavity (Su et al., 2017). The lasing behavior was accompanied by a sharp increase of emission intensity near the zero in-plane wavevector, abrupt linewidth narrowing, and blue shift of the lasing energy. Moreover, long-range spatial coherence was measured by a Michelson interferometer. By replacing the nanoplatelets with nanowires, additional spatial confinement was created, and ballistic propagation of polariton condensate accompanied with long-range coherence was observed (Su et al., 2018). Recently, polariton lattices were also created by patterning the cavity spacer layer of PMMA (Su et al., 2020) (Fig. 15C). Bloch bands with band gap opening were observed. The band gap of 13 meV was one order of magnitude higher than that of inorganic polariton systems (Fig. 15D).

4.4 Transition metal dichalcogenide (TMDC) monolayers

Atomically thin Van der Waals semiconductors have emerged as a new platform to explore novel exciton physics in 2D. Monolayer TMDCs are direct bandgap semiconductors featuring very robust excitons that interact strongly with photons, with binding energies two orders of magnitude larger than those of GaAs and oscillator strengths one to two orders of magnitude higher. Moreover, 2D Van der Waals heterostructures (vdWHs) made of different 2D materials can be created and integrated with other solids, all without lattice matching constraints, allowing unprecedented flexibility for engineering the electronic properties and light-matter coupling. Lastly, many unique properties of vdWHs, such as spin-valley locking, strong phonon–charge interactions, and moiré lattice in twisted bilayers, promise a wealth of novel phenomena and opportunities of new polariton devices.

4.4.1 Excitons in monolayer TMDCs

When thinned down to a monolayer, TMDCs become direct bandgap semiconductors. Due to the strong spin-orbital coupling, both the conduction and valence band split into two bands with opposite electron spins. In addition, the broken inversion symmetry of monolayer TMDs results in the

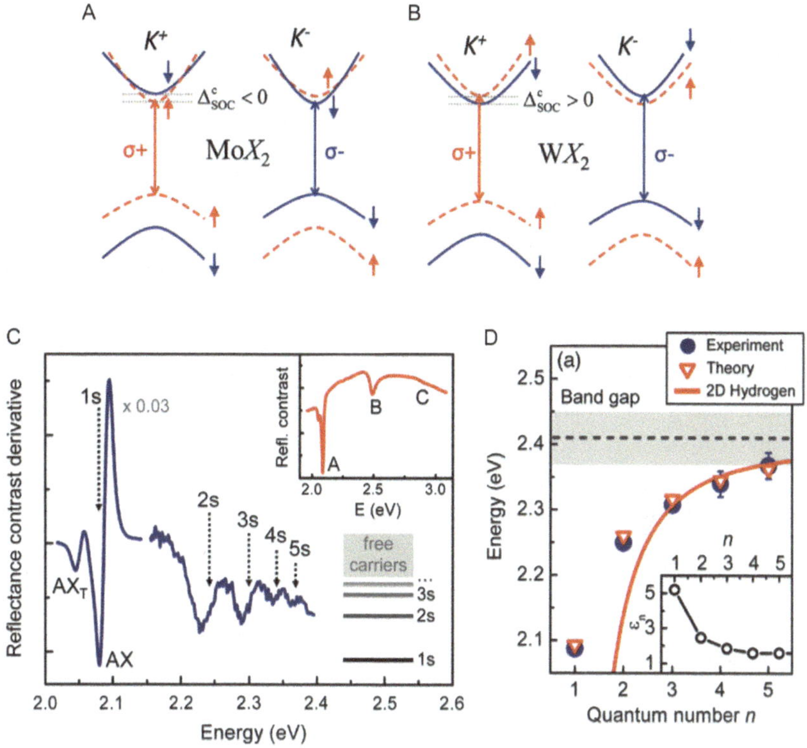

Fig. 16 Excitons in monolayer TMDCs. (A) and (B) Schematic illustrations in a single-particle band picture show that the order of the conduction bands is opposite in MoX_2 (A) and WX_2 (B). (C) and (D) Exciton states of monolayer WS_2 on a SiO_2/Si substrate from reflectance contrast measurements. The extracted transition energies of the states and the inferred band gap position are presented in the *right panel. Panels (A) and (B): From Wang, G., Chernikov, A., Glazov, M.M., Heinz, T.F., Marie, X., Amand, T., Urbaszek, B., 2017b. Excitons in atomically thin transition metal dichalcogenides. Rev. Mod. Phys. 90 (2), 21001. https://doi.org/10.1103/RevModPhys.90.021001; Panels (C) and (D): From Chernikov, A., Berkelbach, T.C., Hill, H.M., Rigosi, A., Li, Y., Aslan, O.B., Reichman, D.R., Hybertsen, M.S., Heinz, T.F., 2014. Exciton binding energy and nonhydrogenic Rydberg series in monolayerWS2. Phys. Rev. Lett. 113 (7), 076802. https://doi.org/10.1103/PhysRevLett.113.076802.*

opposite spin configuration between neighboring K valleys in the reciprocal space. This spin valley locking effect makes valley a new degree of freedom, which can be addressed optically by circularly polarized light (Fig. 16A and B). As a result of strong Coulomb interaction, the tightly bounded exciton dominates the optical response of TMDCs. In the following we focus on several key properties of TMDC excitons for polariton studies.

4.4.1.1 Binding energy

Due to the reduced dielectric screening effect, monolayer TMDC excitons exhibit very large binding energies in the range of 0.2–0.6 eV depending on the dielectric environment. So the excitons are stable above the room temperature. The binding energies of different monolayer TMDCs have been extracted experimentally using one photon linear absorption (Chernikov et al., 2014) (Fig. 16C and D), two photon absorption (He et al., 2014; Ye et al., 2014), and diamagnetic response (Stier et al., 2018).

4.4.1.2 Oscillator strength

The strong Coulomb interaction in TMDC monolayers leads to a large binding energy, a small exciton Bohr radius $a_B \sim 1$nm for the 1s state, as well as a large radiative emission rate. Radiative lifetime is estimated to be $\lesssim$1ps (Glazov et al., 2014; Wang et al., 2016a). The large radiative rate corresponds to a large oscillator strength. An atomically thin layer of TMDCs can absorb 10% $\sim$ 20% of the incident light at the exciton resonance.

4.4.1.3 Radiative linewidth and inhomogeneous broadening.

The linewidth γ of TMDC excitons includes an inhomogeneous linewidth γ_{inh} and a homogeneous linewidth γ_{hom}. γ_{hom} in turn includes a radiative linewidth γ_{rad} and nonradiative linewidth γ_{nr}. Nonlinear spectroscopy can separate γ_{inh} and γ_{hom} Moody et al. (2015). γ_{inh} is typically 1 $\sim$ 10 meV in exfoliated TMDC monolayers, possibly due to inhomogeneity in strain, defect distributions and the environment. In CVD grown monolayers, γ_{inh} is often much broader ($\sim$40 meV), which could be due to physio- or chemisorption of the sample surface (Shree et al., 2019). Encapsulating the TMDCs with few layers of hexagonal boron nitride has shown to be effective in suppressing γ_{inh}, reducing γ to the homogeneous limit (Cadiz et al., 2017). γ_{hom} has been measured to be to about 0.5–2 meV, giving an upper bound of γ_{rad}.

4.4.2 Strong coupling between TMDC excitons and cavity photons

TMDCs have been successfully integrated with various photonic architectures and strong coupling have been demonstrated.

4.4.2.1 Dielectric DBR cavity

Exciton resonances of the four widely studied TMDCs semiconductors (MoS_2, $MoSe_2$, WS_2, WSe_2) are in the wavelength range of 600 nm-800 nm, the high quality DBR mirror for GaAs system grown by MOCVD or MBE, whose bandgaps are below TMDCs, cannot be used. Instead, a DBR in the visible wavelength range consisting of amorphous dielectric materials, such as SiO_2, Si_3N_4, TiO_2, Ta_2O_5, are grown by PECVD, electric beam evaporation, or sputtering. The bare cavity quality factor is often limited to < 1000 but often already corresponds to a cavity linewidth narrower than the exciton homogeneous linewidth.

In 2014, the first demonstration of strong coupling in TMDCs was reported (Liu et al., 2014). The cavity was composed of SiO_2/Si_3N_4 DBR mirrors with a quality factor $\lesssim 400$. The active material is MoS_2 grown using the CVD method, with a relatively broad linewidth. Directly depositing a DBR mirror on the top of monolayer TMDCs may also lower the quantum yield of the exciton emission and induce local strain (Fig. 17A). To overcome some of these problems, an open cavity was developed to create TMDC polaritons (Krórl et al., 2019; Sidler et al., 2017), where the top DBR mirror is mounted on an electrically controlled piezo stage without directly contacting the active material. The cavity length can be continuously tuned (Fig. 17B and C). Such open cavities enjoy the flexibility of a continuously tunable cavity length controlled by the gate voltage, yet difficulty in aligning the top mirror leads to a relatively long cavity length $L_c \sim 2\mu m$ and thus a larger mode volume of the optical field and smaller Rabi splitting. Strong coupling is lost at the room temperature. Most recently, a transferred DBR-DBR cavity was developed, which combines the benefits of smaller mode volume of the compact cavity and the deposition free design of the open cavity. The DBR mirror is peeled off from the as grown substrate and transferred onto the TMDs/bottom DBR and form a cavity by Van der Waals bonding. A bare cavity Q of 5000 has been demonstrated, with Rabi splitting twice as large as that of the open cavity system (Lundt et al., 2019).

4.4.2.2 Plasmonic cavity

To attain larger Rabi splitting, the field mode volume need to be further reduced, which can be implemented by using a metal cavity. The simplest design is replacing one or both the DBR mirrors with a metal mirror (Lundt et al., 2016). The Rabi splitting of TMDCs embedded in full metal FP cavity can be as high as 100 meV (Wang et al., 2016b). Taking advantage of the extremely small mode volume (down to the diffraction limit), a single plasmonic resonator reached the highest reported Rabi splitting 163 meV in a monolayer TMD (Kleemann et al., 2017; Qin et al., 2020). However, a metal cavity will suffer from a much lower Q, which is typically below 100. Although the single plasmonic resonator can couple coherently with the diffraction mode of plasmonic array with higher Q, the reported Q in such devices is still below 100 due to the imperfections in the fabrication (Liu et al., 2016) (Fig. 17D).

4.4.2.3 Dielectric photonic crystal

Slab photonic crystals (PC) have been well established as a versatile optical structure for engineering the properties, confinement, and coupling of the optical modes. Meanwhile, a low loss dielectric photonic crystal combines the high quality of the diffraction mode from the periodic lattice and the small mode volume of the evanescent waveguide mode, providing an attractive platform for strong coupling studies. In the strong coupling regime, controlling the optical properties directly translates to controlling the polariton properties, opening a door to novel polariton quantum many-body phenomena and device application. However, integrating photonic crystals with conventional materials is challenging as they are sensitive to free surface and lattice mismatch with photonic crystal materials. Van der Waals semiconductors are uniquely compatible with photonic crystal, as they do not have dangling bonds on the surface and do not require lattice matching with the substrate. Pristine vdWCs can be directly laid on top of the PC without further processing.

The first demonstration of a PC-TMDCs system in the strong coupling regime up to the room temperature was implemented on a one-dimensional silicon nitride grating (Zhang et al., 2018), as shown by Fig. 17E and F. Recently, the topological boundary in continuum (BIC) modes were also found in the same type of PC, which were predicted to possess extremely high Q (Lee et al., 2012). Strong coupling between a BIC mode and TMD excitons has been demonstrated (Kravtsov et al., 2020).

4.4.3 Valleytronics in TMDC polaritons

As a half-light, half-matter quasiparticle, the exciton polariton inherits the unique properties of both constituents, including the valley degree of freedom from the TMDC exciton. Three different groups reported the observation of valley-polarized polaritons, which can preserve the valley polarization better than bare excitons (Chen et al., 2017; Dufferwiel et al., 2017; Sun et al., 2017b). Comparing with a bare exciton, the total decay rate of a polariton is enhanced, so the valley-polarized state recombines faster than the intervalley scattering rate (Chen et al., 2017; Sun et al., 2017b). At the same time, the spatially extended wavefunctions (micrometer scale) of

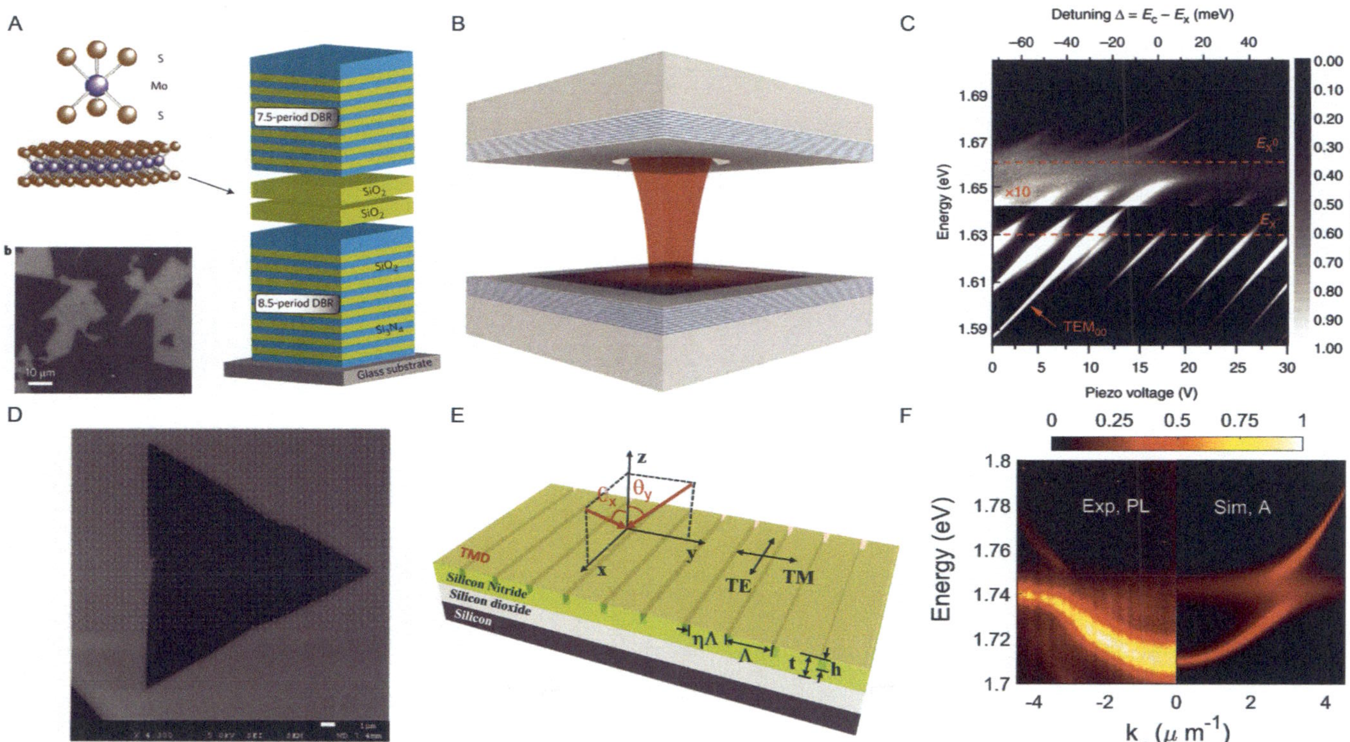

Fig. 17 See legend on next page.

polaritons can reduce the valley depolarization rate induced by defects of nanometer scale (Dufferwiel et al., 2017). As a result, the valley polarization is better preserved in the polariton emission. Similarly, valley coherence as a result of coherent superposition of the K and K′ exciton state, can also be transferred to polariton states (Dufferwiel et al., 2018; Qiu et al., 2019). The robust polariton valley polarization and valley coherence will encourage future exploration of valleytronic devices such as polaritonic circuit elements and optical spin switches.

4.4.4 Perspective

The TMDCs are extremely attractive for polariton research. They are atomically thin crystals insensitive to surface effects or substrate lattice matching, allowing unprecedented flexibility for engineering of the cavity modes in compatible structures. Valley degrees of freedom, moiré lattices and possibilities of stable higher order excitons are all unique features. The large exciton binding energy and oscillator strength of TMDCs lead to strong coupling readily achieved even at room temperature. However, many years after the first demonstration of TMDC polaritons, polariton lasing has yet to be demonstrated in TMDCs. Future research will need to understand and optimize key factors controlling carrier dynamics in TMDC system, such as the exciton and polariton lifetimes, nonradiative processes, and interactions among different excitations in the system.

Fig. 17 Exciton polaritons in TMDC monolayers. Different photonic architectures to demonstrate strong coupling in TMDC monolayers including compact CVD grown DBR-DBR cavity in (A), open cavity in (B) and (C), plasmonic array in (D), and dielectric photonic crystal in (E) and (F). (C) and (F) Clear upper and lower polariton branches with anticrossing behavior from PL measurements. *Panel (A): From Liu, X., Galfsky, T., Sun, Z., Xia, F., Lin, E.C., Lee, Y.H., Kéna-Cohen, S., Menon, V.M., 2014. Strong light-matter coupling in two-dimensional atomic crystals. Nat. Photon. 9 (1), 30–34. https://doi.org/10.1038/nphoton.2014.304; Panels (B) and (C): From Dufferwiel, S., Schwarz, S., Withers, F., Trichet, A.A.P., Li, F., Sich, M., Del Pozo-Zamudio, O., Clark, C., Nalitov, A., Solnyshkov, D. D., Malpuech, G., Novoselov, K.S., Smith, J.M., Skolnick, M.S., Krizhanovskii, D.N., Tartakovskii, A.I., 2015b. Exciton-polaritons in van der Waals heterostructures embedded in tunable microcavities. Nature communications 6, 8579. https://doi.org/10.1038/ncomms9579; Panel (D): From Liu, W., Lee, B., Naylor, C.H., Ee, H.S., Park, J., Johnson, A.T. C., Agarwal, R., 2016. Strong exciton-plasmon coupling in MoS_2 coupled with plasmonic lattice. Nano Lett. 16 (2), 1262–1269. https://doi.org/10.1021/acs.nanolett.5b04588; Panels (E) and (F): From Zhang, L., Gogna, R., Burg, W., Tutuc, E., Deng, H., 2018. Photonic-crystal exciton-polaritons in monolayer semiconductors. Nat. Commun. 9 (1), 713. https://doi.org/10.1038/s41467-018-03188-x.*

Compared to other inorganic materials, exciton and polariton lifetimes are very short in TMDCs due to the large exciton oscillator strength. At the same time, polariton–polariton scattering due to exciton exchange interactions is weaker, due to the heavier effective mass of the electron and hole. These may be obstacles to achieving quantum degeneracy in the branch bottom. The radiative limited linewidth of TMDC excitons have been estimated to be on the order of 1 meV (Horng et al., 2020; Martin et al., 2020). The polariton linewidth can be reduced by using a high-Q cavity with very narrow photon linewidth. But the narrowest linewidth of polaritons at zero detuning obtained from PL is ~6 meV even in cavities with a much narrower empty cavity linewidths (Dufferwiel et al., 2015b). This suggests there may still be significant absorption introduced by excitons, leading to a reduced cavity quality factor. Cleaner TMDC crystals with higher quantum yields under nonresonant excitation may help build up a population in the polariton ground state. Also detrimental to condensate formation are nonradiative Auger process, which has been observed in TMDCs monolayer at high pump densities (Kumar et al., 2014; Sun et al., 2014), and relaxation from the LP branch bottom to other lower energy states, such as trions, defect-bound states, and dark excitons. Further studies will evaluate the impact of these effects in various TMDC systems and how to mitigate them.

5. Summary

As interacting Bosons with nonequilibrium nature, exciton polaritons have exhibited a rich variety of macroscopic quantum phenomena, in which BEC, superfluidity, and vortices have been widely studied, and BKT phase and BCS phase are being explored. Control of exciton polaritons have been implemented by external fields or stress, nanopatterning, and optical pumping, providing a solid-state system for quantum simulations. Precise and versatile control over the potential landscape coupled with robust theoretical models are critical for future development of the field. Quantum light emission is still yet to be demonstrated. Room temperature exciton polariton and its lasing have been obtained in wide bandgap semiconductors, organics and perovskite. Two-dimensional monolayer and heterobilayer semiconductors have many unique properties that promise rich new physics and opportunities for practical device applications. Robust strong coupling with 2D materials has been demonstrated in a variety of photonic structure, while better understanding and control of the resonances and their dynamics are being pursued.

References

Abbarchi, M., Amo, A., Sala, V.G., Solnyshkov, D.D., Flayac, H., Ferrier, L., Sagnes, I., Galopin, E., Lemaître, A., Malpuech, G., Bloch, J., 2013. Macroscopic quantum self-trapping and Josephson oscillations of exciton polaritons. Nat. Phys. 9 (5), 275–279. https://doi.org/10.1038/nphys2609.

Akahane, Y., Asano, T., Song, B.-S., Noda, S., 2003. High-Q photonic nanocavity in a two-dimensional photonic crystal. Nature 425 (6961), 944–947. https://doi.org/10.1038/nature02063.

Alexey, K., 2007. Microcavities. Oxford University Press, Oxford. ISBN: 978-0-19-922894-2.

Amo, A., Lefrère, J.A., Pigeon, S., Adrados, C., Ciuti, C., Carusotto, I., Houdré, R., Giacobino, E., Bramati, A., 2009. Superfluidity of polaritons in semiconductor microcavities. Nat. Phys. 5 (11), 805–810. https://doi.org/10.1038/nphys1364.

Amo, A., Liew, T.C., Adrados, C., Houdré, R., Giacobino, E., Kavokin, A.V., Bramati, A., 2010. Exciton-polariton spin switches. Nat. Photon. 4 (6), 361–366. https://doi.org/10.1038/nphoton.2010.79.

Amthor, M., Liew, T.C.H., Metzger, C., Brodbeck, S., Worschech, L., Kamp, M., Shelykh, I.A., Kavokin, A.V., Schneider, C., Höfling, S., 2015. Optical bistability in electrically driven polariton condensates. Phys. Rev. B 91 (8), 081404. https://doi.org/10.1103/PhysRevB.91.081404.

Azzini, S., Gerace, D., Galli, M., Sagnes, I., Braive, R., Lemaître, A., Bloch, J., Bajoni, D., 2011. Ultra-low threshold polariton lasing in photonic crystal cavities. Appl. Phys. Lett. 99 (11), 111106. https://doi.org/10.1063/1.3638469.

Baas, A., Karr, J.P., Eleuch, H., Giacobino, E., 2004. Optical bistability in semiconductor microcavities. Phys. Rev. A 69 (2), 023809. https://doi.org/10.1103/PhysRevA.69.023809.

Baboux, F., Ge, L., Jacqmin, T., Biondi, M., Galopin, E., Lemaître, A., Le Gratiet, L., Sagnes, I., Schmidt, S., Türeci, H.E., Amo, A., Bloch, J., 2016. Bosonic condensation and disorder-induced localization in a flat band. Phys. Rev. Lett. 116 (6), 066402. https://doi.org/10.1103/PhysRevLett.116.066402.

Bajoni, D., Senellart, P., Wertz, E., Sagnes, I., Miard, A., Lemaître, A., Bloch, J., 2008. Polariton laser using single micropillar GaAs—GaAlAs semiconductor cavities. Phys. Rev. Lett. 100 (4), 047401. https://doi.org/10.1103/PhysRevLett.100.047401.

Bajoni, D., Gerace, D., Galli, M., Bloch, J., Braive, R., Sagnes, I., Miard, A., Lemaître, A., Patrini, M., Andreani, L.C., 2009. Exciton polaritons in two-dimensional photonic crystals. Phys. Rev. B 80 (20), 201308. https://doi.org/10.1103/PhysRevB.80.201308.

Balili, R., Hartwell, V., Snoke, D., Pfeiffer, L., West, K., 2007. Bose-Einstein condensation of microcavity polaritons in a trap. Science 316 (5827), 1007. https://doi.org/10.1126/science.1140990.

Ballarini, D., De Giorgi, M., Cancellieri, E., Houdré, R., Giacobino, E., Cingolani, R., Bramati, A., Gigli, G., Sanvitto, D., 2013. All-optical polariton transistor. Nat. Commun. 4 (1), 1778. https://doi.org/10.1038/ncomms2734.

Ballarini, D., Caputo, D., Dagvadorj, G., Juggins, R., Giorgi, M.D., Dominici, L., West, K., Pfeiffer, L.N., Gigli, G., Szymańska, M.H., Sanvitto, D., 2020. Directional Goldstone waves in polariton condensates close to equilibrium. Nat. Commun. 11 (1), 217. https://doi.org/10.1038/s41467-019-13733-x.

Berezinskiǐ, V., 1972. Destruction of long-range order in one-dimensional and two-dimensional systems possessing a continuous symmetry group. II. Quantum systems. Soviet J. Exp. Theoretical Phys. 34 (3), 610.

Berloff, N.G., Silva, M., Kalinin, K., Askitopoulos, A., Töpfer, J.D., Cilibrizzi, P., Langbein, W., Lagoudakis, P.G., 2017. Realizing the classical XY Hamiltonian in polariton simulators. Nat. Mater. 16 (11), 1120–1126. https://doi.org/10.1038/nmat4971.

Besga, B., Vaneph, C., Reichel, J., Estève, J.A., Reinhard, A., Miguel-Sánchez, J., Imamoğlu, A.A., Volz, T., 2015. Polariton boxes in a tunable fiber cavity. Phys. Rev. Appl. 3 (1), 014008. https://doi.org/10.1103/PhysRevApplied.3.014008.

Bhattacharya, P., Xiao, B., Das, A., Bhowmick, S., Heo, J., 2013. Solid state electrically injected exciton-polariton laser. Phys. Rev. Lett. 110 (20), 206403. https://doi.org/10.1103/PhysRevLett.110.206403.

Bhattacharya, P., Frost, T., Deshpande, S., Baten, M.Z., Hazari, A., Das, A., 2014. Room temperature electrically injected polariton laser. Phys. Rev. Lett. 112 (23), 236802. https://doi.org/10.1103/PhysRevLett.112.236802.

Böer, K.W., Pohl, U.W., 2018. Semiconductor Physics. Springer International Publishing, New York. ISBN: 978-3-319-06540-3.

Bogolyubov, N.N., 1959. The compensation principle and the self-consistent field method. Soviet Phys. Uspekhi 2 (2), 236–254. https://doi.org/10.1070/pu1959v002n02abeh003122.

Boulier, T., Bamba, M., Amo, A., Adrados, C., Lemaitre, A., Galopin, E., Sagnes, I., Bloch, J., Ciuti, C., Giacobino, E., Bramati, A., 2014. Polariton-generated intensity squeezing in semiconductor micropillars. Nat. Commun. 5 (1), 3260. https://doi.org/10.1038/ncomms4260.

Boulier, T., Cancellieri, E., Sangouard, N.D., Glorieux, Q., Kavokin, A.V., Whittaker, D.M., Giacobino, E., Bramati, A., 2016. Injection of orbital angular momentum and storage of quantized vortices in polariton superfluids. Phys. Rev. Lett. 116 (11), 116402. ArXiv: 1509.02680, https://doi.org/10.1103/PhysRevLett.116.116402.

Boulier, T., Jacquet, M.J., Maître, A., Lerario, G., Claude, F., Pigeon, S., Glorieux, Q., Bramati, A., Giacobino, E., Amo, A., Bloch, J., 2020. Microcavity polaritons for quantum simulation. Adv. Quant. Tech. 3 (11), 2000052. http://onlinelibrary.wiley.com/doi/abs/10.1002/qute.202000052.

Brehier, A., Parashkov, R., Lauret, J.S., Deleporte, E., 2006. Strong exciton-photon coupling in a microcavity containing layered perovskite semiconductors. Appl. Phys. Lett. 89 (17), 171110. https://doi.org/10.1063/1.2369533.

Byrnes, T., Horikiri, T., Ishida, N., Yamamoto, Y., 2010. BCS wave-function approach to the BEC-BCS crossover of exciton-polariton condensates. Phys. Rev. Lett. 105 (18), 186402. https://doi.org/10.1103/PhysRevLett.105.186402.

Byrnes, T., Kim, N.Y., Yamamoto, Y., 2014. Exciton–polariton condensates. Nat. Phys. 10 (11), 803–813. https://doi.org/10.1038/nphys3143.

Cadiz, F., Courtade, E., Robert, C., Wang, G., Shen, Y., Cai, H., Taniguchi, T., Watanabe, K., Carrere, H., Lagarde, D., Manca, M., Amand, T., Renucci, P., Tongay, S., Marie, X., Urbaszek, B., 2017. Excitonic linewidth approaching the homogeneous limit in mos_2-based van der Waals heterostructures. Phys. Rev. X 7 (2), 021026. https://doi.org/10.1103/PhysRevX.7.021026.

Chang-Hasnain, C.J., Yang, W., 2012. High-contrast gratings for integrated optoelectronics. Adv. Opt. Photon. 4 (3), 379–440. https://doi.org/10.1364/AOP.4.000379.

Chen, Y.J., Cain, J.D., Stanev, T.K., Dravid, V.P., Stern, N.P., 2017. Valley-polarized exciton-polaritons in a monolayer semiconductor. Nat. Photon. 11 (7), 431–435. ArXiv: 1701.05579, https://doi.org/10.1038/nphoton.2017.86.

Chernikov, A., Berkelbach, T.C., Hill, H.M., Rigosi, A., Li, Y., Aslan, O.B., Reichman, D.R., Hybertsen, M.S., Heinz, T.F., 2014. Exciton binding energy and nonhydrogenic Rydberg series in monolayer WS_2. Phys. Rev. Lett. 113 (7), 076802. https://doi.org/10.1103/PhysRevLett.113.076802.

Christmann, G., Butté, R.A., Feltin, E., Carlin, J.-F.A., Grandjean, N., 2008a. Room temperature polariton lasing in a GaN/AlGaN multiple quantum well microcavity. Appl. Phys. Lett. 93 (5), 051102. https://doi.org/10.1063/1.2966369.

Christmann, G., Butté, R.A., Feltin, E., Mouti, A., Stadelmann, P.A., Castiglia, A., Carlin, J.-F.A., Grandjean, N., 2008b. Large vacuum Rabi splitting in a multiple

quantum well GaN-Based microcavity in the strong-coupling regime. Phys. Rev. B 77 (8), 085310. https://doi.org/10.1103/PhysRevB.77.085310.
Christmann, G., Askitopoulos, A., Deligeorgis, G., Hatzopoulos, Z., Tsintzos, S.I., Savvidis, P.G., Baumberg, J.J., 2011. Oriented polaritons in strongly-coupled asymmetric double quantum well microcavities. Appl. Phys. Lett. 98 (8), 081111. https://doi.org/10.1063/1.3559909.
Christopoulos, S., von Högersthal, G.B.H., Grundy, A.J.D., Lagoudakis, P.G., Kavokin, A.V., Baumberg, J.J., Christmann, G., Butté, R., Feltin, E., Carlin, J.F., Grandjean, N., 2007. Room-Temperature Polariton Lasing in Semiconductor Microcavities. Phys. Rev. Lett. 98 (12), 126405. https://doi.org/10.1103/PhysRevLett.98.126405.
Cristofolini, P., Christmann, G., Tsintzos, S.I., Deligeorgis, G., Konstantinidis, G., Hatzopoulos, Z., Savvidis, P.G., Baumberg, J.J., 2012. Coupling quantum tunneling with cavity photons. Science 336 (6082), 704–707. https://doi.org/10.1126/science.1219010.
Das, A., Heo, J., Jankowski, M., Guo, W., Zhang, L., Deng, H., Bhattacharya, P., 2011. Room Temperature Ultralow Threshold GaN Nanowire Polariton Laser. Phys. Rev. Lett. 107 (6), 066405. https://doi.org/10.1103/PhysRevLett.107.066405.
Daskalakis, K.S., Maier, S.A., Murray, R., Kéna-Cohen, S., 2014. Nonlinear interactions in an organic polariton condensate. Nat. Mater. 13 (3), 271–278. https://doi.org/10.1038/nmat3874.
Delteil, A., Fink, T., Schade, A., Höfling, S., Schneider, C., İmamoğlu, A.A., 2019. Towards polariton blockade of confined exciton-polaritons. Nat. Mater. 18 (3), 219–222. https://doi.org/10.1038/s41563-019-0282-y.
Deng, H., Weihs, G., Santori, C., Bloch, J., Yamamoto, Y., 2002. Condensation of semiconductor microcavity exciton polaritons. Science 298 (5591), 199–202. https://doi.org/10.1126/science.1074464.
Deng, H., Weihs, G., Snoke, D., Bloch, J., Yamamoto, Y., 2003. Polariton lasing vs. photon lasing in a semiconductor microcavity. Proc. Natl. Acad. Sci. 100 (26), 15318–15323. https://doi.org/10.1073/pnas.2634328100.
Deng, H., Press, D., Götzinger, S., Solomon, G.S., Hey, R., Ploog, K.H., Yamamoto, Y., 2006. Quantum Degenerate Exciton-Polaritons in Thermal Equilibrium. Phys. Rev. Lett. 97 (14), 146402. https://doi.org/10.1103/PhysRevLett.97.146402.
Deng, H., Solomon, G.S., Hey, R., Ploog, K.H., Yamamoto, Y., 2007. Spatial Coherence of a Polariton Condensate. Phys. Rev. Lett. 99 (12), 126403. https://doi.org/10.1103/PhysRevLett.99.126403.
Deng, H., Haug, H., Yamamoto, Y., 2010. Exciton-polariton Bose-Einstein condensation. Rev. Mod. Phys. 82 (2), 1489–1537. https://doi.org/10.1103/RevModPhys.82.1489.
Dreismann, A., Ohadi, H., Del Valle-Inclan Redondo, Y., Balili, R., Rubo, Y.G., Tsintzos, S.I., Deligeorgis, G., Hatzopoulos, Z., Savvidis, P.G., Baumberg, J.J., 2016. A sub-femtojoule electrical spin-switch based on optically trapped polariton condensates. Nat. Mater. 15 (10), 1074–1078. https://doi.org/10.1038/nmat4722.
Du, W., Zhang, S., Shi, J., Chen, J., Wu, Z., Mi, Y., Liu, Z., Li, Y., Sui, X., Wang, R., Qiu, X., Wu, T., Xiao, Y., Zhang, Q., Liu, X., 2018. Strong exciton-photon coupling and lasing behavior in all-inorganic $CsPbBr_3$ micro/nanowire Fabry-Pérot cavity. ACS Photonics 5 (5), 2051–2059. https://doi.org/10.1021/acsphotonics.7b01593.
Du, W., Zhang, S., Zhang, Q., Liu, X., 2019. Recent progress of strong exciton-photon coupling in lead halide perovskites. Advanced Materials 31 (45), 1804894. https://doi.org/10.1002/adma.201804894.
Dufferwiel, S., Fras, F., Trichet, A., Walker, P.M., Li, F., Giriunas, L., Makhonin, M.N., Wilson, L.R., Smith, J.M., Clarke, E., Skolnick, M.S., Krizhanovskii, D.N., 2014. Strong exciton-photon coupling in open semiconductor microcavities. Appl. Phys. Lett. 104 (19), 192107. https://doi.org/10.1063/1.4878504.

Dufferwiel, S., Li, F., Cancellieri, E., Giriunas, L., Trichet, A.A.P., Whittaker, D.M., Walker, P.M., Fras, F., Clarke, E., Smith, J.M., Skolnick, M.S., Krizhanovskii, D.N., 2015a. Spin textures of exciton-polaritons in a tunable microcavity with large TE-TM splitting. Phys. Rev. Lett. 115 (24), 246401. https://doi.org/10.1103/PhysRevLett.115.246401.
Dufferwiel, S., Schwarz, S., Withers, F., Trichet, A.A.P., Li, F., Sich, M., Del Pozo-Zamudio, O., Clark, C., Nalitov, A., Solnyshkov, D.D., Malpuech, G., Novoselov, K.S., Smith, J.M., Skolnick, M.S., Krizhanovskii, D.N., Tartakovskii, A.I., 2015b. Exciton-polaritons in van der Waals heterostructures embedded in tunable microcavities. Nat. Commun. 6, 8579. https://doi.org/10.1038/ncomms9579.
Dufferwiel, S., Schwarz, S., Withers, F., Trichet, A.A.P., Li, F., Sich, M., Del Pozo-Zamudio, O., Clark, C., Nalitov, A., Solnyshkov, D.D., Malpuech, G., Novoselov, K.S., Smith, J.M., Skolnick, M.S., Krizhanovskii, D.N., Tartakovskii, A.I., 2015c. Exciton-polaritons in van der Waals heterostructures embedded in tunable microcavities. Nat. Commun. 6 (1), 8579. https://doi.org/10.1038/ncomms9579.
Dufferwiel, S., Lyons, T.P., Solnyshkov, D.D., Trichet, A.A., Withers, F., Schwarz, S., Malpuech, G., Smith, J.M., Novoselov, K.S., Skolnick, M.S., Krizhanovskii, D.N., Tartakovskii, A.I., 2017. Valley-addressable polaritons in atomically thin semiconductors. Nat. Photon. 11 (8), 497–501. ArXiv: 1612.05073, https://doi.org/10.1038/nphoton.2017.125.
Dufferwiel, S., Lyons, T.P., Solnyshkov, D.D., Trichet, A.A.P., Catanzaro, A., Withers, F., Malpuech, G., Smith, J.M., Novoselov, K.S., Skolnick, M.S., Krizhanovskii, D.N., Tartakovskii, A.I., 2018. Valley coherent exciton-polaritons in a monolayer semiconductor. Nat. Commun. 9 (1), 4797. https://doi.org/10.1038/s41467-018-07249-z.
El Daïf, O., Baas, A., Guillet, T., Brantut, J.-P., Kaitouni, R.I., Staehli, J.L., Morier-Genoud, F., Deveaud, B., 2006. Polariton quantum boxes in semiconductor microcavities. Appl. Phys. Lett. 88 (6), 061105. https://doi.org/10.1063/1.2172409.
Estrecho, E., Gao, T., Bobrovska, N., Fraser, M.D., Steger, M., Pfeiffer, L., West, K., Liew, T.C.H., Matuszewski, M., Snoke, D.W., Truscott, A.G., Ostrovskaya, E.A., 2018. Single-shot condensation of exciton polaritons and the hole burning effect. Nat. Commun. 9 (1), 2944. https://doi.org/10.1038/s41467-018-05349-4.
Fieramosca, A., Polimeno, L., Ardizzone, V., De Marco, L., Pugliese, M., Maiorano, V., De Giorgi, M., Dominici, L., Gigli, G., Gerace, D., Ballarini, D., Sanvitto, D., 2019. Two-dimensional hybrid perovskites sustaining strong polariton interactions at room temperature. Sci. Adv. 5 (5), eaav9967. https://doi.org/10.1126/sciadv.aav9967.
Fischer, J., Savenko, I.G., Fraser, M.D., Holzinger, S., Brodbeck, S., Kamp, M., Shelykh, I.A., Schneider, C., Höfling, S., 2014. Spatial coherence properties of one dimensional exciton-polariton condensates. Phys. Rev. Lett. 113 (20), 203902. https://doi.org/10.1103/PhysRevLett.113.203902.
Fraser, M.D., 2017. Coherent exciton-polariton devices. Semicond. Sci. Technol. 32 (9), 093003. https://doi.org/10.1088/1361-6641/aa730c.
Fujita, T., Sato, Y., Kuitani, T., Ishihara, T., 1998. Tunable polariton absorption of distributed feedback microcavities at room temperature. Phys. Rev. B Condens. Matter Mater. Phys. 57 (19), 12428–12434. https://doi.org/10.1103/PhysRevB.57.12428.
Galbiati, M., Ferrier, L., Solnyshkov, D.D., Tanese, D., Wertz, E., Amo, A., Abbarchi, M., Senellart, P., Sagnes, I., Lemaître, A., Galopin, E., Malpuech, G., Bloch, J., 2012. Polariton condensation in photonic molecules. Phys. Rev. Lett. 108 (12), 126403. https://doi.org/10.1103/PhysRevLett.108.126403.
Gao, T., Eldridge, P.S., Liew, T.C.H., Tsintzos, S.I., Stavrinidis, G., Deligeorgis, G., Hatzopoulos, Z., Savvidis, P.G., 2012. Polariton condensate transistor switch. Phys. Rev. B 85 (23), 235102. https://doi.org/10.1103/PhysRevB.85.235102.
Gao, T., Estrecho, E., Bliokh, K.Y., Liew, T.C., Fraser, M.D., Brodbeck, S., Kamp, M., Schneider, C., Höfling, S., Yamamoto, Y., Nori, F., Kivshar, Y.S., Truscott, A.G.,

Dall, R.G., Ostrovskaya, E.A., 2015. Observation of non-Hermitian degeneracies in a chaotic exciton-polariton billiard. Nature 526 (7574), 554–558. https://doi.org/10.1038/nature15522.

Gao, T., Li, G., Estrecho, E., Liew, T., Comber-Todd, D., Nalitov, A., Steger, M., West, K., Pfeiffer, L., Snoke, D., Kavokin, A., Truscott, A., Ostrovskaya, E., 2018. Chiral modes at exceptional points in exciton-polariton quantum fluids. Phys. Rev. Lett. 120 (6), 065301. https://doi.org/10.1103/PhysRevLett.120.065301.

Gessler, J., Baumann, V., Emmerling, M., Amthor, M., Winkler, K., Höfling, S., Schneider, C., Kamp, M., 2014. Electro optical tuning of Tamm-plasmon exciton-polaritons. Appl. Phys. Lett. 105 (18), 181107. https://doi.org/10.1063/1.4901023.

Ghosh, S., Liew, T.C.H., 2020. Quantum computing with exciton-polariton condensates. npj Quantum Inf. 6 (1), 16. https://doi.org/10.1038/s41534-020-0244-x.

Gibbs, H.M., Khitrova, G., Koch, S.W., 2011. Exciton-polariton light-semiconductor coupling effects. Nat. Photon. 5 (5), 273. https://doi.org/10.1038/nphoton.2011.15.

Glauber, R.J., 1963. The quantum theory of optical coherence. Phys. Rev. 130 (6), 2529–2539. https://doi.org/10.1103/PhysRev.130.2529.

Glazov, M.M., Amand, T., Marie, X., Lagarde, D., Bouet, L., Urbaszek, B., 2014. Exciton fine structure and spin decoherence in monolayers of transition metal dichalcogenides. Phys. Rev. B 89 (20), 201302. https://doi.org/10.1103/PhysRevB.89.201302.

Goblot, V., Nguyen, H.S., Carusotto, I., Galopin, E., Lemaître, A., Sagnes, I., Amo, A., Bloch, J., 2016. Phase-controlled bistability of a dark soliton train in a polariton fluid. Phys. Rev. Lett. 117 (21), 217401. https://doi.org/10.1103/PhysRevLett.117.217401.

Goblot, V., Rauer, B., Vicentini, F., Le Boité, A., Galopin, E., Lemaître, A., Le Gratiet, L., Harouri, A., Sagnes, I., Ravets, S., Ciuti, C., Amo, A., Bloch, J., 2019. Nonlinear polariton fluids in a flatband reveal discrete gap solitons. Phys. Rev. Lett. 123 (11), 113901. https://doi.org/10.1103/PhysRevLett.123.113901.

Graf, A., Held, M., Zakharko, Y., Tropf, L., Gather, M.C., Zaumseil, J., 2017. Electrical pumping and tuning of exciton-polaritons in carbon nanotube microcavities. Nat. Mater. 16 (9), 911–917. https://doi.org/10.1038/nmat4940.

Gu, J., Chakraborty, B., Khatoniar, M., Menon, V.M., 2019. A room-temperature polariton light-emitting diode based on monolayer WS2. Nat. Nanotechnol. 14 (11), 1024–1028. https://doi.org/10.1038/s41565-019-0543-6.

Harder, T.H., Egorov, O.A., Beierlein, J., Gagel, P., Michl, J, Emmerling, M., et al., 2020. Exciton-polaritons in flatland: controlling flatband properties in a Lieb lattice. Phys. Rev. B 102, 121302. https://doi.org/10.1103/PhysRevB.102.121302.

Hayat, A., Lange, C., Rozema, L.A., Darabi, A., van Driel, H.M., Steinberg, A.M., Nelsen, B., Snoke, D.W., Pfeiffer, L.N., West, K.W., 2012. Dynamic stark effect in strongly coupled microcavity exciton polaritons. Phys. Rev. Lett. 109 (3), 033605. https://doi.org/10.1103/PhysRevLett.109.033605.

He, K., Kumar, N., Zhao, L., Wang, Z., Mak, K.F., Zhao, H., Shan, J., 2014. Tightly bound excitons in monolayer ws_2. Phys. Rev. Lett. 113 (2), 026803. https://doi.org/10.1103/PhysRevLett.113.026803.

Hopfield, J.J., 1958. Theory of the contribution of excitons to the complex dielectric constant of crystals. Phys. Rev. 112 (5), 1555–1567. https://doi.org/10.1103/PhysRev.112.1555.

Horikiri, T., Yamaguchi, M., Kamide, K., Matsuo, Y., Byrnes, T., Ishida, N., Löffler, A., Höfling, S., Shikano, Y., Ogawa, T., Forchel, A., Yamamoto, Y., 2016. High-energy side-peak emission of exciton-polariton condensates in high density regime. Sci. Rep. 6, 25655. https://doi.org/10.1038/srep25655.

Horng, J., Martin, E.W., Chou, Y.-H., Courtade, E., Chang, T.-C., Hsu, C.-Y., Wentzel, M.-H., Ruth, H.G., Lu, T.-C., Cundiff, S.T., Wang, F., Deng, H., 2020. Perfect absorption by an atomically thin crystal. Phys. Rev. Appl. 14 (2), 024009. https://doi.org/10.1103/PhysRevApplied.14.024009.

Hu, J., Wang, Z., Kim, S., Deng, H., Brodbeck, S., Schneider, C., Höfling, S., Kwong, N.H., Binder, R., 2019. Signatures of a Bardeen-Cooper-Schrieffer polariton laser. ArXiv: 1902.00142, http://arxiv.org/abs/1902.00142.

Huang, R., Yamamoto, Y., André, R., Bleuse, J., Muller, M., Ulmer-Tuffigo, H., 2002. Exciton-polariton lasing and amplification based on exciton-exciton scattering in CdTe microcavity quantum wells. Phys. Rev. B 65 (16), 165314. https://doi.org/10.1103/PhysRevB.65.165314.

Huang, M.C.Y., Zhou, Y., Chang-Hasnain, C.J., 2007. A surface-emitting laser incorporating a high-index-contrast subwavelength grating. Nat. Photon. 1 (2), 119–122. https://doi.org/10.1038/nphoton.2006.80.

Imamogˇlu, A., Ram, R.J., Pau, S., Yamamoto, Y., 1996. Nonequilibrium condensates and lasers without inversion: Exciton-polariton lasers. Phys. Rev. A 53 (6), 4250–4253. https://doi.org/10.1103/PhysRevA.53.4250.

Istrate, E., Sargent, E.H., 2006. Photonic crystal heterostructures and interfaces. Rev. Mod. Phys. 78 (2), 455–481. https://doi.org/10.1103/RevModPhys.78.455.

Jacqmin, T., Carusotto, I., Sagnes, I., Abbarchi, M., Solnyshkov, D.D., Malpuech, G., Galopin, E., Lemaître, A., Bloch, J., Amo, A., 2014. Direct observation of dirac cones and a flatband in a honeycomb lattice for polaritons. Phys. Rev. Lett. 112 (11), 116402. https://doi.org/10.1103/PhysRevLett.112.116402.

Jayaprakash, R., Georgiou, K., Coulthard, H., Askitopoulos, A., Rajendran, S.K., Coles, D.M., Musser, A.J., Clark, J., Samuel, I.D.W., Turnbull, G.A., Lagoudakis, P.G., Lidzey, D.G., 2019. A hybrid organic-inorganic polariton LED. Light Sci. Appl. 8 (1), 81. https://doi.org/10.1038/s41377-019-0180-8.

Kaitouni, R.I., El Daïf, O., Baas, A., Richard, M., Paraiso, T., Lugan, P., Guillet, T., Morier-Genoud, F., Ganière, J.D., Staehli, J.L., Savona, V., Deveaud, B., 2006. Engineering the spatial confinement of exciton polaritons in semiconductors. Phys. Rev. B 74 (15), 155311. https://doi.org/10.1103/PhysRevB.74.155311.

Kamide, K., Ogawa, T., 2010. What determines the wave function of electron-hole pairs in polariton condensates? Phys. Rev. Lett. 105 (5), 056401. https://doi.org/10.1103/PhysRevLett.105.056401.

Karl, N., 1974. Organic Semiconductors. In: Springer, Berlin, Heidelberg.

Karzig, T., Bardyn, C.-E., Lindner, N.H., Refael, G., 2015. Topological polaritons. Phys. Rev. X 5 (3), 031001. https://doi.org/10.1103/PhysRevX.5.031001.

Kasprzak, J., Richard, M., Kundermann, S., Baas, A., Jeambrun, P., Keeling, J.M., Marchetti, F.M., Szymánska, M.H., André, R., Staehli, J.L., Savona, V., Littlewood, P.B., Deveaud, B., Dang, L.S., 2006. Bose-Einstein condensation of exciton polaritons. Nature 443 (7110), 409–414. https://doi.org/10.1038/nature05131.

Kavokin, A., Malpuech, G., Glazov, M., 2005. Optical spin Hall effect. Phys. Rev. Lett. 95 (13), 136601. https://doi.org/10.1103/PhysRevLett.95.136601.

Keeling, J., Berloff, N.G., 2009. Going with the flow. Nature 457 (7227), 273–274. https://doi.org/10.1038/457273a.

Keeling, J., Eastham, P.R., Szymanska, M.H., Littlewood, P.B., 2005. BCS-BEC crossover in a system of microcavity polaritons. Phys. Rev. B 72 (11), 115320. https://doi.org/10.1103/PhysRevB.72.115320.

Kéna-Cohen, S., Forrest, S.R., 2010. Room-temperature polariton lasing in an organic single-crystal microcavity. Nat. Photon. 4 (6), 371–375. https://doi.org/10.1038/nphoton.2010.86.

Kéna-Cohen, S., Davanço, M., Forrest, S.R., 2008. Strong exciton-photon coupling in an organic single crystal microcavity. Phys. Rev. Lett. 101 (11), 116401. https://doi.org/10.1103/PhysRevLett.101.116401.

Kim, N.Y., Kusudo, K., Wu, C., Masumoto, N., Löffler, A., Höfling, S., Kumada, N., Worschech, L., Forchel, A., Yamamoto, Y., 2011. Dynamical d-wave condensation

of exciton-polaritons in a two-dimensional square-lattice potential. Nat. Phys. 7 (9), 681–686. https://doi.org/10.1038/nphys2012.
Kim, N.Y., Kusudo, K., Löffler, A., Höfling, S., Forchel, A., Yamamoto, Y., 2013. Exciton-polariton condensates near the Dirac point in a triangular lattice. New J. Phys. 15 (3), 035032. https://doi.org/10.1088/1367-2630/15/3/035032.
Kim, N.Y., Kusudo, K., Löffler, A., Höfling, S., Forchel, A., Yamamoto, Y., 2014. F -Band condensates in exciton-polariton lattice systems. Phys. Rev. B 89 (8), 085306. https://doi.org/10.1103/PhysRevB.89.085306.
Kim, S., Zhang, B., Wang, Z., Fischer, J., Brodbeck, S., Kamp, M., Schneider, C., Höfling, S., Deng, H., 2016. Coherent polariton laser. Phys. Rev. X 6 (1), 011026. https://doi.org/10.1103/PhysRevX.6.011026.
Kim, S., Wang, Z., Brodbeck, S., Schneider, C., Höfling, S., Deng, H., 2019. Monolithic high-contrast grating based polariton laser. ACS Photonics 6 (1), 18–22. https://doi.org/10.1021/acsphotonics.8b01141.
Kim, S., Rubo, Y.G., Liew, T.C.H., Brodbeck, S., Schneider, C., Höfling, S., Deng, H., 2020. Emergence of microfrequency comb via limit cycles in dissipatively coupled condensates. Phys. Rev. B 101 (8), 085302. https://doi.org/10.1103/PhysRevB.101.085302.
Klaas, M., Flayac, H., Amthor, M., Savenko, I.G., Brodbeck, S., Ala-Nissila, T., Klembt, S., Schneider, C., Höfling, S., 2018. Evolution of temporal coherence in confined exciton-polariton condensates. Phys. Rev. Lett. 120 (1), 017401. https://doi.org/10.1103/PhysRevLett.120.017401.
Kleemann, M.-E., Chikkaraddy, R., Alexeev, E.M., Kos, D., Carnegie, C., Deacon, W., de Pury, A.C., Große, C., de Nijs, B., Mertens, J., Tartakovskii, A.I., Baumberg, J.J., 2017. Strong-coupling of WSe_2 in ultra-compact plasmonic nanocavities at room temperature. Nat. Commun. 8 (1), 1296. https://doi.org/10.1038/s41467-017-01398-3.
Klembt, S., Harder, T.H., Egorov, O.A., Winkler, K., Ge, R., Bandres, M.A., Emmerling, M., Worschech, L., Liew, T.C.H., Segev, M., Schneider, C., Höfling, S., 2018. Exciton-polariton topological insulator. Nature 562 (7728), 552–556. https://doi.org/10.1038/s41586-018-0601-5.
Kozuma, M., Suzuki, Y., Torii, Y., Sugiura, T., Kuga, T., Hagley, E.W., Deng, L., 1999. Phase-coherent amplification of matter waves. Science 286 (5448), 2309–2312. https://doi.org/10.1126/science.286.5448.2309.
Król, M., Lekenta, K., Mirek, R., Łempicka, K., Stephan, D., Nogajewski, K., Molas, M.R., Babiński, A., Potemski, M., Szczytko, J., Piętka, B., 2019. Valley polarization of exciton-polaritons in monolayer WSe_2 in a tunable microcavity. Nanoscale 11 (19), 9574–9579. ArXiv: 1809.09571, https://doi.org/10.1039/c9nr02038a.
Kravtsov, V., Khestanova, E., Benimetskiy, F.A., Ivanova, T., Samusev, A.K., Sinev, I.S., Pidgayko, D., Mozharov, A.M., Mukhin, I.S., Lozhkin, M.S., Kapitonov, Y.V., Brichkin, A.S., Kulakovskii, V.D., Shelykh, I.A., Tartakovskii, A.I., Walker, P.M., Skolnick, M.S., Krizhanovskii, D.N., Iorsh, I.V., 2020. Nonlinear polaritons in a monolayer semiconductor coupled to optical bound states in the continuum. Light Sci. Appl. 9 (1), 56. https://doi.org/10.1038/s41377-020-0286-z.
Kumar, N., Cui, Q., Ceballos, F., He, D., Wang, Y., Zhao, H., 2014. Exciton-exciton annihilation in $MoSe_2$ monolayers. Phys. Rev. B 89 (12), 125427. https://doi.org/10.1103/PhysRevB.89.125427.
Kusudo, K., Kim, N.Y., Löffler, A., Höfling, S., Forchel, A., Yamamoto, Y., 2013. Stochastic formation of polariton condensates in two degenerate orbital states. Phys. Rev. B 87 (21), 214503. https://doi.org/10.1103/PhysRevB.87.214503.
Lagoudakis, K.G., Wouters, M., Richard, M., Baas, A., Carusotto, I., André, R., Dang, L.S., Deveaud-Plédran, B., 2008. Quantized vortices in an exciton-polariton condensate. Nat. Phys. 4 (9), 706–710. ArXiv: 0801.1916, https://doi.org/10.1038/nphys1051.

Lai, C.W., Kim, N.Y., Utsunomiya, S., Roumpos, G., Deng, H., Fraser, M.D., Byrnes, T., Recher, P., Kumada, N., Fujisawa, T., Yamamoto, Y., 2007. Coherent zero-state and π-state in an exciton-polariton condensate array. Nature 450 (7169), 529–532. https://doi.org/10.1038/nature06334.

Larionov, A.V., Kulakovskii, V.D., Höfling, S., Schneider, C., Worschech, L., Forchel, A., 2010. Polarized nonequilibrium Bose-Einstein condensates of spinor exciton polaritons in a magnetic field. Phys. Rev. Lett. 105 (25), 256401. https://doi.org/10.1103/PhysRevLett.105.256401.

Lee, J., Zhen, B., Chua, S.-L., Qiu, W., Joannopoulos, J.D., Soljačić, M., Shapira, O., 2012. Observation and differentiation of unique high Q optical resonances near zero wave vector in macroscopic photonic crystal slabs. Phys. Rev. Lett. 109 (6), 067401. https://doi.org/10.1103/PhysRevLett.109.067401.

Lekenta, K., Krórl, M., Mirek, R., Łempicka, K., Stephan, D., Mazur, R., Morawiak, P., Kula, P., Piecek, W., Lagoudakis, P.G., Piętka, B., Szczytko, J., 2018. Tunable optical spin Hall effect in a liquid crystal microcavity. Light Sci. Appl. 7 (1), 74. https://doi.org/10.1038/s41377-018-0076-z.

Lerario, G., Fieramosca, A., Barachati, F., Ballarini, D., Daskalakis, K.S., Dominici, L., De Giorgi, M., Maier, S.A., Gigli, G., Kéna-Cohen, S., Sanvitto, D., 2017. Room-temperature superfluidity in a polariton condensate. Nat. Phys. 13 (9), 837–841. ArXiv: 1609.03153, https://doi.org/10.1038/nphys4147.

Leyder, C., Romanelli, M., Karr, J.P., Giacobino, E., Liew, T.C.H., Glazov, M.M., Kavokin, A.V., Malpuech, G., Bramati, A., 2007. Observation of the optical spin Hall effect. Nat. Phys. 3 (9), 628–631. https://doi.org/10.1038/nphys676.

Li, F., Orosz, L., Kamoun, O., Bouchoule, S., Brimont, C., Disseix, P., Guillet, T., Lafosse, X., Leroux, M., Leymarie, J., Mexis, M., Mihailovic, M., Patriarche, G., Réveret, F., Solnyshkov, D., Zuniga-Perez, J., Malpuech, G., 2013. From excitonic to photonic polariton condensate in a ZnO-based microcavity. Phys. Rev. Lett. 110 (19), 196406. https://doi.org/10.1103/PhysRevLett.110.196406.

Lidzey, D.G., Bradley, D.D., Skolnick, M.S., Virgili, T., Walker, S., Whittaker, D.M., 1998. Strong exciton-photon coupling in an organic semiconductor microcavity. Nature 395 (6697), 53–55. https://doi.org/10.1038/25692.

Liew, T.C.H., Kavokin, A.V., Shelykh, I.A., 2008. Optical circuits based on polariton neurons in semiconductor microcavities. Phys. Rev. Lett. 101 (1), 016402. https://doi.org/10.1103/PhysRevLett.101.016402.

Liu, X., Galfsky, T., Sun, Z., Xia, F., Lin, E.C., Lee, Y.H., Kéna-Cohen, S., Menon, V.M., 2014. Strong light-matter coupling in two-dimensional atomic crystals. Nat. Photon. 9 (1), 30–34. https://doi.org/10.1038/nphoton.2014.304.

Liu, W., Lee, B., Naylor, C.H., Ee, H.S., Park, J., Johnson, A.T.C., Agarwal, R., 2016. Strong exciton-plasmon coupling in MoS_2 coupled with plasmonic lattice. Nano Lett. 16 (2), 1262–1269. https://doi.org/10.1021/acs.nanolett.5b04588.

Lu, T.-C., Lai, Y.-Y., Lan, Y.-P., Huang, S.-W., Chen, J.-R., Wu, Y.-C., Hsieh, W.-F., Deng, H., 2012. Room temperature polariton lasing vs photon lasing in a ZnO-based hybrid microcavity. Opt. Express 20 (5), 5530. https://doi.org/10.1364/oe.20.005530.

Lundt, N., Klembt, S., Cherotchenko, E., Betzold, S., Iff, O., Nalitov, A.V., Klaas, M., Dietrich, C.P., Kavokin, A.V., Höfling, S., Schneider, C., 2016. Room-temperature Tamm-plasmon exciton-polaritons with a WSe2 monolayer. Nat. Commun. 7 (1), 13328. https://doi.org/10.1038/ncomms13328.

Lundt, N., Dusanowski, Ł., Sedov, E., Stepanov, P., Glazov, M.M., Klembt, S., Klaas, M., Beierlein, J., Qin, Y., Tongay, S., Richard, M., Kavokin, A.V., Höfling, S., Schneider, C., 2019. Optical valley Hall effect for highly valley-coherent exciton-polaritons in an atomically thin semiconductor. Nat. Nanotechnol. 14 (8), 770–775. https://doi.org/10.1038/s41565-019-0492-0.

Ma, X., Berger, B., Aßmann, M., Driben, R., Meier, T., Schneider, C., Höfling, S., Schumacher, S., 2020. Realization of all-optical vortex switching in exciton-polariton condensates. Nat. Commun. 11 (1), 1–7. https://doi.org/10.1038/s41467-020-14702-5.

Martin, E.W., Horng, J., Ruth, H.G., Paik, E., Wentzel, M.-H., Deng, H., Cundiff, S.T., 2020. Encapsulation narrows and preserves the excitonic homogeneous linewidth of exfoliated monolayer $MoSe_2$. Phys. Rev. Appl. 14 (2), 021002. https://doi.org/10.1103/PhysRevApplied.14.021002.

Masumoto, N., Kim, N.Y., Byrnes, T., Kusudo, K., Löffler, A., Höfling, S., Forchel, A., Yamamoto, Y., 2012. Exciton-polariton condensates with flat bands in a two-dimensional kagome lattice. New J. Phys. 14 (6), 065002. https://doi.org/10.1088/1367-2630/14/6/065002.

Moody, G., Kavir Dass, C., Hao, K., Chen, C.-H., Li, L.-J., Singh, A., Tran, K., Clark, G., Xu, X., Berghäuser, G., Malic, E., Knorr, A., Li, X., 2015. Intrinsic homogeneous linewidth and broadening mechanisms of excitons in monolayer transition metal dichalcogenides. Nat. Commun. 6 (1), 8315. https://doi.org/10.1038/ncomms9315.

Muñoz Matutano, G., Wood, A., Johnsson, M., Vidal, X., Baragiola, B.Q., Reinhard, A., Lemaître, A., Bloch, J., Amo, A., Nogues, G., Besga, B., Richard, M., Volz, T., 2019. Emergence of quantum correlations from interacting fibre-cavity polaritons. Nat. Mater. 18 (3), 213–218. https://doi.org/10.1038/s41563-019-0281-z.

Nalitov, A., Solnyshkov, D., Malpuech, G., 2015. Polariton Z topological insulator. Phys. Rev. Lett. 114 (11), 116401. https://doi.org/10.1103/PhysRevLett.114.116401.

Nardin, G., Léger, Y., Pietka, B., Morier-Genoud, F.A., Deveaud-Plédran, B., 2010. Phase-resolved imaging of confined exciton-polariton wave functions in elliptical traps. Phys. Rev. B 82 (4), 045304. https://doi.org/10.1103/PhysRevB.82.045304.

Nitsche, W.H., Kim, N.Y., Roumpos, G., Schneider, C., Kamp, M., Höfling, S., Forchel, A., Yamamoto, Y., 2014. Algebraic order and the Berezinskii-Kosterlitz-Thouless transition in an exciton-polariton gas. Phys. Rev. B 90 (20), 205430. https://doi.org/10.1103/PhysRevB.90.205430.

Ohadi, H., Dreismann, A., Rubo, Y., Pinsker, F., del Valle-Inclan Redondo, Y., Tsintzos, S., Hatzopoulos, Z., Savvidis, P., Baumberg, J., 2015. Spontaneous spin bifurcations and ferromagnetic phase transitions in a spinor exciton-polariton condensate. Phys. Rev. X 5 (3), 031002. https://doi.org/10.1103/PhysRevX.5.031002.

Panzarini, G., Andreani, L.C., Armitage, A., Baxter, D., Skolnick, M.S., Astratov, V.N., Roberts, J.S., Kavokin, A.V., Vladimirova, M.R., Kaliteevski, M.A., 1999. Exciton-light coupling in single and coupled semiconductor microcavities: polariton dispersion and polarization splitting. Phys. Rev. B 59 (7), 5082–5089. https://doi.org/10.1103/PhysRevB.59.5082.

Pieczarka, M., Estrecho, E., Boozarjmehr, M., Bleu, O., Steger, M., West, K., Pfeiffer, L.N., Snoke, D.W., Levinsen, J., Parish, M.M., Truscott, A.G., Ostrovskaya, E.A., 2020. Observation of quantum depletion in a non-equilibrium exciton-polariton condensate. Nat. Commun. 11 (1), 429. https://doi.org/10.1038/s41467-019-14243-6.

Qiao, P., Yang, W., Chang-Hasnain, C.J., 2018. Recent advances in high-contrast metastructures, metasurfaces, and photonic crystals. Adv. Opt. Photon. 10 (1), 180. https://doi.org/10.1364/AOP.10.000180.

Qin, J., Chen, Y.H., Zhang, Z., Zhang, Y., Blaikie, R.J., Ding, B., Qiu, M., 2020. Revealing strong plasmon-exciton coupling between nanogap resonators and two-dimensional semiconductors at ambient conditions. Phys. Rev. Lett. 124 (6), 63902. https://doi.org/10.1103/PhysRevLett.124.063902.

Qiu, L., Chakraborty, C., Dhara, S., Vamivakas, A.N., 2019. Room-temperature valley coherence in a polaritonic system. Nat. Commun. 10 (1), 1513. https://doi.org/10.1038/s41467-019-09490-6.

Qu, Y., Hou, S., Forrest, S.R., 2020. Temperature-dependence of an amorphous organic thin film polariton laser. ACS Photonics 7 (4), 867–872. https://doi.org/10.1021/acsphotonics.9b01656.

Rahimi-Iman, A., Schneider, C., Fischer, J., Holzinger, S., Amthor, M., Höfling, S., Reitzenstein, S., Worschech, L., Kamp, M., Forchel, A., 2011. Zeeman splitting and diamagnetic shift of spatially confined quantum-well exciton polaritons in an external magnetic field. Phys. Rev. B 84 (16), 165325. https://doi.org/10.1103/PhysRevB.84.165325.

Rechcińska, K., Król, M., Mazur, R., Morawiak, P., Mirek, R., Łempicka, K., Bardyszewski, W., Matuszewski, M., Kula, P., Piecek, W., Lagoudakis, P.G., Piętka, B., Szczytko, J., 2019. Engineering spin-orbit synthetic Hamiltonians in liquid-crystal optical cavities. Science 366 (6466), 727–730. https://doi.org/10.1126/science.aay4182.

Richard, M., Kasprzak, J., Romestain, R., André, R., Dang, L.S., 2005. Spontaneous coherent phase transition of polaritons in CdTe microcavities. Phys. Rev. Lett. 94 (18), 187401. https://doi.org/10.1103/PhysRevLett.94.187401.

Rosenberg, I., Mazuz-Harpaz, Y., Rapaport, R., West, K., Pfeiffer, L., 2016. Electrically controlled mutual interactions of flying waveguide dipolaritons. Phys. Rev. B 93 (19), 195151. https://doi.org/10.1103/PhysRevB.93.195151.

Rosenberg, I., Liran, D., Mazuz-Harpaz, Y., West, K., Pfeiffer, L., Rapaport, R., 2018. Strongly interacting dipolar-polaritons. Sci. Adv. 4 (10), eaat8880. https://doi.org/10.1126/sciadv.aat8880.

Roumpos, G., Fraser, M.D., Löffler, A., Höfling, S., Forchel, A., Yamamoto, Y., 2011. Single vortex-antivortex pair in an exciton-polariton condensate. Nat. Phys. 7 (2), 129–133. https://doi.org/10.1038/nphys1841.

Roumpos, G., Lohse, M., Nitsche, W.H., Keeling, J., Szymańska, M.H., Littlewood, P.B., Löffler, A., Höfling, S., Worschechh, L., Forchel, A., Yamamoto, Y., 2012. Power-law decay of the spatial correlation function in exciton-polariton condensates. Proc. Natl. Acad. Sci. U.S.A. 109 (17), 6467–6472. https://doi.org/10.1073/pnas.1107970109.

Sala, V.G., Solnyshkov, D.D., Carusotto, I., Jacqmin, T., Lemaître, A., Terças, H., Nalitov, A., Abbarchi, M., Galopin, E., Sagnes, I., Bloch, J., Malpuech, G., Amo, A., 2015. Spin-orbit coupling for photons and polaritons in microstructures. Phys. Rev. X 5 (1), 011034. https://doi.org/10.1103/PhysRevX.5.011034.

Sanvitto, D., Kéna-Cohen, S., 2016. The road towards polaritonic devices. Nat. Mater. 15 (10), 1061–1073. https://doi.org/10.1038/nmat4668.

Sanvitto, D., Pigeon, S., Amo, A., Ballarini, D., De Giorgi, M., Carusotto, I., Hivet, R., Pisanello, F., Sala, V.G., Guimaraes, P.S., Houdré, R., Giacobino, E., Ciuti, C., Bramati, A., Gigli, G., 2011. All-optical control of the quantum flow of a polariton condensate. Nat. Photon. 5 (10), 610–614. https://doi.org/10.1038/nphoton.2011.211.

Savona, V., Andreani, L.C., Schwendimann, P., Quattropani, A., 1995. Quantum well excitons in semiconductor microcavities: unified treatment of weak and strong coupling regimes. Solid State Commun. 93 (9), 733–739. https://doi.org/10.1016/0038-1098(94)00865-5.

Savvidis, P.G., Baumberg, J.J., Stevenson, R.M., Skolnick, M.S., Whittaker, D.M., Roberts, J.S., 2000. Angle-resonant stimulated polariton amplifier. Phys. Rev. Lett. 84 (7), 1547–1550. https://doi.org/10.1103/PhysRevLett.84.1547.

Scafirimuto, F., Urbonas, D., Scherf, U., Mahrt, R.F., Stöferle, T., 2018. Room-temperature exciton-polariton condensation in a tunable zero-dimensional microcavity. ACS Photonics 5 (1), 85–89. https://doi.org/10.1021/acsphotonics.7b00557.

Schneider, C., Rahimi-Iman, A., Kim, N.Y., Fischer, J., Savenko, I.G., Amthor, M., Lermer, M., Wolf, A., Worschech, L., Kulakovskii, V.D., Shelykh, I.A., Kamp, M., Reitzenstein, S., Forchel, A., Yamamoto, Y., Höfling, S., 2013. An electrically pumped polariton laser. Nature 497 (7449), 348–352. https://doi.org/10.1038/nature12036.

Schneider, C., Winkler, K., Fraser, M.D., Kamp, M., Yamamoto, Y., Ostrovskaya, E.A., Höfling, S., 2017. Exciton-polariton trapping and potential landscape engineering. Rep. Prog. Phys. 80 (1), 016503. https://doi.org/10.1088/0034-4885/80/1/016503.

Semond, F., Sellers, I.R., Natali, F., Byrne, D., Leroux, M., Massies, J., Ollier, N., Leymarie, J., Disseix, P., Vasson, A., 2005. Strong light-matter coupling at room temperature in simple geometry GaN microcavities grown on silicon. Appl. Phys. Lett. 87 (2), 021102. https://doi.org/10.1063/1.1994954.

Shimada, R., Xie, J., Avrutin, V., Özgür, U., Morkoč, H., 2008. Cavity polaritons in ZnO-based hybrid microcavities. Appl. Phys. Lett. 92 (1), 011127. https://doi.org/10.1063/1.2830022.

Shree, S., George, A., Lehnert, T., Neumann, C., Benelajla, M., Robert, C., Marie, X., Watanabe, K., Taniguchi, T., Kaiser, U., Urbaszek, B., Turchanin, A., 2019. High optical quality of MoS_2 monolayers grown by chemical vapor deposition. 2D Materials 7 (1), 015011. https://doi.org/10.1088/2053-1583/ab4f1f.

Sidler, M., Back, P., Cotlet, O., Srivastava, A., Fink, T., Kroner, M., Demler, E., Imamoglu, A., 2017. Fermi polaron-polaritons in charge-tunable atomically thin semiconductors. Nat. Phys. 13 (3), 255–261. ArXiv: 1603.09215, https://doi.org/10.1038/nphys3949.

Srinivasan, K., Borselli, M., Painter, O., Stintz, A., Krishna, S., 2006. Cavity Q, mode volume, and lasing threshold in small diameter AlGaAs microdisks with embedded quantum dots. Opt. Express 14 (3), 1094–1105. https://doi.org/10.1364/OE.14.001094.

Stier, A., Wilson, N., Velizhanin, K., Kono, J., Xu, X., Crooker, S., 2018. Magnetooptics of exciton Rydberg states in a monolayer semiconductor. Phys. Rev. Lett. 120 (5), 057405. https://doi.org/10.1103/PhysRevLett.120.057405.

Sturm, C., Tanese, D., Nguyen, H., Flayac, H., Galopin, E., Lemaître, A., Sagnes, I., Solnyshkov, D., Amo, A., Malpuech, G., Bloch, J., 2014. All-optical phase modulation in a cavity-polariton Mach-Zehnder interferometer. Nat. Commun. 5 (1), 3278. https://doi.org/10.1038/ncomms4278.

Su, R., Diederichs, C., Wang, J., Liew, T.C.H., Zhao, J., Liu, S., Xu, W., Chen, Z., Xiong, Q., 2017. Room-temperature polariton lasing in all-inorganic perovskite nanoplatelets. Nano Lett. 17 (6), 3982–3988. https://doi.org/10.1021/acs.nanolett.7b01956.

Su, R., Wang, J., Zhao, J., Xing, J., Zhao, W., Diederichs, C., Liew, T.C.H., Xiong, Q., 2018. Room temperature long-range coherent exciton polariton condensate flow in lead halide perovskites. Sci. Adv. 4 (10), eaau0244. https://doi.org/10.1126/sciadv.aau0244.

Su, R., Ghosh, S., Wang, J., Liu, S., Diederichs, C., Liew, T.C.H., Xiong, Q., 2020. Observation of exciton polariton condensation in a perovskite lattice at room temperature. Nat. Phys. 16 (3), 301–306. https://doi.org/10.1038/s41567-019-0764-5.

Suchomel, H., Klembt, S., Harder, T.H., Klaas, M., Egorov, O.A., Winkler, K., Emmerling, M., Thomale, R., Höfling, S., Schneider, C., 2018. Platform for electrically pumped polariton simulators and topological lasers. Phys. Rev. Lett. 121 (25), 257402. https://doi.org/10.1103/PhysRevLett.121.257402.

Sun, L., Chen, Z., Ren, Q., Yu, K., Bai, L., Zhou, W., Xiong, H., Zhu, Z.Q., Shen, X., 2008. Direct observation of whispering gallery mode polaritons and their dispersion in a ZnO tapered microcavity. Phys. Rev. Lett. 100 (15), 156403. https://doi.org/10.1103/PhysRevLett.100.156403.

Sun, D., Rao, Y., Reider, G.A., Chen, G., You, Y., Brézin, L., Harutyunyan, A.R., Heinz, T.F., 2014. Observation of rapid exciton-exciton annihilation in monolayer molybdenum disulfide. Nano Lett. 14 (10), 5625–5629. https://doi.org/10.1021/nl5021975.

Sun, Y., Wen, P., Yoon, Y., Liu, G., Steger, M., Pfeiffer, L.N., West, K., Snoke, D.W., Nelson, K.A., 2017a. Bose-Einstein Condensation of Long-Lifetime Polaritons in Thermal Equilibrium. Phys. Rev. Lett. 118 (1), 016602. https://doi.org/10.1103/PhysRevLett.118.016602.

Sun, Z., Gu, J., Ghazaryan, A., Shotan, Z., Considine, C.R., Dollar, M., Chakraborty, B., Liu, X., Ghaemi, P., Kéna-Cohen, S., Menon, V.M., 2017b. Optical control of room temperature valley polaritons. Nat. Photon. 11 (8), 491–496. https://doi.org/10.1038/nphoton.2017.121.

Szymanska, M.H., Littlewood, P.B., Simons, B.D., 2003. Polariton condensation and lasing in optical microcavities: The decoherence-driven crossover. Phys. Rev. A 68 (1), 013818. https://doi.org/10.1103/PhysRevA.68.013818.

Tanaka, T., Zhang, Z., Nishioka, M., Arakawa, Y., 1996. Magnetic field dependence of exciton oscillator strength by measurements of magnetoexciton-polariton mode splitting in quantum wells with a microcavity. Appl. Phys. Lett. 69 (7), 887–889. https://doi.org/10.1063/1.117977.

Tanese, D., Flayac, H., Solnyshkov, D., Amo, A., Lemaître, A., Galopin, E., Braive, R., Senellart, P., Sagnes, I., Malpuech, G., Bloch, J., 2013. Polariton condensation in solitonic gap states in a one-dimensional periodic potential. Nat. Commun. 4 (1), 1749. https://doi.org/10.1038/ncomms2760.

Tartakovskii, A.I., Kulakovskii, V.D., Forchel, A., Reithmaier, J.P., 1998. Exciton-photon coupling in photonic wires. Phys. Rev. B 57 (12), R6807–R6810. https://doi.org/10.1103/PhysRevB.57.R6807.

Tassone, F., Piermarocchi, C., Savona, V., Quattropani, A., Schwendimann, P., 1997. Bottleneck effects in the relaxation and photoluminescence of microcavity polaritons. Phys. Rev. B 56 (12), 7554–7563. https://doi.org/10.1103/PhysRevB.56.7554.

Togan, E., Lim, H.T., Faelt, S., Wegscheider, W., Imamoglu, A., 2018. Enhanced interactions between dipolar polaritons. Phys. Rev. Lett. 121 (22), 227402. https://doi.org/10.1103/PhysRevLett.121.227402.

Torres, J.A., Kaner, R.B., 2014. Graphene synthesis: graphene closer to fruition. Nat. Mater. 13 (4), 328–329. https://doi.org/10.1038/nmat3925.

Tosi, G., Christmann, G., Berloff, N.G., Tsotsis, P., Gao, T., Hatzopoulos, Z., Savvidis, P.G., Baumberg, J.J., 2012. Sculpting oscillators with light within a nonlinear quantum fluid. Nat. Phys. 8 (3), 190–194. https://doi.org/10.1038/nphys2182.

Trichet, A., Sun, L., Pavlovic, G., Gippius, N., Malpuech, G., Xie, W., Chen, Z., Richard, M., Dang, L.S., 2011. One-dimensional ZnO exciton polaritons with negligible thermal broadening at room temperature. Phys. Rev. B 83 (4), 041302. https://doi.org/10.1103/PhysRevB.83.041302.

Tsintzos, S.I., Tzimis, A., Stavrinidis, G., Trifonov, A., Hatzopoulos, Z., Baumberg, J.J., Ohadi, H., Savvidis, P.G., 2018. Electrical tuning of nonlinearities in exciton-polariton condensates. Phys. Rev. Lett. 121 (3), 37401. https://doi.org/10.1103/PhysRevLett.121.037401.

Tsotsis, P., Tsintzos, S., Christmann, G., Lagoudakis, P., Kyriienko, O., Shelykh, I., Baumberg, J., Kavokin, A., Hatzopoulos, Z., Eldridge, P., Savvidis, P., 2014. Tuning the energy of a polariton condensate via bias-controlled Rabi splitting. Phys. Rev. Appl. 2 (1), 014002. https://doi.org/10.1103/PhysRevApplied.2.014002.

Utsunomiya, S., Tian, L., Roumpos, G., Lai, C.W., Kumada, N., Fujisawa, T., Kuwata-Gonokami, M., Löffler, A., Höfling, S., Forchel, A., Yamamoto, Y., 2008. Observation of Bogoliubov excitations in exciton-polariton condensates. Nat. Phys. 4 (9), 700–705. https://doi.org/10.1038/nphys1034.

Vörös, Z., Snoke, D.W., Pfeiffer, L., West, K., 2006. Trapping excitons in a two-dimensional in-plane harmonic potential: experimental evidence for equilibration of indirect excitons. Phys. Rev. Lett. 97 (1), 016803. https://doi.org/10.1103/PhysRevLett.97.016803.

Walck, S.N., Reinecke, T.L., 1998. Exciton diamagnetic shift in semiconductor nanostructures. Phys. Rev. B 57 (15), 9088–9096. https://doi.org/10.1103/PhysRevB.57.9088.

Wang, Z., Zhang, B., Deng, H., 2015. Dispersion engineering for vertical microcavities using subwavelength gratings. Phys. Rev. Lett. 114 (7), 073601. https://doi.org/10.1103/PhysRevLett.114.073601.

Wang, H., Zhang, C., Chan, W., Manolatou, C., Tiwari, S., Rana, F., 2016a. Radiative lifetimes of excitons and trions in monolayers of the metal dichalcogenide mos_2. Phys. Rev. B 93 (4), 045407. https://doi.org/10.1103/PhysRevB.93.045407.

Wang, S., Li, S., Chervy, T., Shalabney, A., Azzini, S., Orgiu, E., Hutchison, J.A., Genet, C., Samorì, P., Ebbesen, T.W., 2016b. Coherent coupling of WS2 monolayers with metallic photonic nanostructures at room temperature. Nano Lett. 16 (7), 4368–4374. ArXiv: 1604.06656, https://doi.org/10.1021/acs.nanolett.6b01475.

Wang, Z., Gogna, R., Deng, H., 2017a. What is the best planar cavity for maximizing coherent exciton-photon coupling. Appl. Phys. Lett. 111 (6), 061102. https://doi.org/10.1063/1.4997171.

Weisbuch, C., Nishioka, M., Ishikawa, A., Arakawa, Y., 1992. Observation of the coupled exciton-photon mode splitting in a semiconductor quantum microcavity. Phys. Rev. Lett. 69 (23), 3314–3317. https://doi.org/10.1103/PhysRevLett.69.3314.

Wertz, E., Ferrier, L., Solnyshkov, D.D., Johne, R., Sanvitto, D., Lemaître, A., Sagnes, I., Grousson, R., Kavokin, A.V., Senellart, P., Malpuech, G., Bloch, J., 2010. Spontaneous formation and optical manipulation of extended polariton condensates. Nat. Phys. 6 (11), 860–864. https://doi.org/10.1038/nphys1750.

Wertz, E., Amo, A., Solnyshkov, D.D., Ferrier, L., Liew, T.C.H., Sanvitto, D., Senellart, P., Sagnes, I., Lemaître, A., Kavokin, A.V., Malpuech, G., Bloch, J., 2012. Propagation and amplification dynamics of 1D polariton condensates. Phys. Rev. Lett. 109 (21), 216404. https://doi.org/10.1103/PhysRevLett.109.216404.

Whittaker, C.E., Cancellieri, E., Walker, P.M., Gulevich, D.R., Schomerus, H., Vaitiekus, D., Royall, B., Whittaker, D.M., Clarke, E., Iorsh, I.V., Shelykh, I.A., Skolnick, M.S., Krizhanovskii, D.N., 2018. Exciton polaritons in a two-dimensional Lieb lattice with spin-orbit coupling. Phys. Rev. Lett. 120 (9), 097401. https://doi.org/10.1103/PhysRevLett.120.097401.

Winkler, K., Schneider, C., Fischer, J., Rahimi-Iman, A., Amthor, M., Forchel, A., Reitzenstein, S., Höfling, S., Kamp, M., 2013. Electroluminescence from spatially confined exciton polaritons in a textured microcavity. Appl. Phys. Lett. 102 (4), 041101. https://doi.org/10.1063/1.4777564.

Winkler, K., Fischer, J., Schade, A., Amthor, M., Dall, R., Geßler, J., Emmerling, M., Ostrovskaya, E.A., Kamp, M., Schneider, C., Höfling, S., 2015. A polariton condensate in a photonic crystal potential landscape. New J. Phys. 17 (2), 023001. https://doi.org/10.1088/1367-2630/17/2/023001.

Winkler, K., Egorov, O.A., Savenko, I.G., Ma, X., Estrecho, E., Gao, T., Müller, S., Kamp, M., Liew, T.C.H., Ostrovskaya, E.A., Höfling, S., Schneider, C., 2016. Collective state transitions of exciton-polaritons loaded into a periodic potential. Phys. Rev. B 93 (12), 121303. https://doi.org/10.1103/PhysRevB.93.121303.

Wouters, M., Carusotto, I., 2007. Excitations in a nonequilibrium Bose-Einstein condensate of exciton polaritons. Phys. Rev. Lett. 99 (14), 140402. https://doi.org/10.1103/PhysRevLett.99.140402.

Xie, W., Dong, H., Zhang, S., Sun, L., Zhou, W., Ling, Y., Lu, J., Shen, X., Chen, Z., 2012. Room-temperature polariton parametric scattering driven by a one-dimensional polariton condensate. Phys. Rev. Lett. 108 (16), 166401. https://doi.org/10.1103/PhysRevLett.108.166401.

Yamaguchi, M., Kamide, K., Nii, R., Ogawa, T., Yamamoto, Y., 2013. Second thresholds in BEC-BCS-laser crossover of exciton-polariton systems. Phys. Rev. Lett. 111 (2), 026404. https://doi.org/10.1103/PhysRevLett.111.026404.

Ye, Z., Cao, T., O'Brien, K., Zhu, H., Yin, X., Wang, Y., Louie, S.G., Zhang, X., 2014. Probing excitonic dark states in single-layer tungsten disulphide. Nature 513 (7517), 214–218. https://doi.org/10.1038/nature13734.

Zasedatelev, A.V., Baranikov, A.V., Urbonas, D., Scafirimuto, F., Scherf, U., Stöferle, T., Mahrt, R.F., Lagoudakis, P.G., 2019. A room-temperature organic polariton transistor. Nat. Photon. 13 (6), 378–383. https://doi.org/10.1038/s41566-019-0392-8.

Zhang, B., Wang, Z., Brodbeck, S., Schneider, C., Kamp, M., Höfling, S., Deng, H., 2014. Zero-dimensional polariton laser in a subwavelength grating-based vertical microcavity. Light Sci. Appl. 3 (1), e135. https://doi.org/10.1038/lsa.2014.16.

Zhang, B., Brodbeck, S., Wang, Z., Kamp, M., Schneider, C., Höfling, S., Deng, H., 2015a. Coupling polariton quantum boxes in sub-wavelength grating microcavities. Appl. Phys. Lett. 106 (5), 051104. https://doi.org/10.1063/1.4907606.

Zhang, L., Xie, W., Wang, J., Poddubny, A., Lu, J., Wang, Y., Gu, J., Liu, W., Xu, D., Shen, X., Rubo, Y.G., Altshuler, B.L., Kavokin, A.V., Chen, Z., 2015b. Weak lasing in one-dimensional polariton superlattices. Proc. Natl. Acad. Sci. U.S.A 112 (13), E1516–E1519. https://doi.org/10.1073/pnas.1502666112.

Zhang, L., Gogna, R., Burg, W., Tutuc, E., Deng, H., 2018. Photonic-crystal exciton-polaritons in monolayer semiconductors. Nat. Commun. 9 (1), 713. https://doi.org/10.1038/s41467-018-03188-x.

Further reading

Vahala, K.J., 2003. Optical microcavities. Nature 424 (6950), 839–846. https://doi.org/10.1038/nature01939.

Wang, G., Chernikov, A., Glazov, M.M., Heinz, T.F., Marie, X., Amand, T., Urbaszek, B., 2017b. Excitons in atomically thin transition metal dichalcogenides. Rev. Mod. Phys. 90 (2), 21001. https://doi.org/10.1103/RevModPhys.90.021001.

CHAPTER THREE

Ultrastrong light–matter coupling in semiconductors

Nicolas Marquez Peraca[a], Andrey Baydin[b], Weilu Gao[c], Motoaki Bamba[d,e], and Junichiro Kono[a,b,f,*]

[a]Department of Physics and Astronomy, Rice University, Houston, TX, United States
[b]Department of Electrical and Computer Engineering, Rice University, Houston, TX, United States
[c]Department of Electrical and Computer Engineering, University of Utah, Salt Lake City, UT, United States
[d]Department of Physics, Kyoto University, Kyoto, Japan
[e]PRESTO, Japan Science and Technology Agency, Saitama, Japan
[f]Department of Materials Science and NanoEngineering, Rice University, Houston, TX, United States
*Corresponding author: e-mail address: kono@rice.edu

Contents

1. Introduction

The main scientific question this chapter deals with is: *What is the nature of the ground state of a condensed matter system that is ultrastrongly coupled with vacuum electromagnetic fields inside a cavity?* There are a variety of

Semiconductors and Semimetals, Volume 105
ISSN 0080-8784
https://doi.org/10.1016/bs.semsem.2020.10.002

theoretically predicted many-body cavity quantum electrodynamics (QED) effects, states, and phases in *the ultrastrong coupling (USC) regime* (Forn-Díaz et al., 2019; Kockum et al., 2019), and it is our goal to review these hitherto-unobserved phenomena. Such phenomena can also be utilized for constructing unique protocols for ultrafast gates and ultrasecure state preparation for quantum information processing. Specifically, we provide an overview of the current state of theoretical and experimental studies of cavity QED systems in the USC regime based on semiconductor quantum wells (QWs), graphene, carbon nanotubes (CNTs), and strongly correlated materials.

1.1 What is the ultrastrong coupling regime?

In a cavity embedding matter, there are four quantities that jointly characterize different light–matter coupling regimes: ω_c, g, κ, and γ. Here, ω_c is the cavity mode photon angular frequency, and the parameter g is the coupling strength (or constant or rate). The parameter κ is the photon decay rate of the cavity; $\tau_{cav} = \kappa^{-1}$ is the photon lifetime of the cavity, and the cavity quality factor is defined as $Q = \omega_c \tau_{cav}$. The parameter γ is the nonradiative matter decay rate which is usually the decoherence rate in the case of solids.

The simplest Hamiltonian one can write down for this system is

$$\mathcal{H} = \hbar\omega_c \hat{a}^\dagger \hat{a} + \hbar\omega_a \hat{b}^\dagger \hat{b} + i\hbar g(\hat{a}^\dagger \hat{b} - \hat{b}^\dagger \hat{a}). \tag{1}$$

Here, $\hat{a}$ ($\hat{a}^\dagger$) and $\hat{b}$ ($\hat{b}^\dagger$) are the annihilation (creation) operators for cavity photons and matter excitations, respectively. When the matter resonance frequency, ω_a, is equal to the cavity frequency, ω_c, i.e., $\omega_a = \omega_c \equiv \omega_0$, the eigen-frequencies of the coupled system are obtained as $\omega_\pm = \omega_0 \pm g$, corresponding to the lower polariton (LP) and upper polariton (UP). The frequency difference between them, $2g$, is called the vacuum Rabi splitting (VRS).

Strong coupling (SC) is achieved (Khitrova et al., 2006) when the VRS, $2g$, is larger than the linewidth, $(\kappa + \gamma)/2$ (see Fig. 1). USC is achieved when g becomes a considerable fraction (say, 10%) of ω_0, and some authors define the deep strong coupling (DSC) regime as $g/\omega_0 > 1$ (Casanova et al., 2010). The two standard figures of merit to measure the coupling strength are

$$C \equiv 4g^2/(\kappa\gamma), \tag{2}$$

$$\eta \equiv g/\omega_0. \tag{3}$$

Here, C is referred to as the *cooperativity* parameter, which is also the determining factor for the onset of optical bistability through resonant absorption

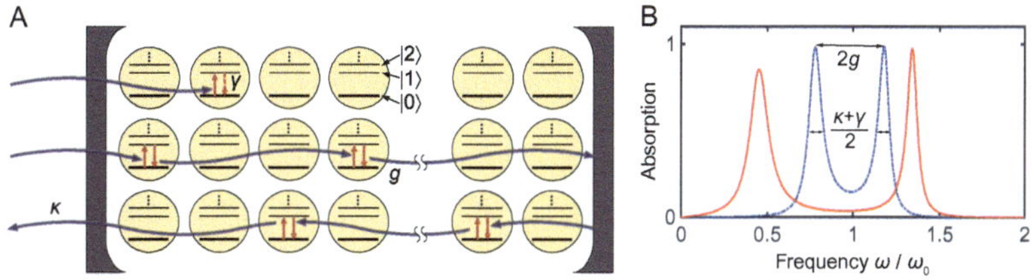

Fig. 1 The parameters that define different light–matter coupling regimes. (A) Sketch of a system consisting of an ensemble of atoms (with multiple levels $|0\rangle$, $|1\rangle$, $|2\rangle$, ...) and a lossy cavity of light. Each atom can be excited by absorbing a photon. A photon is reemitted through relaxation of an atom, while no photon is sometimes emitted by a nonradiative matter decay. g: light–matter coupling rate, γ: matter decay rate, and κ: photon decay rate. (B) Absorption spectra of the cavity system in the SC (*blue*) and USC (*red*) regimes in the case of zero detuning $\omega_a = \omega_c = \omega_0$. The SC is defined as $2g > (\gamma + \kappa)/2$, when the VRS is larger than the linewidth. Furthermore, the USC regime arises when g is a significant fraction of the resonance frequency, ω_0.

saturation (Bonifacio and Lugiato, 1982), and η is the normalized coupling strength. Using these parameters, the three regimes of light–matter coupling are defined as follows (Forn-Díaz et al., 2019; Kockum et al., 2019):

$$\boxed{\textbf{SC}: 2g > (\kappa + \gamma)/2 \quad (\text{or} \quad C > 1), \textbf{USC}: \eta > 0.1, \text{and} \quad \textbf{DSC}: \eta > 1}$$

Note that the above Hamiltonian, Eq. (1), includes only the so-called co-rotating coupling terms, for which the total number of photons and matter excitations is conserved. As is discussed in more detail in Section 2, this Hamiltonian is derived under the rotating-wave approximation (RWA). In general, counter-rotating terms (CRTs), $i\hbar g(\hat{a}^\dagger \hat{b}^\dagger - \hat{a}\hat{b})$, also appear in the Hamiltonian and become nonnegligible in the USC and DSC regimes, where they play essential roles in inducing nonintuitive physical phenomena.

In Table 1, we summarize the commonly used light–matter coupling models (or Hamiltonians). In the Jaynes–Cummings model and the quantum-Rabi model, matter, approximated as a single two-level atom (or a two-level quantum dot), couples with a single photonic mode. The difference between the two models is that the former adopts the RWA, i.e., neglects the CRTs, while the latter retains the CRTs. When the matter is instead approximated as multiple two-level atoms, these models are called the Tavis–Cummings model and the Dicke model, respectively. On the other hand, when the matter is approximated as a single bosonic excitation as in Eq. (1), the model becomes the Hopfield model, irrespective of whether the RWA is used.

Table 1 Light–matter coupling Hamiltonians.

	1 two-level atom	*N* two-level atoms	Bosonic excitations
Within RWA	Jaynes–Cummings model	Tavis–Cummings model	Hopfield model
Beyond RWA	quantum-Rabi model	Dicke model	Hopfield model

Under the RWA, the CRTs are neglected in the Hamiltonian.

1.2 Why ultrastrong light–matter coupling?

It is important to note that in cavity QED systems in the USC regime the "light field" that the matter strongly couples with is not an external laser field but the vacuum fluctuation field in the cavity. This fact makes USC physics distinctly different from ordinary nonlinear optical phenomena which require a strong *external* field and thus inevitably involve excited and/or nonequilibrium matter states. In contrast, USC phenomena arise from the properties of the ground state of the matter-vacuum hybrid *in equilibrium*. This new ground state is a matter-vacuum entangled state (Ashhab and Nori, 2010; Ciuti et al., 2005; Felicetti et al., 2015) which has characteristics that neither the original matter ground state nor the usual vacuum possesses.

There is a long history of theoretical studies of USC physics. Early studies focused on the breakdown of the RWA (Allen and Eberly, 1975; Bloch and Siegert, 1940; Cohen-Tannoudji et al., 1973; De Zela et al., 1997; Shirley, 1965). After the 1973 discovery of a new type of phase transition, *the superradiant phase transition (SRPT)*, based on the Dicke Hamiltonian (Hepp and Lieb, 1973; Wang and Hioe, 1973), attention has been paid to the properties of the ground state and the resultant nonintuitive quantum phenomena: e.g., quantum vacuum radiation (Auer and Burkard, 2012; Ciuti and Carusotto, 2006; Ciuti et al., 2005; De Liberato et al., 2007; Hagenmüller, 2016) (similar to the Unruh-Hawking radiation from black-hole explosions (Hawking, 1974, 1975; Unruh, 1974, 1976; Yablonovitch, 1989)) induced by the dynamic Casimir effect (Fulling and Davies, 1976; Kardar and Golestanian, 1999; Moore, 1970) or through spontaneous conversion (Stassi et al., 2013), unusual photon counting statistics (Ridolfo et al., 2012), the breakdown of the Purcell effect or light–matter "decoupling" (De Liberato, 2014; Jaako et al., 2016), single-photon nonlinear optics (Sanchez-Burillo et al., 2014), quantum chaos (Emary and Brandes, 2003a,b), spontaneous parity-symmetry breaking (Bamba et al., 2016), and ground-state electroluminescence (Cirio et al., 2016, 2019).

For applications, solid-state cavity or circuit QED systems in the USC regime have been recognized to be promising for quantum simulations, simulating, e.g., the quantum-Rabi model (Ballester et al., 2012; Hwang et al., 2015; Mezzacapo et al., 2014; Puebla et al., 2017), Dirac equation physics (Pedernales et al., 2013), and Dicke physics (Lamata, 2017; Mezzacapo et al., 2014). Furthermore, the many-body entangled ground states (Ashhab and Nori, 2010; Felicetti et al., 2015) can be used to construct quantum information processing protocols (Wang et al., 2010) e.g., cat-state-based quantum error correction (Ofek et al., 2016). Also, ultrafast two-qubit quantum gates have been considered (Kyaw et al., 2015a,b; Romero et al., 2012; Wang et al., 2016); in particular, Romero et al. (2012) used *ab initio* calculations to demonstrate quantum gates in the USC regime that can be performed at subnanosecond time scales. Finally, Nataf and Ciuti showed that the qubit coherence time and fidelity of a universal set of quantum gates can be dramatically improved in an optimal regime of USC (Nataf and Ciuti, 2011).

Another possible application of USC is the "quantum battery" where quantum coherence or entanglement enhances the stored energy density or the charging power (stored energy over charging time). Ferraro et al. (2018) theoretically suggested that, in the Dicke model, the charging power can be enhanced by factor $\sqrt{N}$ for $N \gg 1$ due to the Dicke cooperativity (interatom entanglement mediated by the photonic mode), i.e., the charging time is shortened by $\sqrt{N}$. Further, they also suggested that the stored energy density can be enhanced by rearrangement of charges in the superradiant phase obtained by USC (Ferraro et al., 2019). These studies are reviewed by Ferraro et al. (2020).

Finally, it has been proposed that USC may be useful for dark matter detection. An axion haloscope for detecting galactic axions can be improved by the use of magnon–photon USC in a GHz-frequency cavity embedding a yttrium iron garnet (YIG) sphere (Flower et al., 2019a,b). While many ongoing axion haloscope projects are based on the expected axion–photon interaction, an interaction between axions and electron spins may also exist. Flower et al. (2019a) reported an experimental setup to detect signals of magnon polaritons excited by axions. The magnon–photon USC enlarges the frequency (energy) range of the haloscope (Flower et al., 2019b), which is essential for narrowing the parameter space where axions may exist.

1.3 Why solid-state cavity QED systems for USC studies?

In order to maximize C and η, one should construct a cavity QED setup thatcombines a large dipole moment (i.e., large g), a small decoherence rate (i.e.,small γ), a large cavity Q factor (i.e., small κ), and a small resonance frequency ω_0. Condensed matter (or solid-state) systems have a particular advantage over atomic and molecular systems in achieving large g values.

This is due to the cooperative $\sqrt{N}$-fold enhancement of light–matter coupling, i.e., Dicke cooperativity, expected for an N-body system (Amsüss et al., 2011; Dicke, 1954; Kaluzny et al., 1983; Tabuchi et al., 2014; Zhang et al., 2014b), combined with colossal dipole moments available in solids.

Semiconductor QWs can create clean and tunable solid-state environments with quantum-engineered optical properties. Microcavity exciton polaritons in a QW—a strongly coupled light–condensed-matter system—have exhibited a diverse array of quantum many-body phenomena (Deng et al., 2010; Gibbs et al., 2011; Khitrova et al., 1999). However, due to the relatively large values of ω_0, corresponding to the near-infrared and visible ranges, and relatively small dipole moments for interband transitions, it is not practical to achieve large values of $\eta = g/\omega_0$ using QW exciton polaritons.

As detailed in Section 3.1, *intraband* transitions, including intersubband transitions (ISBTs) (Helm, 2000; Paiella, 2006) and cyclotron resonance (CR) (Hilton et al., 2012; Kono, 2001; Lax and Mavroides, 1960; McCombe and Wagner, 1975), are much better suited for exploring USC phenomena because of their small ω_0, typically in the midinfrared (MIR) and THz range, and large dipole moments. Liu (1997) first proposed and analyzed ISB polaritons theoretically; Ciuti et al. (2005) discussed the nonclassical nature of the ground-state properties of ISB polaritons in the USC regime. The first experimental observation of polariton splitting of ISBTs was reported by Dini et al. (2003), followed by a similar observation by Dupont et al. (2003); these early studies already reported η values >0.01. Progressively higher values of η have since been reported by various experimental groups (Anappara et al., 2005, 2006, 2007, 2009; Askenazi et al., 2014, 2017; Delteil et al., 2012; Dietze et al., 2013; Dupont et al., 2007; Geiser et al., 2010, 2012; Günter et al., 2009; Jeannin et al., 2019; Jouy et al., 2011; Laurent et al., 2017; Porer et al., 2012; Sapienza et al., 2007, 2008; Todorov et al., 2009, 2010a; Zanotto et al., 2010, 2012).

CR, i.e., inter-Landau-level transitions (ILLTs), has also been used to study strong coupling in two-dimensional electron gases (2DEGs) in semiconductor QWs (Bayer et al., 2017; Maissen et al., 2014, 2017; Muravev et al., 2013; Scalari et al., 2012; Zhang et al., 2016a) and on the surface of liquid helium (Abdurakhimov et al., 2016). In some of these cases, extremely high values of η (= 0.87 (Maissen et al., 2014) and = 1.43 (Bayer et al., 2017)) have been achieved. More recently, clear evidence has been obtained for the breakdown of the RWA (Li et al., 2018b), which resulted in a shift in the resonance frequency known as *the Bloch-Siegert (B–S) shift* (Bloch and Siegert, 1940) (see Fig. 2). The lowest-order term in the B–S shift is on the order of g^2/ω_0 (Allen and Eberly, 1975). Higher-order terms of

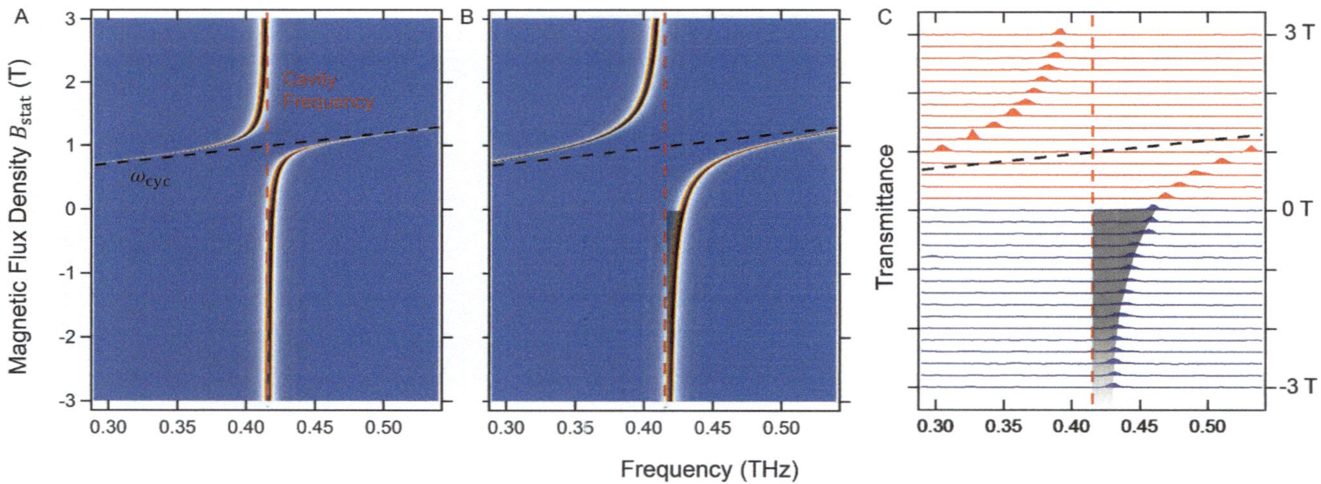

Fig. 2 Vacuum Bloch–Siegert shift of Landau polaritons with ultrahigh cooperativity (Bamba et al., 2019). Numerically calculated THz transmission for the CR–cavity system for interaction strength (A) $g/2\pi = 37.5$ GHz, (B) $g/2\pi = 75.0$ GHz, and (C) experimental data (interaction strength is estimated as $g/2\pi = 150.1$ GHz) as a function of frequency $\omega/2\pi$ and external DC magnetic field B_{DC}. *Dashed red* and *black lines* show bare cavity and CR frequencies, respectively, without considering their interaction. At $B_{DC} > 0$, the CR and a circularly polarized probe THz wave are corotating and the VRS (anticrossing) is obtained. At $B_{DC} < 0$, the vacuum B–S shift appears due to the counterrotating coupling between the CR and the circularly polarized cavity field as highlighted by the *gray shaded areas*.

the B–S shift have also been calculated quantum mechanically (Cohen-Tannoudji et al., 1973; Shirley, 1965), which have to be taken into account in order to correctly estimate the magnetic moment values from magnetic resonance experiments (Abragam, 1961). It should be emphasized that this is a *vacuum* B–S shift in a Landau-polariton system, which occurs when the average photon number inside the cavity is much less than one. The shift in this case is caused by the USC of a Landau-quantized 2DEG with the counter-rotating component of the vacuum fluctuation field inside the cavity.

2. Theory of ultrastrong light–matter coupling

2.1 Hamiltonians and resonance frequencies

When we do not consider the spin degree of freedom, the Hamiltonian describing the electromagnetic fields and charged particles (electrons and ions) is generally expressed by the so-called minimal-coupling Hamiltonian (Cohen-Tannoudji et al., 1989)

$$\hat{\mathcal{H}}_v = \int \mathrm{d}\boldsymbol{r} \left[\frac{\varepsilon_0 \hat{\boldsymbol{E}}_\perp(\boldsymbol{r})^2}{2} + \frac{\hat{\boldsymbol{B}}(\boldsymbol{r})^2}{2\mu_0} \right] + \sum_{j=1}^{N} \frac{\left[\hat{\boldsymbol{p}}_j - e_j \hat{\boldsymbol{A}}(\boldsymbol{r}_j)\right]^2}{2m_j} + \hat{V}(\{\hat{\boldsymbol{r}}_j\}). \quad (4)$$

This is derived in the Coulomb gauge by the Legendre transformation of a Lagrangian describing Maxwell's equations and Newton's equations of motion with the Lorentz force. The first and second terms are the energies of the transverse component of the electric field $\hat{\boldsymbol{E}}_\perp(\boldsymbol{r}) = -\hat{\boldsymbol{\Pi}}(\boldsymbol{r})/\varepsilon_0$ and of the magnetic field $\hat{\boldsymbol{B}}(\boldsymbol{r}) = \nabla \times \hat{\boldsymbol{A}}(\boldsymbol{r})$, respectively. $\hat{\boldsymbol{A}}(\boldsymbol{r})$ is the vector potential, and $\hat{\boldsymbol{\Pi}}(\boldsymbol{r})$ is its conjugate momentum satisfying $[\hat{\boldsymbol{A}}(\boldsymbol{r}), \hat{\boldsymbol{\Pi}}(\boldsymbol{r}')] = \boldsymbol{\delta}_\perp(\boldsymbol{r} - \boldsymbol{r}')$, where $\boldsymbol{\delta}_\perp(\boldsymbol{r}) = \delta(\boldsymbol{r})\mathbf{1} - (2\pi)^{-3} \int \mathrm{d}\boldsymbol{k}\, (\boldsymbol{k}\boldsymbol{k}/k^2)\mathrm{e}^{\mathrm{i}\boldsymbol{k}\cdot\boldsymbol{r}}$ is the dyadic transverse delta function (Cohen-Tannoudji et al., 1989). The third term in Eq. (4) represents the kinetic energy of the charged particles. $\hat{\boldsymbol{r}}_j$ and $\hat{\boldsymbol{p}}_j$ are, respectively, the position and momentum of the jth particle with a mass m_j and a charge e_j. They satisfy $[\hat{\boldsymbol{r}}_j, \hat{\boldsymbol{p}}_{j'}] = \mathrm{i}\hbar\delta_{j,j'}\mathbf{1}$. N is the number of the charged particles. The last term in Eq. (4) represents the Coulomb interaction. The interaction between the electromagnetic fields and charged particles is obtained by expanding the third term in Eq. (4).

We can transform the minimal-coupling Hamiltonian, Eq. (4), into different forms. Here, we define the electric polarization as $\hat{\boldsymbol{P}}(\boldsymbol{r}) \equiv \sum_{j=1}^{N} e_j \hat{\boldsymbol{r}}_j \delta(\boldsymbol{r} - \boldsymbol{R}_j)$, where $\boldsymbol{R}_j$ is the position of a lattice site to which the jth particle belongs.

This expression is valid in the long-wavelength approximation, i.e., the distance $|\hat{r}_j - \boldsymbol{R}_j|$ between the particle and the lattice site is much shorter than the wavelength of interest. By using an unitary operator $\hat{U} \equiv \exp\left[(\mathrm{i}\hbar)^{-1} \int \mathrm{d}\boldsymbol{r}\, \hat{\boldsymbol{A}}(\boldsymbol{r}) \cdot \hat{\boldsymbol{P}}(\boldsymbol{r})\right]$, we get a different form of Hamiltonian $\hat{\mathcal{H}}_l \equiv \hat{U}\hat{\mathcal{H}}_v\hat{U}^\dagger$ as

$$\hat{\mathcal{H}}_l = \int \mathrm{d}\boldsymbol{r} \left\{ \frac{[\hat{\boldsymbol{D}}(\boldsymbol{r}) - \hat{\boldsymbol{P}}_\perp(\boldsymbol{r})]^2}{2\varepsilon_0} + \frac{\hat{\boldsymbol{B}}(\boldsymbol{r})^2}{2\mu_0} \right\} + \sum_{j=1}^{N} \frac{\hat{\boldsymbol{p}}_j^{\,2}}{2m_j} + V(\{\hat{\boldsymbol{r}}_j\}), \quad (5)$$

where we used the following relations: $\hat{U}\hat{\boldsymbol{\Pi}}(\boldsymbol{r})\hat{U}^\dagger = \hat{\boldsymbol{\Pi}}(\boldsymbol{r}) + \hat{\boldsymbol{P}}_\perp(\boldsymbol{r})$ and $\hat{U}\hat{\boldsymbol{p}}_j\hat{U}^\dagger = \hat{\boldsymbol{p}}_j + e_j\hat{\boldsymbol{A}}(\boldsymbol{R}_j)$ for the transverse component of the electric polarization $\hat{\boldsymbol{P}}_\perp(\boldsymbol{r}) \equiv \int \mathrm{d}\boldsymbol{r}\, \boldsymbol{\delta}_\perp(\boldsymbol{r} - \boldsymbol{r}') \cdot \hat{\boldsymbol{P}}(\boldsymbol{r}')$. In this form, the interaction between the electromagnetic fields and charged particles is obtained by expanding the first term in Eq. (5), which still represents the energy of the transverse electric field but is described by the electric displacement field $\hat{\boldsymbol{D}}(\boldsymbol{r}) = -\hat{\boldsymbol{\Pi}}(\boldsymbol{r})$ and the transverse electric polarization $\hat{\boldsymbol{P}}_\perp(\boldsymbol{r})$. The transformation beyond the long-wavelength approximation is known as the Power–Zienau–Woolley transformation (Cohen-Tannoudji et al., 1989).

Eqs. (4) and (5) are said to be in the velocity and length forms, respectively, or sometimes called the Coulomb gauge and the dipole gauge, respectively. These two forms in principle give the same physics in the long-wavelength approximation. However, in calculations of specific systems, we usually truncate some of the matter levels and electromagnetic modes. For example, as shown in Fig. 3, there are in general more than two subbands in the conduction and valence bands and many cavity modes in a Fabry–Pérot (FP) cavity. We usually focus on just two subbands and one cavity mode in a frequency range of interest, and neglect the other levels (subbands) and cavity modes. However, if we perform such truncation, $\mathcal{H}_v$ and $\mathcal{H}_l$ are no longer related to each other through a unitary transform; i.e., they are not equivalent and thus give different dynamics. Such truncation is justified (and thus the two Hamiltonians become approximately equivalent) only when the neglected levels and modes exist far out of the frequency range of interest. However, in the USC regime, the frequency range of interest becomes extremely large, and the influence of truncation frequently becomes crucial (Bamba and Ogawa, 2016; De Bernardis et al., 2018).

This truncation issue is related to the so-called *gauge ambiguity* problem, which has attracted much attention in recent years, especially in deriving the

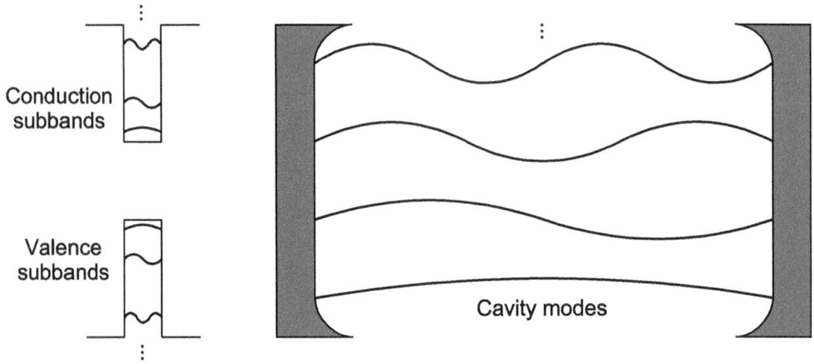

Fig. 3 Sketches of subbands in a semiconductor QW and cavity modes in a FP cavity. In real systems, there can be many subbands in the conduction and valence bands and many cavity modes (as indicated by the *three dots in each*). However, in theoretical analyses, one usually focuses on some of the subbands and cavity modes only (i.e., truncate the other bands and modes). Such simplification invalidates the unitary transformation between the Hamiltonians in the velocity and length forms.

Jaynes-Cummings and quantum-Rabi models. We can derive these models from both forms of the Hamiltonian by truncation. However, when we compare the eigenfrequencies between the truncated and original Hamiltonians, one form can show better agreement than the other. Then, one faces the question of which form gives a better approximation after truncation is done. As we will see in Section 2.1.2, the length form usually provides a better approximation in the case of electric dipole transitions. It has recently been pointed out that the velocity form can also give the same level of approximation if we properly transform the Coulomb potential V under truncation (Di Stefano et al., 2019; Garziano et al., 2020). On the other hand, it has also been suggested that a much better approximation can be obtained when one performs truncation in an intermediate form between the velocity and length forms (Roth et al., 2019; Stokes and Nazir, 2019a). The case of time-dependent light–matter interactions has also been discussed (Stokes and Nazir, 2019b).

In the following, instead of discussing the ambiguity issues in more detail, we will derive simple Hamiltonians for three cases. In Section 2.1.1, we provide a general discussion on the quantization of electromagnetic fields. In Section 2.1.2, ISBTs in semiconductor QWs in the USC regime are theoretically described. In Section 2.1.3, the physics of optical phonons and excitons in bulk semiconductors in the USC regime is examined. In these two cases, it is known that the length form gives a better convergence than the velocity form for matter-level truncation. In Section 2.1.4, CR in a 2DEG in a cavity, or Landau polaritons, is described. In this case, the velocity form is suitable for deriving a simple Hamiltonian, and it well explains the experimental results presented in Section 3 (see also Fig. 2).

2.1.1 Sum rule and quantization of electromagnetic fields

As a general discussion common to all three cases (ISBTs, phonons/excitons, and CR), let us first define the Hamiltonians of the electromagnetic fields and charged particles, respectively, as

$$\hat{\mathcal{H}}_{\mathrm{em}} = \int \mathrm{d}\boldsymbol{r} \left\{ \frac{\hat{\boldsymbol{\Pi}}(\boldsymbol{r})^2}{2\varepsilon_0\varepsilon_{\mathrm{bg}}} + \frac{[\nabla \times \hat{\boldsymbol{A}}(\boldsymbol{r})]^2}{2\mu_0} \right\}, \tag{6a}$$

$$\hat{\mathcal{H}}_{\mathrm{mat}} = \sum_{j=1}^{N} \frac{\hat{\boldsymbol{p}}_j^{\,2}}{2m_j} + \hat{V}(\{\hat{\boldsymbol{r}}_j\}). \tag{6b}$$

Here, we consider a part of the matter as a dielectric medium with a frequency-independent (nondispersive) relative permittivity $\varepsilon_{\mathrm{bg}}$ and put it into $\hat{\mathcal{H}}_{\mathrm{em}}$. Then, the speed of light in the $\hat{\mathcal{H}}_{\mathrm{em}}$ system is $v \equiv c/\sqrt{\varepsilon_{\mathrm{bg}}}$. The total Hamiltonians in Eqs. (4) and (5) can then be rewritten as

$$\hat{\mathcal{H}}_v = \hat{\mathcal{H}}_{\mathrm{em}} + \hat{\mathcal{H}}_{\mathrm{mat}} + \sum_{j=1}^{N} \left[-\frac{e_j\hat{\boldsymbol{p}}_j \cdot \hat{\boldsymbol{A}}(\boldsymbol{r}_j)}{m_j} + \frac{e_j^{\,2}\hat{\boldsymbol{A}}(\boldsymbol{r}_j)^2}{2m_j} \right], \tag{7a}$$

$$\hat{\mathcal{H}}_l = \hat{\mathcal{H}}_{\mathrm{em}} + \hat{\mathcal{H}}_{\mathrm{mat}} + \int \mathrm{d}\boldsymbol{r} \left[-\frac{\hat{\boldsymbol{D}}(\boldsymbol{r}) \cdot \hat{\boldsymbol{P}}_\perp(\boldsymbol{r})}{\varepsilon_0\varepsilon_{\mathrm{bg}}} + \frac{\hat{\boldsymbol{P}}_\perp(\boldsymbol{r})^2}{2\varepsilon_0\varepsilon_{\mathrm{bg}}} \right]. \tag{7b}$$

In each expression, the third term represents the light–matter coupling. The last term in Eq. (7a) (Eq. 7b) is called the A^2 (P^2) term or the diamagnetic term.

From the matter Hamiltonian $\hat{\mathcal{H}}_{\mathrm{mat}}$, Eq. (6b), together with the relation $(m_j/\hbar^2)[\hat{\boldsymbol{r}}_j, [\hat{\mathcal{H}}_{\mathrm{mat}}, \hat{\boldsymbol{r}}_j]] = \mathbf{1}$, we can derive the following general relation called the Thomas–Reiche–Kuhn (TRK) sum rule (Luttinger and Kohn, 1955):

$$\sum_n f_{n,i} = 1, \tag{8}$$

where $f_{n,i}$ is the oscillator strength for the transition from an initial state (ith state) to the nth state and is defined as

$$f_{n,i} \equiv \frac{2m_j\omega_{n,i}}{e_j{}^2\hbar}|\boldsymbol{d}_{n,i}|^2. \tag{9}$$

Here, $\omega_{n,i}$ is the frequency difference between the two states, and $\boldsymbol{d}_{n,i} \equiv \langle n|e_j\hat{\boldsymbol{r}}_j|i\rangle$ is the transition dipole moment.

Next, let us rewrite the Hamiltonian $\hat{\mathcal{H}}_{\text{em}}$ of the electromagnetic fields in terms of the annihilation and creation operators of photons. Following the procedure of quantization of the electromagnetic wave in an dielectric medium described by Glauber and Lewenstein (1991), we can write the operators of the vector potential and its conjugate momentum as

$$\hat{\boldsymbol{A}}(\boldsymbol{r}) = \sum_k \sqrt{\frac{\hbar}{2\varepsilon_0\varepsilon_{\text{bg}}\omega_k}}\left[\boldsymbol{\psi}_k^*(\boldsymbol{r})\hat{a}_k^\dagger + \boldsymbol{\psi}_k(\boldsymbol{r})\hat{a}_k\right], \tag{10}$$

$$\hat{\boldsymbol{\Pi}}(\boldsymbol{r}) = \sum_k \mathrm{i}\sqrt{\frac{\varepsilon_0\varepsilon_{\text{bg}}\hbar\omega_k}{2}}\left[\boldsymbol{\psi}_k^*(\boldsymbol{r})\hat{a}_k^\dagger - \boldsymbol{\psi}_k(\boldsymbol{r})\hat{a}_k\right]. \tag{11}$$

Here, $\hat{a}_k$ is the photon annihilation operator for the kth mode, satisfying $[\hat{a}_k, \hat{a}_{k'}^\dagger] = \delta_{k,k'}$, ω_k is the eigenfrequency, and $\boldsymbol{\psi}_k(\boldsymbol{r})$ is the wavefunction including the polarization direction. The Hamiltonian of the electromagnetic wave is then rewritten as

$$\hat{\mathcal{H}}_{\text{em}} = \sum_k \hbar\omega_k\left(\hat{a}_k^\dagger\hat{a}_k + \frac{1}{2}\right). \tag{12}$$

2.1.2 Intersubband transitions

Let us consider an ISBT in a semiconductor QW embedded in a microcavity. From a one-body potential in the QW through the Coulomb interaction $\hat{V}$, subbands for electrons (holes) can be obtained within the conduction (valence) band. We focus on two subbands with a frequency difference of ω_{a} and describe the jth electron or hole by Pauli operators: $\hat{\sigma}_j^z \equiv |1_j\rangle\langle 1_j| - |0_j\rangle\langle 0_j|$, $\hat{\sigma}_j^+ \equiv |1_j\rangle\langle 0_j|$, $\hat{\sigma}_j^- \equiv |0_j\rangle\langle 1_j|$, $\hat{\sigma}_j^x = \hat{\sigma}_j^+ + \hat{\sigma}_j^-$, and $\hat{\sigma}_j^y = -\mathrm{i}(\hat{\sigma}_j^+ - \hat{\sigma}_j^-)$, where $|0_j\rangle$ and $|1_j\rangle$ represent the lower and upper subband states of the jth particle, respectively. We assume that the QW is placed in a FP cavity with perfect mirrors, distance L, and area S, for which $\omega_{\text{c}} = v(n\pi/L)$ and $\boldsymbol{\psi} = \boldsymbol{e}_\xi(2/SL)^{1/2}\sin[(n\pi/L)z]\mathrm{e}^{\mathrm{i}\boldsymbol{k}_\parallel\cdot\boldsymbol{r}_\parallel}$ for the nth cavity mode with in-plane wavevector $\boldsymbol{k}_\parallel$ polarized in the ξ direction.

By neglecting many-body Coulomb interactions (nonlinear responses) in $\hat{V}$ as well as in-plane energy dispersions, we can write the matter Hamiltonian, Eq. (6b), as

$$\hat{\mathcal{H}}_{\text{mat}} = \sum_{j=1}^{N} \frac{\hbar\omega_{\text{a}}}{2}\left(\hat{\sigma}_j^z + \frac{1}{2}\right) = \hbar\omega_{\text{a}}\left(\hat{S}_z + \frac{N}{2}\right), \tag{13}$$

where $\hat{S}_\xi \equiv \sum_{j=1}^{N} \hat{\sigma}^\xi_j/2$ is a $\frac{N}{2}$-spin operator representing the ensemble of particles. When we focus on just one cavity mode, the Hamiltonians in the two forms, Eqs. (7a) and (7b), are rewritten as

$$\hat{\mathcal{H}}_v/\hbar \approx \omega_{\text{c}}\hat{a}^\dagger\hat{a} + \omega_{\text{a}}\left(\hat{S}_z + \frac{N}{2}\right) + \frac{2\bar{g}}{\sqrt{N}}\hat{S}_y(\hat{a}^\dagger + \hat{a}) + D(\hat{a}^\dagger + \hat{a})^2, \tag{14a}$$

$$\hat{\mathcal{H}}_l/\hbar \approx \omega_{\text{c}}\hat{a}^\dagger\hat{a} + \omega_{\text{a}}\left(\hat{S}_z + \frac{N}{2}\right) + \frac{\text{i}2\tilde{g}}{\sqrt{N}}\hat{S}_x(\hat{a}^\dagger - \hat{a}) + \frac{4\tilde{g}^2}{N\omega_{\text{c}}}\hat{S}_x{}^2. \tag{14b}$$

The coupling strengths appearing in these equations are defined as $\bar{g} \equiv (\omega_{\text{a}}/\omega_{\text{c}})^{1/2}g$ and $\tilde{g} \equiv (\omega_{\text{c}}/\omega_{\text{a}})^{1/2}g$, where

$$g \equiv \sqrt{\frac{\omega_{\text{a}}N}{2\hbar\varepsilon_0\varepsilon_{\text{bg}}}}\,|\boldsymbol{d}_{1,0}\cdot\boldsymbol{\psi}(z_{\text{QW}})|. \tag{15}$$

Here z_{QW} is the z position of the QW. We find that in the velocity form, since $\bar{g} \propto \omega_{\text{a}}$, the coupling strength is larger for a higher subband, which is usually neglected due to truncation in simple analyses. Then, the matter-level truncation is crucial in the velocity form, and the length form, where $\tilde{g}$ is independent of ω_{a}, usually shows better convergence.

Using the sum rule, Eq. (8), we can show that $D > \bar{g}^2/\omega_{\text{a}}$, where D is the coefficient of the A^2 term (see Garziano et al. (2020) for more details). The existence of a lower bound for the value of D is an important issue in the discussion of no-go theorems for the superradiant phase transition, as we discuss in Section 2.3.

In the limit of weak excitation (linear optical response), we can bozonise the spin operators through the Holstein–Primakoff transformation (Emary and Brandes, 2003a,b), i.e., $\hat{S}_z \to \hat{b}^\dagger\hat{b} - N/2$, $\hat{S}_+ \equiv \hat{b}^\dagger(N - \hat{b}^\dagger\hat{b})^{1/2}$, and $\hat{S}_- \equiv (N - \hat{b}^\dagger\hat{b})^{1/2}\hat{b}$. In the lowest-order transformation, we get

$$\hat{\mathcal{H}}_v/\hbar \to \omega_c\hat{a}^\dagger\hat{a} + \omega_a\hat{b}^\dagger\hat{b} + i\overline{g}(\hat{a}^\dagger + \hat{a})(\hat{b}^\dagger - \hat{b}) + D(\hat{a}^\dagger + \hat{a})^2, \tag{16a}$$

$$\hat{\mathcal{H}}_l/\hbar \to \omega_c\hat{a}^\dagger\hat{a} + \omega_a\hat{b}^\dagger\hat{b} + i\widetilde{g}(\hat{a}^\dagger - \hat{a})(\hat{b}^\dagger + \hat{b}) + \frac{\widetilde{g}^2}{\omega_c}(\hat{b}^\dagger + \hat{b})^2. \tag{16b}$$

By deriving the Heisenberg equations of motion (HEMs) for $\hat{a}$ and $\hat{b}$ (also $\hat{a}^\dagger$ and $\hat{b}^\dagger$), Fourier transforming them from the time domain (t) to the frequency domain (ω), and solving those HEMs in the frequency domain, we can get an equation for $\hat{a}$ (or $\hat{a} + \hat{a}^\dagger$, $i(\hat{a} - \hat{a}^\dagger)$, $\hat{b}$, …) as $F(\omega)\hat{a} = 0$, where $F(\omega)$ is a coefficient function of ω. By rewriting $F(\omega) = 0$, which means that $\hat{a}$ can be excited by a small resonant perturbation, we get

$$\frac{{\omega_c}^2}{\omega^2} = 1 + \frac{4g^2}{{\omega_a}^2 - (\omega + i0^+)^2}. \tag{17}$$

From this, we get two positive ω that correspond to the lower and upper polariton frequencies. As we show in Section 2.1.3, the second term on the right-hand side corresponds to the optical susceptibility. By replacing 0^+ with a matter damping rate, $\gamma/2$, and ω_c with $\omega_c - i\kappa/2$ (cavity loss rate κ), we can phenomenologically introduce the effect of dissipation (line broadening).

Note that the two forms of the ISBT Hamiltonian, Eqs. (16a) and (16b), are equivalent to each other. They are related through a unitary transformation. Therefore, Eq. (17) can also be obtained from Eq. (16a), i.e., the Hamiltonian in the velocity form. However, this is possible only if we can set $D = \overline{g}^2/\omega_a$, i.e., if the oscillator strength is concentrated to the two levels. However, such a truncation procedure cannot be justified in most cases in the velocity form. For more detailed studies of the ISBT Hamiltonian, see a series of papers by Todorov and Sirtori (2012), Todorov (2014), and Todorov and Sirtori (2014).

2.1.3 Optical phonons and excitons in bulk semiconductors

In this section, we consider optical phonons and excitons in bulk semiconductors that exhibit USC with light (Bamba and Imoto, 2016; Hopfield, 1958). While excitons are not pure bosons, they can be treated as bosons in the weak excitation limit (Combescot et al., 2008). In the same manner as in the ISBT case, focusing on just one excitation level, we can express the Hamiltonian in two different forms:

$$\frac{\hat{\mathcal{H}}_v}{\hbar} \approx \sum_{\boldsymbol{k}} \times \left[v|\boldsymbol{k}|\hat{a}_{\boldsymbol{k}}^{\dagger}\hat{a}_{\boldsymbol{k}} + \omega_{\mathrm{a},\boldsymbol{k}}\hat{b}_{\boldsymbol{k}}^{\dagger}\hat{b}_{\boldsymbol{k}} + \overline{g}_{\boldsymbol{k}}(\hat{a}_{-\boldsymbol{k}}^{\dagger} + \hat{a}_{\boldsymbol{k}})(\hat{b}_{\boldsymbol{k}}^{\dagger} + \hat{b}_{-\boldsymbol{k}}) + D_{\boldsymbol{k}}(\hat{a}_{-\boldsymbol{k}}^{\dagger} + \hat{a}_{\boldsymbol{k}})(\hat{a}_{\boldsymbol{k}}^{\dagger} + \hat{a}_{-\boldsymbol{k}}) \right], \tag{18a}$$

$$\frac{\hat{\mathcal{H}}_l}{\hbar} \approx \sum_{\boldsymbol{k}} \times \left[v|\boldsymbol{k}|\hat{a}_{\boldsymbol{k}}^{\dagger}\hat{a}_{\boldsymbol{k}} + \omega_{\mathrm{a},\boldsymbol{k}}\hat{b}_{\boldsymbol{k}}^{\dagger}\hat{b}_{\boldsymbol{k}} + \mathrm{i}\widetilde{g}_{\boldsymbol{k}}(\hat{a}_{-\boldsymbol{k}}^{\dagger} - \hat{a}_{\boldsymbol{k}})(\hat{b}_{\boldsymbol{k}}^{\dagger} + \hat{b}_{-\boldsymbol{k}}) + \frac{\widetilde{g}_{\boldsymbol{k}}^{\,2}}{v|\boldsymbol{k}|}(\hat{b}_{-\boldsymbol{k}}^{\dagger} + \hat{b}_{\boldsymbol{k}})(\hat{b}_{\boldsymbol{k}}^{\dagger} + \hat{b}_{-\boldsymbol{k}}) \right]. \tag{18b}$$

Here, the coupling strengths are defined as $\overline{g}_{\boldsymbol{k}} \equiv [\omega_{\mathrm{a}}/(v|\boldsymbol{k}|)]^{1/2} g_{\boldsymbol{k}}$ and $\widetilde{g}_{\boldsymbol{k}} \equiv (v|\boldsymbol{k}|/\omega_{\mathrm{a}})^{1/2} g_{\boldsymbol{k}}$ for $g_{\boldsymbol{k}} \equiv \sqrt{\omega_{\mathrm{a},\boldsymbol{k}}\rho/(2\hbar\varepsilon_0\varepsilon_{\mathrm{bg}})}|\boldsymbol{d}_{1,0}|$ and particle density ρ. The coefficient of the A^2 term satisfies $D_{\boldsymbol{k}} > \overline{g}_{\boldsymbol{k}}^{\,2}/\omega_{\mathrm{a},\boldsymbol{k}}$. From the Heisenberg equations of motion (for $D_{\boldsymbol{k}} = \overline{g}_{\boldsymbol{k}}^2/\omega_{\mathrm{a},\boldsymbol{k}}$ in the velocity form), we get the following dispersion relation determining the resonance frequency, ω, as (Hopfield, 1958)

$$\frac{v^2|\boldsymbol{k}|^2}{\omega^2} = 1 + \frac{4g^2}{\omega_{\mathrm{a},\boldsymbol{k}}^{\,2} - (\omega + \mathrm{i}0^+)^2}. \tag{19}$$

The right-hand side corresponds to the dielectric function divided by $\varepsilon_{\mathrm{bg}}$, and the second term corresponds to the optical susceptibility. We can rewrite its numerator as $4g^2 = f_{1,0}\omega_{\mathrm{plasma}}^{\,2}/\varepsilon_{\mathrm{bg}}$ with the oscillator strength $f_{1,0}$ and the plasma frequency $\omega_{\mathrm{plasma}} \equiv \sqrt{e^2\rho/(m\varepsilon_0)}$. For a fixed $\boldsymbol{k}$, we can find two positive ω, which correspond to the lower and upper polariton frequencies.

2.1.4 Cyclotron resonance (or inter-Landau-level transitions)

In this section, we describe CR (or ILLTs) in a 2DEG interacting with cavity photons in the USC regime. We consider the Hamiltonian $\hat{\mathcal{H}}_v$ only in the velocity form, Eq. (4), since the calculation is simpler than in the length form. Due to Kohn's theorem (Kohn, 1961) for electron–electron interactions, we need not consider the Coulomb interaction $\hat{V}$ for discussing linear optical responses. In the nonlinear regime, contributions beyond Kohn's theorem have been demonstrated (Maag et al., 2016) to emerge from the electron–ion Coulomb coupling.

We have two separate electromagnetic fields in this problem—(i) a DC magnetic field $\boldsymbol{B}_{\rm DC} = \boldsymbol{\nabla} \times \boldsymbol{A}_{\rm DC}$ represented by a static vector potential $\boldsymbol{A}_{\rm DC}$ and (ii) a dynamical electromagnetic field described by a dynamic vector potential $\boldsymbol{A}(t)$. We include the former contribution to the matter Hamiltonian $\hat{\mathcal{H}}_{\rm mat}$. Then, the matter system is the 2DEG in a DC magnetic field along the z direction as $\hat{\mathcal{H}}_{\rm mat} = \sum_{i=1}^{N} \sum_{\xi=x,y} \hat{\pi}_{i,\xi}{}^2/(2m)$. Here, we defined $\hat{\boldsymbol{\pi}}_i \equiv \hat{\boldsymbol{p}}_i + e\boldsymbol{A}_{\rm DC}$. $N = \rho_{\rm 2DEG} S$ is the total number of electrons, when the surface density of the 2DEG is $\rho_{\rm 2DEG}$.

Introducing the CR frequency $\omega_{\rm cyc} = e|B_{\rm DC}|/m$ and the lowering operator $\hat{c}_i \equiv (\hat{\pi}_{i,y} + {\rm i}\hat{\pi}_{i,x})/\sqrt{2m\hbar\omega_{\rm cyc}}$ between Landau levels (LLs) satisfying $[\hat{c}_i, \hat{c}^{\dagger}_j] = \delta_{i,j}$ (Yoshioka, 2002), the matter Hamiltonian is rewritten as $\hat{\mathcal{H}}_{\rm mat} = \sum_{i=1}^{N} \hbar\omega_{\rm cyc}(\hat{c}_i^{\dagger}\hat{c}_i + 1/2)$. Note that CR excitations (i.e., ILLTs) are purely bosonic excitations, and matter levels are not truncated at this stage. Then, the truncation problem does not arise even in the velocity form. In the following, we consider only a bosonic operator $\hat{b} \equiv \sum_{i=1}^{N} \hat{c}_i/\sqrt{N}$ of the collective excitation mode of the 2DEG coherently interacting with the electromagnetic field (Bamba et al., 2019; Li et al., 2018b). Since the other collective modes are dark (their oscillator strengths are zero) for photons with zero in-plane wavevector, we can truncate such modes, and the matter Hamiltonian is approximated as $\hat{\mathcal{H}}_{\rm mat} \approx \hbar\omega_{\rm cyc}\hat{b}^{\dagger}\hat{b} + {\rm const.}$

As for photons, for simplicity, we focus on only one cavity resonance frequency $\omega_{\rm c}$ but explicitly consider the two circular polarizations $\xi = \pm$. We define the annihilation operator of the $\pm$ circularly polarized photon as $\hat{a}_{\pm} \equiv (\hat{a}_x \mp {\rm i}\hat{a}_y)/\sqrt{2}$, where $\hat{a}_{x/y}$ is that of the linearly polarized photon in the x/y direction. Then, when we define the coupling strength as $\overline{g} \equiv (\omega_{\rm cyc}/\omega_{\rm c})^{1/2} g$ and $g \equiv [e^2 \rho_{\rm 2DEG}/(2\varepsilon_0 \varepsilon_{\rm bg} m)]^{1/2} \psi(z_{\rm 2DEG})$ for the position, $z_{\rm 2DEG}$, of the 2DEG inside the cavity, the total Hamiltonian is obtained from Eq. (7a) as

$$\begin{aligned} \hat{\mathcal{H}}_v \approx & \sum_{\xi=\pm} \hbar\omega_{\rm c}\hat{a}_{\xi}^{\dagger}\hat{a}_{\xi} + \hbar\omega_{\rm cyc}\hat{b}^{\dagger}\hat{b} + {\rm i}\hbar\overline{g}\left[\hat{b}^{\dagger}(\hat{a}_{+} + \hat{a}_{-}^{\dagger}) - (\hat{a}_{-} + \hat{a}_{+}^{\dagger})\hat{b}\right] \\ & + \frac{\hbar\overline{g}^2}{\omega_{\rm cyc}}(\hat{a}_{-} + \hat{a}_{+}^{\dagger})(\hat{a}_{+} + \hat{a}_{-}^{\dagger}). \end{aligned} \tag{20}$$

In this way, the CR excitation $\hat{b}$ interacts with the cocircularly polarized photon $\hat{a}_{+}$ in the corotating manner as ${\rm i}\hbar g(\hat{b}^{\dagger}\hat{a}_{+} - \hat{a}_{+}^{\dagger}\hat{b})$, while it interacts with the counter-circularly polarized photon $\hat{a}_-$ in the counterrotating manner as $i\hbar g(\hat{b}^\dagger \hat{a}_-^\dagger - \hat{a}_- \hat{b})$.

By solving the Heisenberg equations of motion, the resonance frequencies, $\omega_\pm$, of the coupled modes (polariton modes) with the $\pm$ circular polarization can be obtained as

$$\frac{{\omega_c}^2}{{\omega_\pm}^2} = 1 - \frac{2g^2}{\omega_\pm(\omega_\pm \mp \omega_{cyc} + i0^+)}. \tag{21a}$$

More detailed calculations can be found in Li et al. (2018b) and Bamba et al. (2019).

2.2 Virtual photons and two-mode squeezed vacuum

As pointed out by Ciuti et al. (2005), owing to the presence of the CRTs, the expectation value of the number of photons, $\langle g|\hat{a}^\dagger \hat{a}|g\rangle$, is nonzero even in the ground-state $|g\rangle$ in the USC regime. Such photons are called virtual photons. It has been theoretically suggested that the ground state of the light–matter coupled system (polariton system) is expressed as a two-mode squeezed vacuum state of the original photon and excitation modes (Artoni and Birman, 1989, 1991; Quattropani and Schwendimann, 2005; Schwendimann and Quattropani, 1992a,b) (a similar discussion has been given for nondispersive dielectric media (Abram, 1987; Glauber and Lewenstein, 1991)). Virtual photons (and virtual excitations) compose such an intrinsic squeezed vacuum. The number of virtual photons and degree of squeezing become a considerable value in the USC regime. While the vacuum B–S shift (Bloch and Siegert, 1940) has been clearly demonstrated as a hallmark of the CRTs in the CR system by Li et al. (2018b), the presence of virtual photons has not yet been proven experimentally. Since the early studies of USC by Ciuti et al. (2005), many proposals for observing virtual photons have been reported.

One of the main strategies is dynamically modulating the coupling strength $g(t)$ or other parameters, by which photons or photon pairs are emitted from the cavity to the outside world (De Liberato et al., 2007; De Liberato et al., 2009; Dodonov et al., 2008; Garziano et al., 2014; Shapiro et al., 2015). Such radiation is called quantum vacuum radiation, and the emission process is similar to the dynamical Casimir effect (Moore, 1970). The modulation frequency should be of the order of the resonance frequencies, and such an experiment has been demonstrated in superconducting circuits (Wilson et al., 2011).

For more explicit evidence of the presence of virtual photons, one can suddenly turn off the coupling, making *g* zero. Then, the virtual photons can escape from the cavity, and the correlation functions of those emitted photons should reflect the state of the virtual photons in the ground state (Ashhab and Nori, 2010; Auer and Burkard, 2012; Garziano et al., 2013; Hagenmüller, 2016). Such schemes have been proposed for ISBT polaritons (Auer and Burkard, 2012) as well as for Landau polaritons (Hagenmüller, 2016).

As shown in Fig. 4A, spontaneous emission of virtual photons has been proposed by Stassi et al. (2013, 2016) by considering a three-level system embedded in a cavity, which is similar to the ISBT system (two conduction subbands and one valence subband) (Günter et al., 2009). When the transition between the two excited levels is ultrastrongly coupled with the cavity mode, the virtual photons are emitted by the spontaneous relaxation of the carrier to its most stable level, because the USC diminishes due to the disappearance of the carrier from the above two levels. In a similar system but with driving between the lower two levels (e.g., between a conduction subband and a valence subband), the (stimulated) emission of virtual photons has been proposed (Carusotto et al., 2012; Huang and Law, 2014; Stefano et al., 2017).

Further, in a two-level or two-(sub)band system embedded in a cavity, as shown in Fig. 4B, as carriers pass through just one of the levels or (sub)bands, the emission of photons is expected while the carriers escape from the system (and then the light–matter coupling diminishes). It is called ground state electroluminescence (Cirio et al., 2016, 2019). It has also been proposed that the signature of virtual photons can be observed as the Lamb shift of an ancillary probe qubit coupled to the ultrastrongly coupled system (Lolli et al., 2015). Further, time-modulated optomechanical coupling has also been proposed by Cirio et al. (2017) for amplifying a mechanical probe of virtual photons.

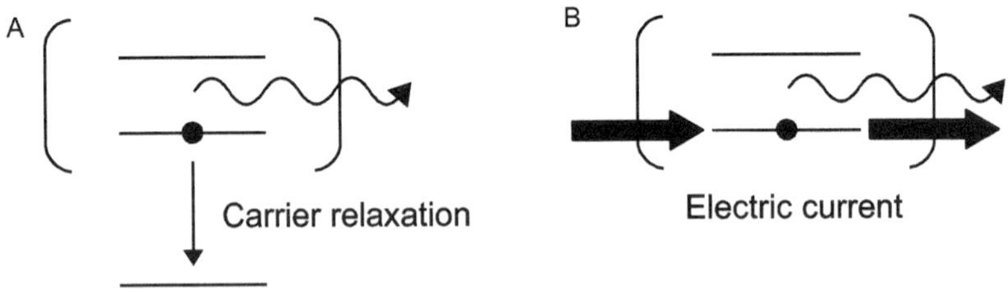

Fig. 4 Sketches of virtual photon emission processes. Virtual photons are released as real photons through (A) spontaneous emission in a three-level system and (B) ground state electroluminescence.

2.3 Superradiant phase transition

Hepp and Lieb (1973) showed that there exists a second-order phase transition in a system of N two-level atoms (with resonance frequency ω_a) interacting with a single mode of light (with frequency ω_c) described by

$$\hat{\mathcal{H}}_{\text{Dicke}}/\hbar \equiv \omega_c \hat{a}^\dagger \hat{a} + \omega_a \left(\hat{S}_z + \frac{N}{2} \right) + \frac{2\bar{g}}{\sqrt{N}} (\hat{a}^\dagger + \hat{a}) \hat{S}_y. \tag{22}$$

Wang and Hioe (1973) also showed the existence of such a phase transition using a different, simpler method. The above Hamiltonian was first used by Dicke (1954) in the context of superradiance, and the predicted phase transition has come to be known as the superradiant phase transition (SRPT) or Dicke phase transition. In the thermodynamic limit ($N \to \infty$), the condition for the superradiant phase to arise is $\bar{g}^2 > \omega_c \omega_a / 4$, i.e., the SRPT occurs in the USC regime (see Fig. 5). Below a critical temperature, the expectation values of the photon annihilation operator $\langle \hat{a} \rangle$ and spin operator $\langle \hat{S}_y \rangle$ become nonzero, signaling a spontaneous appearance of an electromagnetic field and polarization (or electric current) in thermal equilibrium.

However, Rzażewski et al. (1975) pointed out that the A^2 term, $D(\hat{a}^\dagger + \hat{a})^2$, is neglected in the above Hamiltonian used by Hepp and Lieb (1973) (see, e.g., Eq. (14a) for the full Hamiltonian for ISBTs including the A^2 term). Rzażewski et al. (1975) showed that the A^2 term becomes an additional energy cost and prevents the SRPT. In the case of the length form, Eq. (14b), it can be shown that the P^2 term (instead of the A^2 term) prevents the SRPT (Yamanoi and Takatsuji, 1978). The importance of the A^2 or P^2 term in the USC regime has been extensively discussed in relation

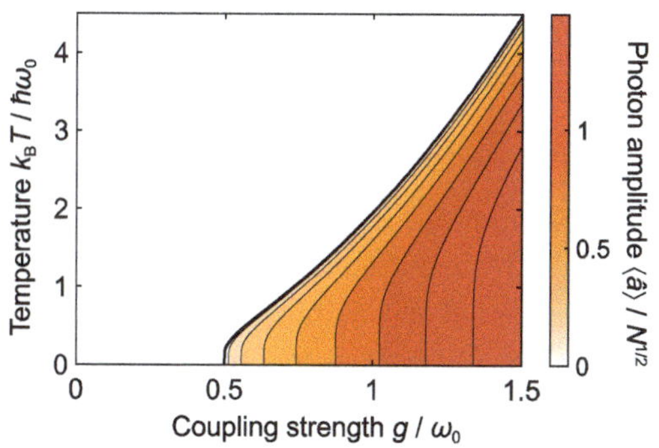

Fig. 5 Phase diagram for the superradiant phase transition predicted by Hepp and Lieb (1973) and Wang and Hioe (1973). The superradiant phase exists in the *shaded area*, where a finite electromagnetic field and polarization arises in thermal equilibrium.

with the gauge invariance (Bamba and Ogawa, 2014; Grießer et al., 2016; Keeling, 2007; Vukics and Domokos, 2012; Vukics et al., 2014, 2015; Woolley, 1976; Yamanoi, 1976; Yamanoi and Takatsuji, 1978). The influence of gauge choice on the SRPT has also been discussed by Stokes and Nazir (2020).

Through a Bogoliubov transformation, $\hat{a}' \equiv (\hat{a} + \zeta\hat{a}^\dagger)/\sqrt{1-\zeta^2}$, with $\zeta \equiv (\sqrt{1+4D/\omega_c} - 1)/(\sqrt{1+4D/\omega_c} + 1)$, we can eliminate the A^2 term, and thus, Eq. (14a) can be transformed to Eq. (22), with replacing $\hat{a} \to \hat{a}'$, $\omega_c \to \omega_c' \equiv \sqrt{\omega_c(\omega_c + 4D)}$, $\bar{g} \to \bar{g}' \equiv \sqrt{(1-\zeta)/(1+\zeta)}\bar{g}$. However, $\bar{g}'^2 > \omega_c'\omega_a/4$ cannot be satisfied for $D > \bar{g}^2/\omega_a$, which is obtained by the sum rule, Eq. (8). Thus, the presence of the A^2 term itself does not prevent the SRPT, but the minimum magnitude of its coefficient (set by the TRK sum rule) does. A similar calculation can be performed also for the P^2 term.

Starting from the minimal-coupling Hamiltonian $\hat{\mathcal{H}}_v$, Eq. (4), without truncation of matter levels nor photonic modes (performed for deriving Eq. (22)), a more general no-go theorem was suggested by Knight et al. (1978) through classical treatment of the electromagnetic fields and polarization. Furthermore, through quantum treatment of the matter but classical treatment of the electromagnetic fields, more general no-go theorems were proposed in the long-wavelength approximation by Bialynicki-Birula and Rzażewski (1979) and in a more general case by Gawędzki and Rzążewski (1981). The classical treatment of the electromagnetic fields can be justified under some conditions, as reviewed by Bamba and Imoto (2017).

Since the no-go theorem has not yet been completely proven even for the minimal-coupling Hamiltonian, proposals of counterexamples and indications of calculation mistakes have been repeated. For example, longitudinal dipole–dipole interaction (Bamba and Ogawa, 2014; Keeling, 2007; Vukics et al., 2014, 2015), graphene (Chirolli et al., 2012; Hagenmüller and Ciuti, 2012), and the excitonic insulator (Andolina et al., 2019; Mazza and Georges, 2019; Nataf et al., 2019) have been discussed. Currently, the possibility of the SRPT by a short-range depolarizing interaction between atoms proposed by Grießer et al. (2016) has not been denied yet.

In a driven-dissipative situation, a scheme for implementing the Dicke model with $\bar{g} > \omega_c\omega_a/4$ was proposed by Dimer et al. (2007). A SRPT-like transition (a nonequilibrium critical phenomenon) was experimentally demonstrated in cold atoms driven by laser light by Baumann et al. (2010). They were able to tune the coupling strength, $\bar{g}$, by changing the power of the laser light, and the transition was observed by the change in $\bar{g}$. While it was not driven by quantum fluctuations (Larson and Irish, 2017), the transition was called a "quantum" phase transition since the coupling term with the tunable $\bar{g}$ was not commutable with the rest of the Hamiltonian terms. In thermal equilibrium, the SRPT has not been observed. While softening (lowering) of a resonance frequency has been experimentally observed in a CR system by Keller et al. (2020), it remains unclear whether it is a SRPT signature.

As pointed out by Knight et al. (1978), for systems beyond the minimal-coupling Hamiltonian, we can circumvent the above-mentioned no-go theorems even in thermal equilibrium. A SRPT analog in superconducting circuits has been discussed by Nataf and Ciuti (2010), Viehmann et al. (2011), Ciuti and Nataf (2012), and Jaako et al. (2016), and one proposal by Bamba et al. (2016) has not been repudiated. Further, a 2DEG with Rashba spin-orbit coupling has been proposed by Nataf et al. (2019), which has also not been invalidated. To date, no-go theorems have not been proposed for magnetic materials.

2.4 Predictions for materials beyond semiconductors

USC between light and various types of quasiparticles in semiconductors has been demonstrated experimentally, as discussed in detail in Section 3. In this section, we focus on recent theoretical proposals predicting fascinating cavity-enabled phenomena in quantum materials beyond traditional semiconductors. In particular, some of the recent proposals on cavity control of properties of materials such as superconductors, excitonic insulators, Mott insulators, topological insulators, and other quantum materials are reviewed.

2.4.1 Superconductors in cavities

Superconductivity is one of the fascinating macroscopic manifestations of quantum mechanics. A superconducting state, exhibiting zero resistance and expelling of magnetic fields, occurs below a critical temperature, T_c. There are currently many efforts to control superconducting properties of materials, particularly their T_c. For example, through a static external stimulus such as pressure, the T_c of a material has been raised to 250 K (Drozdov et al., 2019). Another approach is based on light illumination, especially utilizing femtosecond laser pulses (see, e.g., Kaiser, 2017). Nonequilibrium, light-driven superconductivity and Eliashberg enhancement of superconductivity are usually carried out in free space with an external light source.

Light–matter interactions in cavities in the USC regime offer another exciting method for modifying superconductivity properties. There have been several theoretical studies on physical mechanisms affecting superconductivity properties driven by vacuum fluctuations in cavities, as described in the following.

Curtis et al. (2019) studied a 2D BCS superconductor placed in single and multi-mode cavities. The physical picture is based on the Eliashberg effect (Eliashberg, 1970), i.e., a redistribution of the quasiparticles into a more favorable nonthermal distribution due to an applied electromagnetic field, which was generalized to include both quantum and thermal fluctuations. An enhancement of the superconducting gap by a few percent was predicted (see Fig. 6). Cavity Bardasis-Schrieffer mode polaritons and cavity Higgs polaritons were also predicted to exist. The Bardasis-Schrieffer modes are exciton-like states of the superconducting order parameter (Allocca

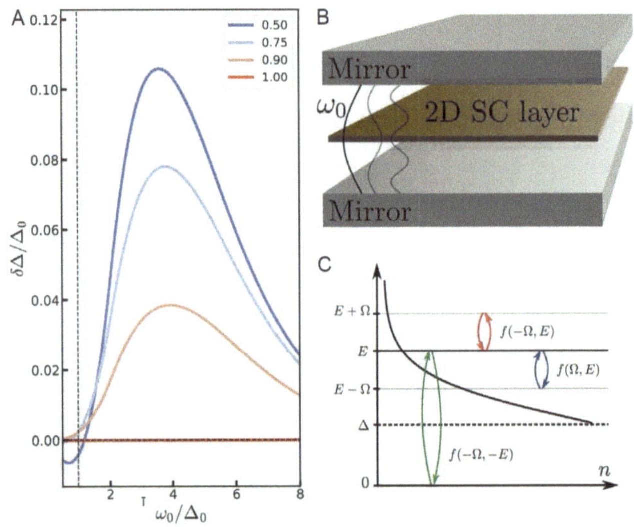

Fig. 6 Enhancement of superconductivity through the cavity quantum Eliashberg mechanism proposed by Curtis et al. (2019). (A) Relative enhancement of the gap function as a function of cavity frequency ω_0. *Curves* are colored and labeled according to the ratio between the photon and quasiparticle temperatures. Enhancement is seen to set in after the cavity frequency surpasses the pair-breaking energy, $2\Delta_0$. (B) Schematic picture of the system used for the calculation. The lowest cavity resonator mode with cutoff frequency ω_0 is shown, as is the 2D superconducting layer. (C) Depiction of various processes that contribute to the quasiparticle collision integral, plotted against the equilibrium quasiparticle distribution $n(E)$. The *blue arrows* depict the downscattering terms captured by $f(\Omega, E)$, the *red arrows* depict the upscattering terms captured by $f(-\Omega, E)$, and the *green arrows* represent the pair processes captured by $f(-\Omega, -E)$.

et al., 2019), while Higgs polaritons are hybridized states between cavity photons and the amplitude mode of a superconductor (Raines et al., 2020). Although none of these modes couple linearly to light, they can be observed and controlled by applying a super-current or observed with THz near-field techniques (Sun et al., 2020).

In another study, Sentef et al. (2018) performed Migdal-Eliashberg simulations to explore modifications of electron–phonon interactions arising from the formation of phonon polaritons at the 2D interface of FeSe/$SrTiO_3$. The observed enhancement in the electron–phonon coupling constant, however, did not result in an enhancement of T_c for the forward-scattering pairing mechanism. This negative effect was explained by the quasilinear dependence of T_c on the electron–phonon coupling constant in their model in contrast to BCS exponential scaling. Most recently, there has been an experimental study reporting a 50% increase of T_c in a BCS-type superconductor (Rb_3C_{60}) coupled to surface plasmon polaritons (SPPs) while showing decreasing T_c for $YBa_2Cu_3O_{7-x}$ (Thomas et al., 2019). In this study, a powder of superconductors was embedded in a polymer matrix and placed on a gold mirror. Qualitatively, the enhancement of T_c is attributed to cooperative SC between phonons in the superconductor and the surface plasmon polaritons via an auxiliary coupler (polymer).

Schlawin et al. (2019) have predicted superconductivity in a semiconductor 2DEG induced by USC with THz cavity photons. By using realistic parameters of a GaAs 2DEG inside a THz cavity, they predicted that superconductivity with T_c in the low-Kelvin regime should be realized. Virtual photons, in this case, act as the glue for pairing (Schlawin et al., 2019). Moreover, in a similar setting, Gao et al. (2020) predicted that driving the cavity might stabilize electron pairing even at higher temperatures. Clearly, cavity QED offers a variety of novel approaches to both manipulation of superconducting materials and investigation of physics behind high-T_c superconductivity.

2.4.2 Quantum materials in cavities

As described in Section 2.3, the Dicke SRPT is one of the most profound collective phenomena in quantum optics. If it is realized in a solid in equilibrium, e.g., a 2DEG (Hagenmüller and Ciuti, 2012; Nataf et al., 2019), it can lead to real-world applications of macroscopic quantum coherence. For so-called *quantum materials*, which host a plethora of collective phases (Tokura et al., 2017), novel, emergent light–matter collective phenomena have been proposed. Because of the interplay between short-range

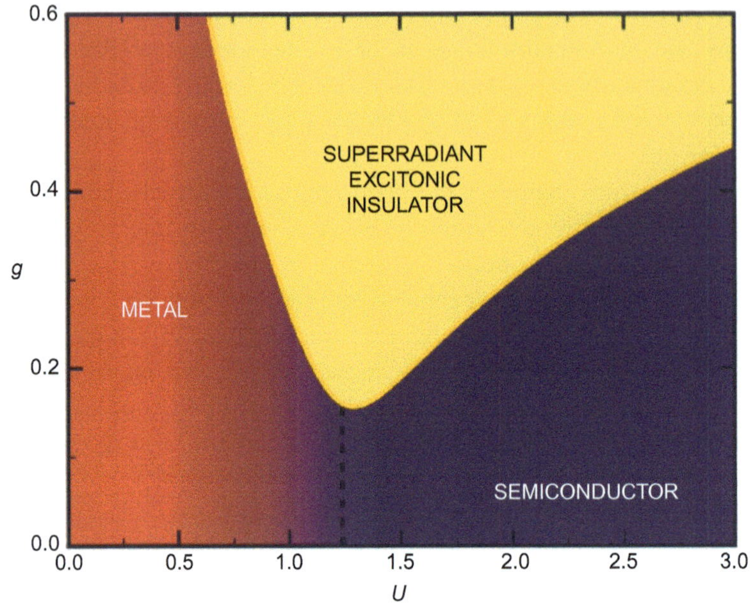

Fig. 7 Phase diagram for interacting electrons strongly coupled with cavity photons proposed by Mazza and Georges (2019). A superradiant excitonic insulator phase exists at certain strengths of electron–electron interactions, *U*, and light–matter coupling, *g*. The *red and blue* intensities reflect band populations. In the metal region (*red*), both orbitals are occupied, whereas in the semiconductor region (*blue*) only the valence band is completely filled.

electrostatic interactions and the nonperturbative coupling to a common cavity mode, several new phases that do not have direct counterparts in the collective Dicke models or solid-state spin systems have been shown to exist through numerical calculations by Schuler et al. (2020).

Another example is the superradiant excitonic insulator (SXI), studied by Mazza and Georges (2019). Fig. 7 shows its phase diagram as a function of the strength of electronic interactions, U, and the light–matter coupling strength, g. The SXI phase is characterized by equilibrium superradiance in the photon field and condensation of excitons in the electronic system happening simultaneously. However, this conclusion is disputed, since the model used by Mazza and Georges (2019) does not guarantee gauge invariance, as pointed out by Andolina et al. (2019) and Nataf et al. (2019).

More recently, Ashida et al. (2020) studied the possibility of observing a SRPT in systems that are naturally close to a spontaneous electric polarization phase. They considered a quantum paraelectric material sandwiched

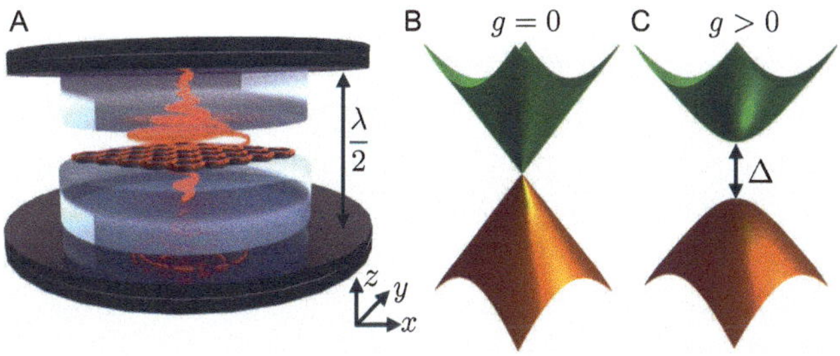

Fig. 8 2D material inside a chiral cavity studied by Wang et al. (2019). (A) Setup for monolayer graphene between cavity mirrors a distance $\lambda/2$ apart, where λ is the wavelength of the fundamental cavity photon mode. The *red spiral* indicates the circular photon polarization. The 2D material is encapsulated in a dielectric medium (*glassy region*). (B) Dirac cone of a 2D Dirac material at electron–photon coupling $g = 0$. (C) Energy gap Δ due to time-reversal symmetry breaking for $g > 0$.

between metallic cladding layers acting as cavity mirrors. In contrast to previous studies, the polariton excitations consist of infrared-active phonons in the quantum paraelectric, electromagnetic fields in the cavity, and plasmons in the metallic electrodes. Hence, the enhancement of the ferroelectric phase in quantum paraelectric materials due to cavity-induced optical phonon softening is predicted.

In general, cavities offer novel ways to manipulate and control material properties. In addition to the proposals mentioned above, other examples include manipulation of macroscopic magnetic and electronic properties of strongly correlated electron systems (Kiffner et al., 2019), nonlinear phononics (Juraschek et al., 2019), control of excitonic optical spectra of van der Waals materials and heterostructures (Latini et al., 2019), and QED Chern insulators (Wang et al., 2019). Fig. 8 demonstrates an opening of an energy gap when monolayer graphene is placed into a cavity, predicted by Wang et al. (2019). Here, photon-dressed Dirac fermions show a quantized Hall response with predicted conductance of $2e^2/h$ that can be characterized by an integer Chern number. Overall, the burgeoning field of condensed-matter cavity QED promises the discoveries of new fascinating physics, in addition to those discussed in the next section, paving the way for unprecedented control of material properties.

3. Experimental demonstrations of ultrastrong coupling

Diverse experimental platforms have emerged during the last decade that exhibit USC. These systems include not only traditional light-matter hybrid systems, such as exciton and phonon polaritons in bulk semiconductors, but also intraband transitions (ISBTs and CR), plasmons, phonons, and magnons in low-dimensional systems. In this section, we review some of the pioneering studies as well as the latest experimental observations of USC in these systems.

3.1 Intraband transitions

3.1.1 Intersubband-plasmon polaritons

ISBTs, also known as ISB plasmons, are defined as resonant optical transitions between two subbands within the conduction or valence band of a semiconductor QW (see, e.g., Helm, 2000; Paiella, 2006). They occur at low photon energies, typically in the MIR or THz range, with enourmous dipole moments (West and Eglash, 1985), which have been advantageously utilized to achieve USC (Anappara et al., 2005, 2006, 2007, 2009; Askenazi et al., 2014, 2017; Delteil et al., 2012; Dietze et al., 2013; Dini et al., 2003; Dupont et al., 2003, 2007; Geiser et al., 2010, 2012; Günter et al., 2009; Jeannin et al., 2019; Jouy et al., 2011; Laurent et al., 2017; Porer et al., 2012; Sapienza et al., 2007, 2008; Todorov et al., 2009, 2010a; Zanotto et al., 2010, 2012). An ISBT is a plasmon excitation because it is a collective response of a 2DEG in the QW to a resonant light field, polarized along the growth (z) direction, with transition frequency $\hbar\omega_{12}$ (see Fig. 9A). In contrast to interband transitions whose optical properties are largely determined by the fixed band gap energy of the semiconductor, the use of band-structure engineering provides ISBTs with ample tunability of the transition frequency and oscillator strength, which is of fundamental importance for quantum devices like the quantum cascade laser (Faist et al., 1994) and the QW infrared photodetector (see, e.g., Choi, 1997; Schneider and Liu, 2007).

Much of the initial work on ISB plasmon microcavity-polaritons in the USC regime (reviewed by Forn-Díaz et al. (2019)) was explored in the MIR range. Todorov et al. (2009) successfully extended the range to the THz band using a metal-dielectric-metal microcavity. This approach, schematically depicted in Fig. 10A, realized subwavelength confinement without significant ohmic losses (Todorov et al., 2010b), which led to USC with a large cooperativity value. Fig. 10B shows low-temperature (4.5 K) reflectivity spectra for different values of cavity detuning with respect to the ISBT. A clear anticrossing behavior is observed for the lowest cavity mode ($K = 1$). In Fig. 10C, the two polaritons branches are plotted as a function of inverse slit-width, $1/s$, which determines $2g/2\pi = 0.8$ THz and $\eta = 0.11$. Todorov et al. (2010a) further increased the number of QWs to 25 and used a zero-dimensional resonator to go deeper into the USC regime.

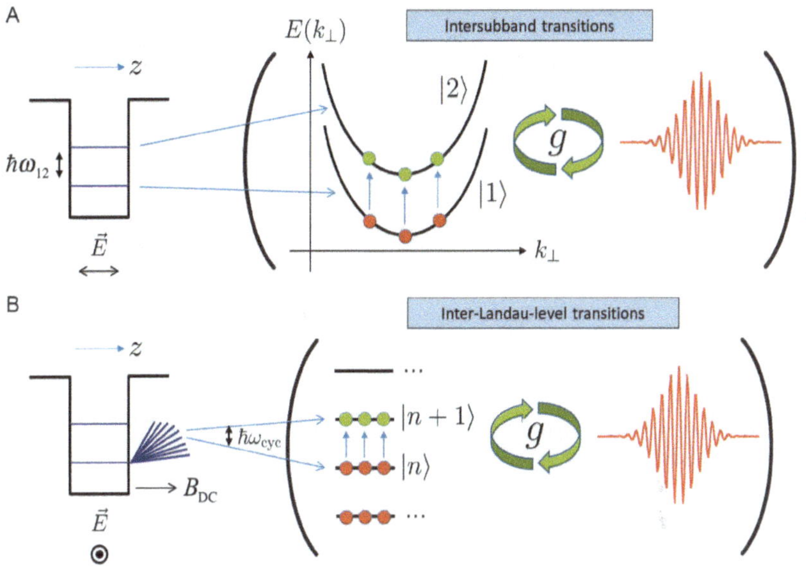

Fig. 9 Cavity polaritons based on intraband transitions in semiconductor quantum wells. (A) ISBTs resonantly coupled at a rate g to a light field $\vec{E}$ polarized along the growth (z) direction. (B) CR, or ILLTs, coupled to a light field $\vec{E}$ polarized in the quantum well ($x-y$) plane. A DC magnetic field B_{DC} applied along the growth (z) direction quantizes the in-plane ($x-y$) motion into discrete states (Landau levels) with energy separation $\hbar\omega_{cyc}$.

The photonic structure consisted of square-shaped microcavities formed by an array of metallic pads above and a planar metallic plate below, as shown in Fig. 10D. Using this configuration, the authors achieved $2g/2\pi = 1.41$ THz at 4.5 K and a record high value at that time for the normalized coupling strength: $\eta = 0.24$. In addition, it was shown that the polariton splitting exhibited a nonlinear behavior arising from the A^2 term in the light–matter interaction Hamiltonian (see Fig. 10E). Finally, the opening of a polaritonic gap of 330 GHz was also observed by plotting the polariton frequencies as a function of cavity frequency (see Fig. 10F).

Since 1 THz ~ 4 meV ~ 50 K, THz resonances in two-level (or two-subband) systems are generally observable only at cryogenic temperatures. Geiser et al. (2010) used parabolic QWs, which enables an observation of THz ISBTs even at room temperature since the equally separated energy levels in this multilevel system makes the resonance frequency and strength independent of the electron distribution. The authors placed the QWs in an electronic feedback microcavity resonator consisting of an inductor-capacitor circuit on top and a gold ground plane below (see Fig. 11A).

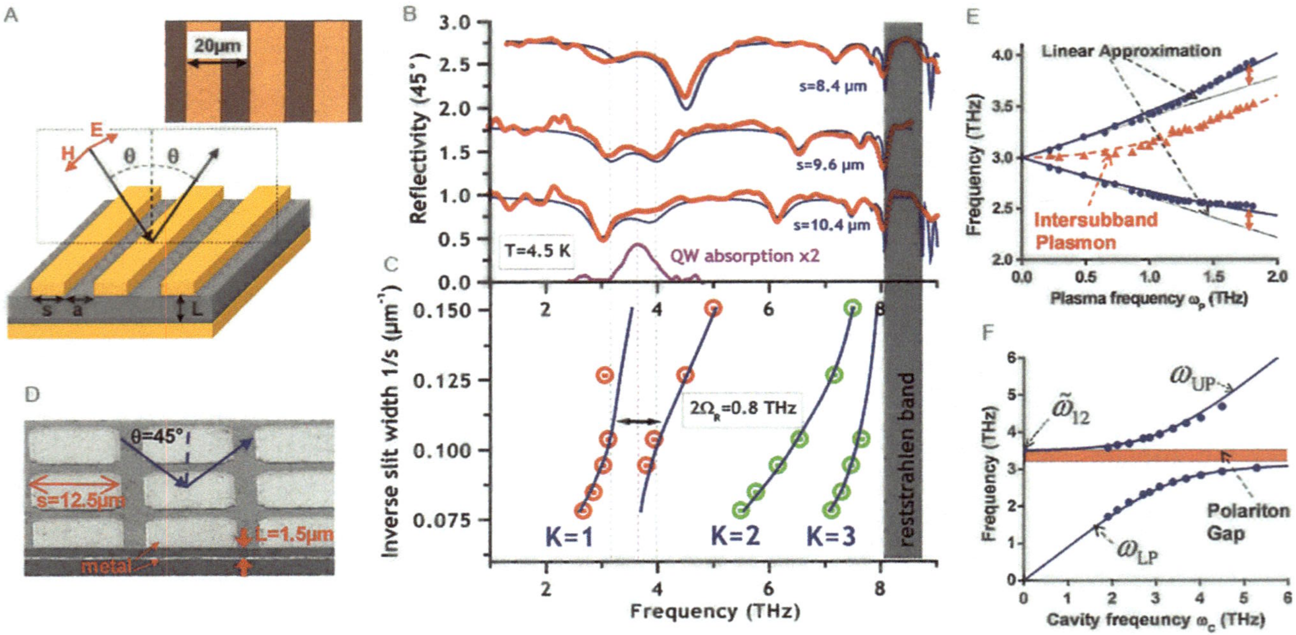

Fig. 10 Intersubband-plasmon polaritons in metal-dielectric microcavities in the USC regime at terahertz frequencies. (A) Schematic of the photonic structure used by Todorov et al. (2009). *Top*: top view of the metallic lamellar grating. *Bottom*: schematic of the geometry used for reflectivity measurements and side view of the grating. (B) Reflectivity spectra at 4.5 K for different values of the strip width parameter *s*. (C) Polariton branches plotted as a function of inverse slit-width, 1/*s*. The lowest cavity mode ($K = 1$) shows an anticrossing behavior with the ISBT. (D) Electron microscopy image of the metal-dielectric-metal photonic structure used by Todorov et al. (2010a). (E) LP and UP peaks (*blue dots*) plotted as a function of the plasma frequency. The ISB plasmon energy is also plotted (*red triangles*). Deviations from the linear approximation are clearly observed. (F) Polariton resonances plotted as a function of cavity frequency. The opening of a polaritonic gap is seen. *Adapted from Todorov, Y., Andrews, A.M., Sagnes, I., Colombelli, R., Klang, P., Strasser, G., Sirtori, C., 2009. Strong light-matter coupling in subwavelength metal dielectric microcavities at terahertz frequencies. Phys. Rev. Lett. 102, 186402; Todorov, Y., Andrews, A.M., Colombelli, R., De Liberato, S., Ciuti, C., Klang, P., Strasser, G., Sirtori, C., 2010a. Ultrastrong light-matter coupling regime with polariton dots. Phys. Rev. Lett. 105, 196402; Todorov, Y., Tosetto, L., Teissier, J., Andrews, A.M., Klang, P., Colombelli, R., Sagnes, I., Strasser, G., Sirtori, C., 2010b. Optical properties of metal dielectric-metal microcavities in the THz frequency range. Opt. Express 18, 13886–13907.*

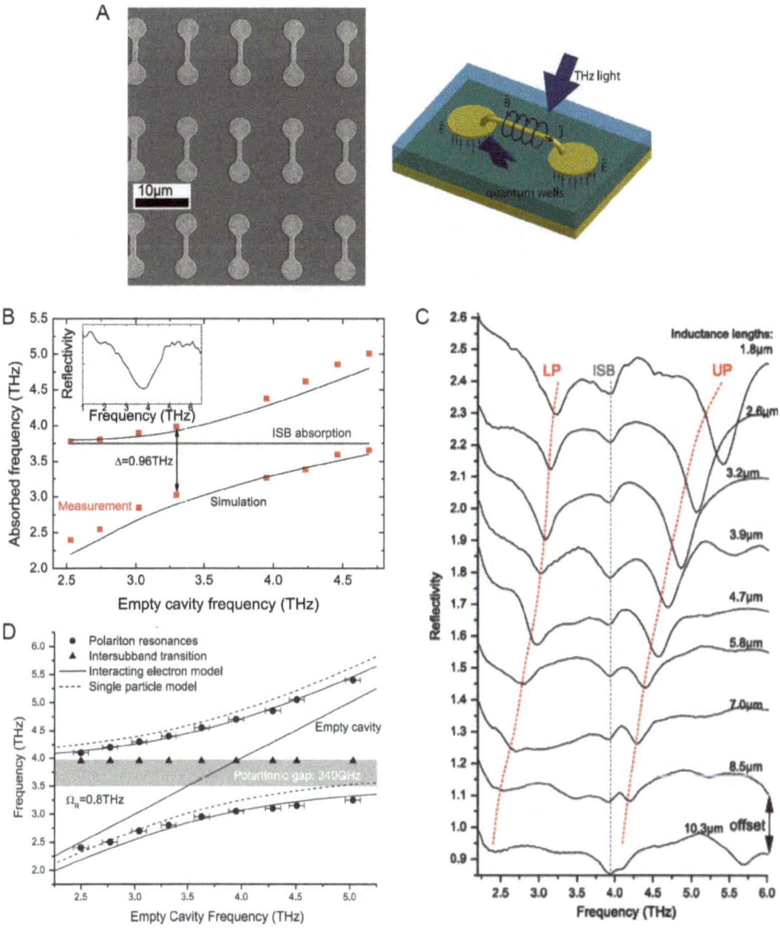

Fig. 11 Intersubband-plasmon polaritons in the USC regime at room temperature and terahertz frequencies using parabolic semiconductor QWs studied by Geiser et al. (2010) and Geiser et al. (2012). (A) *Left*: scanning electron microscopy image of the cavity showing an array of inductor-capacitor resonators. *Right*: schematic of the electronic microcavity coupled to the semiconductor QWs. (B) Anticrossing curve at 300 K. The frequencies of the reflection minima (*red squares*) are plotted as a function of cavity frequency. *Inset*: ISB absorption spectrum at 300 K. (C) Reflection spectra from samples with different inductive lengths. The LP and UP branches as well as the ISB absorption are indicated. (D) Polariton resonances (*black dots*) as a function of bare cavity frequency, showing the opening of a polaritonic gap (*grey shaded area*). *Solid and dashed lines* are a fit to experimental data using interacting and single-particle electron models, respectively. The empty cavity frequency line and the ISB absorption (*black triangles*) are also indicated. *Adapted from Geiser, M., Walther, C., Scalari, G., Beck, M., Fischer, M., Nevou, L., Faist, J., 2010. Strong light-matter coupling at terahertz frequencies at room temperature in electronic LC resonators. Appl. Phys. Lett. 97, 191107; Geiser, M., Castellano, F., Scalari, G., Beck, M., Nevou, L., Faist, J., 2012. Ultrastrong coupling regime and plasmon polaritons in parabolic semiconductor quantum wells. Phys. Rev. Lett. 108, 106402.*

The cavity mode frequency was tuned by varying the wire (i.e., inductor) length that connects the circular capacitive elements. Fig. 11B shows an anticrossing curve at 300 K, together with an ISBT spectrum (inset) and energy (horizontal gray line). The observed Rabi splitting, $2g/2\pi = 0.96$ THz, corresponded to $\eta = 0.14$. Importantly, these results were shown to be temperature independent in the 10–300 K range.

Furthermore, Geiser et al. (2012) studied the role of electron–electron interactions in the USC regime. In a parabolic QW, due to Kohn's theorem (Brey et al., 1989; Kohn, 1961), the depolarization field is not expected to be a limiting factor for η, since its effect is compensated by strong electron–electron interactions. For a highly doped sample, the authors obtained $2g/2\pi = 2.4$ THz, corresponding to $\eta = g/\tilde{\omega} = 0.27$ (see Fig. 11C). Here, $\tilde{\omega}$ is the ISB plasmon frequency, including electron–electron interaction effects. The authors used both a single-particle model and an interacting electron model (Ciuti et al., 2005; Todorov et al., 2010a) (see Fig. 11D) to fit the polariton frequencies. It was shown that even though the different electron–electron interactions cancel out each other as a consequence of Kohn's theorem and lead to $\tilde{\omega} = \omega_{12}$, the plasma frequency remains an important physical quantity in the USC regime, explaining the opening of a polaritonic gap.

3.1.2 Landau polaritons (or cyclotron resonance polaritons)

A magnetic field applied in the growth (z) direction of a QW leads to the quantization of the electronic orbital motion into a ladder of equally separated LLs with a spacing $\hbar\omega_{\text{cyc}} = eB_{\text{DC}}/m^*$, where e is the electronic charge, B_{DC} is the applied external magnetic field, and m^* is the effective carrier mass (see Fig. 9B) (see, e.g., Hilton et al., 2012; Kono, 2001; Lax and Mavroides, 1960; McCombe and Wagner, 1975). To couple with the optical transition between the highest occupied and the lowest unoccupied Landau levels, the THz electric field polarization must be in the QW plane. From an experimental point of view, realizing ILLTs is easier than ISBTs because normal incidence transmission measurements can be made. In addition, since the resonance frequency of ILLTs linearly depends on B_{DC}, ample tunability of the transition frequency for the matter subsystem can be achieved. In this section, we discuss some of the seminal and most recent experimental studies in this field (Abdurakhimov et al., 2016; Bayer et al., 2017; Keller et al., 2017, 2018, 2020; Li et al., 2018b; Maissen et al., 2014, 2017; Muravev et al., 2011, 2013; Paravicini-Bagliani et al., 2017, 2018; Rajabali et al., 2019; Scalari et al., 2012, 2013; Zhang et al., 2016a).

Muravev et al. (2011) observed cavity polaritons in an AlGaAs/GaAs QW structure using coplanar microresonators. These hybrid modes arose from the USC between magnetoplasmons (CR) in the 2DEG system and the photon modes of the microresonator. By placing the sample in a helium cryostat at the center of a superconducting solenoid and performing microwave transmission measurements as a function of B_{DC}, the authors were able to resolve both of the magnetodispersion branches (see Fig. 12A). In addition, the advantages of working with ILLTs was experimentally demonstrated as well, where by changing the electron density or applying an external magnetic field the authors showed how both the Rabi frequency and the dispersion of the cavity polaritons can be tuned over a wide range of frequencies (see Fig. 12B).

Zhang et al. (2016a) studied a high-mobility 2DEG placed at the center of a high-Q 1D photonic crystal cavity (PCC) using THz magnetospectroscopy. The maximum value of the electric field was located at the 2DEG position inside the cavity. The 1D PCC was fabricated using a series of alternating dielectric materials with a large refraction index contrast in the THz range. For this work, the authors used vacuum and Si slabs as the low and high index of refraction materials, respectively. The contrast in their indices reduced the number of layers needed to achieve a high-Q cavity

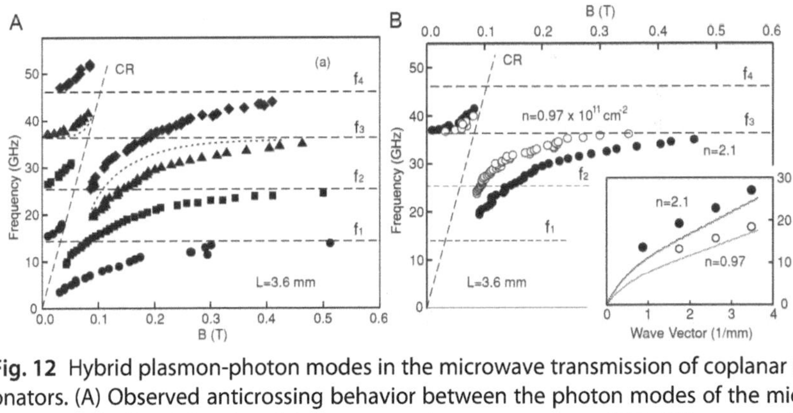

Fig. 12 Hybrid plasmon-photon modes in the microwave transmission of coplanar resonators. (A) Observed anticrossing behavior between the photon modes of the microresonator (f_N, $N = 1, 2, \ldots$) and the magnetoplasmon resonance (CR). Different symbols correspond to different polariton modes. (B) Electron density (*n*) dependence of the third polariton mode's anticrossing. Curves can be seen shifting closer to the photon dispersion line when the electron density is decreased. *Inset*: VRS (denoted as ΔF here) as a function of polariton wavevector for two values of *n*. *Dashed lines* are theoretical results. *Adapted from Muravev, V.M., Andreev, I.V., Kukushkin, I.V., Schmult, S., Dietsche, W., 2011. Observation of hybrid plasmon-photon modes in microwave transmission of coplanar microresonators. Phys. Rev. B 83, 075309.*

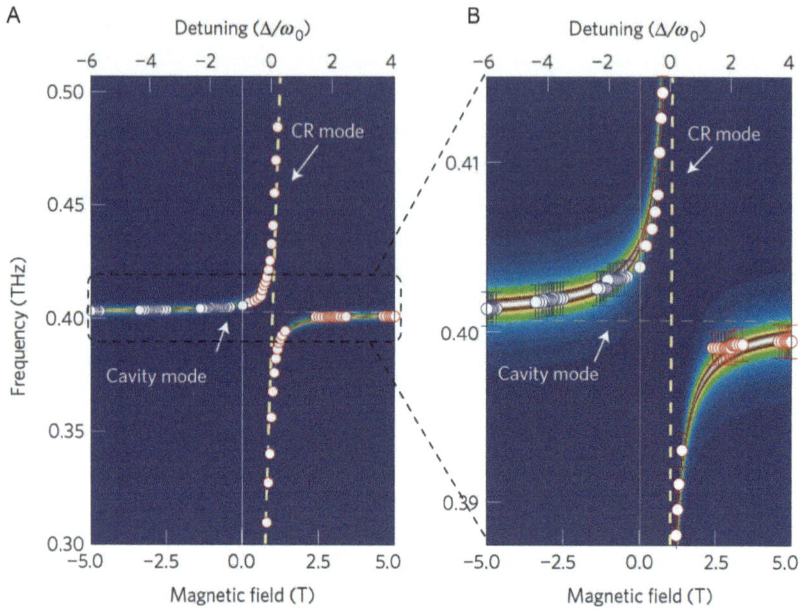

Fig. 13 Cooperative USC of the ILLT of 2D electrons with photons in a high-*Q* THz cavity. (A) LP and UP branches observed in the vicinity of the anticrossing region. The applied DC magnetic field tunes the cyclotron frequency ω_{cyc} with respect to the (fixed) cavity mode frequency, ω_0. *Open circles* are experimental data and *color lines* are THz transmission spectra obtained from electromagnetic field simulations. (B) Zoom-in of the dashed area shown in (A). Deviations from the cold cavity frequency limit persist at negative magnetic fields and can be observed even for a large off-resonant detuning. *Adapted from Zhang, Q., Lou, M., Li, X., Reno, J.L., Pan, W., Watson, J.D., Manfra, M.J., Kono, J., 2016a. Collective non-perturbative coupling of 2D electrons with high-quality-factor terahertz cavity photons. Nat. Phys. 12, 10051011.*

to just four (Chen et al., 2014a; Yee and Sherwin, 2009). The authors simultaneously achieved high values of cooperativity and coupling strength, putting the system in the USC regime with $\eta \sim 0.1$, $C > 300$, and $Q \sim 10^3$. This unique combination of parameters allowed the experimental observation of Rabi oscillations in the time-domain traces, persistence of an ultrastrongly coupled state even when the detuning was off-resonant (see Fig. 13A and B), a $\sqrt{n}$ dependence of the VRS with the electron density, n, and a suppression of the superradiant decay of CR (Zhang et al., 2014a, 2016a).

Paravicini-Bagliani et al. (2018) studied magnetotransport properties of a 2DEG coupled to a subwavelength electronic resonator in the USC regime. The authors prepared two samples with different coupling strengths and resonant frequencies ($\eta = 0.2$ and 0.3 and $\omega_0 = 205$ and 140 GHz, respectively) both of which consisted of a Hall bar placed in the capacitive gap of a complementary LC resonator (see Fig. 14A). The current was along the source-drain channel (parallel to the x-axis direction, see Fig. 14B), and the longitudinal resistivity, ρ_{xx}, was measured, as indicated in Fig. 14A. A significant reduction of the amplitude of the Shubnikov-de Haas oscillations was observed (see Fig. 14C), which was theoretically shown to arise from the strong interaction with the cavity's vacuum field fluctuations (Bartolo and Ciuti, 2018). In addition, the authors measured the irradiation-induced change in ρ_{xx} by illuminating the sample with a single-frequency, tunable, subterahertz source. The changes observed in ρ_{xx} were strongly dependent on whether the filling factor, ν, was an integer or half-integer (see Fig. 14D).

Keller et al. (2020) studied Landau polaritons in highly nonparabolic 2DEGs in the USC regime. The authors used InSb QWs and strained Ge QWs with a nonparabolic heavy-hole band. The CR of the 2D electron or hole gas was coupled to a THz metamaterial resonator that presented subwavelength confinement of the electric field (see Fig. 15A). By increasing the coupling strength via lithographical adjustment of the cavity mode to lower frequencies, the authors were able to observe significant deviations from the standard Hopfield model and a more consistent agreement of the data with a Hopfield model that had a reduced diamagnetic (A^2) term (see Fig. 15B). Possible explanations of the observed behavior include strain, spin-orbit coupling effects, nonparabolic band dispersions, and the enhanced magnetic coupling in the system that arises from the subwavelength electric field confinement. An open question from this work is whether this method can be extended and used to realize the Dicke SRPT in thermal equilibrium, which is expected to occur when the lower polariton branch becomes gapless.

3.2 Plasmon–phonon polaritons

The coupling of collective motion of conduction electrons driven by external electromagnetic waves and lattice vibrations results in a hybrid polariton, which is a mixture of a plasmon polariton and a phonon polariton. Such *plasmon–phonon polaritons* were first identified in doped bulk polar semiconductors as a hybrid of longitudinal optical (LO) phonons and bulk plasmons. Later, the advent of nanomaterials and nanophotonics has enabled confinement and manipulation of light to dimensions much smaller than the wavelength of light, i.e., beyond the diffraction limit. This has made plasmon–phonon polaritons tunable and low-loss nano-objects useful for

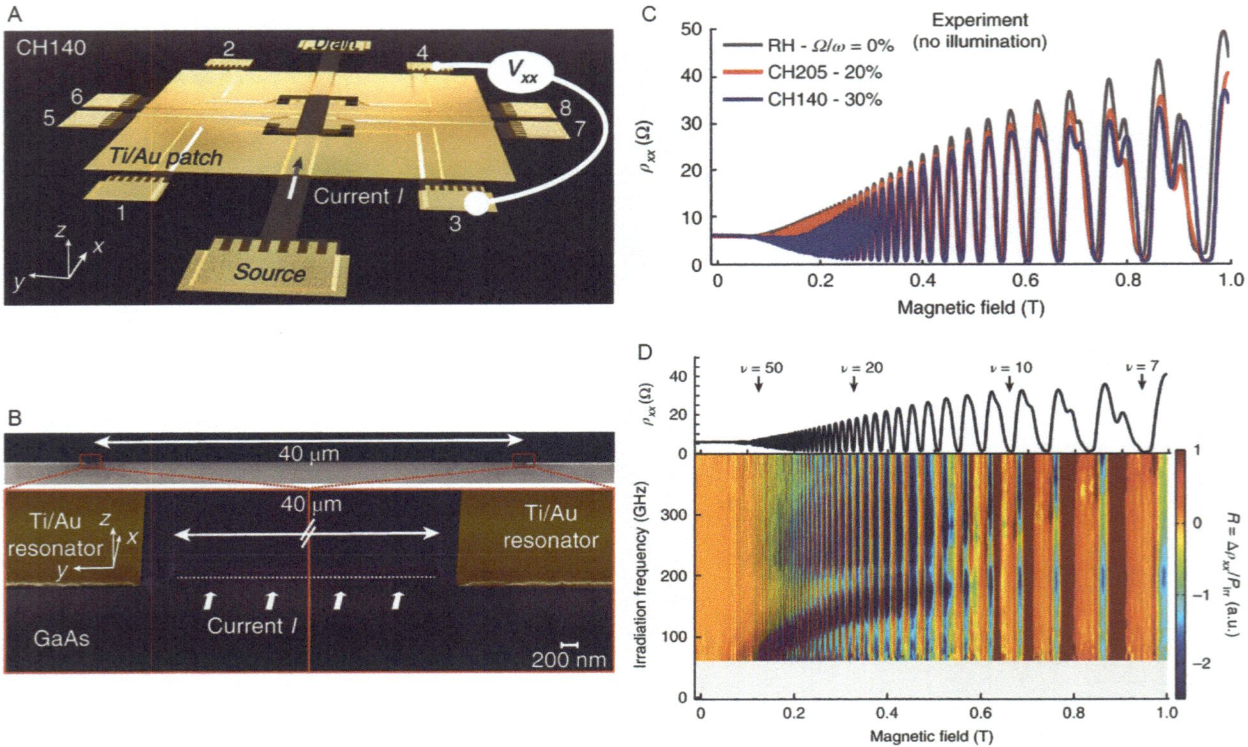

Fig. 14 Magnetotransport in a 2DEG controlled by THz cavity photons. (A) *Sample*: a Hall bar (*center*, *gray*) placed in the capacitive gap of a subwavelength LC resonator (*gold structure*). (B) SEM image showing a *y-z* cut of the Hall bar and the Ti/Au electronic resonator. (C) Shubnikov-de Haas oscillations for three samples ($\eta = 0$, 0.2, and 0.3, respectively) showing a reduction of the modulation amplitude as the coupling strength is increased. (D) Longitudinal resistance change $\Delta\rho_{xx} = \rho_{xx}{}^{illu} - \rho_{xx}{}^{dark}$ when illuminated with a narrow-band subterahertz source normalized to the irradiation power P_{irr}. Strong dependence on the spectrum with the filling factor, ν, is observed. *Adapted from Paravicini-Bagliani, G.L., Appugliese, F., Richter, E., Valmorra, F., Keller, J., Beck, M., Bartolo, N., Rossler, C., Ihn, T., Ensslin, K., Ciuti, C., Scalari, G., Faist, J., 2018. Magneto-transport controlled by Landau polariton states. Nat. Phys. 15, 186–190.*

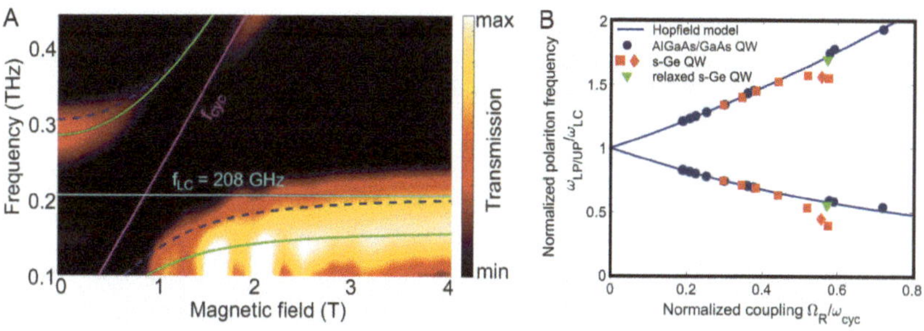

Fig. 15 Landau polaritons in nonparabolic 2DEGs in the USC regime. (A) THz transmission of a strained Ge QW at high filling factors as a function of B_{DC} at 3 K. *Solid cyan and magenta lines* indicate the cavity frequency and CR of the QW, respectively. The polariton dispersions are fit using a Hopfield model with a reduced diamagnetic (or A^2) term (*solid green lines*) and without it (*solid blue lines*). As B_{DC} increases and the CR becomes far-detuned, the cold cavity frequency limit is not recovered, and a LP gap opening is observed. (B) Normalized UP and LP frequencies at the position of minimal splitting plotted as a function of η. *Solid blue lines* indicate predictions by the Hopfield model, which agree with the experimental results for the standard GaAs QWs and a strain-relaxed Ge QW sample. Deviation from this model are seen for the strained Ge QWs as η increases. *Adapted from Keller, J., Scalari, G., Appugliese, F., Rajabali, S., Beck, M., Haase, J., Lehner, C.A., Wegscheider, W., Failla, M., Myronov, M., Leadley, D.R., Lloyd-Hughes, J., Nataf, P., Faist, J., 2020. Landau polaritons in highly non-parabolic two-dimensional gases in the ultrastrong coupling regime. Phys. Rev. B 101, 075301.*

MIR and THz technology. Furthermore, ingenious designs of nanophotonic cavities have pushed the coupling strength into the USC regime, which has opened up new opportunities for plasmon–phonon polaritons, such as control of chemical reactivity and quantum metrology with high-resolution spectroscopy.

This section is organized as follows: a brief historical review of LO phonon–plasmon coupling in doped polar semiconductors is presented, followed by several exciting examples of the coupling between SPPs and surface phonon polaritons (SPhPs), which is uniquely manifested in nanophotonic structures and nanomaterials. Finally, efforts to enhance the coupling strength into the USC regime are reviewed.

3.2.1 Plasmon–phonon polaritons in bulk semiconductors

Polar semiconductors, such as GaAs, AlAs, and InAs, display a narrow high-reflectivity spectral range that is referred to as the Reststrahlen band. This range is bounded by the transverse optical (TO) and LO phonons, whose quantitative expression is characterized through the Lyddane-Sachs-Teller

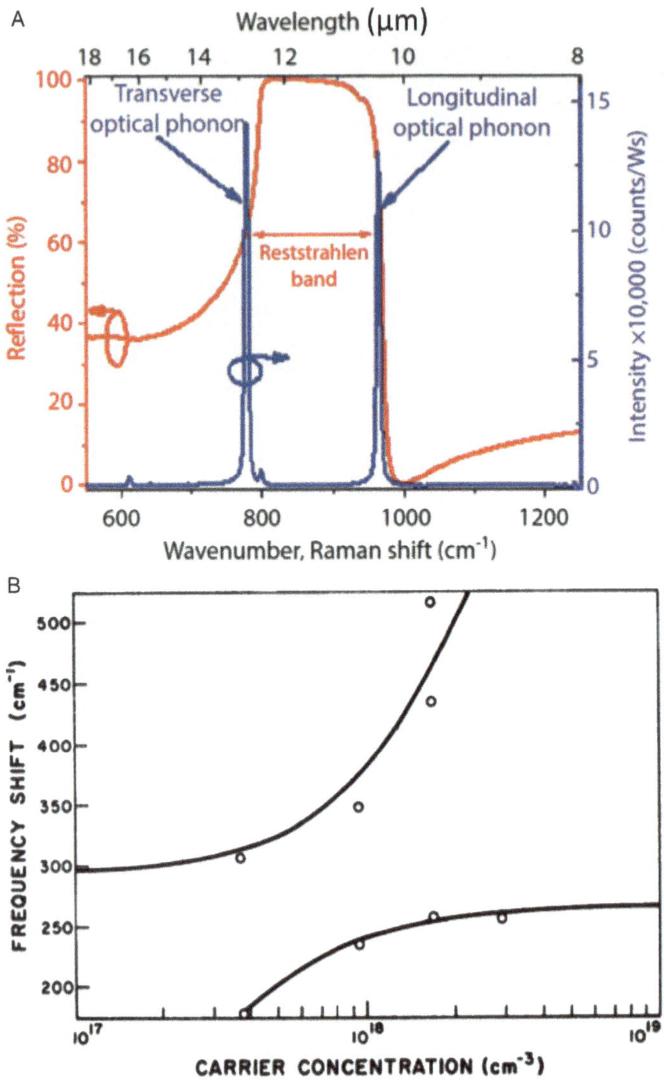

Fig. 16 The Reststrahlen band and manifestations of LO phonon-plasmon coupling in doped bulk semiconductors. (A) Reflectivity (*red trace*) and Raman (*blue trace*) spectra of SiC obtained by Caldwell et al. (2015). (B) Anticrossing in Raman scattering of GaAs observed by Mooradian and Wright (1966).

relation (Fig. 16A) (Caldwell et al., 2015). As the free carrier concentration of these semiconductors increases, the plasma frequency shifts toward higher frequencies, approaching the LO phonon frequency. This causes a spectral blueshift and a shallower slope of the LO phonon edge of the Reststrahlen band.

The dielectric function, $\varepsilon(\omega)$, is then modified through the addition of a Drude term describing free carriers to the Lorentz term describing the TO phonon resonance. The eigenfrequencies of supported longitudinal modes are the solutions to the equation $\varepsilon(\omega) = 0$, and instead of a single LO phonon frequency, there are now two solutions signaling the hybridization of plasmons and phonons. The theory of plasmon–phonon polaritons (Singwi and Tosi, 1966; Varga, 1965; Yokota, 1961) was verified in Raman scattering experiments of bulk doped GaAs (Mooradian and McWhorter, 1967; Mooradian and Wright, 1966). Mooradian and Wright (1966) provided a clear experimental demonstration of these coupled modes, where anticrossing behavior of Raman lines was observed (Fig. 16B). Similar avoid crossing behaviors were also observed in infrared reflectivity measurements (Chandrasekhar and Ramdas, 1980; Gaur, 1976; Kukharskii, 1973; Olson and Lynch, 1969; Perkowitz and Thorland, 1975) and many other doped bulk polar semiconductors, such as AlGaAs (Kim and Spitzer, 1979), SiC (Chafai et al., 2001; Harima et al., 1995; Klein et al., 1972), CdTe (Perkowitz and Thorland, 1975), GaN (Kozawa et al., 1994), InP (Artús et al., 1999), InSb (Gaur, 1976), and InAs (Gaur, 1976; Hasselbeck et al., 2002).

3.2.2 Surface plasmon–phonon polaritons

The advent of nanophotonics has created a new playground for exploring light–matter interactions at the nanoscale. Manipulation of light beyond the diffraction limit has been the persistent focus of nanophotonics, where photons can strongly couple with either plasmons (SPPs) or phonons (SPhPs) (Caldwell et al., 2016). Initial research started with SPPs in metals, including Au, Ag, and Al in the ultraviolet to the near-infrared, which has recently been extended to the MIR and THz using alternative plasmonic materials by Hoffman et al. (2007). Doped semiconductors are attractive because of their ability to tune the spectral range of resonance frequency (Law et al., 2012; Sachet et al., 2015; Zhong et al., 2015). In this regard, graphene arguably has the most dynamic and tunable range, while possessing extreme optical confinement of SPPs (Chen et al., 2012; Fei et al., 2012; Gao et al., 2012, 2013, 2014; Grigorenko et al., 2012; Jablan et al., 2009; Ju et al., 2011; Koppens et al., 2011). However, these materials suffer from short lifetimes (10s of fs) and propagation distances, and broad resonance features (Khurgin, 2015), limiting their deployment for applications.

On the other hand, SPhPs supported in dielectric polar crystals are promising alternatives with long lifetimes (10s of ps). Candidate materials include polar semiconductors, such as SiC and InP, and polar insulators, such as hexagonal boron nitride (hBN) and SiO_2, with their Reststrahlen bands in the MIR and THz. Their nanostructures thus exhibit much sharper resonance features (with Q factors 10s and 100s) (Autore et al., 2018; Caldwell et al., 2014; Chen et al., 2014b; Wang et al., 2013), compared to those in SPP resonators (with $Q < 10$). Graphene nanoribbons have been demonstrated to support strong localized plasmon resonances, whose frequencies can be broadly tuned (Fig. 17A) (Ju et al., 2011). Nanoribbons made of hBN sheets can instead support strong SPhPs in its Reststrahlen band (see Fig. 17B). As a comparison, the experimental Q of hBN ribbons is ≈ 70 while graphene ribbons have $Q \approx 1 - 2$. Despite the benefits of low loss, propagating SPhPs still have short propagation distances mainly due to the slow velocities (Caldwell et al., 2015; Yoxall et al., 2015), and the operation frequency window is limited to a narrow range. Coupled SPPs and SPhPs thus can provide a platform with advantages of individual constituents, if two materials are judiciously combined with sufficient spectral, spatial, and mode symmetry overlap, which will lead to broadband, tunable, and long-lifetime polaritons for MIR and THz optics.

Among many demonstrations of the hybrid materials combination, the coupling of SPPs in graphene with SPhPs in hBN has attracted much interest (see, e.g., Brar et al., 2014; Dai et al., 2015). Especially SPhPs supported in thin layers of hBN are hyperbolic phonon polaritons (HPhPs), where hyperbolicity is defined as an extreme birefringence with permittivities along orthogonal crystal axes possessing opposite signs. Hyperbolic polaritons allow electromagnetic modes with large momenta to propagate within the Reststrahlen band (Ferrari et al., 2015; Poddubny et al., 2013). Brar et al. (2014) reported hybrid SPP-SPhP polaritons in coupled graphene and hBN ribbons. The dispersion relations of the coupled graphene SPPs/hBN SPhPs were derived from measurements of nanoresonators with varying widths, where the graphene SPP mode exhibits an anticrossing behavior near the energy of the hBN optical phonon (see Fig. 17C). Later, Jia et al. (2015) engineered the coupling by varying the hBN thickness in a layer-by-layer manner (Fig. 17D). A systematic experimental study of this coupling using scanning near-field optical microscopy revealed the dispersion relation of SPP-HPhP in graphene/hBN heterostructures and pure HPhP in hBN (see Fig. 17E). The polariton wavelength within graphene/hBN heterostructures could be varied by $> 20\%$ compared with hBN, by changing the free carrier density in graphene (Fig. 17F), and it is also sensitive to the thickness of hBN (Fig. 17G pink line) while SPP waves in graphene were independent of thickness (Fig. 17G blue lines).

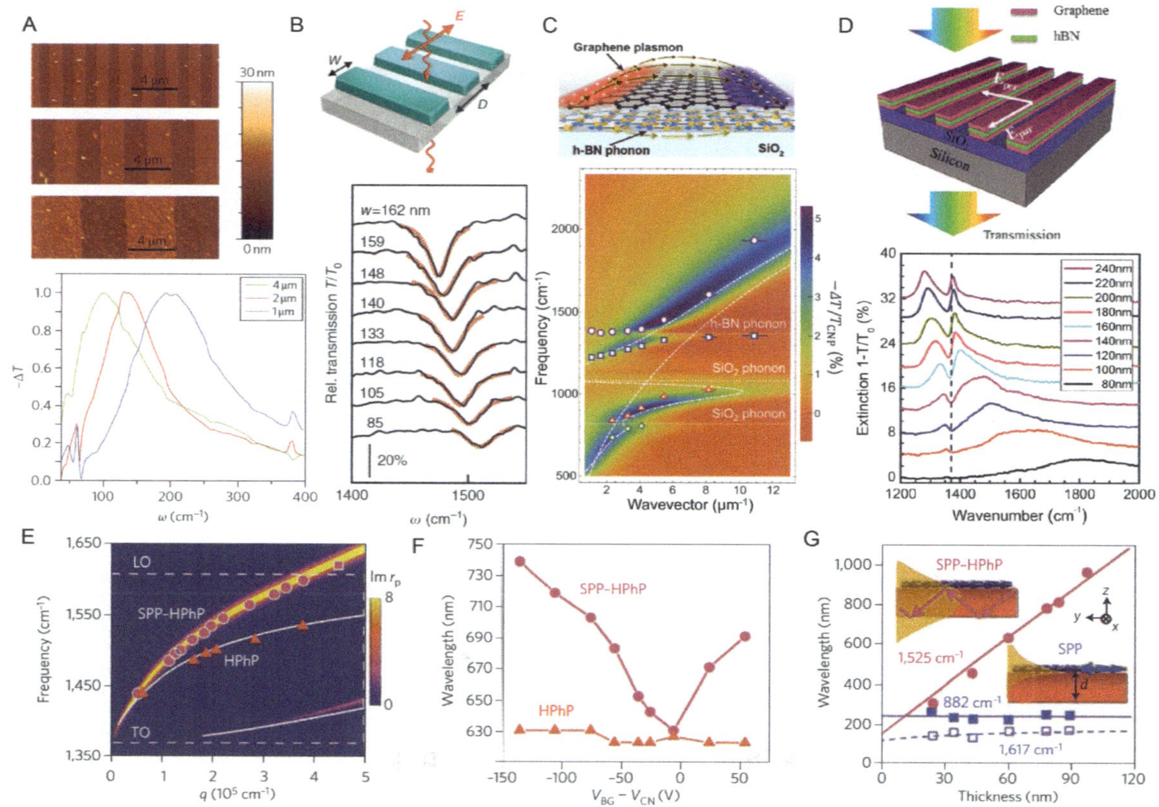

Fig. 17 See figure legend on next page.

3.2.3 Ultrastrong coupling of SPP and SPhP

Despite much success in understanding and demonstrating the physics of plasmon–phonon polaritons in a variety of materials, the coupling strength has remained quite limited due to the intrinsically weaker oscillator strengths of vibrational modes compared to electronic transitions. In addition, the SPP waves are more weakly bonded to the surface at long wavelengths, leading to a significantly higher mode volume. Although engineering phonon properties in materials can be challenging, nanophotonic tools have enabled ingenious design of structures possessing ultrasmall mode volumes, which can greatly enhance light–matter interaction strengths.

Fig. 17 SPPs and SPhP nanoresonators in graphene and hBN. (A) Graphene ribbons and electric gating to change the Fermi level. Infrared spectra of graphene ribbons with varying widths. (B) hBN ribbons and infrared spectra of ribbons with different widths. (C) Graphene/hBN nanoribbons together with their calculated (*white dashed lines*) and experimental dispersion (*colored markers*). (D) Multilayer graphene/hBN ribbons and their infrared spectra. (E) Calculated and corresponding experimental dispersion relationship of the hybridized graphene SPPs and hBN SPhPs (*pink circles and square*) and pure hBN (*red triangles*), which is visualized in terms of the imaginary part of the reflection coefficient. (F) Dependence of the polariton wavelength on the applied bias, with respect to the charge neutrality point, of graphene within the graphene/hBN (*pink circles*) and hBN (*red triangles*). (G) hBN thickness dependence of polariton wavelength for coupled SPP-HPhPs within graphene/hBN at 1525 cm^{-1} (*pink circles*), and of graphene SPP modes at 882 and 1617 cm^{-1} (*filled and open blue squares*, respectively). Schematics of the SPP-HPhP and SPP modes are provided as insets with graphene represented by the blue top layer and hBN by the red slab with thickness. *Panel A: Adapted from Ju, L., Geng, B., Horng, J., Girit, C., Martin, M., Hao, Z., Bechtel, H.A., Liang, X., Zettl, A., Shen, Y.R., Wang, F., 2011. Graphene plasmonics for tunable terahertz metamaterials. Nat. Nanotechnol. 6, 630–634. Panel B: Adapted from Autore, M., Li, P., Dolado, I., Alfaro-Mozaz, F.J., Esteban, R., Atxabal, A., Casanova, F., Hueso, L.E., Alonso-Gonzalez, P., Aizpurua, J., et al., 2018. Boron nitride nanoresonators for phonon-enhanced molecular vibrational spectroscopy at the strong coupling limit. Light Sci. Appl. 7, 17172–17172. Panel C: Adapted from Brar, V.W., Jang, M.S., Sherrott, M., Kim, S., Lopez, J.J., Kim, L.B., Choi, M., Atwater, H., 2014. Hybrid surface-phonon-plasmon polariton modes in graphene/monolayer h-BN heterostructures. Nano Lett. 14, 3876–3880. Panel D: Adapted from Jia, Y., Zhao, H., Guo, Q., Wang, X.,Wang, H., Xia, F., 2015. Tunable plasmonphonon polaritons in layered graphene-hexagonal boron nitride heterostructures. ACS Photonics. 2, 907–912; Panel G: Adapted from Dai, S., Ma, Q., Liu, M.K., Andersen, T., Fei, Z., Goldflam, M.D., Wagner, M., Watanabe, K., Taniguchi, T., Thiemens, M., et al., 2015. Graphene on hexagonal boron nitride as a tunable hyperbolic metamaterial. Nat. Nanotechnol. 10, 682–686; Caldwell, J.D., Vurgaftman, I., Tischler, J.G., Glembocki, O.J., Owrutsky, J.C., Reinecke, T.L., 2016. Atomic-scale photonic hybrids for mid-infrared and terahertz nanophotonics. Nat. Nanotechnol. 11, 9–15.*

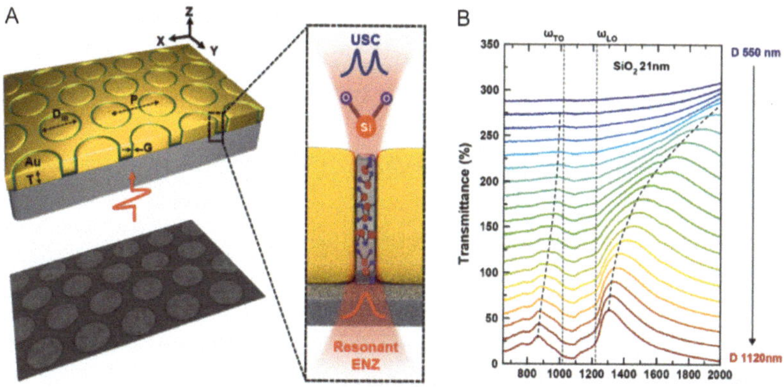

Fig. 18 Ultrastrong plasmon–phonon coupling in the MIR. (A) Schematics and scanning electron microscope images of coaxial nanocavities filled with SiO_2. (B) Infrared transmission spectra of coaxial nanocavities filled with SiO_2 of varying diameters. *Adapted from Yoo, D., de León-Pérez, F., Lee, I.-H., Mohr, D.A., Pelton, M., Raschke, M.B., Caldwell, J.D., Martín-Moreno, L., Oh, S.-H., 2020. Ultrastrong coupling of polar phonons and effective epsilon-near-zero modes in coaxial nanocavities. AU8 arXiv:2003.00136.*

Very recently, Yoo et al. (2020) utilized the epsilon-near-zero (ENZ) properties of coaxial nanocavities possessing extreme light confinement in a few-nm gap to achieve USC of vibrational modes in SiO_2. The designed cavities exhibited a transmission resonance that can be understood as either an ENZ resonance or a zeroth-order FP resonance, and the resonance frequency can be shifted by adjusting geometric parameters while maintaining strong optical confinement (Fig. 18A). As the resonance happens near the frequency where the dielectric constant is close to zero, the electric field is approximately spatially uniform with a very long effective wavelength to collectively connect a wide range of oscillators for boosting the coupling strength. Thus, as the dimension of the designed cavities is continuously changed and the resonance frequency is swept across the SiO_2 LO phonons, a clear anticrossing in transmission spectra is observed (Fig. 18B) and detailed analysis reveals that the Rabi splitting is 50 % of the resonance frequency, or $\eta \sim 0.25$.

3.3 Exciton polaritons

Excitons resonantly interacting with photons in a microcavity—microcavity exciton polaritons—have been studied for many years since the pioneering work on GaAs QWs by Weisbuch et al. (1992). GaAs QWs are sandwiched between two epitaxially grown distributed Bragg reflectors (DBRs) that form a cavity (Bloch et al., 1998; Deng et al., 2002; Skolnick et al., 1998; Weisbuch et al., 1992). The formed cavity can display a sharp resonance in reflectivity, and the electric or magnetic field can be maximized at the position where QWs are located. When the exciton energy is on resonance with a cavity mode frequency, two dips or peaks appear in a reflectivity or transmittance spectrum with a clear anticrossing behavior. However, the oscillator strength of Wannier excitons in GaAs QWs is typically small, leading to small η, typically less than 10^{-3}. Furthermore, the exciton binding energy is comparable to the ambient thermal fluctuations, and thus, the strong coupling limit, $C > 1$, is achieved only at cryogenic temperatures. Wannier excitons in other inorganic semiconductors, including GaN (Christmann et al., 2008) and ZnO (Chen et al., 2011; Guillet et al., 2011; van Vugt et al., 2006), with larger exciton binding energies and oscillator strengths, have been utilized to push the operation temperature to room temperature and achieve larger values of η, but USC has never been obtained for excitons in inorganic semiconductors.

Screening is typically weak in organic materials because of their small dielectric constants, which leads to the formation of Frenkel excitons with Bohr radii of the same order as the size of the unit cell. These excitons possess large binding energies and oscillator strengths and generally display, when in cavities, larger values of η than Wannier excitons. Some exemplary materials include tetra-(2,6-t-butyl)phenol-porphyrin zinc (4TBPPZn) ($\eta \approx 0.03$) (Lidzey et al., 1998), J aggregates ($\eta \approx 0.09$) (Wei et al., 2013), 2,7-bis [9,9-di(4-methylphenyl)-fluoren-2-yl]-9,9-di (4-methylphenyl)fluorene ($\eta \approx 0.14$) (Kéna-Cohen and Forrest, 2010), and squaraine ($\eta \approx 0.27$) (Gambino et al., 2014). Fig. 19A displays angle-dependent transmission measurements with TE polarized light for the full cavity containing squaraine (contour plot) and a bare cavity (circular dots). Both the upper and lower polariton branches display an almost dispersionless behavior and agree with theoretical results based on the transfer matrix method, which intrinsically takes into account the CRTs. Fig. 19B shows the upper and lower polariton energies normalized to the exciton energy as a function of coupling strength. Linear relation is only approximately valid if $\eta < 0.2$, while large deviation from linear dependence when $\eta > 0.2$ can be excellently corrected using the full Hopfield model, suggesting that USC is achieved. Also, a polariton gap is observed in the dispersion for a system with $\eta = 0.54$, which is another evidence of USC (see Fig. 19C).

Emerging nanomaterials, such as semiconducting single-wall carbon nanotubes (SWCNTs) (Gao et al., 2018; Graf et al., 2016, 2017), transition metal

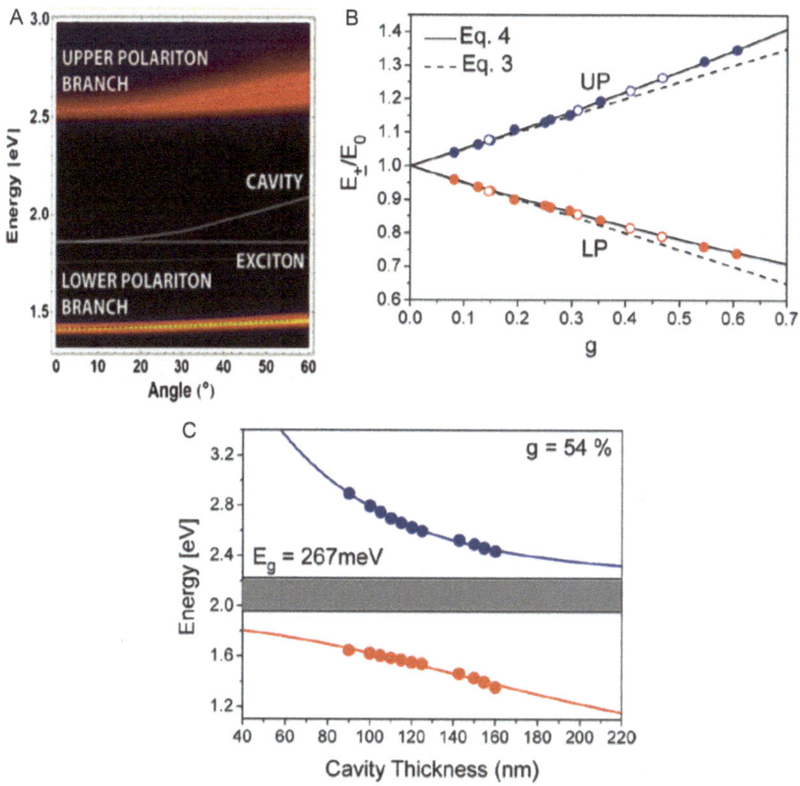

Fig. 19 Ultrastrong coupling of excitons and photons in organic semiconductor squaraine in a microcavity. (A) Contour plots of angle-resolved transmission spectra for a 140-nm-thick microcavity entirely filled with squaraine. (B) Polariton peak energies of UPs and LPs normalized to the transition energy as a function of coupling strength. Both *solid and empty dots* are from experiments. *Dashed lines* are calculated eigenfrequencies of the approximated Hamiltonian ignoring CRTs, while *solid lines* are calculated eigenfrequencies of the full Hamiltonian. (C) Polariton peak energies as a function of cavity thickness (detuning) at $\eta = 0.54$. *Solid dots* are experimental data and *solid lines* are calculation based on the Hopfield Hamiltonian. *Shaded area* indicates a polariton gap. *Adapted from Gambino, S., Mazzeo, M., Genco, A., Di Stefano, O., Savasta, S., Patanè, S., Ballarini, D., Mangione, F., Lerario, G., Sanvitto, D., et al., 2014. Exploring light-matter interaction phenomena under ultrastrong coupling regime. ACS Photonics 1, 1042–1048.*

dichalcogenides (TMDs) (Chen et al., 2017; Dufferwiel et al., 2015, 2017; Lee et al., 2017; Liu et al., 2015, 2016; Sun et al., 2017), and low-dimensional perovskites (Bao et al., 2019; Shang et al., 2018; Su et al., 2017, 2018, 2020; Wang et al., 2018; Zhang et al., 2018), have characteristics of both Wannier and Frenkel excitons, where the exciton wavefunction has a wsubstantial spatial extent while the screening from the lattice is weak. Due to strong quantum confinement, excitons in these materials usually have large oscillator strengths and binding energies and are stable even at room temperature.

Graf et al. (2016) demonstrated SC between photons in a FP cavity and excitons in a semiconducting SWCNT-polymer composite, both in reflectivity and photoluminescence, as shown in Fig. 20A. In addition to a large VRS exceeding 100 meV, a cooperative enhancement effect with $g \propto \sqrt{N}$ was observed as the SWCNT occupation ratio was systematically adjusted, where N is the number of nanotubes. Graf et al. (2017) further implemented a cavity-integrated light-emitting field effect transistor consisting of a semiconducting SWCNT-polymer composite, which allowed ambipolar charge transport. Under strong current injection, electroluminescence from the LP branch was observed from the low wavevector region of the dispersion, suggesting efficient relaxation. By changing the gate voltage to adjust charge density, the ground state was bleached and overall oscillator strength was reduced. As a result, the VRS decreased, and the system continuously transitioned from strong to weak coupling.

Gao et al. (2018) employed *aligned* SWCNT films also inside a FP cavity and observed a polarization-dependent VRS (Fig. 20B), suggesting that the exciton-polaritons inherited anisotropic properties from the SWCNT excitons. Detailed analysis revealed the existence of *exceptional points* in the polariton dispersion surfaces (Fig. 20C), where the LP branch and UP branch collapse at the exciton energy. The largest VRS in the thickest device was 329 meV, corresponding to $\eta \approx 0.13$ (see Fig. 20D). Furthermore, an increase of the film thickness displayed *cooperative enhancement* of g, i.e., VRS was proportional to the square root of the film thickness (see Fig. 20E).

3.4 Magnon polaritons, spin–magnon coupling, and magnon–magnon coupling

Hybrid quantum systems have seen an increased interest in recent years due to their potential use in quantum communication platforms such as quantum memories, quantum transducers, and quantum information processors (Bittencourt et al., 2019; Zhang et al., 2016b, 2020). In these systems, coupling occurs between two excitations coming from distinctly different constituents, e.g., molecules embedded in optical cavities, exciton-polaritons in a microcavity, and magnons in microwave cavities. In this section, we briefly review some of the recent research efforts on SC and USC phenomena involving magnons.

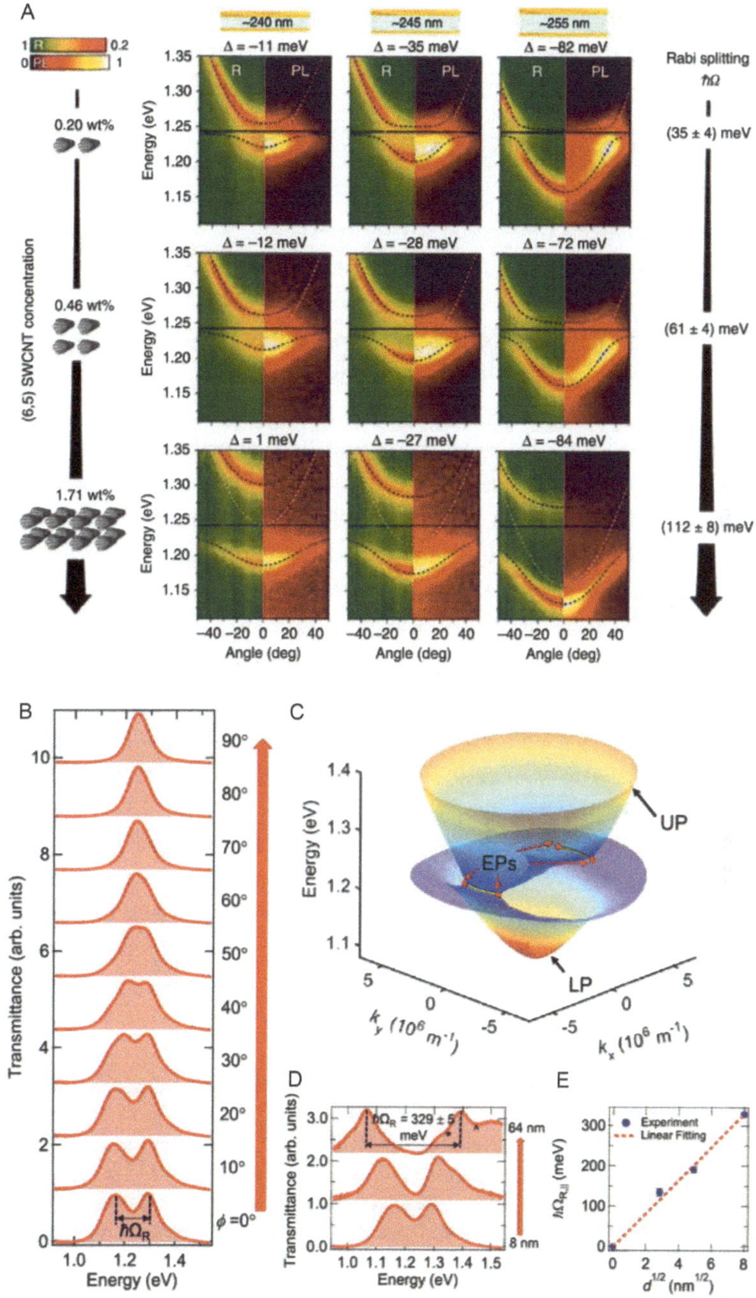

Fig. 20 SC of excitons and photons in semiconducting SWCNTs. (A) Angle-resolved reflectivity and photoluminescence spectra for (6,5) SWCNT microcavity exciton polaritons with increasing nanotube concentration (from *top to bottom*) and increasing cavity thickness and detuning from (*left to right*). (B) Transmittance spectra for a cavity

(Continued)

Cavity optomagnonics, where a set of spin waves (or magnons) collectively couples to the confined electromagnetic field of a cavity (Kusminskiy, 2019a,b; Parvini et al., 2020), has recently seen some of the first experimental observations of coupling between magnons in the ferrimagnetic insulator YIG to microwave photons both in the SC (Haigh et al., 2015; Huebl et al., 2013; Tabuchi et al., 2014; Zhang et al., 2014b) and USC (Goryachev et al., 2014) regimes. Goryachev et al. (2014) observed USC between the magnon resonances of a submillimeter YIG sphere and the bright cavity mode of a double-post 3D microwave reentrant cavity at mK temperatures. Using a novel multipost design (Tobar and Goryachev, 2019), $C \sim 10^5$ was obtained, together with $\eta \sim 0.1$. The cavity transmission as a function of B_{DC} showed a clear anticrossing between the bright cavity mode $f_{\uparrow\downarrow}$ and the magnon mode M_1 (see Fig. 21A) and between the dark cavity mode $f_{\uparrow\uparrow}$ and the magnetostatic magnon modes M_2 and M_3. The latter anticrossing showed SC with C reaching 10^3 (see Fig. 21B). In addition, the authors provided straightforward modifications to the cavity design that could potentially reach $C \sim 10^7$.

Recent years have also seen the development of an exciting area of research that focuses on antiferromagnets (AFMs) instead of ferromagnets/ferrimagnets. In AFMs, magnonic excitations typically occur in the THz frequency range and can couple with another degree of freedom coming from the same magnetic system rather than excitations supplied by an external source. For example, a spin–magnon system studied by Li et al. (2018a) exhibited USC between the electron paramagnetic resonance (EPR) of Er^{3+} spins and the quasi-ferromagnetic (q_{FM}) mode of the Fe^{3+} magnons in the rare-earth orthoferrite, $Er_xY_{1-x}FeO_3$. An applied external DC magnetic field tuned the EPR with respect to the fixed q_{FM} magnon mode frequency, and a clear anticrossing behavior was resolved in the THz range (see Fig. 22A).

Fig. 20—cont'd containing aligned (6,5) SWCNTs at zero detuning for various polarization angles. (C) The dispersion surfaces of the UP and LP for the device in (B) showing exceptional points (EPs). (D) Transmittance spectra for parallel polarization at zero detuning for devices containing aligned SWCNT films of different thicknesses. The largest VRS is 329 meV for the thickest film. (E) VRS for parallel polarization at zero detuning as a function of the square root of the film thickness, demonstrating collective enhancement. *Panel A: Adapted from Graf, A., Tropf, L., Zakharko, Y., Zaumseil, J., Gather, M.C., 2016. Nearinfrared exciton-polaritons in strongly coupled single-walled carbon nanotube microcavities. Nat. Commun. 7, 13078; Panel E: Adapted from Gao, W., Li, X., Bamba, M., Kono, J., 2018. Continuous transition between weak and ultrastrong coupling through exceptional points in carbon nanotube microcavity exciton-polaritons. Nat. Photon. 12, 362–367.*

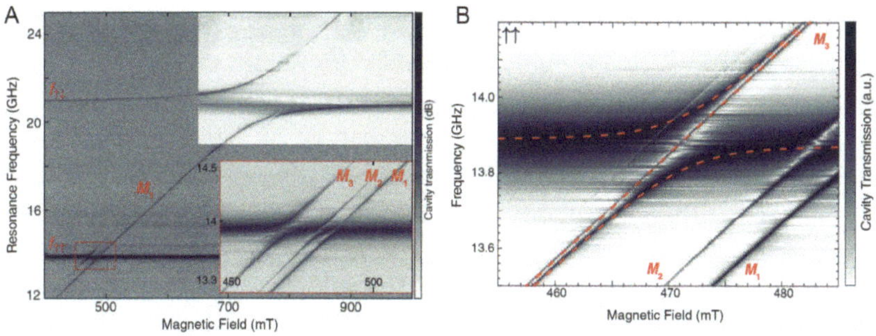

Fig. 21 High-cooperativity cavity QED in the USC regime with magnons at microwave frequencies. (A) Cavity microwave transmission as a function of applied magnetic field, B_{DC}. Anticrossings between the cavity modes $f_{\uparrow\downarrow}$ (*bright*), $f_{\uparrow\uparrow}$ (*dark*) and the magnon modes M_1, M_2, and M_3 are observed close to $B_{DC} = 0.743$ T and $B_{DC} = 0.471$ T, respectively. *Inset*: zoom-in of the red dashed area shown on the left. (B) Three-mode anticrossing spectra and interacting model fit (red dashed red lines). First-order cancelation of the coupling prohibits the magnon mode M_1 to interact with the dark cavity mode. *Adapted from Goryachev, M., Farr, W.G., Creedon, D.L., Fan, Y., Kostylev, M., Tobar, M.E., 2014. High-cooperativity cavity QED with magnons at microwave frequencies. Phys. Rev. Appl. 2, 054002.*

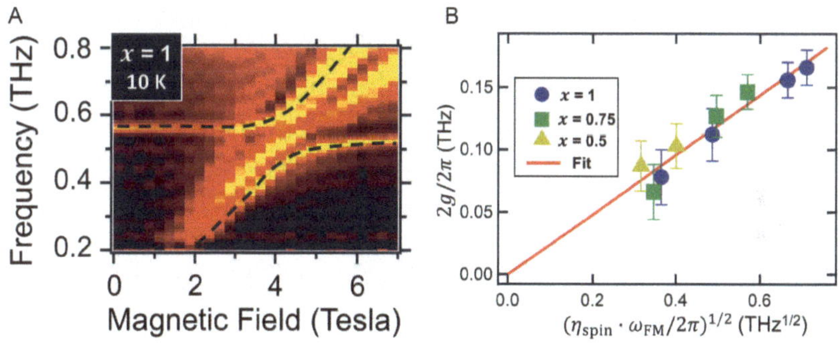

Fig. 22 Dicke cooperativity observed in spin–magnon interactions in the USC regime. (A) Anticrossing behavior at 10 K in the THz frequency range between the Er^{3+} EPR and the Fe^{3+} q_{FM} magnon mode of the rare-earth orthoferrite $Er_xY_{1-x}FeO_3$. An applied magnetic field tunes the EPR with respect to the fixed magnon mode. (B) Cooperative square-root-dependence of the coupling strength g on the number of EPR-contributing spins. *Adapted from Li, X., Bamba, M., Yuan, N., Zhang, Q., Zhao, Y., Xiang, M., Xu, K., Jin, Z., Ren, W., Ma, G., Cao, S., Turchinovich, D., Kono, J., 2018a. Observation of Dicke cooperativity in magnetic interactions. Science 361 (6404), 794–797.*

Moreover, a $g \propto \sqrt{N}$ behavior (Dicke cooperativity), where N is the net density of EPR-contributing spins, was obtained (see Fig. 22B). The authors estimated a maximum $\eta = 0.18$ in one of the configurations, putting the system in the USC regime. Another example falling in this category is *magnon–magnon coupling* inside a magnetically ordered material (Liensberger et al., 2019; MacNeill et al., 2019; Makihara et al., 2020). In these studies, coupling occurred between different magnon modes, and *g* was tunable by changing the applied magnetic field direction. Makihara et al. (2020) reported USC between the q_{FM} and quasi-antiferromagnetic (q_{AFM}) magnon modes in $YFeO_3$. The authors were able to map the spin model of this system into an *anisotropic* Hopfield Hamiltonian in which the CRTs dominated the co-rotating terms, leading to giant vacuum BS shifts that dominated the shifts due to VRS.

4. Summary and outlook

In this chapter, we reviewed the theory behind USC and some of the groundbreaking experiments that have taken place since the first theoretical proposal by Ciuti et al. (2005). These pioneering studies have unquestionably shown that the USC regime can be achieved in diverse experimental systems, particularly engineered semiconductor systems. Given the fact that the first experimental observation of USC was made only 11 years ago at the time of this writing, the fast-paced and continuous progress this field has exhibited in the recent years is truly remarkable. Not only the coupling strengths that can be achieved in experiments have shown an order of magnitude increase during this past decade, but also new and unexplored regimes of light–matter interaction such as the DSC regime have emerged (Bayer et al., 2017; Mueller et al., 2020; Yoshihara et al., 2017). The USC regime is now readily accessible using equipment found in most university laboratories around the world.

However, research in this field and our current understanding about it are far from complete. Open questions regarding how to explain some of the experimental observations made in the USC regime remain, and there is much more that can be done on the experimental front to push the boundaries even further to reveal some of the exotic properties predicted for the ground states of systems in the USC regime. In addition, in the same way as applications of light–matter interactions that occur in cavities in the weak and SC regimes advanced the development of solid-state lasers, quantum emitters, and precision metrology, the present and future applications of the USC regime are now starting to surface and have found immediate applicability in the long-sought dream of realizing quantum technologies such as quantum computers and quantum simulators. The authors look forward to seeing where this field will be in 5, 10, and 20 years from now and what new exciting discoveries will emerge.

Acknowledgments

The authors thank Takuma Makihara for proofreading the manuscript. J.K. acknowledges support from the U.S. Army Research Office (W911NF-17-1-0259), the U.S. National Science Foundation (NSF MRSEC DMR-1720595), the U.S. Department of Energy (DEFG02-06ER46308), and the Robert A. Welch Foundation (C-1509). M.B. acknowledges support from the Japan Science and Technology Agency (JST) PRESTO program (grant JPMJPR1767).

References

Abdurakhimov, L.V., Yamashiro, R., Badrutdinov, A.O., Konstantinov, D., 2016. Strong coupling of the cyclotron motion of surface electrons on liquid helium to a microwave cavity. Phys. Rev. Lett. 117, 056803.

Abragam, A., 1961. Principles of Nuclear Magnetism. Oxford University Press, Oxford.

Abram, I., 1987. Quantum theory of light propagation: linear medium. Phys. Rev. A 35, 4661–4672.

Allen, L., Eberly, J.H., 1975. Optical Resonance and Two-Level Atoms. John Wiley & Sons, New York.

Allocca, A.A., Raines, Z.M., Curtis, J.B., Galitski, V.M., 2019. Cavity superconductor-polaritons. Phys. Rev. B 99, 020504.

Amsüss, R., Koller, C., Nöbauer, T., Putz, S., Rotter, S., Sandner, K., Schneider, S., Schramböck, M., Steinhauser, G., Ritsch, H., Schmiedmayer, J., Majer, J., 2011. Cavity QED with magnetically coupled collective spin states. Phys. Rev. Lett. 107, 060502.

Anappara, A.A., Tredicucci, A., Biasiol, G., Sorba, L., 2005. Electrical control of polariton coupling in intersubband microcavities. Appl. Phys. Lett. 87, 051105.

Anappara, A.A., Tredicucci, A., Beltram, F., Biasiol, G., Sorba, L., 2006. Tunnel-assisted manipulation of intersubband polaritons in asymmetric coupled quantum wells. Appl. Phys. Lett. 89, 171109.

Anappara, A.A., Barate, D., Tredicucci, A., Devenson, J., Teissier, R., Baranov, A., 2007. Giant intersubband polariton splitting in InAs/AlSb microcavities. Solid State Commun. 142, 311–313.

Anappara, A.A., De Liberato, S., Tredicucci, A., Ciuti, C., Biasiol, G., Sorba, L., Beltram, F., 2009. Signatures of the ultrastrong light-matter coupling regime. Phys. Rev. B 79, 201303(R).

Andolina, G.M., Pellegrino, F.M.D., Giovannetti, V., MacDonald, A.H., Polini, M., 2019. Cavity quantum electrodynamics of strongly correlated electron systems: a No-Go theorem for photon condensation. Phys. Rev. B 100, 121109.

Artoni, M., Birman, J.L., 1989. Polariton squeezing: theory and proposed experiment. Quant. Opt. 1, 91–97.

Artoni, M., Birman, J.L., 1991. Quantum-optical properties of polariton waves. Phys. Rev. B 44, 3736–3756.

Artús, L., Cuscó, R., Ibáñez, J., Blanco, N., González-Díaz, G., 1999. Raman scattering by LO phonon-plasmon coupled modes in *n*-type InP. Phys. Rev. B 60, 5456.

Ashhab, S., Nori, F., 2010. Qubit-oscillator systems in the ultrastrong-coupling regime and their potential for preparing nonclassical states. Phys. Rev. A 81, 042311.

Ashida, Y., Imamoglu, A., Faist, J., Jaksch, D., Cavalleri, A., Demler, E., 2020. Quantum electrodynamic control of matter: cavity-enhanced ferroelectric phase transition. Phys. Rev. X 10, 041027.

Askenazi, B., Vasanelli, A., Delteil, A., Todorov, Y., Andreani, L.C., Beaudoin, G., Sagnes, I., Sirtori, C., 2014. Ultra-strong light-matter coupling for designer Reststrahlen band. New J. Phys. 16, 043029.

Askenazi, B., Vasanelli, A., Todorov, Y., Sakat, E., Greffet, J.-J., Beaudoin, G., Sagnes, I., Sirtori, C., 2017. Midinfrared ultrastrong light-matter coupling for THz thermal emission. ACS Photonics 4, 2550–2555.

Auer, A., Burkard, G., 2012. Entangled photons from the polariton vacuum in a switchable optical cavity. Phys. Rev. B 85, 235140.

Autore, M., Li, P., Dolado, I., Alfaro-Mozaz, F.J., Esteban, R., Atxabal, A., Casanova, F., Hueso, L.E., Alonso-González, P., Aizpurua, J., et al., 2018. Boron nitride nanoresonators for phonon-enhanced molecular vibrational spectroscopy at the strong coupling limit. Light Sci. Appl. 7, 17172. 17172.

Ballester, D., Romero, G., García-Ripoll, J.J., Deppe, F., Solano, E., 2012. Quantum simulation of the ultrastrong-coupling dynamics in circuit quantum electrodynamics. Phys. Rev. X 2, 021007.

Bamba, M., Imoto, N., 2016. Maxwell boundary conditions imply non-Lindblad master equation. Phys. Rev. A 94, 033802.

Bamba, M., Imoto, N., 2017. Circuit configurations which may or may not show superradiant phase transitions. Phys. Rev. A 96, 053857.

Bamba, M., Ogawa, T., 2014. Stability of polarizable materials against superradiant phase transition. Phys. Rev. A 90, 063825.

Bamba, M., Ogawa, T., 2016. Laser under ultrastrong light-matter interaction: qualitative aspects and quantitative influences by level and mode truncations. Phys. Rev. A 93, 033811.

Bamba, M., Inomata, K., Nakamura, Y., 2016. Superradiant phase transition in a superconducting circuit in thermal equilibrium. Phys. Rev. Lett. 117, 173601. https://doi.org/10.1103/PhysRevLett.117.173601.

Bamba, M., Li, X., Kono, J., 2019. Terahertz strong-field physics without a strong external terahertz field. In: Betz, M., Elezzabi, A.Y. (Eds.), Ultrafast Phenom. Nanophotonics XXIII, SPIE, p. 5. 1901.06749.

Bao, W., Liu, X., Xue, F., Zheng, F., Tao, R., Wang, S., Xia, Y., Zhao, M., Kim, J., Yang, S., et al., 2019. Observation of Rydberg exciton polaritons and their condensate in a perovskite cavity. Proc. Natl. Acad. Sci. U.S.A. 116, 20274–20279.

Bartolo, N., Ciuti, C., 2018. Vacuum-dressed cavity magnetotransport of a two-dimensional electron gas. Phys. Rev. B 98, 205301.

Baumann, K., Guerlin, C., Brennecke, F., Esslinger, T., 2010. Dicke quantum phase transition with a superfluid gas in an optical cavity. Nature 464, 1301–1306.

Bayer, A., Pozimski, M., Schambeck, S., Schuh, D., Huber, R., Bougeard, D., Lange, C., 2017. Terahertz light-matter interaction beyond unity coupling strength. Nano Lett. 17, 6340–6344.

Bialynicki-Birula, I., Rzażewski, K., 1979. No-Go theorem concerning the superradiant phase transition in atomic systems. Phys. Rev. A 19, 301–303.

Bittencourt, V.A.S.V., Feulner, V., Kusminskiy, S.V., 2019. Magnon heralding in cavity optomagnonics. Phys. Rev. A 100, 013810.

Bloch, F., Siegert, A., 1940. Magnetic resonance for nonrotating fields. Phys. Rev. 57, 522–527.

Bloch, J., Freixanet, T., Marzin, J.Y., Thierry-Mieg, V., Planel, R., 1998. Giant Rabi splitting in a microcavity containing distributed quantum wells. Appl. Phys. Lett. 73, 1694–1696.

Bonifacio, R., Lugiato, L.A., 1982. Theory of optical bistability. In: Bonifacio, R. (Ed.), Dissipative Systems in Quantum Optics. Topics in Current Physics, Springer-Verlag, Berlin, pp. 61–92.

Brar, V.W., Jang, M.S., Sherrott, M., Kim, S., Lopez, J.J., Kim, L.B., Choi, M., Atwater, H., 2014. Hybrid surface-phonon-plasmon polariton modes in graphene/monolayer h-BN heterostructures. Nano Lett. 14, 3876–3880.

Brey, L., Johnson, N.F., Halperin, B.I., 1989. Optical and magneto-optical absorption in parabolic quantum wells. Phys. Rev. B 40, 10647–10649.
Caldwell, J.D., Kretinin, A.V., Chen, Y., Giannini, V., Fogler, M.M., Francescato, Y., Ellis, C.T., Tischler, J.G., Woods, C.R., Giles, A.J., et al., 2014. Sub-diffractional volume-confined polaritons in the natural hyperbolic material hexagonal boron nitride. Nat. Commun. 5, 1–9.
Caldwell, J.D., Lindsay, L., Giannini, V., Vurgaftman, I., Reinecke, T.L., Maier, S.A., Glembocki, O.J., 2015. Low-loss, infrared and terahertz nanophotonics using surface phonon polaritons. Nanophotonics 4, 44–68.
Caldwell, J.D., Vurgaftman, I., Tischler, J.G., Glembocki, O.J., Owrutsky, J.C., Reinecke, T.L., 2016. Atomic-scale photonic hybrids for mid-infrared and terahertz nanophotonics. Nat. Nanotechnol. 11, 9–15.
Carusotto, I., De Liberato, S., Gerace, D., Ciuti, C., 2012. Back-reaction effects of quantum vacuum in cavity quantum electrodynamics. Phys. Rev. A 85, 23805.
Casanova, J., Romero, G., Lizuain, I., García-Ripoll, J.J., Solano, E., 2010. Deep strong coupling regime of the Jaynes-Cummings model. Phys. Rev. Lett. 105, 263603.
Chafai, M., Jaouhari, A., Torres, A., Antón, R., Martın, E., Jiménez, J., Mitchel, W.C., 2001. Raman scattering from LO phonon-plasmon coupled modes and Hall-effect in *n*-type silicon carbide 4H-SiC. J. Appl. Phys. 90, 5211–5215.
Chandrasekhar, H.R., Ramdas, A.K., 1980. Nonparabolicity of the conduction band and the coupled plasmon-phonon modes in *n*-GaAs. Phys. Rev. B 21, 1511.
Chen, J.-R., Lu, T.-C., Wu, Y.-C., Lin, S.-C., Hsieh, W.-F., Wang, S.-C., Deng, H., 2011. Characteristics of exciton-polaritons in ZnO-based hybrid microcavities. Opt. Exp. 19, 4101–4112.
Chen, J., Badioli, M., Alonso-González, P., Thongrattanasiri, S., Huth, F., Osmond, J., Spasenović, M., Centeno, A., Pesquera, A., Godignon, P., et al., 2012. Optical nano-imaging of gate-tunable graphene plasmons. Nature 487, 77–81.
Chen, T., Liu, P., Liu, J., Hong, Z., 2014a. A terahertz photonic crystal cavity with high *Q*-factors. Appl. Phys. B 115, 105–109.
Chen, Y., Francescato, Y., Caldwell, J.D., Giannini, V., Maß, T.W.W., Glembocki, O.J., Bezares, F.J., Taubner, T., Kasica, R., Hong, M., et al., 2014b. Spectral tuning of localized surface phonon polariton resonators for low-loss mid-IR applications. ACS Photon. 1, 718–724.
Chen, Y.-J., Cain, J.D., Stanev, T.K., Dravid, V.P., Stern, N.P., 2017. Valley-polarized exciton-polaritons in a monolayer semiconductor. Nat. Photon. 11, 431–435.
Chirolli, L., Polini, M., Giovannetti, V., MacDonald, A.H., 2012. Drude weight, cyclotron resonance, and the Dicke model of graphene cavity QED. Phys. Rev. Lett. 109, 267404.
Choi, K.-K., 1997. The Physics of Quantum Well Infrared Photodetectors. vol. 7 World Scientific, Singapore.
Christmann, G., Butté, R., Feltin, E., Carlin, J.-F., Grandjean, N., 2008. Room temperature polariton lasing in a GaN/AlGaN multiple quantum well microcavity. Appl. Phys. Lett. 93, 051102.
Cirio, M., De Liberato, S., Lambert, N., Nori, F., 2016. Ground state electroluminescence. Phys. Rev. Lett. 116, 113601.
Cirio, M., Debnath, K., Lambert, N., Nori, F., 2017. Amplified optomechanical transduction of virtual radiation pressure. Phys. Rev. Lett. 119, 053601.
Cirio, M., Shammah, N., Lambert, N., De Liberato, S., Nori, F., 2019. Multielectron ground state electroluminescence. Phys. Rev. Lett. 122, 190403.
Ciuti, C., Carusotto, I., 2006. Input-output theory of cavities in the ultrastrong coupling regime: the case of time-independent cavity parameters. Phys. Rev. A 74, 033811.
Ciuti, C., Nataf, P., 2012. Comment on "superradiant phase transitions and the standard description of circuit QED". Phys. Rev. Lett. 109, 179301.

Ciuti, C., Bastard, G., Carusotto, I., 2005. Quantum vacuum properties of the intersubband cavity polariton field. Phys. Rev. B 72, 115303.

Cohen-Tannoudji, C., Dupont-Roc, J., Fabre, C., 1973. A quantum calculation of the higher order terms in the Bloch-Siegert shift. J. Phys. B 6, L214–L217.

Cohen-Tannoudji, C., Dupont-Roc, J., Grynberg, G., 1989. Photons and Atoms: Introduction to Quantum Electrodynamics. John Wiley & Sons, Inc., New York.

Combescot, M., Betbedermatibet, O., Dubin, F., 2008. The many-body physics of composite bosons. Phys. Rep. 463, 215–320.

Curtis, J.B., Raines, Z.M., Allocca, A.A., Hafezi, M., Galitski, V.M., 2019. Cavity quantum Eliashberg enhancement of superconductivity. Phys. Rev. Lett. 122, 167002.

Dai, S., Ma, Q., Liu, M.K., Andersen, T., Fei, Z., Goldflam, M.D., Wagner, M., Watanabe, K., Taniguchi, T., Thiemens, M., et al., 2015. Graphene on hexagonal boron nitride as a tunable hyperbolic metamaterial. Nat. Nanotechnol. 10, 682–686.

De Bernardis, D., Pilar, P., Jaako, T., De Liberato, S., Rabl, P., 2018. Breakdown of gauge invariance in ultrastrong-coupling cavity QED. Phys. Rev. A 98, 053819.

De Liberato, S., 2014. Light-matter decoupling in the deep strong coupling regime: the breakdown of the Purcell effect. Phys. Rev. Lett. 112, 016401.

De Liberato, S., Ciuti, C., Carusotto, I., 2007. Quantum vacuum radiation spectra from a semiconductor microcavity with a time-modulated vacuum Rabi frequency. Phys. Rev. Lett. 98, 103602.

De Liberato, S., Gerace, D., Carusotto, I., Ciuti, C., 2009. Extracavity quantum vacuum radiation from a single qubit. Phys. Rev. A 80, 53810.

De Zela, F., Solano, E., Gago, A., 1997. Micromaser without the rotating-wave approximation: the Bloch-Siegert shift and related effects. Opt. Commun. 142, 106–118.

Delteil, A., Vasanelli, A., Todorov, Y., Feuillet Palma, C., Renaudat St-Jean, M., Beaudoin, G., Sagnes, I., Sirtori, C., 2012. Charge-induced coherence between intersubband plasmons in a quantum structure. Phys. Rev. Lett. 109, 246808.

Deng, H., Weihs, G., Santori, C., Bloch, J., Yamamoto, Y., 2002. Condensation of semiconductor microcavity exciton polaritons. Science 298 (5591), 199–202.

Deng, H., Haug, H., Yamamoto, Y., 2010. Exciton-polariton Bose-Einstein condensation. Rev. Mod. Phys. 82, 1489–1537.

Di Stefano, O., Settineri, A., Macrì, V., Garziano, L., Stassi, R., Savasta, S., Nori, F., 2019. Resolution of gauge ambiguities in ultrastrong-coupling cavity QED. Nat. Phys. 15, 803–808.

Dicke, R.H., 1954. Coherence in spontaneous radiation processes. Phys. Rev. 93, 99–110.

Dietze, D., Andrews, A.M., Klang, P., Strasser, G., Unterrainer, K., Darmo, J., 2013. Ultrastrong coupling of intersubband plasmons and terahertz metamaterials. Appl. Phys. Lett. 103, 201106.

Dimer, F., Estienne, B., Parkins, A.S., Carmichael, H.J., 2007. Proposed realization of the Dicke-model quantum phase transition in an optical cavity QED system. Phys. Rev. A 75, 013804.

Dini, D., Köhler, R., Tredicucci, A., Biasiol, G., Sorba, L., 2003. Microcavity polariton splitting of intersubband transitions. Phys. Rev. Lett. 90, 116401.

Dodonov, A.V., Celeri, L.C., Pascoal, F., Lukin, M.D., Yelin, S.F., 2008. Photon generation from vacuum in non-stationary circuit QED. arXiv:0806.4035v3. 0806.4035v3.

Drozdov, A.P., Kong, P.P., Minkov, V.S., Besedin, S.P., Kuzovnikov, M.A., Mozaffari, S., Balicas, L., Balakirev, F.F., Graf, D.E., Prakapenka, V.B., et al., 2019. Superconductivity at 250 k in lanthanum hydride under high pressures. Nature 569, 528–531.

Dufferwiel, S., Schwarz, S., Withers, F., Trichet, A.A.P., Li, F., Sich, M., Del Pozo-Zamudio, O., Clark, C., Nalitov, A., Solnyshkov, D.D., et al., 2015. Exciton-polaritons in van der Waals heterostructures embedded in tunable microcavities. Nat. Commun. 6, 1–7.

Dufferwiel, S., Lyons, T.P., Solnyshkov, D.D., Trichet, A.A.P., Withers, F., Schwarz, S., Malpuech, G., Smith, J.M., Novoselov, K.S., Skolnick, M.S., et al., 2017. Valley-addressable polaritons in atomically thin semiconductors. Nat. Photon. 11, 497–501.

Dupont, E., Liu, H.C., SpringThorpe, A.J., Lai, W., Extavour, M., 2003. Vacuum-field Rabi splitting in quantum-well infrared photodetectors. Phys. Rev. B 68, 245320.

Dupont, E., Gupta, J.A., Liu, H.C., 2007. Giant vacuum-field Rabi splitting of intersubband transitions in multiple quantum wells. Phys. Rev. B 75, 205325.

Eliashberg, G.M., 1970. Film superconductivity stimulated by a high-frequency field. JETP Lett. 11, 114–116.

Emary, C., Brandes, T., 2003a. Chaos and the quantum phase transition in the Dicke model. Phys. Rev. E 67, 066203.

Emary, C., Brandes, T., 2003b. Quantum chaos triggered by precursors of a quantum phase transition: the dicke model. Phys. Rev. Lett. 90, 044101.

Faist, J., Capasso, F., Sivco, D.L., Sirtori, C., Hutchinson, A.L., Cho, A.Y., 1994. Quantum cascade laser. Science 264, 553–556.

Fei, Z., Rodin, A.S., Andreev, G.O., Bao, W., McLeod, A.S., Wagner, M., Zhang, L.M., Zhao, Z., Thiemens, M., Dominguez, G., Fogler, M.M., Castro Neto, A.H., Lau, C.N., Keilmann, F., Basov, D.N., 2012. Gate-tuning of graphene plasmons revealed by infrared nano-imaging. Nature 487, 82–85.

Felicetti, S., Douce, T., Romero, G., Milman, P., Solano, E., 2015. Parity-dependent state engineering and tomography in the ultrastrong coupling regime. Sci. Rep. 5, 11818.

Ferrari, L., Wu, C., Lepage, D., Zhang, X., Liu, Z., 2015. Hyperbolic metamaterials and their applications. Prog. Quantum Electron. 40, 1–40.

Ferraro, D., Campisi, M., Andolina, G.M., Pellegrini, V., Polini, M., 2018. High-power collective charging of a solid-state quantum battery. Phys. Rev. Lett. 120, 117702.

Ferraro, D., Andolina, G.M., Campisi, M., Pellegrini, V., Polini, M., 2019. Quantum supercapacitors. Phys. Rev. B 100, 075433.

Ferraro, D., Campisi, M., Andolina, G.M., Pellegrini, V., Polini, M., 2020. Quantum resources for energy storage. EPJ Web Conf. 230, 00003.

Flower, G., Bourhill, J., Goryachev, M., Tobar, M.E., 2019a. Broadening frequency range of a ferromagnetic axion haloscope with strongly coupled cavity-magnon polaritons. Phys. Dark Universe 25, 100306.

Flower, G., Goryachev, M., Bourhill, J., Tobar, M.E., 2019b. Experimental implementations of cavity-magnon systems: from ultrastrong coupling to applications in precision measurement. New J. Phys. 21, 095004.

Forn-Díaz, P., Lamata, L., Rico, E., Kono, J., Solano, E., 2019. Ultrastrong coupling regimes of light-matter interaction. Rev. Mod. Phys. 91, 025005. https://doi.org/10.1103/RevModPhys.91.025005.

Fulling, S.A., Davies, P.C.W., 1976. Radiation from a moving mirror in two dimensional space-time: conformal anomaly. Proc. Roy. Soc. London A 348, 393–414.

Gambino, S., Mazzeo, M., Genco, A., Di Stefano, O., Savasta, S., Patanè, S., Ballarini, D., Mangione, F., Lerario, G., Sanvitto, D., et al., 2014. Exploring light-matter interaction phenomena under ultrastrong coupling regime. ACS Photonics 1, 1042–1048.

Gao, W., Shu, J., Qiu, C., Xu, Q., 2012. Excitation of plasmonic waves in graphene by guided-mode resonances. ACS Nano 6, 7806–7813.

Gao, W., Shi, G., Jin, Z., Zhang, Q., Ajayan, P.M., Kono, J., Xu, Q., 2013. Excitation and active control of propagating surface plasmon polaritons in graphene. Nano Lett. 13, 3698–3702.

Gao, W., Shu, J., Reichel, K., Nickel, D., He, X., Shi, G., Vajtai, R., Ajayan, P.M., Kono, J., Mittleman, D.M., Xu, Q., 2014. High-contrast terahertz wave modulation by gated graphene enhanced by extraordinary transmission through ring apertures. Nano Lett. 14, 1242–1248.

Gao, W., Li, X., Bamba, M., Kono, J., 2018. Continuous transition between weak and ultrastrong coupling through exceptional points in carbon nanotube microcavity exciton-polaritons. Nat. Photon. 12, 362–367.
Gao, H., Schlawin, F., Buzzi, M., Cavalleri, A., Jaksch, D., 2020. Photoinduced electron pairing in a driven cavity. Phys. Rev. Lett. 125, 053602.
Garziano, L., Ridolfo, A., Stassi, R., Di Stefano, O., Savasta, S., 2013. Switching on and off of ultrastrong light-matter interaction: photon statistics of quantum vacuum radiation. Phys. Rev. A 88, 63829.
Garziano, L., Stassi, R., Ridolfo, A., Di Stefano, O., Savasta, S., 2014. Vacuum-induced symmetry breaking in a superconducting quantum circuit. Phys. Rev. A 90, 43817.
Garziano, L., Settineri, A., Di Stefano, O., Savasta, S., Nori, F., 2020. Gauge invariance of the Dicke and Hopfield models. Phys. Rev. A 102, 23718.
Gaur, N.K.S., 1976. Longitudinal optical phonon-plasmon coupling in InSb and InAs. Physica B+C 82, 262–270.
Geiser, M., Walther, C., Scalari, G., Beck, M., Fischer, M., Nevou, L., Faist, J., 2010. Strong light-matter coupling at terahertz frequencies at room temperature in electronic LC resonators. Appl. Phys. Lett. 97, 191107.
Gawędzki, K., Rzążewski, K., 1981. No-go theorem for the superradiant phase transition without dipole approximation. Phys. Rev. A 23, 2134–2136.
Geiser, M., Castellano, F., Scalari, G., Beck, M., Nevou, L., Faist, J., 2012. Ultrastrong coupling regime and plasmon polaritons in parabolic semiconductor quantum wells. Phys. Rev. Lett. 108, 106402.
Gibbs, H.M., Khitrova, G., Koch, S.W., 2011. Exciton-polariton light-semiconductor coupling effects. Nat. Photon. 5, 273. 273.
Glauber, R.J., Lewenstein, M., 1991. Quantum optics of dielectric media. Phys. Rev. A 43, 467–491.
Goryachev, M., Farr, W.G., Creedon, D.L., Fan, Y., Kostylev, M., Tobar, M.E., 2014. High-cooperativity cavity QED with magnons at microwave frequencies. Phys. Rev. Appl. 2, 054002.
Graf, A., Tropf, L., Zakharko, Y., Zaumseil, J., Gather, M.C., 2016. Near-infrared exciton-polaritons in strongly coupled single-walled carbon nanotube microcavities. Nat. Commun. 7, 13078.
Graf, A., Held, M., Zakharko, Y., Tropf, L., Gather, M.C., Zaumseil, J., 2017. Electrical pumping and tuning of exciton-polaritons in carbon nanotube microcavities. Nat. Mater. 16, 911–917.
Grießer, T., Vukics, A., Domokos, P., 2016. Depolarization shift of the superradiant phase transition. Phys. Rev. A 94, 033815.
Grigorenko, A.N., Polini, M., Novoselov, K.S., 2012. Graphene plasmonics. Nat. Photon. 6, 749–758.
Guillet, T., Mexis, M., Levrat, J., Rossbach, G., Brimont, C., Bretagnon, T., Gil, B., Butté, R., Grandjean, N., Orosz, L., Réveret, F., Leymarie, J., Zúñiga-Pérez, J., Leroux, M., Semond, F., Bouchoule, S., 2011. Polariton lasing in a hybrid bulk ZnO microcavity. Appl. Phys. Lett. 99, 161104.
Günter, G., Anappara, A.A., Hees, J., Sell, A., Biasiol, G., Sorba, L., De Liberato, S., Ciuti, C., Tredicucci, A., Leitenstorfer, A., Huber, R., 2009. Sub-cycle switch-on of ultrastrong light-matter interaction. Nature 458, 178–181.
Hagenmüller, D., 2016. All-optical dynamical Casimir effect in a three-dimensional terahertz photonic band gap. Phys. Rev. B 93, 235309.
Hagenmüller, D., Ciuti, C., 2012. Cavity QED of the graphene cyclotron transition. Phys. Rev. Lett. 109, 267403.
Haigh, J.A., Lambert, N.J., Doherty, A.C., Ferguson, A.J., 2015. Dispersive readout of ferromagnetic resonance for strongly coupled magnons and microwave photons. Phys. Rev. B 91, 104410.

Harima, H., Nakashima, S.-I., Uemura, T., 1995. Raman scattering from anisotropic LO-phonon-plasmon-coupled mode in *n*-type 4H-and 6H-SiC. J. Appl. Phys. 78, 1996–2005.

Hasselbeck, M.P., Stalnaker, D., Schlie, L.A., Rotter, T.J., Stintz, A., Sheik-Bahae, M., 2002. Emission of terahertz radiation from coupled plasmon-phonon modes in InAs. Phys. Rev. B 65, 233203.

Hawking, S.W., 1974. Black hole explosions? Nature (London) 248, 30–31.

Hawking, S.W., 1975. Particle creation by black holes. Commun. Math. Phys. 43, 199–220.

Helm, M., 2000. The basic physics of intersubband transitions. In: Liu, H.C., Capasso, F. (Eds.), Intersubband Transitions in Quantum Wells: Physics and Device Applications I. Semiconductors and Semimetals, vol. 62. Academic Press, San Diego, pp. 1–99.

Hepp, K., Lieb, E.H., 1973. On the superradiant phase transition for molecules in a quantized radiation field: the Dicke maser model. Ann. Phys. 76, 360–404.

Hilton, D.J., Arikawa, T., Kono, J., 2012. Cyclotron resonance. In: Kaufmann, E.N. (Ed.), Characterization of Materials. John Wiley & Sons, Inc., New York, pp. 1–15.

Hoffman, A.J., Alekseyev, L., Howard, S.S., Franz, K.J., Wasserman, D., Podolskiy, V.A., Narimanov, E.E., Sivco, D.L., Gmachl, C., 2007. Negative refraction in semiconductor metamaterials. Nat. Mater. 6, 946–950.

Hopfield, J.J., 1958. Theory of the contribution of excitons to the complex dielectric constant of crystals. Phys. Rev. 112, 1555–1567.

Huang, J.F., Law, C.K., 2014. Photon emission via vacuum-dressed intermediate states under ultrastrong coupling. Phys. Rev. A 89, 033827.

Huebl, H., Zollitsch, C.W., Lotze, J., Hocke, F., Greifenstein, M., Marx, A., Gross, R., Goennenwein, S.T.B., 2013. High cooperativity in coupled microwave resonator ferrimagnetic insulator hybrids. Phys. Rev. Lett. 111, 127003.

Hwang, M.-J., Puebla, R., Plenio, M.B., 2015. Quantum phase transition and universal dynamics in the Rabi model. Phys. Rev. Lett. 115, 180404.

Jaako, T., Xiang, Z.-L., Garcia-Ripoll, J.J., Rabl, P., 2016. Ultrastrong-coupling phenomena beyond the Dicke model. Phys. Rev. A 94, 033850.

Jablan, M., Buljan, H., Soljačić, M., 2009. Plasmonics in graphene at infrared frequencies. Phys. Rev. B 80, 245435.

Jeannin, M., Mariotti Nesurini, G., Suffit, S., Gacemi, D., Vasanelli, A., Li, L., Davies, A.G., Linfield, E., Sirtori, C., Todorov, Y., 2019. Ultrastrong light-matter coupling in deeply subwavelength THz LC resonators. ACS Photonics 6, 1207–1215.

Jia, Y., Zhao, H., Guo, Q., Wang, X., Wang, H., Xia, F., 2015. Tunable plasmon-phonon polaritons in layered graphene-hexagonal boron nitride heterostructures. ACS Photonics 2, 907–912.

Jouy, P., Vasanelli, A., Todorov, Y., Delteil, A., Biasiol, G., Sorba, L., Sirtori, C., 2011. Transition from strong to ultrastrong coupling regime in mid-infrared metal-dielectric-metal cavities. Appl. Phys. Lett. 98, 231114.

Ju, L., Geng, B., Horng, J., Girit, C., Martin, M., Hao, Z., Bechtel, H.A., Liang, X., Zettl, A., Shen, Y.R., Wang, F., 2011. Graphene plasmonics for tunable terahertz metamaterials. Nat. Nanotechnol. 6, 630–634.

Juraschek, D.M., Neuman, T., Flick, J., Narang, P., 2019. Cavity control of nonlinear phononics. arXiv:1912.00122.

Kaiser, S., 2017. Light-induced superconductivity in high-Tc cuprates. Phys. Scr. 92, 103001.

Kaluzny, Y., Goy, P., Gross, M., Raimond, J.M., Haroche, S., 1983. Observation of self-induced Rabi oscillations in two-level atoms excited inside a resonant cavity: the ringing regime of superradiance. Phys. Rev. Lett. 51, 1175–1178.

Kardar, M., Golestanian, R., 1999. The "friction" of vacuum, and other fluctuation-induced forces. Rev. Mod. Phys. 71, 1233–1245.

Keeling, J., 2007. Coulomb interactions, gauge invariance, and phase transitions of the Dicke model. J. Phys. Condens. Matter 19, 295213.

Keller, J., Scalari, G., Cibella, S., Maissen, C., Appugliese, F., Giovine, E., Leoni, R., Beck, M., Faist, J., 2017. Few-electron ultrastrong light-matter coupling at 300 GHz with nanogap hybrid LC microcavities. Nano Lett. 17, 7410–7415.

Keller, J., Scalari, G., Appugliese, F., Mavrona, E., Rajabali, S., Süess, M.J., Beck, M., Faist, J., 2018. High Tc superconducting THz metamaterial for ultrastrong coupling in a magnetic field. ACS Photonics 5, 3977–3983. https://doi.org/10.1021/acsphotonics.8b01025.

Keller, J., Scalari, G., Appugliese, F., Rajabali, S., Beck, M., Haase, J., Lehner, C.A., Wegscheider, W., Failla, M., Myronov, M., Leadley, D.R., Lloyd-Hughes, J., Nataf, P., Faist, J., 2020. Landau polaritons in highly nonparabolic two-dimensional gases in the ultrastrong coupling regime. Phys. Rev. B 101, 075301.

Kéna-Cohen, S., Forrest, S.R., 2010. Room-temperature polariton lasing in an organic single-crystal microcavity. Nat. Photon. 4, 371–375.

Khitrova, G., Gibbs, H.M., Jahnke, F., Kira, M., Koch, S.W., 1999. Nonlinear optics of normal-mode-coupling semiconductor microcavities. Rev. Mod. Phys. 71, 1591–1639.

Khitrova, G., Gibbs, H.M., Kira, M., Koch, S.W., Scherer, A., 2006. Vacuum Rabi splitting in semiconductors. Nat. Phys. 2, 81–90. https://doi.org/10.1038/nphys227y.

Khurgin, J.B., 2015. How to deal with the loss in plasmonics and metamaterials. Nat. Nanotechnol. 10, 2–6.

Kiffner, M., Coulthard, J.R., Schlawin, F., Ardavan, A., Jaksch, D., 2019. Manipulating quantum materials with quantum light. Phys. Rev. B 99, 085116.

Kim, O.K., Spitzer, W.G., 1979. Study of plasmon-LO-phonon coupling in Te-doped $Ga_{1-x}Al_xAs$. Phys. Rev. B 20, 3258.

Klein, M.V., Ganguly, B.N., Colwell, P.J., 1972. Theoretical and experimental study of Raman scattering from coupled LO-phonon-plasmon modes in silicon carbide. Phys. Rev. B 6, 2380.

Knight, J.M., Aharonov, Y., Hsieh, G.T.C., 1978. Are super-radiant phase transitions possible? Phys. Rev. A 17, 1454–1462.

Kockum, A.F., Miranowicz, A., De Liberato, S., Savasta, S., Nori, F., 2019. Ultrastrong coupling between light and matter. Nat. Rev. Phys. 1, 19–40.

Kohn, W., 1961. Cyclotron resonance and de Haas-van Alphen oscillations of an interacting electron gas. Phys. Rev. 123, 1242.

Kono, J., 2001. Cyclotron resonance. In: Kaufmann, E.N., Abbaschian, R., Bocarsly, A., Chien, C.L., Dollimore, D., Doyle, B., et al. (Eds.), Methods in Materials Research. John Wiley & Sons, New York.

Koppens, F.H.L., Chang, D.E., Garcia de Abajo, F.J., 2011. Graphene plasmonics: a platform for strong light-matter interactions. Nano Lett. 11, 3370–3377.

Kozawa, T., Kachi, T., Kano, H., Taga, Y., Hashimoto, M., Koide, N., Manabe, K., 1994. Raman scattering from LO phonon-plasmon coupled modes in gallium nitride. J. Appl. Phys. 75, 1098–1101.

Kukharskii, A.A., 1973. Plasmon-phonon coupling in GaAs. Solid State Commun. 13, 1761–1765.

Kusminskiy, S.V., 2019a. Cavity optomagnonics. arXiv:1911.11104.

Kusminskiy, S.V., 2019b. Quantum Magnetism, Spin Waves, and Optical Cavities. Springer, Berlin.

Kyaw, T.H., Felicetti, S., Romero, G., Solano, E., Kwek, L.C., 2015a. Scalable quantum memory in the ultrastrong coupling regime. Sci. Rep. 5, 8621.

Kyaw, T.H., Herrera-Martí, D.A., Solano, E., Romero, G., Kwek, L.C., 2015b. Creation of quantum error correcting codes in the ultrastrong coupling regime. Phys. Rev. B 91, 064503.

Lamata, L., 2017. Digital-analog quantum simulation of generalized Dicke models with superconducting circuits. Sci. Rep. 7, 43768.
Larson, J., Irish, E.K., 2017. Some remarks on 'superradiant' phase transitions in light-matter systems. J. Phys. A Math. Theor. 50, 174002.
Latini, S., Ronca, E., De Giovannini, U., Hübener, H., Rubio, A., 2019. Cavity control of excitons in two-dimensional materials. Nano Lett. 19, 3473–3479.
Laurent, T., Manceau, J.M., Monroy, E., Lim, C.B., Rennesson, S., Semond, F., Julien, F.H., Colombelli, R., 2017. Short-wave infrared ($\lambda = 3\ \mu m$) intersubband polaritons in the GaN/AlN system. Appl. Phys. Lett. 110, 131102.
Law, S., Adams, D.C., Taylor, A.M., Wasserman, D., 2012. Mid-infrared designer metals. Opt. Express 20, 12155–12165.
Lax, B., Mavroides, J.G., 1960. Cyclotron resonance. In: Seitz, F., Turnbull, D. (Eds.), Solid State Physics. vol. 11. Academic Press, New York, pp. 261–400.
Lee, B., Liu, W., Naylor, C.H., Park, J., Malek, S.C., Berger, J.S., Johnson, A.T.C., Agarwal, R., 2017. Electrical tuning of exciton-plasmon polariton coupling in monolayer MoS_2 integrated with plasmonic nanoantenna lattice. Nano Lett. 17, 4541–4547.
Li, X., Bamba, M., Yuan, N., Zhang, Q., Zhao, Y., Xiang, M., Xu, K., Jin, Z., Ren, W., Ma, G., Cao, S., Turchinovich, D., Kono, J., 2018a. Observation of Dicke cooperativity in magnetic interactions. Science 361 (6404), 794–797.
Li, X., Bamba, M., Zhang, Q., Fallahi, S., Gardner, G.C., Gao, W., Lou, M., Yoshioka, K., Manfra, M.J., Kono, J., 2018b. Vacuum Bloch-Siegert shift in Landau polaritons with ultra-high cooperativity. Nat. Photon. 12, 324–329.
Lidzey, D.G., Bradley, D.D.C., Skolnick, M.S., Virgili, T., Walker, S., Whittaker, D.M., 1998. Strong exciton-photon coupling in an organic semiconductor microcavity. Nature 395, 53–55.
Liensberger, L., Kamra, A., Maier-Flaig, H., Geprägs, S., Erb, A., Goennenwein, S.T., Gross, R., Belzig, W., Huebl, H., Weiler, M., 2019. Exchange-enhanced ultrastrong magnon-magnon coupling in a compensated ferrimagnet. Phys. Rev. Lett. 123, 117204.
Liu, A., 1997. Rabi splitting of the optical intersubband absorption line of multiple quantum wells inside a Fabry-Pérot microcavity. Phys. Rev. B 55, 7101–7109.
Liu, X., Galfsky, T., Sun, Z., Xia, F., Lin, E.-c., Lee, Y.-H., Kéna-Cohen, S., Menon, V.M., 2015. Strong light-matter coupling in two-dimensional atomic crystals. Nat. Photon. 9, 30–34.
Liu, W., Lee, B., Naylor, C.H., Ee, H.-S., Park, J., Johnson, A.T.C., Agarwal, R., 2016. Strong exciton-plasmon coupling in MoS_2 coupled with plasmonic lattice. Nano Lett. 16, 1262–1269.
Lolli, J., Baksic, A., Nagy, D., Manucharyan, V.E., Ciuti, C., 2015. Ancillary qubit spectroscopy of vacua in cavity and circuit quantum electrodynamics. Phys. Rev. Lett. 114, 183601.
Luttinger, J.M., Kohn, W., 1955. Motion of electrons and holes in perturbed periodic fields. Phys. Rev. 97, 869–883.
Maag, T., Bayer, A., Baierl, S., Hohenleutner, M., Korn, T., Schller, C., Schuh, D., Bougeard, D., Lange, C., Huber, R., Mootz, M., Sipe, J.E., Koch, S.W., Kira, M., 2016. Coherent cyclotron motion beyond kohns theorem. Nature Phys. 12, 119–123.
MacNeill, D., Hou, J.T., Klein, D.R., Zhang, P., Jarillo-Herrero, P., Liu, L., 2019. Gigahertz frequency antiferromagnetic resonance and strong magnon-magnon coupling in the layered crystal $CrCl_3$. Phys. Rev. Lett. 123, 047204.
Maissen, C., Scalari, G., Valmorra, F., Beck, M., Faist, J., Cibella, S., Leoni, R., Reichl, C., Charpentier, C., Wegscheider, W., 2014. Ultrastrong coupling in the near field of complementary split-ring resonators. Phys. Rev. B 90, 205309.
Maissen, C., Scalari, G., Beck, M., Faist, J., 2017. Asymmetry in polariton dispersion as function of light and matter frequencies in the ultrastrong coupling regime. New J. Phys. 19, 043022.

Makihara, T., Hayashida, K., Noe II, G.T., Li, X., Marquez Peraca, N., Ma, X., Jin, Z., Ren, W., Ma, G., Katayama, I., Takeda, J., Nojiri, H., Turchinovich, D., Cao, S., Bamba, M., Kono, J., 2020. Magnonic quantum simulator of antiresonant ultrastrong light-matter coupling. arXiv:2008.10721.
Mazza, G., Georges, A., 2019. Superradiant quantum materials. Phys. Rev. Lett. 122, 017401.
McCombe, B.D., Wagner, R.J., 1975. Intraband magneto-optical studies of semiconductors in the far-infrared. I. In: Marton, L. (Ed.), Advances in Electronics and Electron Physics, 37. vol. 37. Academic Press, New York, pp. 1–78.
Mezzacapo, A., Las Heras, U., Pedernales, J.S., DiCarlo, L., Solano, E., Lamata, L., 2014. Digital quantum Rabi and Dicke models in superconducting circuits. Sci. Rep. 4, 7482.
Mooradian, A., McWhorter, A.L., 1967. Polarization and intensity of Raman scattering from plasmons and phonons in gallium arsenide. Phys. Rev. Lett. 19, 849.
Mooradian, A., Wright, G.B., 1966. Observation of the interaction of plasmons with longitudinal optical phonons in GaAs. Phys. Rev. Lett. 16, 999.
Moore, G.T., 1970. Quantum theory of the electromagnetic field in a variable-length one-dimensional cavity. J. Math. Phys. 11, 2679–2691.
Mueller, N.S., Okamura, Y., Vieira, B.G.M., Juergensen, S., Lange, H., Barros, E.B., Schulz, F., Reich, S., 2020. Deep strong light-matter coupling in plasmonic nanoparticle crystals. Nature 583, 780–784.
Muravev, V.M., Andreev, I.V., Kukushkin, I.V., Schmult, S., Dietsche, W., 2011. Observation of hybrid plasmon-photon modes in microwave transmission of coplanar microresonators. Phys. Rev. B 83, 075309.
Muravev, V.M., Gusikhin, P.A., Andreev, I.V., Kukushkin, I.V., 2013. Ultrastrong coupling of high-frequency two-dimensional cyclotron plasma mode with a cavity photon. Phys. Rev. B 87, 045307.
Nataf, P., Ciuti, C., 2010. No-go theorem for superradiant quantum phase transitions in cavity QED and counter-example in circuit QED. Nat. Commun. 1, 72.
Nataf, P., Ciuti, C., 2011. Protected quantum computation with multiple resonators in ultrastrong coupling circuit QED. Phys. Rev. Lett. 107, 190402.
Nataf, P., Champel, T., Blatter, G., Basko, D.M., 2019. Rashba cavity QED: a route towards the superradiant quantum phase transition. Phys. Rev. Lett. 123, 207402.
Ofek, N., Petrenko, A., Heeres, R., Reinhold, P., Leghtas, Z., Vlastakis, B., Liu, Y., Frunzio, L., Girvin, S.M., Jiang, L., Mirrahimi, M., Devoret, M.H., Schoelkopf, R.J., 2016. Extending the lifetime of a quantum bit with error correction in superconducting circuits. Nature 536, 441–445.
Olson, C.G., Lynch, D.W., 1969. Longitudinal-optical-phonon-plasmon coupling in GaAs. Phys. Rev. 177, 1231.
Paiella, R., 2006. Intersubband Transitions in Quantum Structures. McGraw-Hill, New York.
Paravicini-Bagliani, G.L., Scalari, G., Valmorra, F., Keller, J., Maissen, C., Beck, M., Faist, J., 2017. Gate and magnetic field tunable ultrastrong coupling between a magnetoplasmon and the optical mode of an LC cavity. Phys. Rev. B 95, 205304.
Paravicini-Bagliani, G.L., Appugliese, F., Richter, E., Valmorra, F., Keller, J., Beck, M., Bartolo, N., Rössler, C., Ihn, T., Ensslin, K., Ciuti, C., Scalari, G., Faist, J., 2018. Magneto-transport controlled by Landau polariton states. Nat. Phys. 15, 186–190.
Parvini, T.S., Bittencourt, V.A.S.V., Kusminskiy, S.V., 2020. Antiferromagnetic cavity optomagnonics. Phys. Rev. Res. 2, 022027.
Pedernales, J.S., Di Candia, R., Ballester, D., Solano, E., 2013. Quantum simulations of relativistic quantum physics in circuit QED. New J. Phys. 15, 055008.
Perkowitz, S., Thorland, R.H., 1975. Generalized dielectric function and the plasmon-phonon coupling in GaAs and CdTe. Solid State Commun. 16, 1093–1096.

Poddubny, A., Iorsh, I., Belov, P., Kivshar, Y., 2013. Hyperbolic metamaterials. Nat. Photon. 7, 948–957.
Porer, M., Ménard, J.-M., Leitenstorfer, A., Huber, R., Degl'Innocenti, R., Zanotto, S., Biasiol, G., Sorba, L., Tredicucci, A., 2012. Nonadiabatic switching of a photonic band structure: ultrastrong light-matter coupling and slow-down of light. Phys. Rev. B 85, 081302(R).
Puebla, R., Hwang, M.-J., Casanova, J., Plenio, M.B., 2017. Probing the dynamics of a superradiant quantum phase transition with a single trapped ion. Phys. Rev. Lett. 118, 073001.
Quattropani, A., Schwendimann, P., 2005. Polariton squeezing in microcavities. phys. stat. sol. (b) 242, 2302–2314.
Raines, Z.M., Allocca, A.A., Hafezi, M., Galitski, V.M., 2020. Cavity Higgs polaritons. Phys. Rev. Res. 2, 013143.
Rajabali, S., Scalari, G., Keller, J., Beck, M., Faist, J., 2019. The upper branch broadening in ultrastrongly coupled thz landau polaritons. In: Proc. of the 44th Int. Conf. Infrared, Millimeter, and Terahertz Waves (IRMMW-THz), p. 1. 1.
Ridolfo, A., Leib, M., Savasta, S., Hartmann, M.J., 2012. Photon blockade in the ultrastrong coupling regime. Phys. Rev. Lett. 109, 193602.
Romero, G., Ballester, D., Wang, Y.M., Scarani, V., Solano, E., 2012. Ultrafast quantum gates in circuit QED. Phys. Rev. Lett. 108, 120501.
Roth, M., Hassler, F., DiVincenzo, D.P., 2019. Optimal gauge for the multimode Rabi model in circuit QED. Phys. Rev. Res. 1, 033128.
Rzażewski, K., Wódkiewicz, K., Żakowicz, W., 1975. Phase transitions, two-level atoms, and the A^2 term. Phys. Rev. Lett. 35, 432–434.
Sachet, E., Shelton, C.T., Harris, J.S., Gaddy, B.E., Irving, D.L., Curtarolo, S., Donovan, B.F., Hopkins, P.E., Sharma, P.A., Sharma, A.L., et al., 2015. Dysprosium-doped cadmium oxide as a gateway material for mid-infrared plasmonics. Nat. Mater. 14, 414–420.
Sanchez-Burillo, E., Zueco, D., Garcia-Ripoll, J.J., Martin-Moreno, L., 2014. Scattering in the ultrastrong regime: nonlinear optics with one photon. Phys. Rev. Lett. 113, 263604.
Sapienza, L., Vasanelli, A., Ciuti, C., Manquest, C., Sirtori, C., Colombelli, R., Gennser, U., 2007. Photovoltaic probe of cavity polaritons in a quantum cascade structure. Appl. Phys. Lett. 90, 201101.
Sapienza, L., Vasanelli, A., Colombelli, R., Ciuti, C., Chassagneux, Y., Manquest, C., Gennser, U., Sirtori, C., 2008. Electrically injected cavity polaritons. Phys. Rev. Lett. 100, 136806.
Scalari, G., Maissen, C., Turčinková, D., Hagenmüller, D., De Liberato, S., Ciuti, C., Reichl, C., Schuh, D., Wegscheider, W., Beck, M., Faist, J., 2012. Ultrastrong coupling of the cyclotron transition of a 2D electron gas to a THz metamaterial. Science 335, 1323–1326.
Scalari, G., Maissen, C., Hagenmüller, D., De Liberato, S., Ciuti, C., Reichl, C., Wegscheider, W., Schuh, D., Beck, M., Faist, J., 2013. Ultrastrong light-matter coupling at terahertz frequencies with split ring resonators and inter-Landau level transitions. J. Appl. Phys. 113, 136510.
Schlawin, F., Cavalleri, A., Jaksch, D., 2019. Cavity-mediated electron-photon superconductivity. Phys. Rev. Lett. 122, 133602.
Schneider, H., Liu, H.C., 2007. Quantum well infrared photodetectors. Springer, Berlin.
Schuler, M., De Bernardis, D., Läuchli, A.M., Rabl, P., 2020. The vacua of dipolar cavity quantum electrodynamics. arXiv:2004.13738v3.
Schwendimann, P., Quattropani, A., 1992a. Nonclassical properties of polariton states. Europhys. Lett. 17, 355–358.
Schwendimann, P., Quattropani, A., 1992b. Nonclassical properties of polariton states. Europhys. Lett. 18, 281. 281.

Sentef, M.A., Ruggenthaler, M., Rubio, A., 2018. Cavity quantum-electrodynamical polaritonically enhanced electron-phonon coupling and its influence on superconductivity. Sci. Adv. 4, eaau6969.
Shang, Q., Zhang, S., Liu, Z., Chen, J., Yang, P., Li, C., Li, W., Zhang, Y., Xiong, Q., Liu, X., et al., 2018. Surface plasmon enhanced strong exciton-photon coupling in hybrid inorganic-organic perovskite nanowires. Nano Lett. 18, 3335–3343.
Shapiro, D.S., Zhukov, A.A., Pogosov, W.V., Lozovik, Y.E., 2015. Dynamical Lamb effect in a tunable superconducting qubit-cavity system. Phys. Rev. A 91, 1–10.
Shirley, J.H., 1965. Solution of the Schrödinger equation with a Hamiltonian periodic in time. Phys. Rev. 138, B979–B987.
Singwi, K.S., Tosi, M.P., 1966. Interaction of plasmons and optical phonons in degenerate semiconductors. Phys. Rev. 147, 658.
Skolnick, M.S., Fisher, T.A., Whittaker, D.M., 1998. Strong coupling phenomena in quantum microcavity structures. Semicond. Sci. Technol. 13, 645–669.
Stassi, R., Ridolfo, A., Di Stefano, O., Hartmann, M.J., Savasta, S., 2013. Spontaneous conversion from virtual to real photons in the ultrastrong-coupling regime. Phys. Rev. Lett. 110, 243601.
Stassi, R., Savasta, S., Garziano, L., Spagnolo, B., Nori, F., 2016. Output field-quadrature measurements and squeezing in ultrastrong cavity-QED. New J. Phys. 18, 123005.
Stefano, O.D., Stassi, R., Garziano, L., Kockum, A.F., Savasta, S., Nori, F., 2017. Feynman-diagrams approach to the quantum Rabi model for ultrastrong cavity QED: stimulated emission and reabsorption of virtual particles dressing a physical excitation. New J. Phys. 19, 053010.
Stokes, A., Nazir, A., 2019a. Gauge ambiguities imply Jaynes-Cummings physics remains valid in ultrastrong coupling QED. Nat. Commun. 10, 499.
Stokes, A., Nazir, A., 2019. Ultrastrong time-dependent light-matter interactions are gauge-relative. arXiv:1902.05160v2. 1902.05160v2.
Stokes, A., Nazir, A., 2020. Uniqueness of the phase transition in many-dipole cavity quantum electrodynamical systems. Phys. Rev. Lett. 125, 143603.
Su, R., Diederichs, C., Wang, J., Liew, T.C.H., Zhao, J., Liu, S., Xu, W., Chen, Z., Xiong, Q., 2017. Room-temperature polariton lasing in all-inorganic perovskite nanoplatelets. Nano Lett. 17, 3982–3988.
Su, R., Wang, J., Zhao, J., Xing, J., Zhao, W., Diederichs, C., Liew, T.C.H., Xiong, Q., 2018. Room temperature long-range coherent exciton polariton condensate flow in lead halide perovskites. Sci. Adv. 4, eaau0244.
Su, R., Ghosh, S., Wang, J., Liu, S., Diederichs, C., Liew, T.C., Xiong, Q., 2020. Observation of exciton polariton condensation in a perovskite lattice at room temperature. Nat. Phys. 16, 301–306.
Sun, Z., Gu, J., Ghazaryan, A., Shotan, Z., Considine, C.R., Dollar, M., Chakraborty, B., Liu, X., Ghaemi, P., Kéna-Cohen, S., et al., 2017. Optical control of room-temperature valley polaritons. Nat. Photon. 11, 491.
Sun, Z., Fogler, M.M., Basov, D.N., Millis, A.J., 2020. Collective modes and terahertz near-field response of superconductors. Phys. Rev. Res. 2, 023413.
Tabuchi, Y., Ishino, S., Ishikawa, T., Yamazaki, R., Usami, K., Nakamura, Y., 2014. Hybridizing ferromagnetic magnons and microwave photons in the quantum limit. Phys. Rev. Lett. 113, 083603.
Thomas, A., Devaux, E., Nagarajan, K., Chervy, T., Seidel, M., Hagenmüller, D., Schütz, S., Schachenmayer, J., Genet, C., Pupillo, G., et al., 2019. Exploring superconductivity under strong coupling with the vacuum electromagnetic field. arXiv: 1911.01459.

Tobar, M.E., Goryachev, M., 2019. Microwave frequency magnetic field manipulation systems and methods and associated application instruments, apparatus and system. US Patent 10,290,916.
Todorov, Y., 2014. Dipolar quantum electrodynamics theory of the three-dimensional electron gas. Phys. Rev. B 89, 75115.
Todorov, Y., Sirtori, C., 2012. Intersubband polaritons in the electrical dipole gauge. Phys. Rev. B 85, 45304.
Todorov, Y., Sirtori, C., 2014. Few-electron ultrastrong light-matter coupling in a quantum LC circuit. Phys. Rev. X 4, 041031.
Todorov, Y., Andrews, A.M., Sagnes, I., Colombelli, R., Klang, P., Strasser, G., Sirtori, C., 2009. Strong light-matter coupling in subwavelength metal-dielectric microcavities at terahertz frequencies. Phys. Rev. Lett. 102, 186402.
Todorov, Y., Andrews, A.M., Colombelli, R., De Liberato, S., Ciuti, C., Klang, P., Strasser, G., Sirtori, C., 2010a. Ultrastrong light-matter coupling regime with polariton dots. Phys. Rev. Lett. 105, 196402.
Todorov, Y., Tosetto, L., Teissier, J., Andrews, A.M., Klang, P., Colombelli, R., Sagnes, I., Strasser, G., Sirtori, C., 2010b. Optical properties of metal-dielectric-metal microcavities in the THz fquency range. Opt. Express 18, 13886–13907.
Tokura, Y., Kawasaki, M., Nagaosa, N., 2017. Emergent functions of quantum materials. Nat. Phys. 13, 1056–1068.
Unruh, W.G., 1974. Second quantization in the Kerr metric. Phys. Rev. D 10, 3194–3205.
Unruh, W.G., 1976. Notes on black-hole evaporation. Phys. Rev. D 14, 870–892.
van Vugt, L.K., Rühle, S., Ravindran, P., Gerritsen, H.C., Kuipers, L., Vanmaekelbergh, D., 2006. Exciton polaritons confined in a ZnO nanowire cavity. Phys. Rev. Lett. 97, 147401.
Varga, B.B., 1965. Coupling of plasmons to polar phonons in degenerate semiconductors. Phys. Rev. 137, A1896.
Viehmann, O., von Delft, J., Marquardt, F., 2011. Superradiant phase transitions and the standard description of circuit QED. Phys. Rev. Lett. 107, 113602.
Vukics, A., Domokos, P., 2012. Adequacy of the Dicke model in cavity QED: a counter-no-go statement. Phys. Rev. A 86, 53807.
Vukics, A., Grießer, T., Domokos, P., 2014. Elimination of the *A*-square problem from cavity QED. Phys. Rev. Lett. 112, 073601.
Vukics, A., Grießer, T., Domokos, P., 2015. Fundamental limitation of ultrastrong coupling between light and atoms. Phys. Rev. A 92, 043835.
Wang, Y.K., Hioe, F.T., 1973. Phase transition in the Dicke model of superradiance. Phys. Rev. A 7, 831–836.
Wang, Y.-D., Chesi, S., Loss, D., Bruder, C., 2010. One-step multiqubit Greenberger-Horne-Zeilinger state generation in a circuit QED system. Phys. Rev. B 81, 104524.
Wang, T., Li, P., Hauer, B., Chigrin, D.N., Taubner, T., 2013. Optical properties of single infrared resonant circular microcavities for surface phonon polaritons. Nano Lett. 13, 5051–5055.
Wang, Y., Zhang, J., Wu, C., You, J.Q., Romero, G., 2016. Holonomic quantum computation in the ultrastrong-coupling regime of circuit QED. Phys. Rev. A 94, 012328.
Wang, J., Su, R., Xing, J., Bao, D., Diederichs, C., Liu, S., Liew, T.C., Chen, Z., Xiong, Q., 2018. Room temperature coherently coupled exciton-polaritons in two-dimensional organic-inorganic perovskite. ACS Nano 12, 8382–8389.
Wang, X., Ronca, E., Sentef, M.A., 2019. Cavity quantum electrodynamical Chern insulator: towards light-induced quantized anomalous Hall effect in graphene. Phys. Rev. B 99, 235156.

Wei, H.-S., Jaing, C.-C., Chen, Y.-T., Lin, C.-C., Cheng, C.-W., Chan, C.-H., Lee, C.-C., Chang, J.-F., 2013. Adjustable exciton-photon coupling with giant Rabi-splitting using layer-by-layer J-aggregate thin films in all-metal mirror microcavities. Opt. Express 21, 21365–21373.

Weisbuch, C., Nishioka, M., Ishikawa, A., Arakawa, Y., 1992. Observation of the coupled exciton-photon mode splitting in a semiconductor quantum microcavity. Phys. Rev. Lett. 69, 3314–3317.

West, L.C., Eglash, S.J., 1985. First observation of an extremely large-dipole infrared transition within the conduction band of a GaAs quantum well. Appl. Phys. Lett. 46, 1156–1158.

Wilson, C.M., Johansson, G., Pourkabirian, A., Simoen, M., Johansson, J.R., Duty, T., Nori, F., Delsing, P., 2011. Observation of the dynamical Casimir effect in a superconducting circuit. Nature 479 (7373), 376–379.

Woolley, R.G., 1976. Gauge invariance and the thermodynamics of the electromagnetic field. J. Phys. A. Math. Gen. 9, L15–L16.

Yablonovitch, E., 1989. Accelerating reference frame for electromagnetic waves in a rapidly growing plasma: Unruh-Davies-Fulling-DeWitt radiation and the nonadiabatic Casimir effect. Phys. Rev. Lett. 62, 1742–1745.

Yamanoi, M., 1976. Influence of omitting the A^2 term in the conventional photon-matter-Hamiltonian on the photon-field equation. Phys. Lett. A 58, 437–439.

Yamanoi, M., Takatsuji, M., 1978. Influence of omitting the P^2 term in the multipole photon-matter Hamiltonian on the stability and propagation. In: Mandel, L., Wolf, E. (Eds.), Coherence Quantum Opt. IV Proc. Fourth Rochester Conf. Coherence Quantum Opt, Plenum Press, New York, pp. 839–850.

Yee, C.M., Sherwin, M.S., 2009. High-Q terahertz microcavities in silicon photonic crystal slabs. Appl. Phys. Lett. 94, 154104.

Yokota, I., 1961. On the coupling between optical lattice vibrations and carrier plasma oscillations in polar semiconductors. J. Phys. Soc. Jpn 16, 2075. 2075.

Yoo, D., de León-Pérez, F., Lee, I.-H., Mohr, D.A., Pelton, M., Raschke, M.B., Caldwell, J.D., Martín-Moreno, L., Oh, S.-H., 2020. Ultrastrong plasmon-phonon coupling via epsilon-near-zero nanocavities. arXiv:2003.00136v2.

Yoshihara, F., Fuse, T., Ashhab, S., Kakuyanagi, K., Saito, S., Semba, K., 2017. Superconducting qubit-oscillator circuit beyond the ultrastrong-coupling regime. Nat. Phys. 13, 44–47.

Yoshioka, D., 2002. The Quantum Hall Effect. Springer, Berlin and New York, ISBN: 9783540431152.

Yoxall, E., Schnell, M., Nikitin, A.Y., Txoperena, O., Woessner, A., Lundeberg, M.B., Casanova, F., Hueso, L.E., Koppens, F.H.L., Hillenbrand, R., 2015. Direct observation of ultraslow hyperbolic polariton propagation with negative phase velocity. Nat. Photon. 9, 674–678.

Zanotto, S., Biasiol, G., Degl'Innocenti, R., Sorba, L., Tredicucci, A., 2010. Intersubband polaritons in a one-dimensional surface plasmon photonic crystal. Appl. Phys. Lett. 97, 231123.

Zanotto, S., Degl'Innocenti, R., Xu, J.-H., Sorba, L., Tredicucci, A., Biasiol, G., 2012. Ultrafast optical bleaching of intersubband cavity polaritons. Phys. Rev. B 86, 201302.

Zhang, Q., Arikawa, T., Kato, E., Reno, J.L., Pan, W., Watson, J.D., Manfra, M.J., Zudov, M.A., Tokman, M., Erukhimova, M., Belyanin, A., Kono, J., 2014a. Superradiant decay of cyclotron resonance of two-dimensional electron gases. Phys. Rev. Lett. 113, 047601.

Zhang, X., Zou, C.-L., Jiang, L., Tang, H.X., 2014b. Strongly coupled magnons and cavity microwave photons. Phys. Rev. Lett. 113, 156401.

Zhang, Q., Lou, M., Li, X., Reno, J.L., Pan, W., Watson, J.D., Manfra, M.J., Kono, J., 2016a. Collective non-perturbative coupling of 2D electrons with high-quality-factor terahertz cavity photons. Nat. Phys. 12, 1005–1011.

Zhang, X., Zhu, N., Zou, C.-L., Tang, H.X., 2016b. Optomagnonic whispering gallery microresonators. Phys. Rev. Lett. 117, 123605.

Zhang, S., Shang, Q., Du, W., Shi, J., Wu, Z., Mi, Y., Chen, J., Liu, F., Li, Y., Liu, M., et al., 2018. Strong exciton-photon coupling in hybrid inorganic-organic perovskite micro/nanowires. Adv. Opt. Mater. 6, 1701032.

Zhang, X., Galda, A., Han, X., Jin, D., Vinokur, V.M., 2020. Broadband nonreciprocity enabled by strong coupling of magnons and microwave photons. Phys. Rev. Appl. 13, 044039.

Zhong, Y., Malagari, S.D., Hamilton, T., Wasserman, D.M., 2015. Review of mid-infrared plasmonic materials. J. Nanophoton. 9, 093791.

CHAPTER FOUR

Quantum integrated photonic circuits

Samir Bounouar[a], Marcelo Davanco[b], and Stephan Reitzenstein[a,*]
[a]Technische Universität Berlin, Institute of Solid State Physics, Berlin, Germany
[b]National Institute of Standards and Technology, Gaithersburg, MD, United States
[*]Corresponding author: e-mail address: stephan.reitzenstein@physik.tu-berlin.de

Contents

Semiconductors and Semimetals, Volume 105
ISSN 0080-8784
https://doi.org/10.1016/bs.semsem.2020.09.003

1. Introduction

Integrated quantum photonic circuits (IQPCs) are key elements in the emerging field of quantum technologies. They provide exciting opportunities to realize quantum devices with high functionality in a scalable, robust, energy efficiency, and compact fashion. Similar to the development in of integrated electronic circuits in the 1960s it is expected that IQPCs will boost the progress in the field and lead to high-performance quantum systems to perform for instance quantum computation and quantum simulations (Georgescu et al., 2014; Knill et al., 2001). Here, scaling up to quantum operations with $N \gtrsim 50$ qubits promises to outperform classical computers and to enter the regime of quantum supremacy (Arute et al., 2019; Markov et al., 2018). The overarching motivation for the development of IQPCs is based on the efforts to reduce complex and sensitive laboratory structures to a chip level. As with electronically integrated circuits, this leads to a significantly increased functional density and is the basis for successfully transferring exciting basic experiments into real-world applications of quantum technology with low fabrication and operation costs.

Depending on the intended task, IQPCs have different levels of complexity, but they have in common that the relevant information is transmitted within the circuits by means of individual photons. In the simplest case, they only contain photonic waveguide (WG) structures, beam splitters, and possibly phase shifters to perform basic quantum operations (Knill et al., 2001; Politi et al., 2008). However, much more complex architectures are required for many goals and advanced applications, such as the generation of 2D photonic cluster states, multidimensional quantum entanglement and the realization photonic quantum computers (Briegel and Raussendorf, 2001; Harris et al., 2017; Nielsen, 2004; Raussendorf and Briegel, 2001; Wang et al., 2018b). In addition to simple WG structures, fully integrated IQPCs ideally also include active elements such as single-photon emitters, which in the best case are also excited directly on the chip electrically via carrier injection or optically via integrated micro- or nanolasers. Such circuits quickly become more complex and require sophisticated design concepts and manufacturing processes. In addition, the combination of several material systems in heterogeneous approaches is often required in order to be able to map all the necessary functionalities. For example, it makes sense to combine quantum dots (QDs) in the GaAs material system in hybrid architectures with loss-minimized silicon waveguides in order to, respectively, generate single photons at high rate and to transfer and manipulate them in the silicon waveguides with low optical losses (Davanco et al., 2017; Kim et al., 2017; Schnauber et al., 2019; Zadeh et al., 2016). Similarly, it is attractive to combine superconducting materials with waveguide materials to implement on-chip single-photon detectors (Pernice et al., 2012; Reithmaier et al., 2015a) and recently even a fully integrated Hanbury Brown and Twiss experiment (Schwartz et al., 2018).

The development of IQPCs promises exciting opportunities across the board to implement components with high quantum functionality. In addition to the underlying ideas and sophisticated numerical design methods, advanced manufacturing methods must be developed and applied. In some cases, methods of classic integrated photonics can be used to implement low loss silicon WGs, for example. For the envisioned operation in the quantum regime of individual emitters and individual photons, however, new types of technology concepts have to be established. For example, the targeted and scalable integration of individual quantum emitters in IQPCs represents a major challenge that can only be solved with new approaches in the field of quantum technology and quantum engineering. In this context, deterministic manufacturing processes have been developed in recent years to select individual quantum emitters based on their optical properties and to integrate them into WGs with nanometer accuracy. In addition, methods for spectral control of individual quantum emitters have been established that form the basis for IQPCs whose functionality is based on multiphoton interference, when implementing for instance large-scale boson sampling circuits or scalable quantum computing using the Knill–Laflamme–Milburn (KLM) scheme (Knill et al., 2001; Tillmann et al., 2013). In addition, it has been proposed to implement IQPCs using quantum physical effects such as entanglement to generate 2D photonic cluster states as key resource for universal measurement-based quantum computation (Van den Nest et al., 2006). These concepts and many other approaches in the field of photonic quantum technology benefit decisively from the integration of optical components in functional IQPCs, which explains the high interest and the timely importance of developing such quantum photonic elements.

Against this background, this chapter shows important basics and current developments in the field of IQPCs. The aim is to give undergraduate students, doctoral students and interested scientists from related subject areas a well-founded insight into the topic and also to show the enormous potential and future developments of integrated quantum nanophotonics.

Following the thematic introduction in the present section, Section 2 introduces numerical methods are presented that are used for the design and functional optimization of corresponding quantum optical circuits. Building on this, Section 3 gives an insight into modern manufacturing methods for IQPCs. One focus here is on novel and unconventional approaches that have been developed to integrate quantum emitters in a highly controlled manner as single-photon sources in quantum circuits. Sections 4 and 5 include an overview of the status and current developments in the development of quantum circuits based on homogeneous material compositions or on the basis of heterogeneous component approaches. Homogeneous approaches have the advantage that they are generally easier to implement, but their performance is often limited to individual functionalities enabled by a single type of material. In contrast, heterogeneous concepts have a significantly greater development and optimization potential. For example, quantum emitters, low-loss WGs and efficient single-photon detectors made of different materials can be combined on a chip in front of fully integrated solutions. The chapter closes with an outlook in Section 6 on future developments and challenges to be solved toward fully integrated IQPCs with full quantum functionality and a summary of the chapter.

2. Numeric modeling and optimization

This section discusses numerical methods that are used for the design and optimization of WG systems. The focus is on systems with integrated quantum emitters, in which the coupling efficiency between emitter and WG is usually the figure of merit. For instance, the chiral light–matter interaction can be maximized for a directional light coupling into WGs at certain emitter positions. Further examples of numerical optimization are maximizing the photon extraction efficiency via grating couplers normal to the surface or lateral coupling elements and improving the structural design of nanocavities to maximize their quality (Q) factors. Typical numerical approaches that are used here are the finite difference time domain (FDTD) and the finite element method (FEM) concepts.

2.1 Finite difference time domain

The FDTD method (Taflove and Hagness, 2005) is one of the most popular methods for electromagnetic simulation and is currently available from multiple commercial as well as open sources (Oskooi et al., 2010).

As the name indicates, the method solves Maxwell's equations in time domain in three-dimensional space, allowing straightforward simulation of general, complex electromagnetic and photonic devices with a great degree of detail.

Starting from the differential form of the equations, both time- and space-derivatives are approximated with a finite differences scheme in a spatio-temporal grid. The standard grid, used in most implementations of the method, is generated from the so-called Yee cell, defined such that the electric and magnetic fields are staggered in time and in space, and sampled at spatial locations offset by a half pixel. Such a prescription allows space- and time-derivatives to be expressed in the center-difference scheme (Oskooi et al., 2010; Taflove and Hagness, 2005).

The method allows almost arbitrary geometries to be defined by assigning spatially varying electric permittivity ($\epsilon(\mathbf{r})$) and magnetic permeability ($\mu(\mathbf{r})$) values across the grid. Importantly, because the discretization grid is regular, mesh refinement is not as flexible and efficient as in the FEM approach. Nonetheless, simulation of dispersive, anisotropic, nonlinear, and time-varying materials is possible. The spatially finite computational domain can furthermore support a variety of boundary conditions (e.g., perfect electric or magnetic conductor, symmetry conditions), and open boundaries can be implemented by introduction of perfectly matched layers (PMLs). A PML is a finite, absorptive electromagnetic medium that is placed side-by-side with the computation domain, while being matched to it. The matching condition means that waves emanating from the computation domain are not reflected at the PML interface, such that the latter, being sufficiently thick to absorb the wave completely, emulates an infinite half-space. The FDTD method allows excitation of electromagnetic structures with time-varying, spatially distributed electric or magnetic current sources, as well as with illumination field distributions, such as plane waves, Gaussian, beams, and WG modes. In general, the temporal shape of the excitation source determines the spectral range over which the simulation is valid. Importantly, a minimum time step length must generally be met, depending primarily on the spatial discretization with respect to the wavelength, though also on boundary conditions and material properties, in order to ensure stability of the simulation.

In an FDTD simulation, the electromagnetic field is obtained from the previous time step over the entire computational domain, so that computation of relevant quantities such as electric and magnetic energy densities, Poynting vector, absorbed power, etc., at each time step is also possible.

FDTD is commonly used for the calculation of resonant cavity modes (Vuckovic et al., 1999). Typically In this case, a point dipole source with a short temporal envelope is placed within the cavity, and illuminates it with a broad range of frequencies. If the cavity supports a resonance at a particular frequency within the range, it will be excited by the source. The time-evolution of the cavity field in this case will display a sustained oscillation at the resonance frequency, with an exponential decay associated with energy escape from the cavity, and corresponding to its quality factor Q. In the frequency domain, this decay appears as a Lorentzian peak centered at the resonance frequency.

In contrast with frequency-domain methods, FDTD can allow the determination of the spectral response of a simulated system, within a specified frequency range, with a single run. After completion of the simulation, the time-dependence of the electromagnetic field and derived quantities can be converted to the frequency domain via Fourier transformation and normalization by the time-varying excitation source's spectrum. As an example, Fig. 20C in Section 5 shows the steady-state, simulated electric field amplitude for the on-chip adiabatic mode transformer demonstrated in Davanco et al. (2017), at a single wavelength. The mode transformer consisted of a GaAs ridge stacked on a Si_3N_4 ridge, both of which featured slow width tapers along the propagation direction. The simulation consisted of launching the hybrid WG mode in Fig. 20A at the device input, and calculating the power delivered to the mode in Fig. 20C, at the output. Fig. 1A shows the calculated modal conversion spectrum for the structure, obtained from a single simulation run. As another example, Fig. 1B shows results of a simulation of an electric point dipole radiating at the center of the GaAs ridge in Fig. 20A, which supports the transverse electric (TE) mode in the bottom panel. Fig. 1B shows the fraction β of total emitted dipole power that couples to the TE mode, as a function of wavelength, as well as GaAs ridge width. The wavelength dependence here was obtained by Fourier transforming the time-dependent electromagnetic field, for each dipole position.

2.2 Finite element method

In the case of FEM simulations, the Maxwell equations are solved under boundary conditions that are given by the device geometry, see, e.g., Piprek (2018). Suitable source terms can be taken into account that approximate the emission properties of quantum emitters. The given problem is

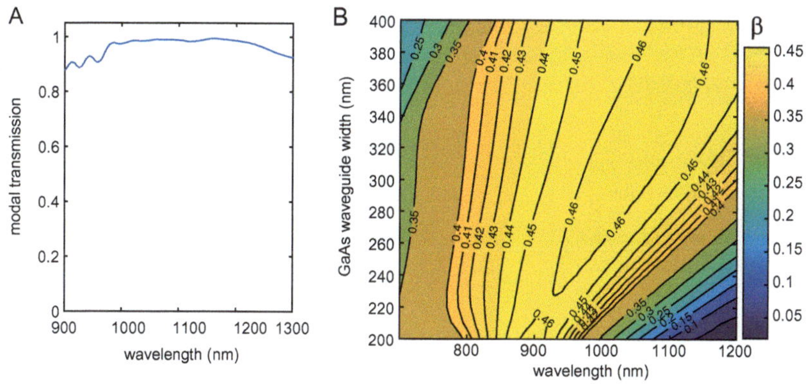

Fig. 1 Examples of FDTD simulation results. (A) Modal transmission for the mode transformer from Davanco et al. (2017), illustrated in Fig. 20C. (B) Fraction β of the total power emitted by a dipole in the hybrid WG of Fig. 20A, coupled into the fundamental TE mode for the hybrid WG of Fig. 20B. *Reprinted from Davanco, M., Liu, J., Sapienza, L., Zhang, C.-Z., De Miranda Cardoso, J.V., Verma, V., Mirin, R., Nam, S.W., Liu, L., Srinivasan, K., 2017. Heterogeneous integration for on-chip quantum photonic circuits with single quantum dot devices. Nat. Commun. 8 (1), 889. https://doi.org/10.1038/s41467-017-00987-6 with permission from Springer Nature.*

subdivided into finite elements, or small, easy-to-handle domains with locally defined boundary conditions and polynomial field-approximating basis functions. Taken together, finite element piecewise-continuous field solutions can be obtained for the entire domain. Common examples of finite elements are triangles and rectangles in 2D and tetrahedrons and cubes in 3D, see, e.g., Fig. 2A and B. In modern FEM solvers, these domains, i.e., the computational meshes, are dynamically adjusted during the calculation in order to adapted to the multiscale properties of the photonic structures and to minimize the necessary computing time for a given computational accuracy. Before calculating solutions of the problem these domains are pieced together in a way that tangential continuity of the electric and magnetic field across the boundaries of neighboring patches are ensured. Noteworthy, while FEM applies a discretization of the geometry it does not use any approximation to Maxwell's equations itself. In fact, it approximates the solution space in which one seeks a reasonable approximation to the exact solution. Advantages of the FEM concept include the accurate treatment of complex geometrical shapes, a flexible and iterative refinement of the finite element mesh by considering details of intermediate solutions, and the availability of high order approximations to ensure fast convergence.

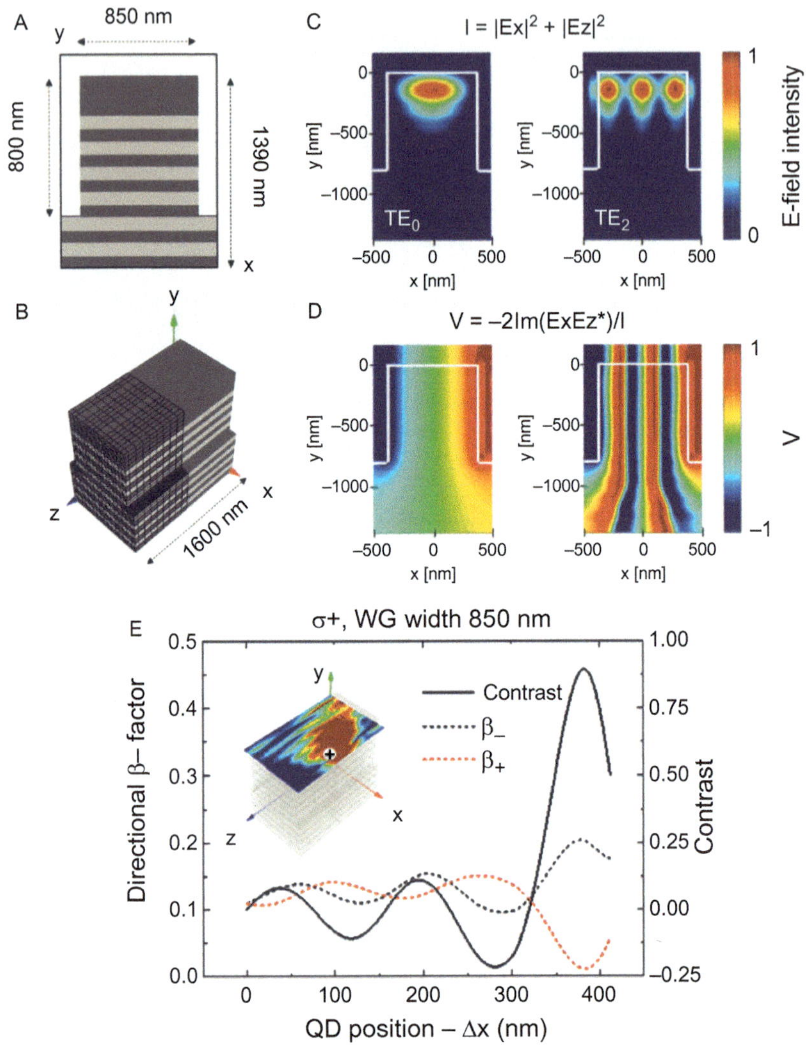

Fig. 2 Numeric device modeling using the FEM approach. (A) Cross section of the considered ridge WG system. The WG has a width of 850 nm and the layer design consists of lower DBR with 8 AlGaAs/GaAs (*dark gray/gray*) mirror pairs followed by 230 nm of GaAs (*dark gray*), which includes the QD layer (not shown). (B) 3D representation of the considered model structure including information about the used computing domains. Calculated intensity distributions (C) and degree of circular polarization (D) for the fundamental and a 2nd order TE-like mode. (E) Calculated directional coupling factor and directionality contrast in dependence of lateral dipole (quantum emitter) displacement Δx. *Inset*: Visualization (cross section) through the electric field distribution for the dipole placed at $\Delta x = 350$ nm. *Reprinted from Mrowinski, P., Schnauber, P., Gutsche, P., Kaganskiy, A., Schall, J., Burger, S., Rodt, S., Reitzenstein, S., 2019. Directional emission of a deterministically fabricated quantum dot-Bragg reflection multimode waveguide system. ACS Photon. 6 (9), 2231–2237 with permission from American Chemical Society.*

Overall, FEM constitutes a likewise flexible and powerful concept to numerically model to optical properties of nanophotonic devices and integrated photonic circuits.

To illustrate and underline the suitability of FEM for the numeric optimization of photonic WG systems we present results on chiral light–matter interaction and directional coupling in QD-based ridge WG circuits. We consider a GaAs-based ridge WG system as an easy-to-fabricate model system for the studying and maximizing the directional emission of a quantum emitter coupled to confined modes in the WG, where the lateral displacement of the emitter Δx acts as parameter. The layer design includes a lower AlGaAs/GaAs distributed Bragg reflector (DBR) to reduce photon losses into the substrate and GaAs capping layer which includes the quantum emitter at the vertical field maximum of the WG modes, see Fig. 3A. The computational geometry is shown in panel (B) which also indicates the used mesh and coordinates. A FEM solver was used to calculate the propagating eigenmodes for a WG width of 850 nm and a height of 800 nm at a wavelength of 930 nm. The corresponding electric field distributions $I = |E_x|^2 + |E_z|^2$ are presented in Fig. 3C for modes TE_0 and TE_2, where the different mode index is clearly reflected in the corresponding mode pattern. Beyond the simple calculation of intensity profiles FEM modeling also provides additional information like on the corresponding degree of polarization. This aspect is presented in panel (D), where the degree of circular polarization $V = -Im(E_x E_z^*)/I$ is plotted in the cross section of the considered WG geometry for modes TE_0 and TE_2. While for the mode TE_0 a continuous transition from negative to positive values of V and oscillatory behavior is observed for and TE_2. It turns out that this oscillatory behavior of the circular polarization together with the peculiar mode profile of the TE_2 leads to pronounced chiral light–matter interaction as quantified in the directional β-factor $\beta_\pm = \frac{\gamma_\pm}{\gamma_+ + \gamma_- + \Gamma_{out}}$ with the decay rates $\gamma_\pm$ into left and right propagation modes and Γ_{out} the residual recombination rate by all other channels including losses due to nonguided modes and nonradiative intrinsic recombination, and the directionality contrast $C = \frac{\gamma_- - \gamma_+}{\gamma_- + \gamma_+}$ (Lodahl et al., 2015). Both parameters are plotted in Fig. 3D for TE_2 as function of the emitter's lateral displacement Δx and show an oscillatory behavior with maximum directionality at an emitter position close to the edge of the WG, i.e., at $\Delta x \approx 380$ nm (Mrowinski et al., 2019). Thus, the numerical modeling here shows the existence of a chiral point near the edges of the WG, for which directional emission is

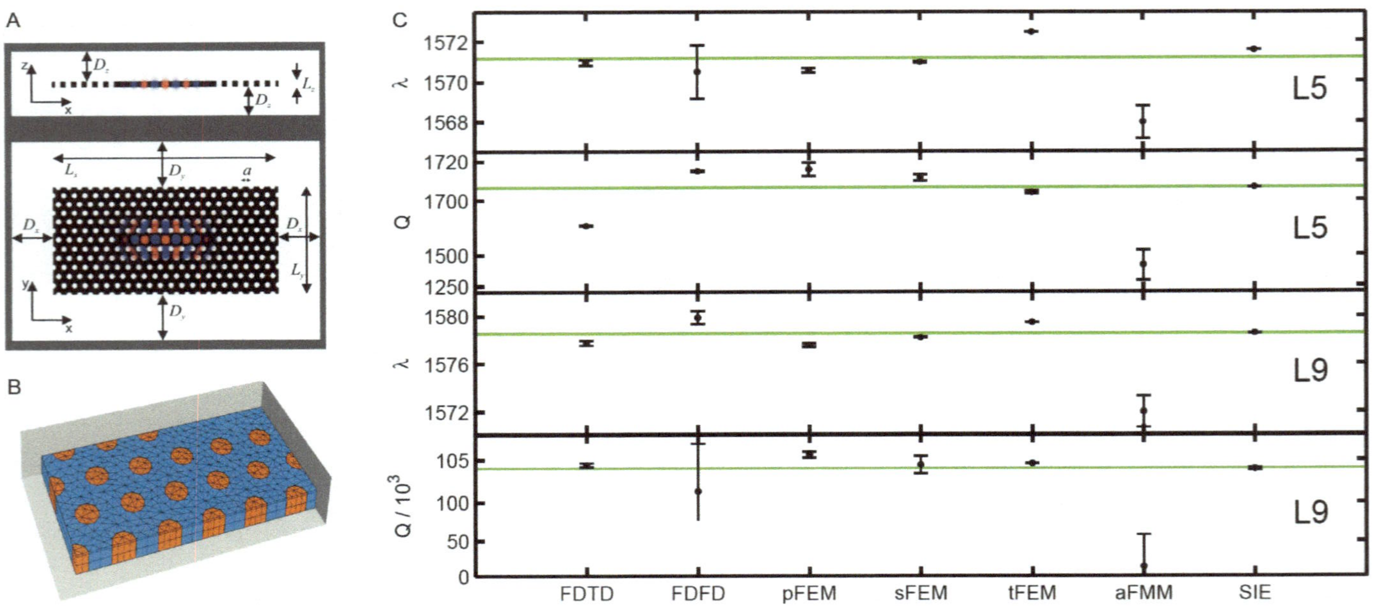

Fig. 3 Schematic illustrating the PhC membrane geometry and the electric field E_y profile for the M1 mode in an L9 cavity in the $y = 0$ (top) and the $z = 0$ (bottom) plane of the structure, respectively. (B) Computational mesh of the L5 PhC geometry used in the scattering analysis FEM (sFEM) simulations. (C) Comparison of the resonance wavelength and the Q-factor for an L5 and an L9 PC cavity. The *green horizontal lines* indicate the averaged values and *vertical bars* the numerical error. The required computational time was in the range of 3 (FDFD) to 146 (aFMM) minutes. *Reprinted from de Lasson, J.R., Frandsen, L.H., Gutsche, P., Burger, S., Kim, O.S., Breinbjerg, O., Ivinskaya, A., Wang, F., Sigmund, O., Häyrynen, T., Lavrinenko, A.V., Mørk, J., Gregersen, N., 2018. Benchmarking five numerical simulation techniques for computing resonance wavelengths and quality factors in photonic crystal membrane line defect cavities. Opt. Express 26 (9), 11366–11392. https://doi.org/10.1364/OE.26.011366 with permission from Optical Society of America.*

maximized, thereby establishing a design rule to maximize the directionality of emission in good agreement with experiment as presented in Section 4. This example nicely illustrates the power of the FEM approach to support the design of quantum devices to obtain optimum performance with regard to the desired functionality. Noteworthy is the fact that FEM simulations are also very beneficial to maximize the photon coupling and outcoupling efficiency of photonic quantum devices (Fischbach et al., 2017; Gschrey et al., 2015c; Schneider et al., 2018; Srocka et al., 2018, 2020).

2.3 Comparison of finite element method and finite difference time domain approaches

Nanoresonators based on PC membranes form an interesting system for comparing different numerical methods. In de Lasson et al. (2018) this was carried out for FDTD and different FEM approaches, as well as for finite-difference frequency-domain (FDFD), the aperiodic Fourier modal method (aFMM) and the surface integral equation (SIE) approach, in which resonance wavelengths and Q-factors for L5 and L9 cavities were calculated and compared. Here in the case of L5 and L9 cavities, which can act as integral parts of IQPCs, 5 and 9 holes are left out of the periodic lattice of holes, see Fig. 3A, in order to localize light in the resulting photonic defects. The calculation of the Q-factors represents a significant computational challenge, which requires a 3D calculation in the system under consideration. Furthermore, the Q-factor describes optical losses to the environment, which cannot be taken into account directly in common models and is usually approximated by absorbing boundary conditions in terms of PLMs (Berenger, 1994), whereby the choice of the computational domain can affect the result in an uncontrolled way (de Lasson et al., 2018).

The main results of comparison between the aforementioned numeric methods are summarized in Fig. 3C which summarizes the calculated resonance wavelengths λ and Q-factors for the considered L5 and L9 cavities. It is seen that all methods, apart from aFMM, approximate the average values (green lines) quite well. However, it is interesting to note that the given error bars are too small to account for the deviation of the different method's results from the average value. In other words, it seems that the used methods underestimate the numeric error, an aspect which one should have in mind when interpreting the obtained results. Moreover, the comparison yields that for the particular problem the SIE method provides the most accurate results.

Overall, the presented numerical methods and the results shown give an impression of their remarkable level of development and of the possibilities that result in particular for the design optimization of nanophotonic elements and especially IQPCs. Current and future developments can use machine learning approaches to achieve a further increase in efficiency and computing accuracy using, e.g., Bayesian algorithms. In addition, it will be interesting to integrate more precise descriptions of the quantum emitters beyond simple dipole models in the simulations and to take quantum effects into account.

3. Fabrication technologies

IQPCs are highly optimized components, the manufacture of which requires sophisticated and precise methods of modern nanotechnology. IQPCs are usually realized on single-crystalline planar semiconductor structures mainly based on silicon and III–V compound semiconductors, as well as combinations thereof. These materials are structured using metal and dielectric layer deposition, lithography and etching processes with nanometer precision in order to achieve the desired functionality. Furthermore, they can be combined with other materials, e.g., via wafer bonding in heterogeneous approaches, so that complex, fully integrated quantum circuits are created. This section provides an overview of common and advanced manufacturing processes for the implementation of highly functional IQPCs. The focus is on modern technology processes such as in situ lithography that were specially developed for structuring IQPCs with integrated quantum emitters. The fabrication aspects presented here are taken up again in Sections 4 and 5, in which the specific requirements and physical-technical properties of the circuits as well as their physical properties are discussed.

3.1 Realization of planar substrates for integrated quantum photonics circuits

Silicon-based IQPCs are usually realized on high-purity single-crystal silicon, as is also used for highly integrated electronic circuits. These are manufactured using crucible processes and are available as planar wafers for the further processing of IQPCs. In this case, the desired functionalization is achieved essentially via the lateral structuring in combination with the deposition of further materials. We refer to May and Sze (2003) for further details on the very established production of silicon-based substrates.

Optically active III–V compound semiconductors with a direct band gap are of high interest for the development of IQPCs. GaAs and InP semiconductors in particular are characterized by excellent optical properties. They have a pronounced optical nonlinearity as the basis, for example, for quantum frequency conversion in customized WG structures (Chang et al., 2018) and integrated photon pair generation (Horn et al., 2012). In addition, self-organized InGaAs QDs form the basis for state-of-the-art single-photon sources with record values in terms of the single-photon emission rate (Schlehahn et al., 2015), the single-photon purity (Schweickert et al., 2018), and the photon indistinguishability (Wei et al., 2014). Corresponding planar wafers are realized on the basis of single-crystal GaAs substrates through the epitaxy of semiconductor heterostructures. For this purpose, the methods of molecular beam epitaxy (MBE) and metal organic chemical vapor deposition (MOCVD) are used. Both make it possible to achieve single-crystal layers with atomically precise interfaces, tailored layer sequences and the highest optical quality. With regard to IQPCs, layer sequences that contain QDs as single-photon emitters are of particular interest. These are usually integrated into the semiconductor heterostructures using the Stranski–Krastanov process. For example, InGaAs QDs are created in a GaAs matrix. The quantum emitters thus achieved form almost ideal two-level systems with narrow emission line preparation in the range of the homogeneous linewidth and quantum efficiencies beyond 90% as the basis for quantum optical applications (Kuhlmann et al., 2015). However, the self-organized manufacturing approach has an adverse effect on the further processing. The random position and spectral position of individual QDs prevent scalable component production using common lithography processes. This has motivated the development of deterministic manufacturing concepts, which are presented in Section 3.2.

3.1.1 Wafer bonding

We next discuss the wafer bonding technique which is of particular importance for the realization of hybrid or heterogeneous photonic circuits to be presented in section 5. While QD lasers produced on silicon substrates through a direct wafer fusion bonding process were demonstrated (Fang et al., 2006), in Davanco et al. (2017) it was shown that direct wafer bonding could be leveraged to produce SiN WG-based photonic circuits incorporating single QDs as single-photon sources. In this method, two wafers of different materials are initially brought together and bonded to form a heterogeneous wafer stack. Typically, the substrate of one of the wafers is removed, leaving behind a thin film of material that can be processed alongside the receiving material substrate. Device fabrication can then follow a single high-throughput process flow with completely top-down techniques, where different material layers are subsequently etched, following high-resolution lithography. The fully top-down fabrication process, which allows high-resolution lithography-based definition of the IQPC geometries of all layers involved, in principle allows fabrication of high-performance geometries.

One of the main advantages of such an approach is in scalability—it allows wafer-scale device production by leveraging the massive parallelism enabled by mature, top-down semiconductor fabrication methods. Indeed, the wafer bonding approach has been employed commercially for the creation of heterogeneous integrated III–V semiconductor/silicon-on-insulator photonic devices, such as lasers and transceivers, with optical gain provided by III–V materials (Komljenovic et al., 2018).

An important issue with the wafer bonding approach is that considerable effort is required for bringing two materials together before device fabrication can even be considered. After the two materials are bonded, suboptimal device performance may still arise due to process incompatibility. Increasing the complexity to more than two types of materials on one chip is furthermore challenging. At least for fast device prototyping, deterministic fabrication methods presented in the following section offer advantages over wafer bonding.

3.2 Deterministic fabrication technologies for quantum circuits with integrated quantum emitters

The implementation of fully functional IQPCs usually requires the monolithic integration of individual quantum emitters as active elements. For example, applications in the field of boson sampling would benefit enormously if the input state based on individual photons could be generated directly on-chip. So far, this has not been possible and external single-photon sources are used in combination with complex multiplexing (Loredo et al., 2017; Wang et al., 2017, 2019). This limitation can be explained by the fact that high-quality quantum emitters such as nitrogen-vacancy (NV) centers in diamond or semiconductor QDs in solid-state systems are usually realized in a self-organized way. As a result, their positions and their emission energies are not predetermined, so that the process yield using conventional nanostructuring methods is very low and prevents a scalable production of

IQPCs. This is mainly explained by the fact that for advanced applications a given number N_{QE} of quantum emitters with the same emission energy on the scale of the homogeneous linewidth (approx. 1 μeV) have to be integrated precisely into the WG.

To clarify this aspect, the table in Fig. 4 shows comparative the expected process yield for the integration of multiple QDs in WG systems using conventional lithography and deterministic in situ lithography. The underlying estimate assumes that the QDs are self-organized with a specified areal density and an inhomogeneous broadening of the ensemble emission of 50 meV which is typical for In(Ga)As QDs. Furthermore, spectral fine-tuning in a range of 1 meV for individual QDs is included, for example, via Stark tuning (Patel et al., 2010), in order to achieve spectral resonance of the QD emission. From these considerations, which are described in Schnauber et al. (2018, Supplementary Information) and summarized in Fig. 4, it becomes clear that deterministic manufacturing processes are far superior to conventional methods and that it is only likely that the necessary scaling can be achieved using them. The process yield of conventional lithography is almost zero for integration of two QDs and a low QD areal density of $10^7 cm^2$. It can be increased by a higher areal density, but even with a QD density of $10^9 cm^2$, the process yield for the integration of 6 quantum emitters is below 0.1%. Areal densities of this size, however, result in a large number of QDs that were not intentionally integrated, which has a disadvantageous effect on the single-photon purity and can also lead to absorption

	Random yield $P_{N,\mathrm{rdm}}$			Deterministic yield P_N		
N \ QD Density (cm^{-2})	10^7	10^8	10^9	10^7	10^8	10^9
2		0.02	0.17	0.99		
3		$<10^{-3}$	0.03	0.94		
4	0		0.005	0.83	1	
5		0	$<10^{-3}$	0.66		
6			$<10^{-3}$	0.49		

Fig. 4 Estimated process yield per write field for fabricating a WG structure with *N* QD nodes that are all within a resonance tuning range of 1 meV. A 50 mm × 50 mm deterministic preselection field and a 100 nm × 10 mm (parallel and normal to the WG) spatial positioning tolerance is assumed. Values $<10^{-4}$ are rounded to 0 and >0.9999 rounded to 1. *Reprinted with permission from Schnauber, P., Schall, J., Bounouar, S., Höhne, T., Park, S.-I., Ryu, G.-H., Heindel, T., Burger, S., Song, J.-D., Rodt, S., Reitzenstein, S., 2018a. Deterministic integration of quantum dots into on-chip multimode interference beamsplitters using in situ electron beam lithography. Nano Lett. 18 (4), 2336–2342. https://doi.org/10.1021/acs.nanolett.7b05218, Supplementary Information.*

losses. In contrast, deterministic methods promise almost perfect process yields at a QD area density of $10^8 cm^2$ and up to 50% at $10^7 cm^2$, which illustrates the enormous advantages of such methods.

Deterministic manufacturing processes for quantum devices essentially include the positioned growth or the positioned implantation of quantum emitters, so-called pick-and-place techniques and deterministic lithography processes (Rodt et al., 2020). In this context, the positioned growth, for example, via nanohole arrays as nucleation sites, and to a certain extent also the positioned implantation of quantum emitters have the disadvantage that the spectral position of the quantum emitters is not controlled, so that there is still no high scalability. In contrast, quantum emitters with desired spectral properties can basically be selected from a large pool of conventionally produced structures in the pick-and-place process. However, this can lead to a large overhead of potentially unusable quantum emitter structures. Alternatively, in the concept of deterministic lithography processes, suitable quantum emitters are selected on the basis of their emission intensity and their spectral properties, in order to then define and produce the desired nanophotonic structure with nanometer accuracy.

3.2.1 Pick-and-place and transfer techniques

In the pick-and-place (Elshaari et al., 2017; Kim et al., 2017; Mouradian et al., 2015; Zadeh et al., 2016) and transfer printing (Katsumi et al., 2018, 2020) approaches, which are of particular importance for heterogenous IQPCs (see section 5), different components of a device, composed of different materials, are produced separately and in parallel through well-developed fabrication techniques, and then brought together through a lower-throughput mechanical transfer process. As an example, in Mouradian et al. (2015) integration of quantum memories based on NV center defects in etched diamond nanobeam WGs onto SiN WGs was demonstrated using a pick-and-place process that employed micro-manipulation of tungsten tips. The same technique was also used in Najafi et al. (2015) to demonstrated the integration of niobium nitride superconducting nanowire single-photon detectors (SNSPDs), produced on SiN membranes, over an SOI-based photonic circuit. Pick-and-place with tungsten tips was also used in Zadeh et al. (2016) and Elshaari et al. (2017) to integrated single-photon sources based on InP nanowires with embedded InAsP QDs into a SiN photonic circuit. In Katsumi et al. (2018), InAs quantum dot-containing GaAs nanobeam photonic crystal cavities evanescently coupled to underlying GaAs WGs were produced through a transfer printing process. Here, a polydimethylsiloxane

(PDMS) stamp was used to lift suspended GaAs nanobeams from a processed GaAs wafer. The destination substrate consisted of GaAs-on-SiO_2 wafer onto which GaAs ridge WGs were etched, covered by a spin-on-glass (SOG) top cladding.

In Kim et al. (2017) the pick-and-place technique was used to realize heterogenous QD-WG systems. Here InGaAs QDs are integrated into photonic membrane structures using conventional lithography methods. Low loss Si WG structures are structured in parallel. Subsequently, the quantum emitters are spectroscopically characterized and membrane structures with the desired optical properties are selected. In order to realize the heterogeneous QD WG structures, suitable QD membrane structures are detached from the carrier material (GaAs) in a focused ion beam (FIB) machine and are transferred using a pick-and-place technique by a microprobe tip with high accuracy to selected positions on the silicon WG structure, see Fig. 5. IQPCs can thus be implemented with high-quality quantum emitters, low optical losses and high design flexibility. However, many QD membrane structures have to be generated and elaborately precharacterized in order to select elements with the desired optical properties. Furthermore, the pick-and-place process is rather tedious, time-consuming and difficult to automate.

Overall, the pick-and-place and transfer printing techniques offer a lower barrier for creating maximally performing devices based on components fabricated with low-yield processes, by allowing prescreening and selection of the best performers within a population. For instance, such capability was leveraged in the diamond NV center quantum memories in Mouradian et al. (2015), given that consistent, desired spectral and spin properties of fabricated nanobeams with inhomogeneous populations of NV centers are difficult to achieve. Similarly, Najafi et al. (2015) report on an integrated photonic circuit chip for photon correlation measurements, containing 10 low-jitter SNSPD detectors with consistent timing jitter and detection efficiencies. With pick-and-place and transfer printing, greater scalability is sacrificed in favor of processing flexibility and fast turn-around. Because finalized heterogeneous photonic chips are processed in separate runs, fewer issues with process compatibility can be expected, in comparison with a wafer bonding. This endows the techniques with a great degree of flexibility, offering a low barrier toward the incorporation of different types of materials together. While the technique is particularly attractive for rapid device prototyping and proof-of-principle demonstrations, agile automation can potentially lead to significant gains in device scalability.

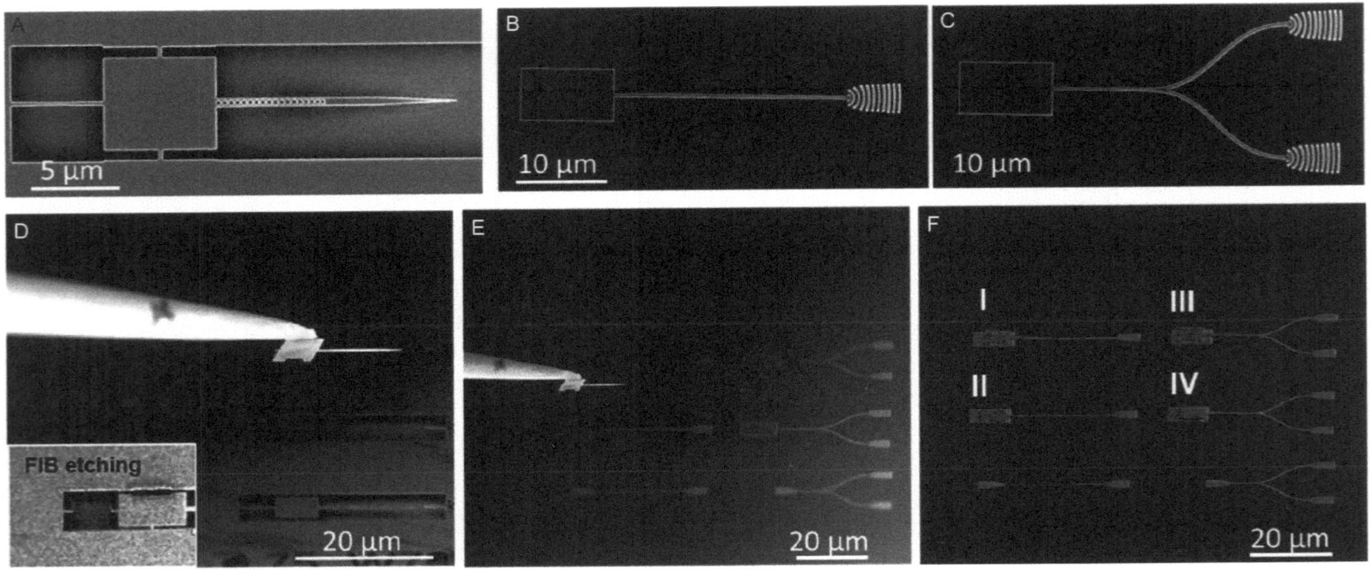

Fig. 5 Pick-and-place technique for the deterministic realization of IQPCs: (A) SEM image of a tapered InP nanobeam containing InAs QDs. (B) and (C) SEM images of WG structures in silicon. The square pads at the nanobeam's left end are used for placing the InP nanobeam. (D)–(F) Presentation of the pick-and-place procedure which uses a microprobe tip combined in an SEM equipped with a FIB unit to place InP nanobeams with InAs QD onto Si WGs. *Reprinted from Kim, J.-H., Aghaeimeibodi, S., Richardson, C.J.K., Leavitt, R.P., Englund, D., Waks, E., 2017. Hybrid integration of solid-state quantum emitters on a silicon photonic chip. Nano Lett. 17 (12), 7394–7400. https://doi.org/10.1021/acs.nanolett.7b03220 with permission from American Chemical Society.*

Importantly, however, the lack of deterministic control over self-assembled QD growth location is, likewise to the wafer bonding approach, a challenging scalability bottleneck for pick-and-place based device fabrication.

3.2.2 In situ lithography techniques

As an alternative concept, deterministic in situ lithography has been developed in the past ten years and is now an important tool in the field of quantum nanophotonics. In 2008, the optical in situ lithography was pioneered (Dousse et al., 2008) and has been used very successfully for the deterministic realization of QD micropillars as single-photon sources (Somaschi et al., 2016; Unsleber et al., 2016). Here, for optimal performance of the single-photon sources, the QD must be in spectral and spatial resonance with the resonator mode. To ensure this, the surface of a planar microcavity sample is first scanned with a low-energy (red) laser in a photoluminescence scan and the positions of QDs with a suitable spectral position are determined, see Fig. 6A. In the following lithography step, the desired structure, here a micropillar, is defined via a green laser in a UV-sensitive resist precisely aligned to the position and with its diameter matched to ensure spectral resonance with the preselected QD. The UV resist is spin-coated in advance and the entire process is carried out at cryogenic temperatures to ensure a sufficiently high emission intensity of the QDs. An attractive variant of this method is based on a two-stage process in which the positions of suitable QDs are first determined relative to alignment markers and then, after the mark detection, the desired structure is defined by means of electron-beam lithography (Liu et al., 2019; Sapienza et al., 2015), see Fig. 6B. This process benefits from the significantly higher resolution of electron-beam lithography (EBL) compared to optical lithography in the original process, where the higher resolution is strictly required for the realization of WG circuits with feature sizes well below the resolution of optical lithography. However, the two-stage process is technically much more complex.

To combine the advantages of in situ lithography and the high flexibility and resolution of EBL a further technique was developed and named in situ EBL (Gschrey et al., 2013). This deterministic nanofabrication technology platform applies low temperature cathodoluminescence (CL) spectroscopy to find select suitable QDs via convenient 2D scanning, and then performs low temperature EBL using the same scanning electron microscope (SEM) system. The structuring benefits from the highly optimized scanning properties of the SEM or CL technology and uses the enormous advantages of the EBL in order to produce quantum devices of the highest quality in a precise and variable manner (Gschrey et al., 2014, 2015a; Kaganskiy et al., 2016;

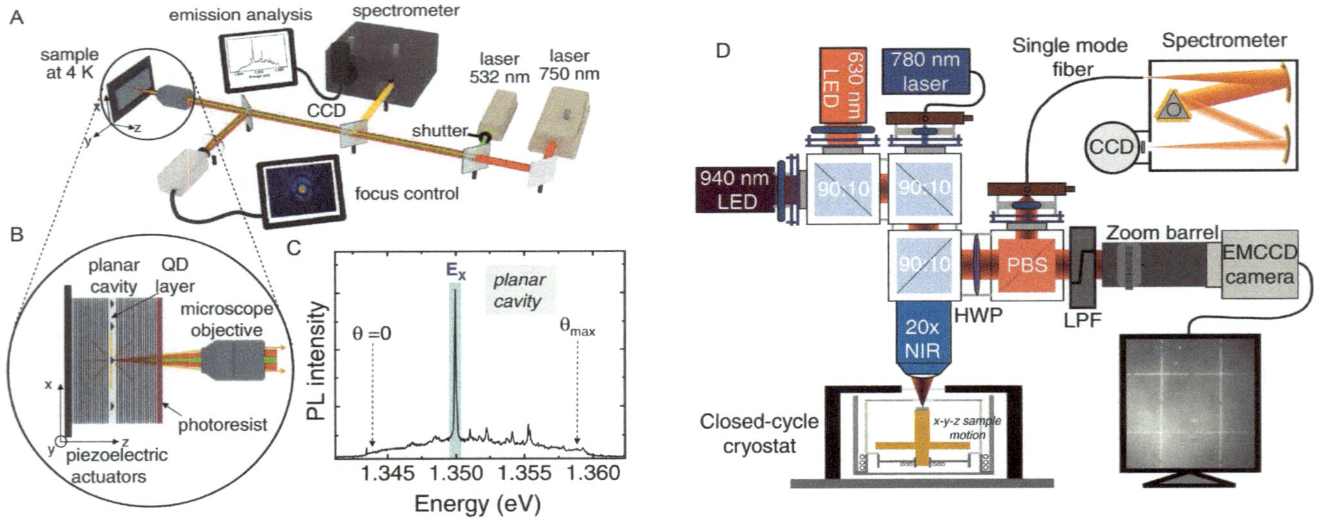

Fig. 6 In situ optical lithography performed at 10 K and photoluminescence (PL) imaging as deterministic tools for QD device processing. (A) Schematic view of the in situ optical lithography setup including two lasers used for PL registering (*red laser*) and optical lithography (*green laser*). (B) Zoom-in view illustrating the selection of suitable QDs via the red laser and the patterning of the device cross section into UV resist with the cross section aligned to the QD position by the green laser. (C) Low temperature PL spectrum of planar microcavity sample. Emission of a selected QD is marked (E_X). (D) Schematic view of the PL imaging setup. The sample is illuminated by an infrared LED and either a red LED or a 780-nm laser is for wide-range and local excitation of QDs, respectively. PL imaging and QD registering relative to alignment markers are performed by detecting the emitted and reflected light by an electron-multiple charge-coupled device (EMCCD) camera. Spectral information is obtained by a grating spectrometer. Subsequent to the PL imaging nanophotonic structures are patterned by conventional EBL using the alignment markers to place these structures with high accuracy preregistered QDs. *Panels (A)–(C): Reprinted from Ellis, D.J.P., Bennett, A.J., Dangel, C., Lee, J.P., Griffiths, J.P., Mitchell, T.A., Paraiso, T.K., Spencer, P., Ritchie, D.A., Shields, A.J., 2018. Independent indistinguishable quantum light sources on a reconfigurable photonic integrated circuit. Appl. Phys. Lett. 112 (21), 211104. https://doi.org/10.1063/1.5028339 with permission from American Physical Society; Panel (D): Reprinted with permission from Sapienza, L., Davanco, M., Badolato, A., Srinivasan, K., 2015. Nanoscale optical positioning of single quantum dots for bright and pure single-photon emission. Nat. Commun. 6 (1), 7833. https://doi.org/10.1038/ncomms8833.*

Schnauber et al., 2016). The in situ EBL technique has been refined more and more in recent years and has been used very successfully and in many ways for the implementation of efficient single-photon sources, for example based on microlenses (Gschrey et al., 2015c; Heindel et al., 2017; Kaganskiy et al., 2015; Schlehahn et al., 2016). With regard to the development of IQPCs, in situ EBL has the great, if not decisive, advantage over in situ optical lithography that, thanks to the extremely flexible and high-resolution EBL, it can also write very complex structures with the finest structural details. In particular, the in situ EBL technology has succeeded in integrating InGaAs QDs deterministically into GaAs WG structures and expanding them into more complex units including e.g., multimode interference (MMI) beam splitters (Mrowinski et al., 2019; Schnauber et al., 2018, 2019). The corresponding process flow is shown in Fig. 7. Panel (A) shows the 2D scanning of the QD sample using CL spectroscopy using a low EBL dose (typically performed at 10 K). The EBL positive resist (PMMA or CSAR) is already on the sample surface and is homogeneously exposed in this step in parallel to finding suitable QDs. Immediately afterwards, the desired WG structure is precisely written into the resist with a high EBL dose as shown in panel (B). Furthermore, elements of the intended IQPS such as beam splitters (here a MMI coupler) and possibly also output ports are defined in the resist (also at 10 K). The EBL dose is chosen so high that the resist is locally cross-linked again in negative-tone resist mode. After development in the clean room the inverted remaining resist acts as an etching mask (see panel C) in a plasma etching step in which the written structure is transferred to the semiconductor (see panel D). The real process flow is further illustrated by panels (E)–(G). Panel (E) shows a 2D CL map recorded in the first in in situ EBL step to select a suitable QD with a suitable spectral features and high CL intensity, which is marked here with QD1. Panel (F) shows an SEM image of a processed QD-WG element, in which the position of the previously selected and deterministically integrated QD (QD1) is marked. The 2D CL map in panel (G) was recorded for process control and demonstrates impressively that the QD was successfully integrated into the QD at the planned position. Furthermore, all other emission centers were eliminated in the etching step that removes the semiconductor material including the QD layer away from the WG. By using the inverted resist as an etching mask, the process described is overall very simple, reproducible and robust. It basically allows a very scalable production of IQPCs and can also be used, for example, to interface superconducting single-photon detector elements with high precision and alignment accuracy with QD WGs.

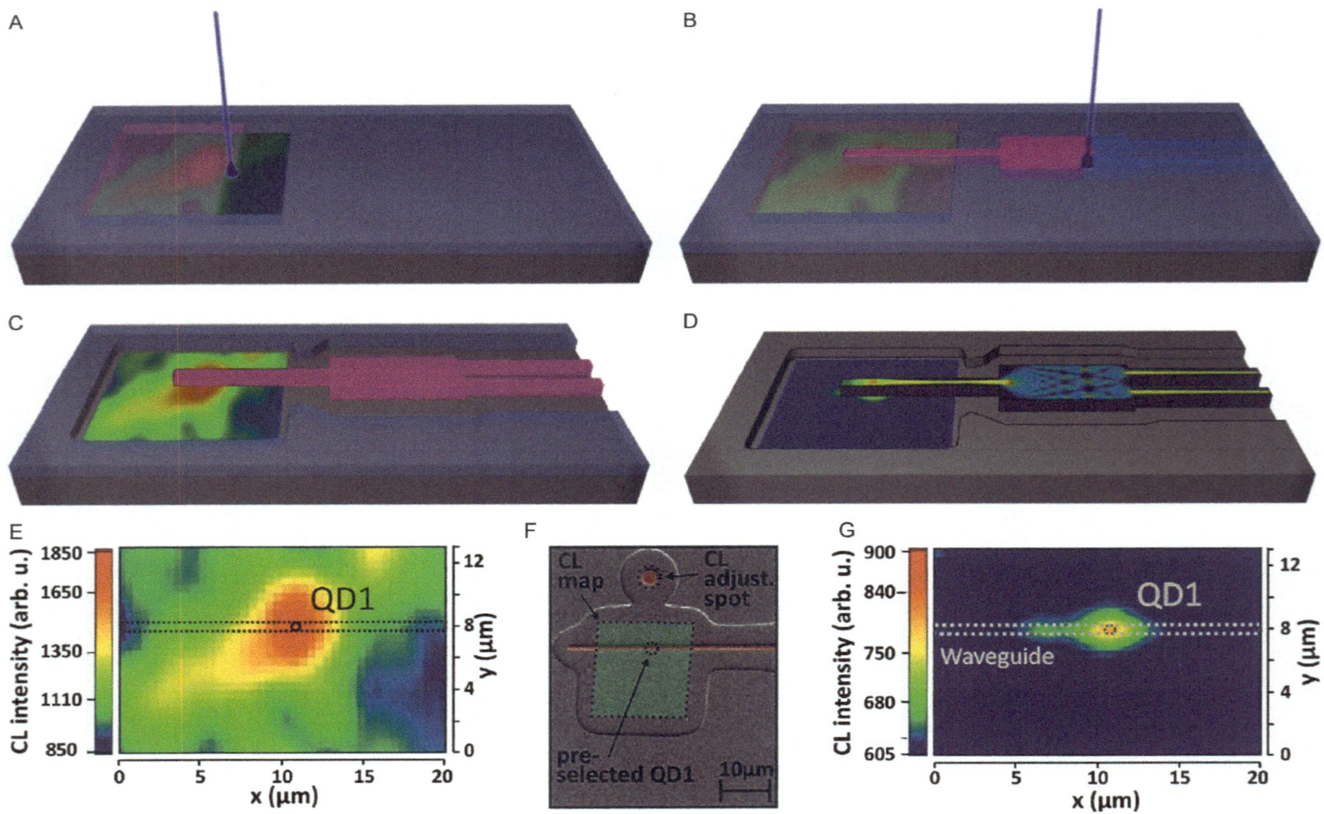

Fig. 7 See figure legend on opposite page.

3.3 Etching and processing technologies for integrated quantum photonic circuits

After the production of the starting materials and the lithography of the nanophotonic elements, the actual structuring of the IQPCs is carried out. In the simplest case, this only involves the etching of WG structures. However, complex manufacturing processes can also be involved, for example, to manufacture heterogeneous IQPCs based on different materials or fully integrated circuits. An overview of the most important process technologies is given below.

3.3.1 Etching of cavity and WG structures

The most common method for realizing photonic WG structures is reactive-ion etching (RIE). Here, the semiconductor material is removed physically and often also chemically supported in areas that are not protected by etching masks. As mentioned above, when describing the in situ EBL process, cross-linked resist can be used as etch mask, however, most often so-called hardmasks, such as Si_3N_4, SiO_2, or metals, are used to improve the selectivity of small features. Here, in contrast to wet chemical etching, dry plasma etching enables a very anisotropic etching pattern, which is essential for WGs. In some cases, combinations of the two etching processes are also used, for example, to implement membrane structures.

As an example in Fig. 8 a GaAs membrane structure with InAs QDs is shown, which contains a grating outcoupler for the light decoupling normal to the surface (Zhou et al., 2018). The structure was grown using MBE and

Fig. 7 In situ electron-beam lithography of IQPCs. The process is performed in a customized SEM which includes a CL unit, a He flow cryostat and an EBL pattern generator. (A) Selection of suitable QDs by CL registering at low electron-beam dose at 10 K. (B) In situ EBL at 10 K to integrate selected QDs into ridge WGs and to pattern also more complex WG structures such as MMIs at electron-beam dose. (C) Schematic of the sample after the subsequent development process in the cleanroom. The locally inverted electron-beam resist (magenta) acts as etch mask in the following plasma etching process (D). CL map obtained during QD registering (E), SEM of a fully processed device (F) and CL map showing emission of a QD deterministically integrated into a ridge WG (G). *Reprinted from Schnauber, P., Schall, J., Bounouar, S., Höhne, T., Park, S.-I., Ryu, G.-H., Heindel, T., Burger, S., Song, J.-D., Rodt, S., Reitzenstein, S., 2018a. Deterministic integration of quantum dots into on-chip multimode interference beamsplitters using in situ electron beam lithography. Nano Lett. 18 (4), 2336–2342. https://doi.org/10.1021/acs.nanolett.7b05218 with permission from American Chemical Society.*

defined using conventional EBL. As can be seen in partial image (B), the layer structure includes the actual (160 nm thick) GaAs membrane structure (with InAs QDs in the center) as well as a 1350 nm thick $Al_{0.75}Ga_{0.25}As$ layer, which acts as sacrificial layer when realizing the membrane structure.

The fabrication process to realize the membrane structure involves two EBL steps followed by dry etching. First, the grating grooves are written and etched using low-power BCl_3/Ar RIE. In the second EBL step the WGs and focusing tapers are defined with high accuracy to the grooves, before $BCl_3/Cl_2/Ar$ supported inductively coupled plasma RIE (ICP-RIE) is used remove the desired semiconductor material. It is important to note that this RIE essentially exposes the structures of the upper GaAs layer and that the actual membrane structure is realized in the following wet chemical process step. In this, the locally exposed $Al_{0.75}Ga_{0.25}$ As sacrificial layer is removed selectively by means of HF etching below the previously dry-chemically etched areas, so that the partially suspended GaAs membrane is formed (Midolo et al., 2015). The latter process is often combined with CO_2 critical point drying in order to reduce mechanical stress on the fragile WG structures (Leclercq et al., 2003). Alternatively, hydrogen fluoride vapor etching can also potentially be used. Such a process has been employed to remove oxide sacrificial layers in Deotare et al., (2009) to create high-quality factor silicon-on-insulator photonic crystal nanobeam cavities. This example shows in an impressive way how the sophisticated interaction between dry-wet chemical etching processes can lead to complex IQPCs.

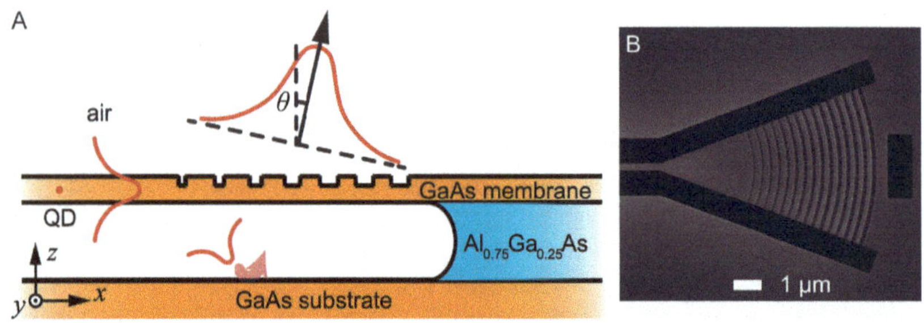

Fig. 8 IQPC based on GaAs membranes. The structure is realized by a combination of dry etching (ICP-RIE) and wet chemical etching. (A) Schematic cross section of the GaAs membrane with integrated shallow-etched grating couplers. (B) SEM image of the fully processed structure. *Reprinted from Zhou, X., Kulkova, I., Lund-Hansen, T., Hansen, S.L., Lodahl, P., Midolo, L., 2018. High-efficiency shallow-etched grating on gaAs membranes for quantum photonic applications. Appl. Phys. Lett. 113 (25), 251103. https://doi.org/10.1063/1.5055622 with permission from American Institute of Physics.*

3.3.2 Direct laser writing of photonic waveguide systems

An alternative method for producing photonic WG structures is direct laser writing (Marshall et al., 2009; Pertsch et al., 2004; Szameit and Nolte, 2010). In this process, 3D WG structures are flexibly and comparatively simply written optically with focused laser pulses in solid-state materials. This means that optical circuits can be generated quickly and efficiently, for example, to emulate quantum mechanics (Pertsch et al., 2004; Szameit and Nolte, 2010), to implement boson sampling (Tillmann et al., 2013) and to study topological photonics (Weimann et al., 2017).

Direct laser writing of WG structures not only makes it possible to implement individual WGs, but, above all, arrays of WGs with high precision—for example, in fused silica (SiO_2). Each WG can be structured individually, and modifications such as defects due to a local change in the refractive index can also be integrated. The basic idea is to first break molecular ring structures in fused SiO_2 locally using focused fs-laser pulses (Szameit and Nolte, 2010; Zhang et al., 2006). The subsequent recombination leads to smaller ring structures, which in turn locally increases the refractive index and thus enables light guidance in WGs with suitable dimensions. The direct laser writing concept is shown in Fig. 9. In practice, WGs are manufactured using this method by continuously moving the

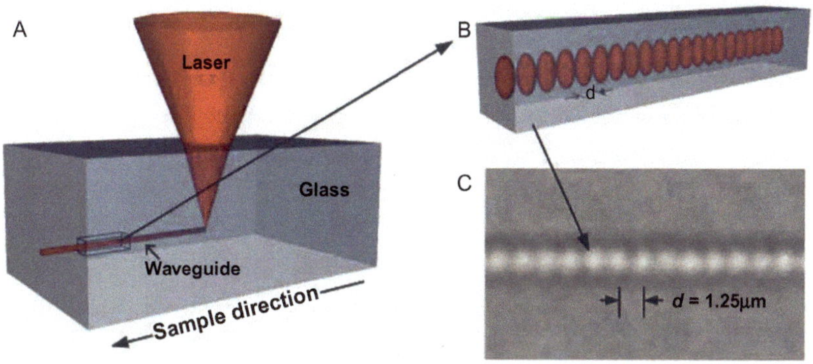

Fig. 9 Illustration of direct 3D fs-laser writing WGs in fused silica. (A) High light intensities of a fs-laser in the focal region of the writing objective break the SiO_2 bonds in fused silica which leads to a locally increased refractive index after densification. (B) In the writing process the WG (with higher refractive index) is created by moving the sample relative to the focus point of the laser in a spot-like fashion, which results in isolated modification volumes forming the desired WG (C). *Reprinted from Zhang, H., Eaton, S.M., Herman, P.R., 2006. Low-loss Type II waveguide writing in fused silica with single picosecond laser pulses. Opt. Express 14 (11), 4826–4834. https://doi.org/10.1364/OE.14.004826 with permission from Optical Society of America.*

material to be structured with the required accuracy relative to a focused laser spot. Here, highly complex 3D WG structures can be written using translations in all three spatial directions, the smallest structure dimensions being essentially given by the size of the laser spot, on the order of a few μm. The refractive index contrast that can be achieved typically reaches values in the range of 10^{-3} and can be controlled by the writing speed. The resulting structures are permanent, require no further postprocessing and have low optical losses in the range of 0.5 dB/cm. However, it is disadvantageous that structures of this type are purely passive and a direct integration of active elements such as quantum emitters and detectors should hardly be possible.

3.4 Concepts for efficient photon outcoupling

The focus of IQPCs is on-chip processing of single photons. Nevertheless, the photon coupling and decoupling is of great importance. As long as there are no fully integrated IQPCs, for example, photons must be coupled in at the input ports and coupled out at the outputs for detection or for transfer to photonic networks. For this purpose, various approaches have been developed, which include grating outcouplers (Coles et al., 2016; Faraon et al., 2008; Liu et al., 2018; Mrowinski et al., 2019) (see, e.g., Fig. 8B) for photon extraction normal to the surface and adiabatic outcouplers for in-plane extraction (Kiršanskė et al., 2017).

The aforementioned approaches to photon coupling are aimed primarily at proof-of-principle experiments on a laboratory scale and free-space optics. For practical applications, however, it is crucial to connect IQPCs directly to optical fibers in order to avoid complex and fault-prone optical paths and to ensure user friendliness. The direct fiber-connection represents a major technological challenge, since IQPC elements and fiber optics, especially in case of single-mode fibers, have to be aligned with each other with sub-μm accuracy and have to be mechanically and thermally very stable due to the cryogenic operation of the sources. In addition, good mode matching must also be ensured for high coupling efficiency between IQPC and fiber. In Daveau et al. (2017) an adiabatic-evanescent coupling between a tapered nanobeam WG and a microfiber was established for this purpose via an in-plane approach as shown in Fig. 10A and C. Here, high chip-to-fiber coupling efficiency was achieved, however, the fiber was not permanently connected to the IQPC. Concepts for permanent coupling between glass fibers and nanophotonic elements have already been established in superconducting detectors and use either high-precision mechanical elements or suitable adhesives.

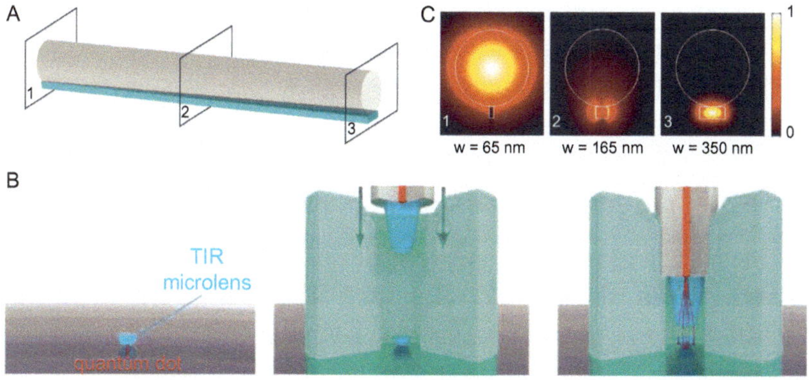

Fig. 10 Direct fiber coupling of nanophotonic elements. (A) Schematic of a fiber-coupled WG system. The microfiber (*gray cylinder*) is brought into contact with the tapered GaAs WG to allow for efficient light coupling. (C) Normalized electric field amplitude $|E|$ plotted in the cross section plane of the coupled WG system different WG widths w. (B) Illustration of the on-chip fiber coupling using 3D printed microoptics (Total Internal Reflection (TIR) microlens) in combination with a corresponding microlens at the fiber-facet and on-chip fiber chuck. *Panels (A) and (C): Reprinted from Daveau, R.S., Balram, K.C., Pregnolato, T., Liu, J., Lee, E.H., Song, J.D., Verma, V., Mirin, R., Nam, S.W., Midolo, L., Stobbe, S., Srinivasan, K., Lodahl, P., 2017. Efficient fiber-coupled single photon source based on quantum dots in a photonic-crystal waveguide. Optica 4 (2), 178–184. https://doi.org/10.1364/OPTICA.4.000178 with permission from Optical Society of America; Panel (B): Reprinted with permission from Bremer, L., Weber, K., Fischbach, S., Thiele, S., Schmidt, M., Kaganskiy, A., Rodt, S., Herkommer, A., Sartison, M., Portalupi, S.L., Michler, P., Giessen, H., Reitzenstein, S., 2020. Quantum dot single-photon emission coupled into single-mode fibers with 3d printed micro-objectives. APL Photon. 5 (10), 106101. https://doi.org/10.1063/5.0014921.*

Current developments in the field of fiber-coupled single-photon sources are very promising and point the way to on-chip fiber coupling of IQPCs as well. The alignment between emitter and fiber is particularly challenging here, since QDs, for example, hardly emit any light at room temperature, which prevents active alignment based on the emitter signal. This issue was circumvented in Schlehahn et al. (2018) by using the local wetting layer signal for alignment and subsequent gluing of a multimode glass fiber. In another very elegant method, the optical interference signal between a laser-illuminated glass fiber and the sample surface was used to determine the surface topography and to couple a single-mode glass fiber with gluing and sub-μm accuracy with a QD structure (Musial et al., 2020; żołnacz et al., 2019). A promising alternative to this, which is also very attractive for the fiber coupling of IQPCs, uses 3D printing of microoptics and on-chip fiber holders to achieve an extremely precise and optically efficient connection between nanophotonic structure and glass fiber (Bogucki et al., 2018; Bremer et al., 2020), as presented in Fig. 10B. The mentioned approaches are very promising for direct fiber coupling to IQPCs with grating outcouplers, toward modular devices for direct integration into quantum networks.

4. Homogenous waveguide systems

The integration of on-chip photonic quantum information processing has been developed on a variety of platforms and has benefited from very productive advances in the past decade. The simplest method was to integrate individual photon sources and elements of linear optics into a system that consists of single-material, or in other words homogeneous WGs. This was the first approach to implementing IQPCs. These successes were achieved using mainly already established manufacturing techniques developed for classical photonic chips. The problems of purity, decoherence, scalability, and versatility of individual photons are very different depending on the material used. We give an overview of the recently developed and investigated platforms for the implementation of a homogeneous IQPC. We discuss the strengths and weaknesses of the individual platforms. In particular, we will highlight the advances in silicon-based platforms, III–V semiconductors and diamond. So far, they are the most reliable and most advanced platforms for integrated quantum electronics with different strengths and weaknesses. In this section we only refer to WG-based structures made of naturally compatible materials (Si/SiO_2 or GaAs/AlGaAs/InGaAs) as opposed to hybrid platforms made of different material types, which have their own challenges and problems and are discussed in section 5. We conclude this section with a synthesis of the platform viability discussed as potential self-sufficient, scalable, and programmable IQPCs.

4.1 Si-based platforms

Historically, silicon-based platforms are the first researched and most successful way to implement IQPCs (Politi et al., 2009). They include structures based on SiO_2 on Si, Si-on-insulator (SOI) and SiN on SiO_2. The very first demonstrations were conducted on silicon-on-silicon WGs, with the first implementation of linear optics based on chip quantum gates (Politi et al., 2008). The large mode size typical for these structures enables simple coupling to fibers and free-space optics, which offers early opportunities for experiments with passive elements on the chip. Silica-on-silicon platforms are still widely used (Masada et al., 2015; Metcalf et al., 2014), but have been increasingly surpassed by SOI-IQPCs (Silverstone et al., 2016), which are better suited for processing quantum information on a large scale. They have several advantages over the other existing alternatives: they are transparent to photons at the telecommunications wavelengths, they have a strong nonlinearity of $\chi^{(3)}$, which enables compact single-photon source designs, and they benefit from the extensively developed manufacturing techniques in the field of silicon electronics and photonics and compatibility with the CMOS industry.

4.2 Single-photon generation and integration

Integrated single-photon sources based on silicon chips typically rely on spontaneous four-wave mixing (SFWM), a third-order nonlinear process in which photon pairs are produced spontaneously from a pump beam, following energy conservation and phase-matching. SFWM sources have been implemented on silica-on-silicon (Spring et al., 2017), SOI (Engin et al., 2013), SiN-on-silica (Lu et al., 2019) platforms (see Fig. 11), in a variety of successful proof-of-principle photonic quantum information demonstrations. The Poissonian statistics of the nonlinear generation process, however, imposes a fundamental limit to the efficiency and scalability of these structures, as the pumping level, and thus the pair-generation rate, must be kept low in order to avoid multiphoton creation events. Time (Kaneda and Kwiat, 2019) and spatial (Collins et al., 2013) multiplexing techniques can further improve source efficiency by up to 66.7% (Collins et al., 2013). It has been recently shown that an array of 18 SFWM single-photon sources in SiO_2 WGs (Spring et al., 2017) can generate highly indistinguishable and highly pure photons with a heralding efficiency of 50%. In addition, the source can be used to generate entangled photons (Grassani et al., 2015). A silicon chip that embeds two sources, frequency demultiplexers and a reconfigurable optics can perform two-photon tomography or Bell CHSH tests on the chip (Silverstone et al., 2015). Recently, high-dimensional entanglement generation has been achieved in a large-area Si chip with 16 SFWM sources (Wang et al., 2018b), and a coherent control over the generated states has also been demonstrated (Kues et al., 2017).

4.3 Large-scale integrated photonics

So far, the vast majority of integrated quantum optical experiments have been based on WG systems that consist of *N* interferometer circuits that perform uniform transformations for *N* spatial modes. In 2001, Knill et al. (2001) showed in their groundbreaking work that uniform transformations or logical operations can be carried out with linear optics and in particular beam splitters.

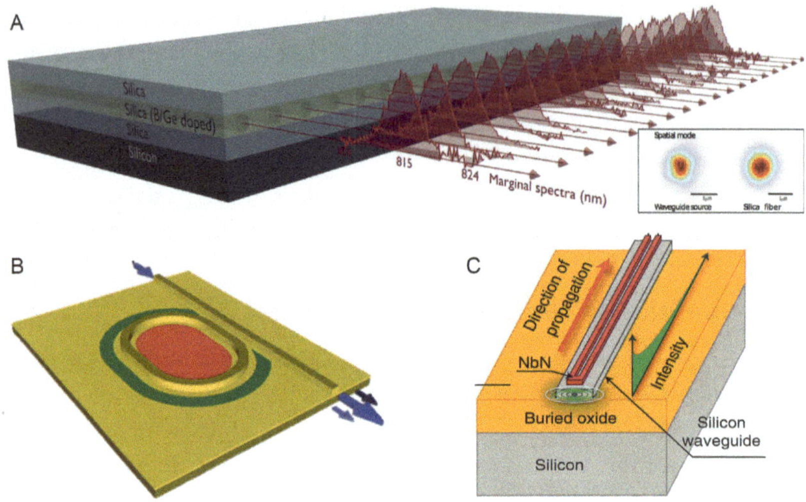

Fig. 11 (A) An array of heralded single-photon sources on a silica photonic chip. (B) Photon pair generation in a Si microring resonator with reverse bias enhancement. (C) Design of the traveling wave SSPD: a sub-wavelength absorbing NbN nanowire is patterned atop a Si WG to detect single photons. *Panel (A): Reprinted from Spring, J.B., Mennea, P.L., Metcalf, B.J., Humphreys, P.C., Gates, J.C., Rogers, H.L., Söller, C., Smith, B.J., Kolthammer, W.S., Smith, P.G.R., Walmsley, I.A., 2017. Chipbased array of near-identical, pure, heralded single-photon sources. Optica 4 (1), 90–96. https://doi.org/10.1364/OPTICA.4.000090 with permission from the Optical Society of America; Panel (B): Reprinted from Engin, E., Bonneau, D., Natarajan, C.M., Clark, A.S., Tanner, M.G., Hadfield, R.H., Dorenbos, S.N., Zwiller, V., Ohira, K., Suzuki, N., Yoshida, H., Iizuka, N., Ezaki, M., O'Brien, J. L., Thompson, M. G., 2013. Photon pair generation in a silicon micro-ring resonator with reverse bias enhancement. Opt. Express 21 (23), 27826–27834. https://doi.org/10.1364/OE.21.027826 with permission from the Optical Society of America; Panel (C): Reprinted from Pernice, W.H.P., Schuck, C., Minaeva, O., Li, M., Goltsman, G.N., Sergienko, A.V., Tang, H.X., 2012. High-speed and high-efficiency travelling wave single-photon detectors embedded in nanophotonic circuits. Nat. Commun. 3 (1), 1325. https://doi.org/10.1038/ncomms2307 with permission from the Springer Nature.*

The setup can be made reprogrammable as desired by implementing reconfigurable Mach–Zehnder interferometers that act as reconfigurable beam splitters (RBS) (Carolan et al., 2015a; Harris et al., 2014).

The RBS is realized with two 50% reflectivity beam splitters with adjustable internal and external phase shifters and corresponds to the linear optical Bogoliubov transformation:

$\begin{pmatrix} e^{i\phi} sin(\theta) & e^{i\phi} cos(\theta) \\ cos(\theta) & -sin(\theta) \end{pmatrix}$ where ϕ is the phase difference between the two outer arms of the Mach–Zehnder interferometer and θ is the phase difference between the two inner arms. Control of the phases enables any rotation in the U(2) group. The ability to generate arbitrary transformations for a large number of spatial modes is at the expense of complexity. For example, any U(20) transformation would require a total of 190 reflecting beam splitter (RBSs), each of which is phased and requires 380 electrical connections (Carolan et al., 2015a; Harris et al., 2014).

4.3.1 Gate based quantum information processing

The setup required for implementing one of the most basic logic operations, called CNOT. This gate was implemented on chip in several silicon quantum integrated circuits (Crespi et al., 2015; Politi et al., 2008; Santagati et al., 2017), see Fig. 12A. With this technology, an on-chip demonstration of the factoring Shor's algorithm with two CNOT gates was realized (Politi et al., 2009). The factorization of 15 was obtained with an accuracy of 99%. The principle of the CNOT gate with linear optics requires the presence of ancillary waveguides which makes them intrinsically probabilistic. However, the demonstration of the CNOT operation with the detection of a photon on the auxiliary line can signal that the successfully processed qubits are available for use in the larger architecture. The first heralded CNOT gate was implemented on a SiO_2 chip that could also implement universal linear optical operations (Carolan et al., 2015a). A similar laser-written SiO_2 chip was used as the platform for integrated quantum teleportation with an average fidelity of 89%. The outstanding possibilities of the RBSs lie in the capability of reprogramming a single chip and enabling it to process several quantum tasks and algorithms. In 2012, Shadbolt et al. (2012) reported on the first fully reconfigurable two-qubit quantum processor on a SiO_2 chip. The device consisted of two symmetrically designed entangled CNOT logic gates, which enabled the preparation, manipulation, and processing of quantum states on the chip. The device could create any maximum and not maximum entangled states, perform thousands of different high-fidelity quantum experiments related to the violation of the Bell inequality and the calculation of ground state molecular energy (Peruzzo et al., 2014). Further work on other silicon platforms confirmed these possibilities (Flamini et al., 2015; Qiang et al., 2018), in particular with the implementation of 98 different logic gates on the same programmable chip with an average fidelity of 93%. Carolan et al. (2015a) proposed recently first universal linear-optic chip and fully reprogrammable and demonstrated its ability to perform several key applications such as heralded quantum logics, entangling gates, boson sampling with verification tests and six-dimensional complex Hadamard operations. A later publication demonstrated the outstanding proficiency of the device by showing its ability to perform quantum

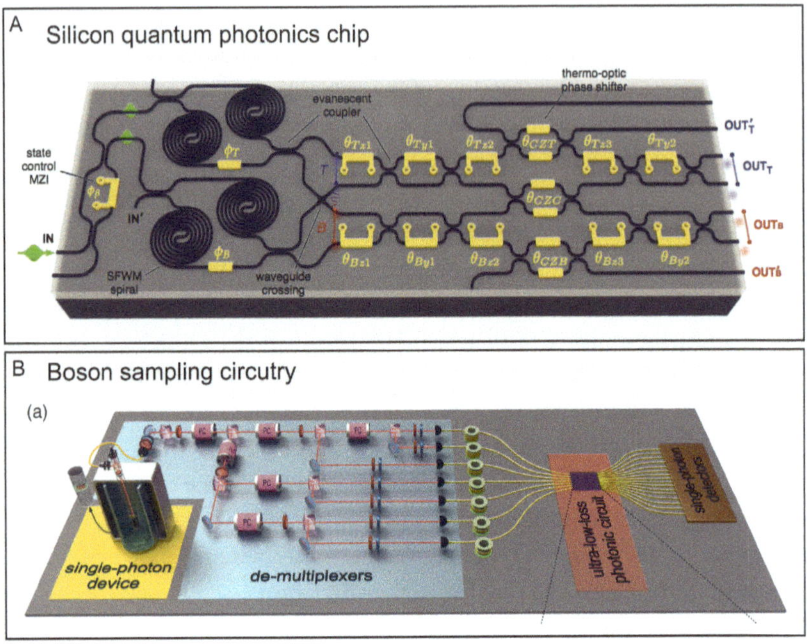

Fig. 12 (A) Silicon quantum photonics chip integrating four-photon pair sources with a reconfigurable six-mode interferometer. It can generate on-chip entangled or separable states on-demand. (B) Experimental setup for the boson sampling using a deterministic single-photon source to prepare the input state. Only the interferometer part is integrated. Single-photon source (semiconductor QDs), demultiplexing setup, and detectors are off-chip. *Panel (A): Reprinted from Santagati, R., Silverstone, J.W., Strain, M.J., Sorel, M., Miki, S., Yamashita, T., Fujiwara, M., Sasaki, M., Terai, H., Tanner, M.G., Natarajan, C.M., Hadfield, R.H., O'Brien, J.L., Thompson, M.G., 2017. Silicon photonic processor of two-qubit entangling quantum logic. J. Opt. 19 (11), 114006. https://doi.org/10.1088/2040-8986/aa8d56 with permission; Panel (B): Reprinted from Wang, H., Li, W., Jiang, X., He, Y.M., Li, Y.H., Ding, X., Chen, M.C., Qin, J., Peng, C.Z., Schneider, C., Kamp, M., Zhang, W.J., Li, H., You, L.X., Wang, Z., Dowling, J.P., Höfling, S., Lu, C.-Y., Pan, J.-W., 2018a. Toward scalable boson sampling with photon loss. Phys. Rev. Lett. 120 (23), 230502. https://doi.org/10.1103/PhysRevLett.120.230502 with permission from the American Physical Society.*

simulation on several vibrational molecule processes (Sparrow et al., 2018). A universal linear optical circuit has also recently been described in a silicon nitride-on-silicon platform (Taballione et al., 2019).

4.3.2 Boson sampling

A direct and appealing application of the IQPC is boson sampling. Taking into account M indistinguishable photons that are injected into a linear optical circuit with N modes, the task of boson sampling is to take a sample from the probability distribution of single-photon measurements at the output of the circuit (Broome et al., 2013).

Solving this problem in a classical computer entails the calculation of the permanent of the unitary matrix representing the linear optic circuit, a task that is computationally very demanding, of #P-complete complexity (Wang et al., 2018b). Indeed, it is believed that for a sufficiently large number of photons and optical channels, the problem may be more efficiently solvable in a photonic quantum simulator than by classical algorithms. Whereas this would constitute a first demonstration of quantum supremacy, the number of photons and channels required for such demonstrations is currently a moving target. Boson sampling does not require any ancillary line or entangling operation as in the Knill, Laflamme, and Milburn schemes (Knill et al., 2001). While boson sampling is not universal, it can implement a classically difficult task with far fewer physical resources than a full linear optical quantum computer setup. The technical requirements are the implementation of efficient sources for single photons, linear interferometers and efficient single-photon detectors. As highlighted in the previous sections, these requirements are met by the various Si platforms. Therefore, shortly after the theory was published, several groups simultaneously reported on their experimental implementation (Broome et al., 2013; Carolan et al., 2014; Spring et al., 2013; Tillmann et al., 2013). They were all based on the injection of three photons into a five- or six-mode interferometer-on-chip circuit. Interestingly, the demonstrations with the highest number of photons are based on off-chip single semiconductor QD demultiplexed photons (Loredo et al., 2017; Wang et al., 2018a, b), see Fig. 12C. Due to the probabilistic nature of SPDCs, the probability that the source generates simultaneously N photons decreases exponentially as N is increased. A variant called scattershot boson sampling (Lund et al., 2014), consisting of generating $M > N$ photons in order to increase the probability to effectively have a N photon input, has been proposed to make best use of the SPDC sources for boson sampling. The protocol was recently implemented with on-chip SFWM sources (Paesani et al., 2019). Another variant is the Gaussian boson sampling relies on an input of squeezed light (Hamilton et al., 2017) has been realized on the Si chip (Paesani et al., 2019). In contrast to the general-purpose computer, it is unlikely that boson sampling can be verified exactly by classic computers, especially since they are associated with the estimation of large matrix permanent elements, an unsolvable task for them. This posed the problem that no result of the boson sampling could be certified. However, various approaches were successfully implemented in the integrated optics, which made it possible to distinguish between uniform and boson scanning or between quantum and classic distribution.

4.4 III–V Semiconductor integrated optics

Another promising approach to realizing IQPCs is the use of III–V semiconductor materials. Unlike silicon, such materials feature a direct bandgap, allowing efficient light generation. Of particular relevance to quantum optics is that semiconductor quantum dots can be produced in III–V semiconductors, which can act as deterministic single-photon sources—in contrast to the spontaneous generation of parametric nonlinear processes such as SFWM. This is a considerable advantage over silicon: provided that the deterministically generated photons are efficiently collected into the WG circuits, the likelihood of success of on-chip logic operations is dramatically increased—a critical necessity for large-scale circuits. This also creates the possibility to implement scattershot Boson sampling. In particular, GaAs is a well-known and mature material system with well-established manufacturing techniques and a long history of photonics integration. Its high refractive index enables strong light confinement and the creation of low-loss WGs. The large electro-optical effect in GaAs offers great opportunities for the routing and manipulation of photons. The integration of QDs in nanophotonic GaAs structures that aim at the deterministic generation of individual photons has achieved remarkable success in recent years (Fig. 13). In addition, the implementation of on-chip detectors in GaAs WGs has attracted much community attention and paved the way for the first fully integrated semiconductor IQPCs.

4.5 On-chip single-photon generation

On-chip single-photon sources are realized using the radiation recombination from an excitonic state of a single QD (Michler et al., 2000). QDs proved to be very efficient emitters of single photons and entangled photons (Akopian et al., 2006). They can be electrically driven (Yuan et al., 2002), nonresonantly and resonantly addressed (Muller et al., 2007). The generation of indistinguishable photons on demand requires a resonance laser that addresses the on-chip QD from above for better pump filtering. Efficient photon collection, low-loss waveguiding of light along the circuit and a good connection between the various functionalities are the main requirements for effective semiconductor IQPCs. The integration of QDs in WGs

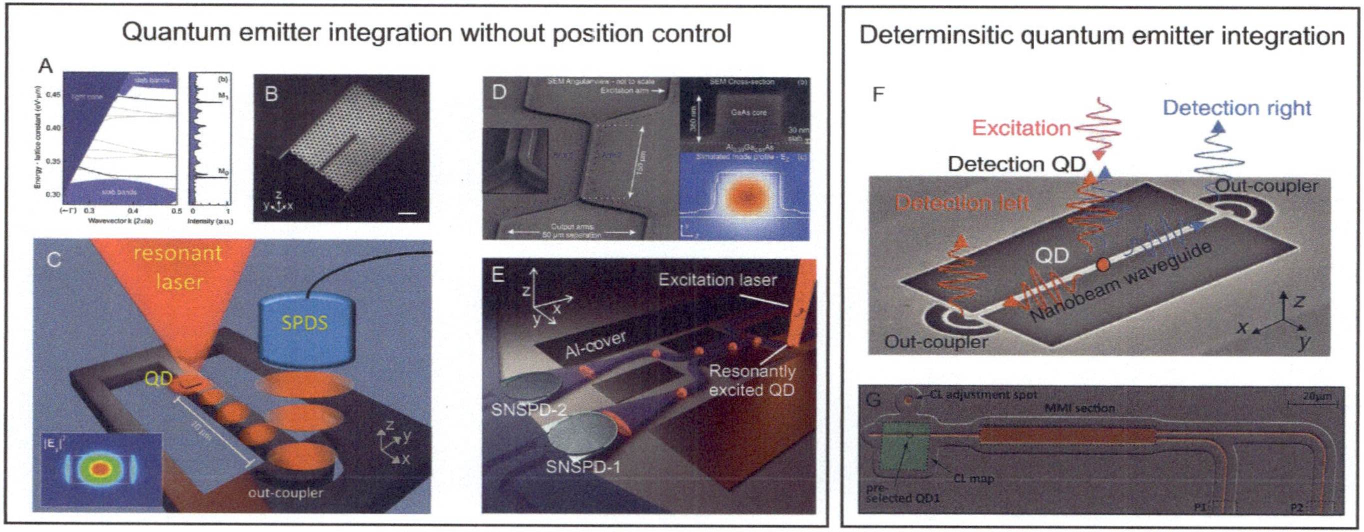

Fig. 13 See figure legend on opposite page.

has proven to be particularly efficient. PhC WGs (Ba Hoang et al., 2012; Laucht et al., 2012; Schwagmann et al., 2011) and ridge WGs are the two most popular approaches for on-chip collection and the transport of photons.

4.5.1 Photonic crystal based quantum circuits

PhC WG structures have the advantage that they offer tight confinement and high coupling efficiencies between quantum emitters and propagating modes (Lecamp et al., 2007; Manga Rao and Hughes, 2007; Rao and Hughes, 2007). A popular type of PhC WG consists of a missing row of

Fig. 13 (A) Calculated band structure for TE modes of an infinite PhC WG. (B) Scanning electron microscopy picture of a PhC WG. (C) Experimental scheme for the observation of resonance fluorescence QD coupled to a WG. (D) Scanning electron microscope image of a beam splitter evanescent field coupler implemented with ridge WGs. (E) Schematic representation of a fully integrated Hanbury Brown and Twiss experiment, comprising the SPS, the beam splitter and two integrated SNSPDs. (F) Nanophotonic waveguide used for demonstration of chiral emission from an embedded quantum dot. (G) Integrated multimode interference beam splitter fabricated using in situ EBL technique. *Panels (A) and (B): Reprinted from Schwagmann, A., Kalliakos, S., Farrer, I., Griffiths, J.P., Jones, G.A.C., Ritchie, D.A., Shields, A.J., 2011. On-chip single photon emission from an integrated semiconductor quantum dot into a photonic crystal waveguide. Appl. Phys. Lett. 99 (26), 261108. https://doi.org/10.1063/1.3672214 with permission from the American Institute of Physics; Panel (C): Reprinted from Makhonin, M.N., Dixon, J.E., Coles, R.J., Royall, B., Luxmoore, I.J., Clarke, E., Hugues, M., Skolnick, M.S., Fox, A.M., 2014. Waveguide coupled resonance fluorescence from on-chip quantum emitter. Nano Lett. 14 (12), 6997–7002. https://doi.org/10.1021/nl5032937 with permission from the American Chemical Society; Panel (D): Reprinted from Rengstl, U., Schwartz, M., Herzog, T., Hargart, F., Paul, M., Portalupi, S.L., Jetter, M., Michler, P., 2015. On-chip beamsplitter operation on single photons from quasi-resonantly excited quantum dots embedded in GaAs rib waveguides. Appl. Phys. Lett. 107 (2), 021101. https://doi.org/10.1063/1.4926729 with permission from the American Institute of Physics; Panel (E): Reprinted from Schwartz, M., Schmidt, E., Rengstl, U., Hornung, F., Hepp, S., Portalupi, S.L., Ilin, K., Jetter, M., Siegel, M., Michler, P., 2018. Fully on-chip single-photon Hanbury-Brown and Twiss experiment on a monolithic semiconductor-superconductor platform. Nano Lett. 18 (11), 6892–6897. https://doi.org/10.1021/acs.nanolett.8b02794 with permission from the American Chemical Society; Panel (F): Reprinted from Coles, R.J., Price, D.M., Dixon, J.E., Royall, B., Clarke, E., Kok, P., Skolnick, M.S., Fox, A.M., Makhonin, M.N., 2016. Chirality of nanophotonic waveguide with embedded quantum emitter for unidirectional spin transfer. Nat. Commun. 7 (1), 11183. https://doi.org/10.1038/ncomms11183 with permission; Panel (G): Reprinted from Schnauber, P., Schall, J., Bounouar, S., Höhne, T., Park, S.-I., Ryu, G.-H., Heindel, T., Burger, S., Song, J.-D., Rodt, S., Reitzenstein, S., 2018a. Deterministic integration of quantum dots into on-chip multimode interference beamsplitters using in situ electron beam lithography. Nano Lett. 18 (4), 2336–2342. https://doi.org/10.1021/acs.nanolett.7b05218 with permission from the American Chemical Society.*

holes in a triangular lattice of holes etched into a semiconductor slab (Lund-Hansen et al., 2008). In such structures, the PhC photonic band gap provides strong lateral confinement for light, leading to the formation of guided modes with strong field concentration within the line-defect. Efficient coupling of quantum emitter radiation to such line-defect modes arises due to the strong field confinement. The coupling efficiency can be further improved if the emitter is, in addition, resonant with a WG slow-light mode, for which propagation velocity is considerably lower than for other WG modes. Efficient routing in PhC WGs has been reported for 900 nm (Laucht et al., 2012) and for photons with a telecommunications wavelength of 1300 nm (Ba Hoang et al., 2012) with efficiencies up to 98% (Arcari et al., 2014; Lund-Hansen et al., 2008). PhC WGs are also very practical for modifying the local density of states and allow manipulation of the emission direction and polarization (see Section 4.7).

4.5.2 Ridge waveguides

The main obstacles to the implementation of all-PhC circuits are the typically high loss rates (of the order of dB/mm) (Sugimoto et al., 2004) and the mechanical fragility of large, free-standing membrane PhC structures. Ridge WGs are an attractive alternative to both PhC and free-standing channel WGs, being mechanically robust, at least one order of magnitude less lossy, though offering a comparatively modest spatial confinement. Light-matter interaction can nonetheless be enhanced though a compatible cavity, such as the Bragg grating cavity WG demonstrated by Hepp et al., (2018). The combination of PhC cavities connected to ridge WGs using a one-step lithography process was reported and showed coupling efficiencies of 70% (Fattah et al., 2013), using a one-step lithography process, a strategy that alleviates the mechanical fragility issue while preserving strong optical confinement for enhanced interaction with QDs. Coupling efficiencies of 70% between ridge and PhC waveguides were achieved, which allowed single-photon count rates of 3.5 MHz, to be achieved, exceeding typical rates obtained for single QDs in bulk. Much effort has also been made to embed QDs directly into free-standing GaAs WGs (Makhonin et al., 2014). Free-standing WGs offer a comparable level of optical confinement and slightly better propagation losses than PhC WGs, with a less complex geometry. This class of WGs was used by Javadi et al., (2018) to create an interface between a single photon and a QD spin with high spin preparation fidelity, and also a single-spin photonic switch. Like with freestanding membrane PhC WGs, however, the mechanical fragility of such structures creates challenges towards the creation of larger circuits.

Overall, ridge WGs have been widely used for the fabrication of more complex structures and paved the way toward the first integrated quantum optics experiments in GaAs. The following subsection describes such developments.

4.6 Beam splitters and on-chip interferometers

GaAs platforms are particularly well positioned for integrating monolithic circuits from beam splitters, interferometers, and detectors, which are required to implement KLM schemes or boson scanning protocols. The possibilities of GaAs structures were first demonstrated in 2014 (Wang et al., 2014) by characterizing an interferometric circuit, while the single-photon sources were outside the chip. The Mach–Zehnder interferometer with two directional couplers and two electro-optical phase shifters showed classic interference visibility of up to 98% and quantum Hong–Ou–Mandel quantum interference visibility of up to 95%. The relative phase control evidenced by the oscillations of the output count rates is a critical achievement because it is the key to fine-tuning the circuit and achieving universal linear computing in GaAs. The integration of quantum emitters in GaAs WG beam splitter systems has been demonstrated several times, with different designs such as rib-WGs (Rengstl et al., 2015) and ridge WGs (Jöns et al., 2015). They showed a successful integrated beam splitting with 50/50 directional couplers and multimode WGs. Another robust option is the production of multimode interference beam splitters, a technology that is already used in Si materials (Peruzzo et al., 2011). It was demonstrated with free-standing evanescent coupling beam splitters (Prtljaga et al., 2014). An important step forward in the production of practical semiconductor IQPCs was made with the implementation of a single-mode on-chip GaAs WG beam splitter. The device was addressed from above in an orthogonal configuration and the resulting laser filtering is very efficient since a multiphoton probability of only 0.18 was measured (Schwartz et al., 2016). The development of the superconducting on-chip detectors in GaAs (Reithmaier et al., 2013, 2015a) enabled the first demonstration of the fully integrated Hanbury Brown and Twiss experiment (Schwartz et al., 2018) (see Fig. 13E).

This series of advances has reached its technological limits due to the randomness of the manufacturing process related to the self-assembled epitaxial growth of the considered QDs. A very large number of devices must be manufactured and characterized before a single one can be selected for use. This represents a significant obstacle to the scalability of the semiconductor IQPCs. It is believed that deterministic manufacturing processes such as in situ optical lithography (Coles et al., 2016) and in situ EBL (Gschrey et al., 2013, 2015c) can overcome this problem, see Section 3. As an example of recent progress in this direction, the integration of QDs into a 50/50 multimode interference beam splitter has been mastered using in situ EBL (Schnauber et al., 2018) (see Fig. 13G). The fabrication technology offers the advantage of nanometer alignment accuracy and leads to IQPCs with excellent quantum optical properties. Thus, deterministic fabrication technologies open up realistic perspectives for the development of complex IQPCs with multiple monolithically integrated quantum emitters. Deterministic methods also prove to be competitive technologies for the manipulation of the emitter's emission directionality and potentially fulfill all the requirements for the complete implementation of on-chip logic quantum gates.

4.7 Chiral light–matter interaction

The WGs discussed above are powerful tools for exploring the interaction between light and matter. In most of these structures, light is confined transversely, in a plane orthogonal to the direction of propagation (Lodahl et al., 2017) and features a significant longitudinal electric field component. At specific, fixed locations (chiral points) within a WG, the electric field vector for a guided mode traveling in one direction rotates on the wafer plane. Locally, the field is circularly polarized, and carries a spin angular momentum. Inversion of the guided wave propagation direction leads to a sign change for the transverse spin component (Lodahl et al., 2017). This means that chiral effects may be explored through the interaction of light with circularly polarized emitters. Specifically, a circularly polarized dipole located at a chiral point only emits guided waves that propagate in one direction, depending on the sense of rotation of the latter's electric field. The resulting emission chirality has been experimentally demonstrate with a variety of emitters and waveguides. Fig. 14 shows examples of typical structures used to realize directional emission in GaAs quantum circuits.

Fig. 14A shows the deterministically produced, free-standing orthogonal GaAs WG structure, which is used to realize circular polarization directionality (Luxmoore et al., 2013a,b). This result was confirmed with another design based on a deterministic in situ EBL approach (Mrowinski et al., 2019) (Fig. 14B) where the lateral position of the emitter was controlled

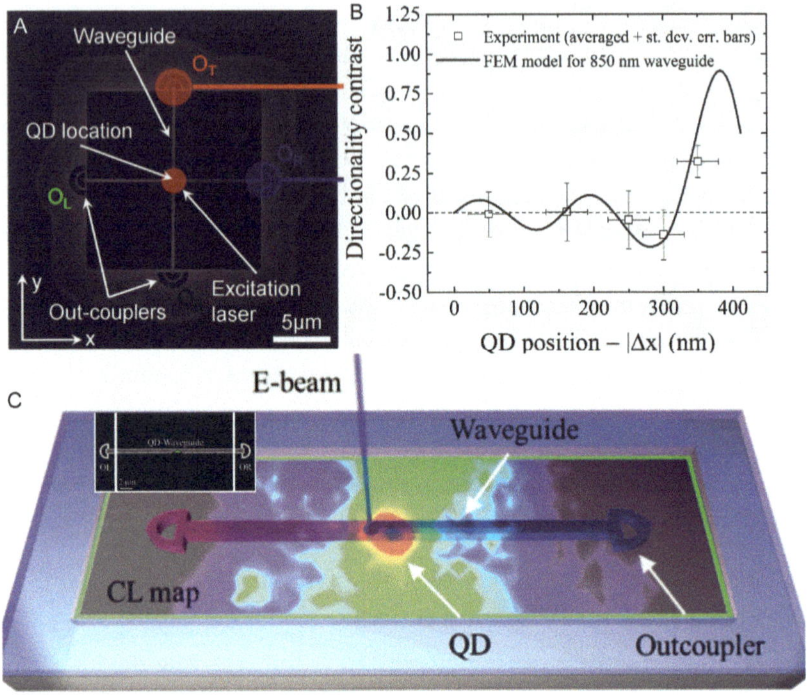

Fig. 14 (A) Spin-photon interface based on two orthogonal WGs where the polarization of the QD-emitted photon is converted into a path-encoded photon. (B) Deterministically fabricated ridge WG as a chiral platform for QD photon emission. On the two extremities, outcoupler gratings have been implemented for vertical efficient detection. (C) Measured directionality of the photon emission as a function of the quantum dot position respective to the WG center. *Panel (A): Reprinted from Luxmoore, I.J., Wasley, N.A., Ramsay, A.J., Thijssen, A.C.T., Oulton, R., Hugues, M., Kasture, S., Achanta, V.G., Fox, A.M., Skolnick, M.S., 2013b, Interfacing spins in an InGaAs quantum dot to a semiconductor waveguide circuit using emitted photons. Phys. Rev. Lett. 110 (3). https://doi.org/10.1103/physrevlett.110.037402 with permission from the American Physical Society; Panels (B) and (C): Reprinted from Mrowinski, P., Schnauber, P., Gutsche, P., Kaganskiy, A., Schall, J., Burger, S., Rodt, S., Reitzenstein, S., 2019. Directional emission of a deterministically fabricated quantum dot-Bragg reflection multimode waveguide system. ACS Photon. 6 (9), 2231–2237 with permission from the American Chemical Society.*

with nanometer accuracy to maximize in chiral interaction with propagating ridge WG modes and to achieve maximum directionality contrast (Fig. 14C). In combination with an external magnetic field, the device enabled a high directionality contrast (90%) of the circularly polarized emission. Directionality with high fidelity can be applied to the path-encoded qubit initialization and implemented directly in integrated quantum logic gates (Sollner et al., 2015). In the same publication, a scheme dealt with two additional beam splitters and an additional WG for the device with two control target photons and three microwave pulses for initialization and manipulation of the QD state within the photonic crystal WG. This may be the long-term goal of a deterministic logic gate operation, as originally proposed, based on the 1D atom principle.

4.8 Diamond integrated quantum photonic systems

Diamond has emerged as a competitive platform for the development of reliable IQPCs thanks to new fabrication techniques (Rath et al., 2015b). It presents several advantageous structural and optical properties: a wide bandgap, high Debye temperature, high isotopic purity, and low free electron concentration. NV centers feature long electron spin coherence times, on the order of microseconds. Single shot spin readout and coherent manipulation have been demonstrated (Robledo et al., 2011). This allowed for the generation of spin photon entanglement (Togan et al., 2010) and for the realization of loophole free tests of the Bell inequalities (Hensen et al., 2015). Silicon vacancy (SiV) centers can also be inserted in diamond and present a reliable alternative to NV centers. SiV centers present an interesting alternative with 90% of its emission into the zero-phonon line, near transform-limited optical linewidths and spin-lattice relaxation times approaching 1 min as well as coherence times approaching 1 s (Rose et al., 2018). Recently further emitters, such as Cr (Aharonovich et al., 2010), Ni- (Gaebel et al., 2004), Ge- (Wang et al., 2014), and Xe- (Iwasaki et al., 2015) emitters, attracted attention for their promising optical properties. Integrated circuits comprising color centers and optical cavities have been proposed as potentially deterministic platforms for generation and manipulation of quantum information (Nemoto et al., 2014).

4.9 Photonic systems in diamond thin films

Suspended diamond films or diamond structures placed on SiO_2 offer the high index contrast required for light confinement. Their development made full use of the techniques already established for the SOI platform. The first integration of multiple optical elements into a diamond film was demonstrated despite the difficulty in producing single-crystal diamond films of uniform thickness. In 2011, Hausmann et al. (2012) reported on the realization of a nanophotonic diamond that embeds a ring resonator with a high Q-factor ($Q = 12{,}600$) that is evanescently coupled to a WG with

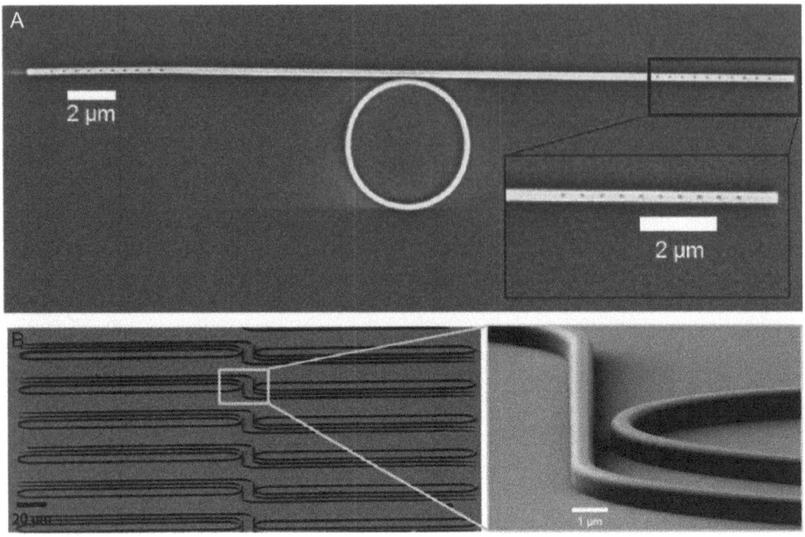

Fig. 15 (A) Diamond integrated nanophotonic network composed of a ring resonator evanescently coupled to a waveguide. (B) Single-crystal diamond racetrack resonators integrated with low-loss diamond waveguides terminated by polymer pads. *Panel (A) Reprinted from Hausmann, B.J.M., Shields, B., Quan, Q., Maletinsky, P., McCutcheon, M., Choy, J.T., Babinec, T.M., Kubanek, A., Yacoby, A., Lukin, M.D., Loncar, M., 2012. Integrated diamond networks for quantum nanophotonics. Nano Lett. 12 (3), 1578–1582. https://doi.org/10.1021/nl204449n with permission from the American Chemical Society; Panels (B) and (C) Reprinted from Hausmann, B.J.M., Bulu, I.B., Deotare, P.B., McCutcheon, M., Venkataraman, V., Markham, M.L., Twitchen, D.J., Loncar, M., 2013. Integrated high-quality factor optical resonators in diamond. Nano Lett. 13 (5), 1898–1902. https://doi.org/10.1021/nl3037454 with permission from the American Chemical Society.*

grating couplers (Fig. 15A). Here, single photons emitted by a NV center were coupled to the resonator mode and routed through the WG with a total extraction efficiency of 10%. This result was reproduced in a very similar device with a remarkably strong improvement in the zero-phonon line in the resonator mode (Faraon et al., 2013). Another device, based on high-quality single-crystal diamond racetrack resonators and operating in the telecom C-band (1550 nm) associated with low loss diamond WGs terminated by polymer spot size converters, showed very high Q-factors ($Q = 250{,}000$) (Hausmann et al., 2013) (Fig. 15B). An integrated optical thermal switch was fabricated with two temperature-tuned WGs coupled to a resonator, reaching switching efficiencies up to 73%. SNSPDs have been successfully implemented in polycrystalline diamond surface grown on silica-on-silicon substrate (Kahl et al., 2016; Rath et al., 2015a).

4.10 Quantum photonic systems in bulk diamond

The first attempts to implement photonic circuits in diamond were carried out in bulk. The latter is preferred because color centers have superior emission properties in a bulky single crystal. The development of bulk diamond chips appears to have been dramatically slowed down by structural instabilities and complicated structuring techniques (Hiscocks et al., 2008). The main problem is the need for free-standing structures. However, the use of triangular ion etching enables the implementation of optical elements such as optical WGs, photonic crystals and cavities in bulk diamond. The recently developed laser writing technique in bulk diamond shows promising potential for the production of integrated flat shares (Sotillo et al., 2016). There are obviously still some challenges and difficulties (which will be discussed in more detail in Section 6) before the first fully functional IQPC can emerge. However, the excellent optical properties of color centers motivate strong research in this direction and make them a potentially competitive solution for the future.

5. Heterogeneous waveguide systems

Hybrid or heterogeneous integration allows the creation of photonic circuits composed of two or more different materials, each offering unique, usually complementary, optical properties. Such a "best of both worlds" approach in principle allows incorporation of desirable functionalities, provided by the different materials, to a single photonic chip.

Silicon-based photonic integrated circuits, for instance, are very promising for large system scaling, as foundry services offer the fabrication of user-designed, high-quality integrated photonic circuits comprising thousands of elements on shared-project wafers (Hochberg and Baehr-Jones, 2010). Furthermore, optical losses in silicon photonic WGs and related linear elements such as beam splitters, phase delays, linear filters, etc., can be made sufficiently low through design and process control, which allows large-scale photonic circuits (Carolan et al., 2015b). Thermal or carrier injection-based refractive index control, in addition, allows tunable phase delays, filters and modulators (Komljenovic et al., 2018), which are important in both classical and quantum photonics. Lastly, silicon also offers pronounced Kerr nonlinearity, which, together with strong field confinement and dispersion control enabled by a high refractive index ($\approx$3.5 for Si, $\approx$2 for SiN), allows efficient nonlinear optical processes (Li et al., 2016). Silicon's indirect bandgap, however, makes it an inefficient optical gain medium, imposing challenges for the creation of photonic circuits with elements such as on-chip electrically pumped tunable lasers, amplifiers and detectors. To circumvent this issue, heterogeneous integration processes started being developed in the mid-2000s, allowing introduction of direct-bandgap, III–V compound semiconductors to silicon-on-insulator (SOI) wafers. Since the demonstration of the first silicon photonic laser with optical gain III–V semiconductor-based quantum wells (Fang et al., 2006), integration of other on-chip elements such as amplifiers, modulators, detectors, followed, with faster growth rate of devices per chip than homogeneous III–V devices (Komljenovic et al., 2018). Importantly, most of the early development of heterogeneous integrated photonic circuits has aimed at classical applications, primarily optical communications. Aiming at scalability, such development has been based on entirely top-down fabrication processes, starting from a bonded wafer stack. As we show below, many such strategies have been applied toward quantum photonic circuits. The main focus of heterogeneous integration for quantum photonics has been the incorporation of single-photon sources and detectors onto otherwise passive, low-loss, and (potentially) reconfigurable integrated photonic circuits. We next describe relevant efforts to date in these two directions, with considerable technical detail.

5.1 Heterogeneous quantum integrated photonic circuits with quantum emitters

The introduction of solid-state quantum emitters (Aharonovich et al., 2016) onto otherwise purely passive integrated photonic circuits creates on-chip capabilities akin to those made possible through the introduction of III–V semiconductor-based optical gain media onto silicon photonics. Quantum emitters can act as high rate, on-demand sources of indistinguishable single photons (Ding et al., 2016; Liu et al., 2019; Somaschi et al., 2016), providing the large on-chip photon fluxes necessary for a variety of quantum photonic systems. These include boson sampling type simulators simulators (Loredo et al., 2017; Wang et al., 2017). Emitters with optically addressable spins may furthermore act as stationary qubits in photonic networks, and, along similar lines, single-photon nonlinearities in single-emitter quantum cavity-electrodynamic systems (Javadi et al., 2015; Sun et al., 2018) may allow networks of deterministic quantum logic gates to be implemented.

A number of types of quantum emitters have been integrated onto photonic circuits to date, primarily using the techniques described in Section 3. We next describe some of the most relevant work demonstrated to date.

The work of Mouradian et al. (2015) demonstrated integration of a SiN photonic circuit with long-lived quantum memories based on NVs in diamond, as shown in Fig. 16A. The diamond quantum nodes consisted of micron-length WGs of ≈200 nm cross section, bridging two low-loss SiN WGs over an air gap. Photonic design was such that 86% of the NV zero-phonon line fluorescence intensity at 638 nm would be launched into

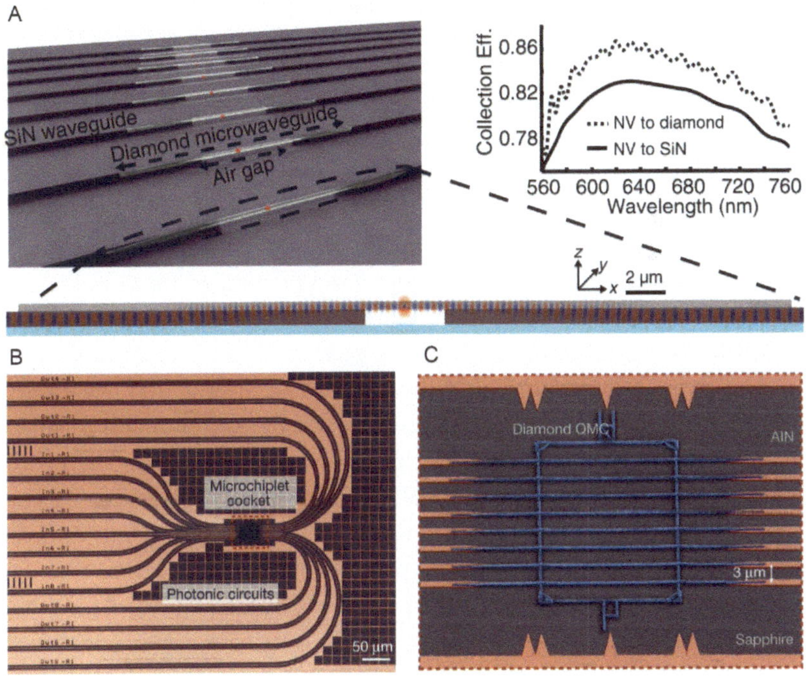

Fig. 16 (A) SiN photonic circuit with integrated diamond WGs containing NV centers. The simulated electric field produce by a single NV center at the WG center can be seen at the bottom of the figure, showing efficient light coupling to the SiN WG. *Top right inset*: dipole coupling efficiency spectrum into the diamond and SiN WGs. (B) AlN photonic circuit for large-scale diamond WG coupling. The micro-chiplet socket indicated at the center receives the diamond WG array chiplet in (C), via a pick-and-place approach. *Panel (A): Reprinted from Mouradian, S.L., Schröder, T., Poitras, C.B., Li, L., Goldstein, J., Chen, E.H., Walsh, M., Cardenas, J., Markham, M.L., Twitchen, D.J., Lipson, M., Englund, D., 2015, Scalable Integration of Long-Lived Quantum Memories into a Photonic Circuit. Phys. Rev. X 5 (3), 031009. https://doi.org/10.1103/PhysRevX.5.031009 with permission from the American Physical Society; Panel (B): Reprinted from Wan, N.H., Lu, T.-J., Chen, K.C., Walsh, M.P., Trusheim, M.E., De Santis, L., Bersin, E.A., Harris, I.B., Mouradian, S.L., Christen, I.R., Bielejec, E.S., Englund, D., 2019. Large-scale integration of near-indistinguishable artificial atoms in hybrid photonic circuits. arXiv:1911.05265 [quant-ph], http://arxiv.org/abs/1911.05265 with permission.*

the diamond WG, in both directions, for an optimally positioned NV center. The diamond WGs also featured adiabatic mode transformers at the ends, consisting of linear width tapers down to 100 nm over 4 μm in length, for highly efficient light (96%) transfer into SiN WG. The pick-and-place method utilized to experimentally produce the photonic circuits allowed separate fabrication and precharacterization of the diamond WGs separately from the SiN WGs, and ultimately the deterministic integration of diamond WGs containing single, stable, negatively charged NVs onto the circuit. Experimentally, the fabricated devices were investigated at room temperature by confocally exciting single NV devices from the top with a 532 nm pump laser, and collecting the fluorescence both through the objective and through the SiN WGs, using a lensed single-mode fiber. The objective-collected signal showed second-order autocorrelation with $g^{(2)}(0)$ as low as 0.07, indicating good isolation between the NV and other fluorescence sources in the diamond WG. Cross-correlation between the fiber- and objective-collected signals on the other hand yielded $g^{(2)}(0) \approx 0.48$, however, indicating the presence of considerable background emission, which was attributed to fluorescence due to scattering of the excitation laser into the SiN WGs. Nonetheless, an estimated flux of 1.45×10^6 NV-emitted photons/s was collected into one direction of the single-mode SiN WG. Cryogenic photoluminescence excitation (PLE) measurements at 18 K revealed NV zero-phonon linewidths of <393 MHz, approximately 30 times wider than the natural linewidth for the $m_s = 0$ transition in bulk diamond. The broadened spectrum was at least partially due to spectral diffusion of the NV emitter, which was faster than the experimental integration times. Through a Hahn-echo experiment, an electron spin coherence time of $T_2 \approx 120$ μs was measured, which was comparable to that observed in the bulk diamond used to produce the WGs. To circumvent the issue of qubit inhomogeneities in the path toward large-scale quantum photonic integration, work by Wan et al. demonstrated that the pick-and-place of arrays of diamond nano-WGs at a time onto an AlN photonic circuit, as shown in Fig. 16B (Wan et al., 2019). The process allowed inclusion of a 72-channel array of germanium-vacancy (GeV) and silicon-vacancy (SiV) centers embedded in the nanodiamond WGs. Photoluminescence spectroscopy at below 4 K temperatures revealed close to lifetime-limited emission linewidths for the two types of color centers. Strain-based tuning of the emission wavelengths was also demonstrated, via electrostatic actuation of the suspended nano-WGs. Although the demonstration in this work only involved single-photon launching into the AlN, the material offers a number of desirable optical properties, such as $\chi^{(2)}$ nonlinearities, as well as electro-optic and piezoelectric properties that can be used to create a number of on-chip functional elements (Xiong et al., 2012b).

The first report of hybrid integration of epitaxial QD-based single-photon sources on a low-loss silicon-based photonic circuit was by Murray et al. (2015). In this work, a bare strip of a GaAs wafer containing InAs QDs was orthogonally bonded to the cleaved facet of a Silicon Oxynitride (SiON) WG-based Mach–Zehnder interferometer (MZI), as shown in Fig. 17A. Coupling of QD single photons into one of the MZI input WGs relied on the chance spatial alignment of an individual dot to the former; without a proper GaAs geometry to help funnel QD emission into the input WG, a theoretical maximum collection efficiency of about 3% was predicted. The MZI featured integrated, nickel-chromium alloy heaters that allowed tuning of the its coupling ratio, from ≈50% to ≈100% upon application of voltages up to 20 V. Operating at 4 K, the MZI was used as a splitter in a Hanbury Brown and Twiss (HBT) measurement, yielding $g^{(2)}(0) \approx 0.09$. In a later publication (Ellis et al., 2018), the same group combined an array of tuneable InGaAs/GaAs QD single photon sources with the SiON WG circuit. The same bonding approach as in Murray et al. (2015) was used here, though the QD wafer featured a regular array of emitters, with defined electric contacts, at a compatible spacing with the SiON WGs (Fig. 17B). The QDs were embedded in a diode heterostructure comprised of a weak planar cavity with 4 and 10 pairs of GaAs/AlGaAs distributed Bragg reflectors above and below a $\lambda/2$ spacer. A coupling efficiency of ≈8% into the SiON was predicted for such source, and spectral tunability was enabled through the DC Stark effect. Each WG in the SiON circuit could be addressed by a separate, electrically controlled QD-containing diode. Emission from neighboring diodes were independently tuned to degeneracy, and allowed observation of Hong–Ou–Mandel (HOM)-type interference with a visibility of ≈80%, limited by source purity. The authors also used electrically pumped wetting layer light, guided by the planar cavity, to pump individual QDs in separate diodes, demonstrating the possibility of an on-chip pump for single-photon emission.

In the work by Zadeh et al., the pick-and-place technique was used to produce hybrid devices as shown in Fig. 18A, composed of InP nanowires (NW) containing single InAsP QDs embedded in SiN WGs (Zadeh et al., 2016). The high index contrast between the InP and SiN helped promote high collection efficiency for QD-emitted photons into a confined InP WG mode, and a slight tapering of the NW tip, with geometry controllable

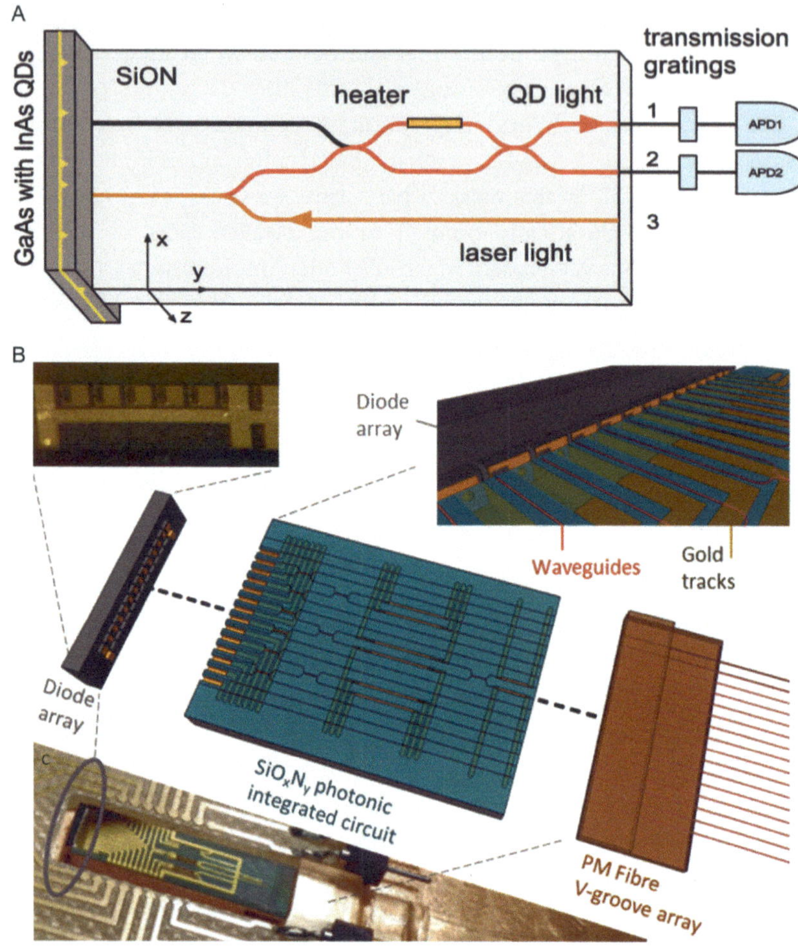

Fig. 17 (A) SiON photonic circuit with side-bonded GaAs wafer containing InAs QDs. The QDs here were in bulk GaAs. (B) SiON photonic circuit with side-bonded GaAs wafer containing InAs QD single-photon sources based on planar microcavities. The low-loss SiON photonic circuit was used to demonstrate interference of photons produced by two independent QD sources. *Panel (A): Reprinted from Murray, E., Ellis, D.J.P., Meany, T., Floether, F.F., Lee, J.P., Griffiths, J.P., Jones, G.A.C., Farrer, I., Ritchie, D.A., Bennett, A.J., Shields, A.J., 2015. Quantum photonics hybrid integration platform. Appl. Phys. Lett. 107 (17), 171108 https://doi.org/10.1063/1.4935029 with permission from the American Institute of Physics; Panel (B): Reprinted from Ellis, D.J.P., Bennett, A.J., Dangel, C., Lee, J.P., Griffiths, J.P., Mitchell, T.A., Paraiso, T.K., Spencer, P., Ritchie, D.A., Shields, A.J., 2018. Independent indistinguishable quantum light sources on a reconfigurable photonic integrated circuit. Appl. Phys. Lett. 112 (21), 211104. https://doi.org/10.1063/1.5028339 with permission from the American Institute of Physics.*

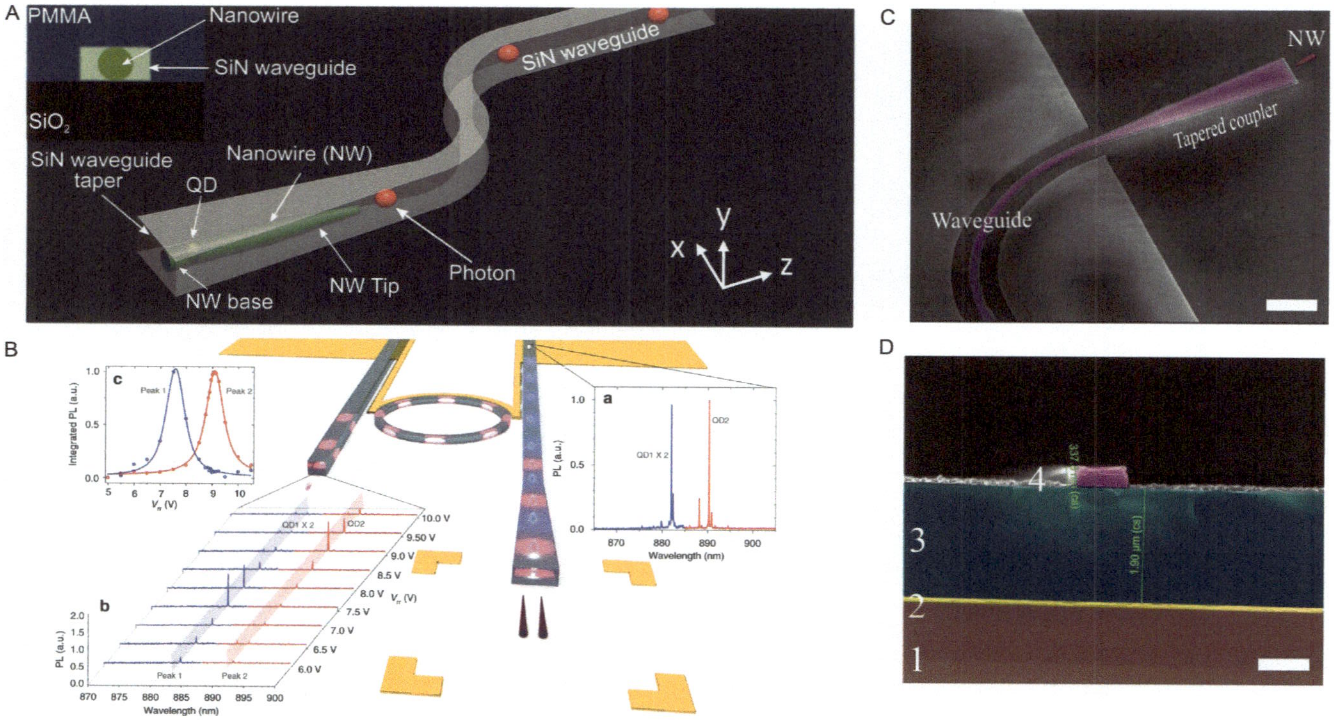

Fig. 18 See figure legend on opposite page.

through growth parameters, allowed efficient launch of dot emitted photons into the SiN WG, with a theoretical maximum of 36%. The III–V nanowire and QD were grown by a selective-area and vapor–liquid–solid epitaxy process. These microns-long, ≈100 nm diameter wurtzite nanocrystals were plucked from the growth substrate using a tungsten tip mounted on a 3-axis movable stage, and transferred to a silicon chip, with <500 nm positioning precision. The receiving chip consisted of a Si wafer covered by a 2.4 μm SiO_2 film, which served as a low-index substrate layer, and featured prefabricated markers used to align the photonic circuits with respect to the nanowires. The sample was covered in SiN grown by plasma-enhanced chemical deposition (PECVD), and lithography and etching of the SiN were used to define the photonic circuits. The work in Zadeh et al. (2016) showed coupling of single photons from the InAsP QDs into a low-loss (≈2.5 dB/cm at 880 nm) SiN WGs with an efficiency ≈24%. The single-photon nature of the emission was verified through an HBT measurement of the QD photoluminescence signal under nonresonant excitation, collected from the WG facet, and showed multiphoton probability $g^{(2)}(0) \approx 0.07$. Importantly, the QD emission linewidth was of ≈3.65 GHz.

Fig. 18 (A) Hybrid SiN WG with InP-based nanowire/QD-based single-photon source. (B) Schematic of SiN microring resonator with implemented add-drop filter, the circuit incorporates nanowire/QD single-photon sources as in (A). Two single-photon sources, with different spectral characteristics, are pictured, coupling into the microring resonator via a bus WG on the right-hand side. The microring drops signals resonant with its whispering-gallery modes, into the bus WG to the left-hand side. Microheaters, depicted in *yellow*, allow the microring add-drop filter to be spectrally tuned. (C) Scanning electron micrograph of SiN WG and InP nanowire (NW) single-photon source, where NW emission could be tuned via strain induced from the magnesium niobate–lead titanate (PMN–PT) substrate on which the circuit was produced. (D) False-color SEM of the cross section of the SiN WG in (C). Regions labeled 1, 2, 3, and 4 are the piezoelectric PMN-PT substrate, gold electrode, silicon oxide substrate and silicon nitride WG, respectively. *Panel (A): Reprinted from Zadeh, I.E., Elshaari, A.W., Jöns, K.D., Fognini, A., Dalacu, D., Poole, P.J., Reimer, M.E., Zwiller, V., 2016. Deterministic integration of single photon sources in silicon based photonic circuits. Nano Lett. 16 (4), 2289–2294. https://doi.org/10.1021/acs.nanolett.5b04709 with permission from the American Chemical Society; Panel (B): From Elshaari, A.W., Zadeh, I.E., Fognini, A., Reimer, M.E., Dalacu, D., Poole, P.J., Zwiller, V., Jöns, K.D., 2017. On-chip single photon filtering and multiplexing in hybrid quantum photonic circuits. Nat. Commun. 8 (1), 379. doi:10.1038/s41467-017-00486-8. The license is Creative Commons; Panels (C, D): Reprinted from Elshaari, A.W., Büyüközer, E., Zadeh, I.E., Lettner, T., Zhao, P., Schöll, E., Gyger, S., Reimer, M.E., Dalacu, D., Poole, P.J., Jöns, K.D., Zwiller, V., 2018. Strain-tunable quantum integrated photonics. Nano Lett. 18 (12), 7969–7976. https://doi.org/10.1021/acs.nanolett.8b03937 with permission from the American Chemical Society.*

While this first demonstration was mostly concerned with outcoupling single photons into the low-loss SiN WG, Elshaari et al. (2017) later used the same basic platform to demonstrate filtering and multiplexing of single QD emission by on-chip SiN microring resonator-based add-drop filters, as shown in Fig. 18B. The SiN add-drop filters, designed to for critical coupling to SiN bus WGs, had a free spectral range (FSR) $\approx$0.96 nm with full width at half-maximum (FWHM) of $\approx$0.13 nm at the QD emission wavelength with, and incorporated metal microheaters which allowed thermo-optic wavelength tuning. The ring resonator's PMMA top cladding displayed a pronounced negative thermo-optic coefficient, $\approx 10 \times$ that of SiN, which allowed tuning of the whispering-gallery modes by 120% of the FSR with $\approx$15 V applied to the electrode heaters, and covering $\approx$40 nm of QD spectral range. Importantly, the photonic circuit design sought to minimize the thermal coupling between the filter electrodes and the QDs, the emission of which red-shifted slightly due to thermal coupling. Nonetheless, the tunable filter was used to drop one of the QD's trion emission line into an output WG, eliminating both the QD's exciton line and the nonresonant, 532 nm wavelength pump from the original WG-coupled signal, with suppression of $\approx$95 dB. The resulting $g^{(2)}(0)$ $\approx$0.4 for the dropped trion line demonstrated the effectiveness of the ring resonator as an on-chip spectral filter. A similar add-drop structure was also used to multiplex emission from two spectrally distinct QD sources placed at two of the microring inputs, as shown in Fig. 18B. In Elshaari et al. (2018), the same group demonstrated a SiN photonic circuit on piezoelectric magnesium niobate–lead titanate (PMN–PT) substrate, and used their microprobe-based pick-and place technique to incorporate QD nanowire single-photon sources. The piezoelectric substrate allowed voltage-controlled, mechanical strain-based spectral tuning of the QD emission, with a tuning rate of as much as 1.33 pm/V, as well as tuning of a microring resonator filter by a compatible rate of 0.96 pm/V.

In the work by Kim et al. (2017), pick-and-place was done in a focused ion beam (FIB), yielding hybrid devices that consisted of an InP nanobeam containing InAs QDs (emitting in the 1300 nm telecom band), on top of Silicon-on-insulator (SOI) photonic WGs. In contrast with the nanowire-based platform of Zadeh et al. (2016) and Elshaari et al. (2017), QD-containing nanobeam geometries were produced through lithography and etching steps. This allowed the creation of a photonic crystal back-reflector on the nanobeam, consisting of a periodic array of $\approx$100 nm diameter etched holes, which was used to direct the QD emission to one direction. Lithography also allowed the creation of mode transformers at the end of the nanobeam, for efficient power transfer into the silicon WG. An outcoupling efficiency of ≈32% was theoretically predicted into the Si WG, for an ideally positioned QD. The theoretical efficiency breakdown was $\beta = 71\%$ of QD emission into the nanobeam's fundamental TE guided mode, and coupling efficiency of 45% from such guided mode to the silicon WG, through an adiabatic mode transformer. An on-chip silicon photonic 50:50 beam splitter was implemented, which directed photoluminescence from a single QD between to WGs that were terminated into two free-space grating outcouplers. Emission lines from the same QD could be seen in light collected from the two grating couplers, and an HBT measurement revealed $g^{(2)} \approx 0.25$ for one of the lines, proving single-photon emission. In a more recent development, using the same FIB microprobe-based pick-and-place technique, the same group integrated InP nano-WGs containing InAs QDs onto $LiNbO_3$ WGs (Aghaeimeibodi et al., 2018), and demonstrated HBT single-photon correlation measurements with an on-chip beam splitter and off-chip detection. Although the electro-optic and nonlinear optical properties of $LiNbO_3$ were not explored, this work showed that efficient coupling of QD emission into $LiNbO_3$ materials is possible, which creates many opportunities for fast on-chip single-photon switching and wave-packet manipulation (Rakher and Srinivasan, 2011).

In the work by Katsumi et al. (2018), InAs QD-containing GaAs nanobeam photonic crystal cavities evanescently coupled to underlying GaAs WGs were produced through a transfer printing process. Here, a polydimethylsiloxane (PDMS) stamp was used to lift suspended GaAs nanobeams from a processed GaAs wafer. The destination substrate consisted of GaAs-on-SiO_2 wafer onto which GaAs ridge WGs were etched, covered by a spin-on-glass (SOG) top cladding. The small mode volume (≈0.5 cubic wavelengths) of the designed nanobeam photonic crystal cavity's fundamental mode allows β-factors of ≈99% to be achieved, primarily through Purcell radiative rate enhancement, even for loaded quality factors Q in the few thousands. Importantly, Q is dictated by the coupling rate between the cavity and the WG, which is determined by the vertical distance d between the two. Although the coupling efficiency into the WG can be limited by parasitic losses to radiative modes as the distance d between guide and cavity goes below a certain point, efficiencies in excess of 99% are still achievable theoretically. Experimentally, the estimated emitter–cavity coupling efficiency was $\beta \approx 87\%$, the cavity-WG coupling efficiency was $\eta \approx 72\%$, for a total single-photon launch efficiency of 63% into the GaAs WG. Similar performance was also predicted for Si_3N_4 and Si WGs (in fact, integration of

GaAs/InAs QD single-photon sources onto glass-clad silicon photonic WG processed by a CMOS foundry was demonstrated in Katsumi et al. (2019), shown in Fig. 19). More recently, the same technique was used to demonstrate thermally tunable QD single-photon sources on CMOS silicon photonic circuits, implemented by adding a thermal heating pad to the GaAs device (Katsumi et al., 2020). Independent QD spectral tuning of up to ≈1 nm was demonstrated, though corresponding to an effective temperature swing of ≈15 K. This capability nonetheless allowed in situ wavelength matching between two dissimilar QD sources integrated on the same silicon chip. Multiplexing of the QD source signals in an on-chip 2 × 2, multimode interference coupler based 50:50 splitter was also demonstrated, though two-photon interference effects were not observed. In work by Osada

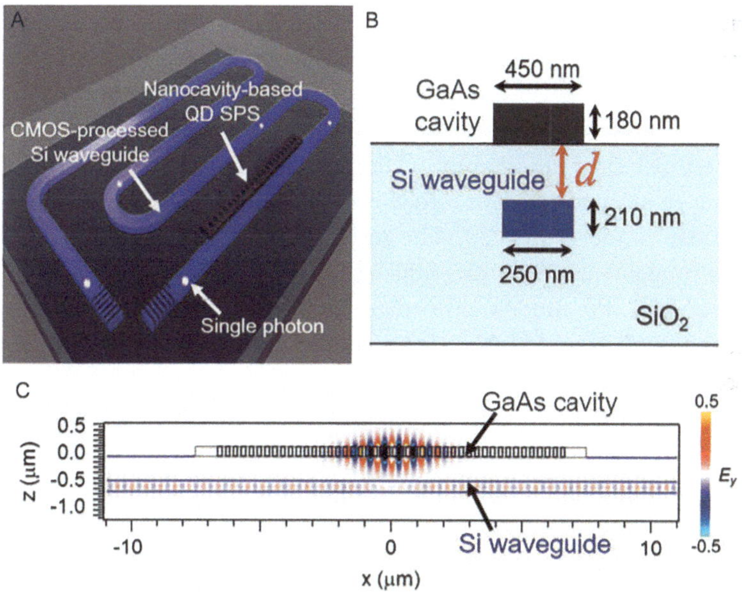

Fig. 19 CMOS-based integrated photonic circuit with on-chip single-photon sources (SPS) based on GaAs nanobeam photonic crystals containing InGaAs QDs (QDs), from Katsumi et al. (2019). (A) Artistic rendering of Si photonic WG with evanescently coupled GaAs photonic crystal nanobeam. (B) Cross section of the circuit in (A), at the nanobeam position. (C) Simulated electric field of photonic crystal cavity mode, showing evanescent coupling to the underlying Si WG. *Reprinted from Katsumi, R., Ota, Y., Osada, A., Yamaguchi, T., Tajiri, T., Kakuda, M., Iwamoto, S., Akiyama, H., Arakawa, Y., 2019. Quantum-dot single-photon source on a CMOS silicon photonic chip integrated using transfer printing. APL Photon. 4 (3), 036105. https://doi.org/10.1063/1.5087263 with permission from the American Institute of Physics.*

et al. (2019), the same group also demonstrated strongly coupled cavity QED systems, based on single InAs QDs in GaAs photonic crystal cavities, on a CMOS silicon photonic chip.

The work by Davanco et al. (2017), showed that direct wafer bonding could be leveraged to produce Si_3N_4 WG-based photonic circuits incorporating single QDs as single-photon sources. In this work, a heterogeneous GaAs/Si_3N_4 stack was produced with a low temperature, oxygen plasma-activated wafer bonding procedure (Fang et al., 2006). The stack consisted of a silicon substrate covered by 3 μm of thermal SiO_2, 550 nm of stoichiometric Si_3N_4 produced through low-pressure chemical vapor deposition (LPCVD), and a ≈200 nm thick layer of GaAs containing InAs QDs at half thickness. After wafer bonding, fabrication proceeded through two subsequent, aligned electron-beam lithography and etching steps, and could produce circuits as shown schematically in Fig. 20A. In these, compact GaAs nanophotonic structures hosted single InAs QDs, and were designed to launch emitted photons efficiently into the Si_3N_4 WGs composing the photonic circuit. The GaAs nanophotonic structure comprised a photon capture WG and a mode transformer, both with nanometer-scale features defined by electron-beam lithography. The photon capture allowed efficient coupling of emitted photons into the fundamental transverse electric (TE) GaAs ridge mode in Fig. 20. The mode transformer converted the latter into the fundamental TE Si_3N_4 ridge mode shown in Fig. 20D. As indicated in Fig. 20A–C, the mode transformer consisted of an adiabatic width taper for both the GaAs and Si_3N_4 ridges. A number of high-resolution geometries were generated in the GaAs layer, such as nano-WGs (Fig. 20A), WG-coupled microring resonators, (Fig. 20A and B all of which were efficiently coupled to the underlying Si_3N_4 WGs via adiabatic mode transformers, as indicated in Fig. 20A. Although the III–V material contained a high density of randomly positioned QDs (>100 μm^2), single-photon emission from single QDs coupled to nano-WGs and microring resonators were observed, by pumping the QDs in the p-shell with a tunable laser. For these devices, the QD single-photon emission (with raw $g^{(2)}(0) < 0.4$ overall) was launched, with >90% efficiency, into the Si_3N_4 WG, via the adiabatic mode transformers. A further, relevant detail is that the emitted single photon streams were collected from the Si_3N_4 WG facets with a lensed optical fiber in endfire configuration, as typically done for testing classical integrated photonic devices. The fiber-to-Si_3N_4 coupling efficiency was of ≈20%, which, while also typical of integrated photonic devices, can potentially be significantly improved by proper Si_3N_4 mode transformer design.

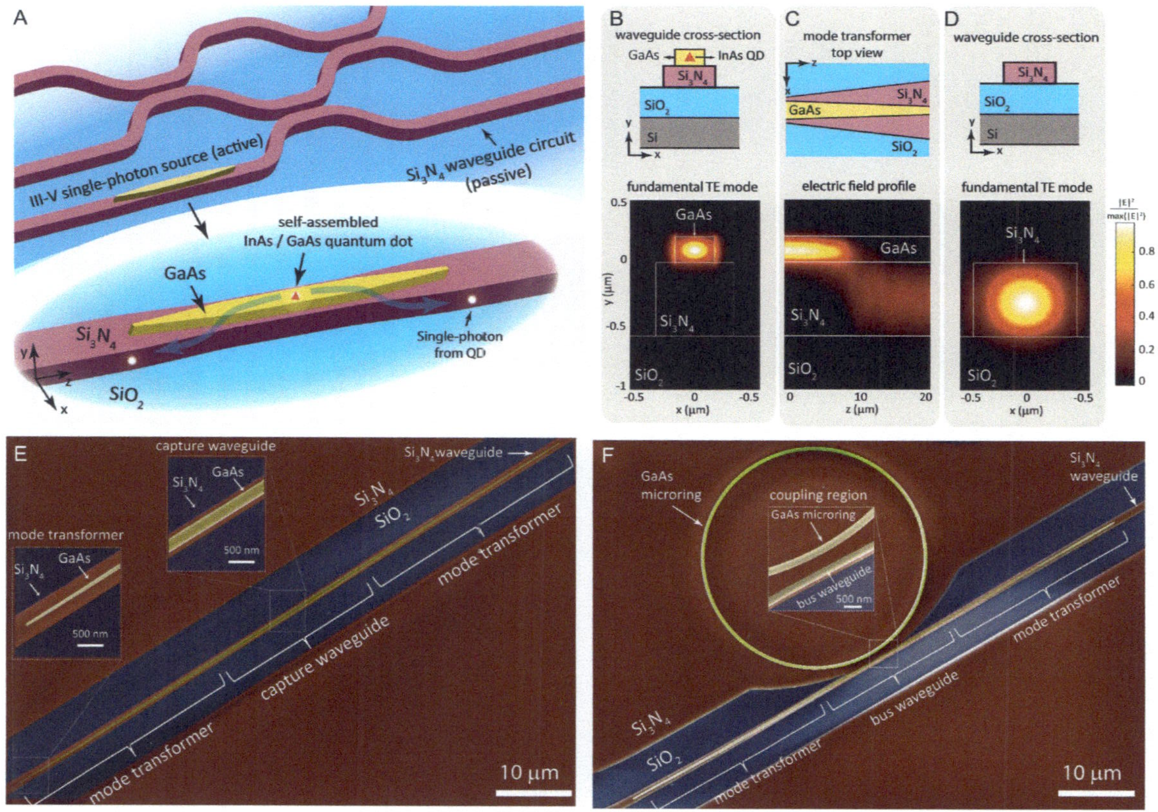

Fig. 20 See figure legend on opposite page.

Regarding β-factors, the nano-WG were designed to feature $\beta = 0.35$ for the fundamental transverse electric (TE) GaAs ridge mode in one direction, and for a QD with transverse dipole moment at the center of the GaAs ridge, even though the structure was not fully optimized (a design with $\beta > 0.45$ is provided in the publication). For a single dot in a representative, fabricated GaAs nano-WG, $\beta \approx 0.20$ was estimated—a value that is not too far from the theoretical maximum. It is worth noting that the QDs were not deterministically positioned in the nano-WG in this case, and so both positions and dipole moment orientations were likely suboptimal. Lifetimes observed in nano-WGs were of $\approx$1.5 ns, close to the "natural" value for an InAs QD in bulk. No radiative rate enhancement was expected here, as nano-WGs are nonresonant. In contrast, in microring resonators, weak-coupling of single dots to whispering-gallery modes of quality factors in the 10^3 to 10^4 range led to significantly reduced radiative lifetimes, with an estimated Purcell factor $F_p \approx 4$. This value was not too far from the theoretically predicted based on the microring Q-factor and mode volume $V_{\text{eff}} \approx 75(\lambda/n)^3$. This work extended the application space of a mature, scalable, top-down heterogeneous photonic integrated circuit platform into the quantum realm. The technique is unique in allowing independent, flexible, and high-resolution tailoring of both active and passive photonic circuit elements with precise and repeatable, sub-50 nm alignment defined strictly by lithography. The underlying Si_3N_4 WGs demonstrated here provide not only a low-loss medium for single-photon propagation on-chip, but also the possibility of

Fig. 20 Wafer bonding-based heterogeneous Si_3N_4 photonic circuits with single-photon sources based on InAs QDs in GaAs nanophotonic geometries. (A) Idealized Si_3N_4 photonic circuit showing passive WGs in pink and local III–V semiconductor-based single-photon sources. The source depicted comprises a GaAs ridge WG on top of a Si_3N_4, with cross section as shown in (B), *top*. The GaAs ridge contains a single QD, which emits single photons into the transverse electric (TE) mode in (B), with high efficiency. (C) *Top*: Adiabatic mode transformer, implemented through linear width tapers for both GaAs and Si_3N_4 ridges, allows efficient transfer of emitted QD light from the mode in (B) to the Si_3N_4 mode in (D). (E) False-color scanning electron micrograph of a fabricated nano-WG with mode transformers, as depicted in (A). The insets show details of the photon capture and mode transformer sections of the device. (F) False-color scanning electron micrograph of a fabricated, hybrid microring resonator evanescently coupled to a bus WG. The WG has mode transformers that allow the access to the Si_3N_4 WG. *Reprinted from Davanco, M., Liu, J., Sapienza, L., Zhang, C.-Z., De Miranda Cardoso, J.V., Verma, V., Mirin, R., Nam, S.W., Liu, L., Srinivasan, K., 2017. Heterogeneous integration for on-chip quantum photonic circuits with single quantum dot devices. Nat. Commun. 8 (1), 889 https://doi.org/10.1038/s41467-017-00987-6 with permission from Springer Nature.*

using nonlinear optical processes for functionality. An example is four-wave-mixing-based quantum frequency conversion, which has been demonstrated in Si_3N_4 microring resonators (Li et al., 2016; Singh et al., 2019) made out of WGs with comparable dimensions and the same materials as in Davanco et al. (2017). The introduction of elements such as on-chip delay lines, high-quality Si_3N_4-based filters, and microring add-drops, can also be envisioned, as well as integration with WG-based superconducting single-photon nanowire detectors (Pernice et al., 2012). Finally, the fabrication process can be adapted for materials such as AlN and $LiNbO_3$, which may enable active electro-optic phase control.

While the work from Davanco et al. (2017) featured devices with randomly positioned QDs, in Schnauber et al. (2019) heterogeneous nano-WGs containing precisely positioned single QDs were deterministically produced via the in situ EBL technique of Gschrey et al. (2015b,c), Schnauber et al. (2018), and Mrowinski et al. (2019), see Section 3. To fabricate the devices, single QDs were initially located through cathodoluminescence spectroscopy and imaging, on a heterogeneous bonded wafer. Immediately after localization, proximity-corrected gray scale in situ EBL was performed at 7 K with unchanged beam column settings to define the GaAs WG taper patterns and alignment marker for the Si_3N_4 layer, both aligned to the identified QDs. The Si_3N_4 WG patterns were defined as in Davanco et al. (2017), using the GaAs alignment markers. An important result of this work was that, for a QD located sufficiently far from the GaAs etched sidewalls, postselection indistinguishable single-photon emission could be observed. The QD here was located in the middle of a $\approx$800 nm GaAs ridge, and was pumped in the p-shell. High-resolution spectra of one of the QD's emission lines (likely a trion) revealed a Voigt lineshape with a linewidth of $\approx$2.2 GHz full width at half-maximum (FWHM). The Lorentzian component of $\approx$1 GHz FWHM suggested homogeneous broadening beyond the Fourier limit of $\approx$0.1 GHz, likely due to dephasing from phonon interactions. The Gaussian component of width $\approx$1.5 GHz FWHM suggested inhomogeneous linewidth broadening due to spectral diffusion. A Hong–Ou–Mandel (HOM) measurement of the single-photon stream revealed clear evidence of two-photon interference with a visibility of $\approx$89%, and a coherence time $\tau_c \approx 0.33$ ns. The coherence time here indicates a postselection time window where indistinguishable photons are available. Precise alignment of the EBL patterns with respect to the QDs is essential to avoid excessive proximity to etched sidewalls, which may lead to degradation of quantum efficiency and, especially, coherence (Liu et al., 2018). The observation of a 2.2 GHz linewidth from a positioned QD emission line, and subsequent demonstration of two-photon interference, suggested that the required precision can be met in a heterogeneous integration platform, via the chosen the positioning method.

Monolayer transition metal dichalcogenides and (TMDCs) and monolayer and multilayer hexagonal boron nitride (hBN) have been shown to support spatially localized single photon emitters, producing light at visible and near-infrared wavelengths. Peyskens et al. (2019) have demonstrated integration of monolayer WSe_2 single-photon emitters with SiN WGs. The monolayer WSe_2 here was produced though an exfoliation process, and the transfer onto the SiN WG led to formation of localized emitters at stress points at the top edges of the SiN ridge. The emitter launched single photons into the SiN, as verified by HBT measurements. The maximum theoretical single-photon coupling efficiency was $\approx$8%, relatively low due to a lack of an optimized photonic structure to capture and launch photons into the WG. A strategy to improve this efficiency through cavity coupling is discussed in the publication. Single emitters based on WSe_2 monolayers were also integrated onto a Titanium in-diffused lithium niobate directional coupler in work by White et al. (2019), however very low single-photon collections efficiencies ($\approx$0.1%) prevented observation of the HBT effect.

The work of Khasminskaya et al. (2016), demonstrated a nanoscale, electrically driven single-photon emitter using WG-coupled semiconducting single-walled carbon nanotubes (sc-SWCNTs), coupled to a silicon photonic circuit. The circuit consisted of a Si_3N_4 WG hosting, at its center, a CNT that was contacted by gold electrodes, for carrier injection. The circuit also featured niobium nitride (NbN) superconducting nanowire single-photon detectors (SNSPDs) defined on top and at the two ends of the Si_3N_4 WG. The SNSPD detectors comprised a single meander with 80 nm width and 100 nm gap size, a critical temperature $T_c = 10.7$ K, and a superconducting transition width $T_c = 0.5$ K. Interestingly, because the nanotube emitter was located at the center of the WG, emission toward the two detectors at the WG ends was equally likely, so the entire configuration was equivalent to an HBT setup. Generation of single photons from sc-SWCNTs was postulated to take place via radiative recombination of defect-localized excitons, after carrier injection through a Schottky contact. At 1.6 K, the nanotube's electroluminescence spectrum revealed a peak around 1370 nm, with a FWHM of $\approx$30 nm. Running the HBT measurement at various injection current levels yielded antibunching curves that indicated nonclassical light generation from the nanotube. Antibunching with $g^{(2)}(0) < 0.5$ was observed (within experimental noise) only for the lowest current injection level, though increased currents led to decreased levels of single-photon purity.

5.2 Heterogeneous quantum integrated photonic circuits with single-photon detectors

Detector integration is one of the necessary building blocks on the way to a fully integrated device. On-chip detection enables single-photon counting correlations photon number resolving measurements directly on chip, alongside single-photon generation. In addition, it allows one to avoid the significant photonic losses, electronic delays and wiring complexity involved in off-chip detection. Superconducting nanowire single-photon detectors (SNSPDs) have been extensively explored in quantum integrated photonic circuits. With quantum efficiencies of over 90%, response times in the picosecond regime, negligible dark counts, detection rates of over 100 MHz over a broad spectral range, and potential for photon-number resolution (Gerrits et al., 2011), SNSPDs are highly attractive for quantum photonics applications in general, albeit at the expense of the necessity of cryogenic temperature operation. Single-photon detection by an SNSPDs is based on the formation of a transient resistive barrier across an initially superconducting wire, following absorption of a single photon, which produces a measurable voltage pulse. The resistive barrier is formed by breaking of Cooper-pairs by the incident photon, which leads to formation of a hotspot across the wire. Superconductivity is later restored after healing of the hotspot, in time scales of typically tens of picoseconds. Critical for high-performance SNSPD operation are the quality of the superconducting material film, as well as with nanowire geometry and imperfections. The former consideration is associated with deposition methods as well as substrate quality, and the latter with high-resolution lithography and etching process quality. Both considerations are additional challenges in the creation of waveguide-integrated detectors. The work discussed below covers significant efforts to date, toward SNSPD integration within integrated photonic circuits.

In 2011, Sprenger et al. demonstrated NbN SNSPDs integrated onto GaAs ridge waveguides, following the same working principle as non-waveguided, superconducting nanowire detectors demonstrated up to that point. The detectors were deposited and patterned on top of GaAs ridge waveguides, and sensed the evanescent tail of the modal field on the surface. A detection efficiency of $\approx$20% telecom wavelengths was reported, with response time in the nanosecond range, and a timing accuracy of ≈60 ps, all at <4 K operating temperatures.

Around the same time, Pernice et al. (2012) demonstrated traveling wave niobium nitride (NbN) SNSPDs atop silicon nanophotonic WGs on a circuit, and have shown that such integration can lead to drastic increase of the absorption length for guided photons, as compared to normal incidence. An on-chip single-photon detection efficiency up to 91% at telecom wavelengths was achieved. Dark counts in the Hz range were achieved without compromising the efficiency, as well as a timing jitter of 18 ps. The high temporal resolution allowed observation of ballistic photon transport in silicon ring resonators produced on the same chip as the detectors.

In work by Schuck et al. (2016), sputter-deposited NbTiN SNSPDs were produced onto a Si_3N_4 photonic circuit that formed part of a Hong–Ou–Mandel (HOM) setup, as shown in Fig. 21B, featuring a directional coupler to combine incoming photons and route them toward two on-chip detectors. An on-chip detection efficiency of 11.5% and a typical dark count rate in the Hz range when biasing two detectors close to the critical current were achieved for operation at 1.7 K. These characteristics allowed the observation of two-photon interference with a visibility of 97% from an off-chip spontaneous parametric down-conversion photon-pair source.

Reithmaier et al. (2013) demonstrated integration of NbN SNSPDs on top of GaAs-on-AlGaAs WGs containing InAs QDs, with a detection efficiency of ≈0.1%, and a timing jitter of 72 ps. The low efficiency here was associated with the 10 nm NbN thickness. Nonetheless, in later work, a similar type of device was used to perform time-resolved excitation spectroscopy on single InAs QDs and demonstrate resonance fluorescence with a linewidth of ≈10 μeV, by temporally filtering the time-resolved luminescence signal (Reithmaier et al., 2015a).

On a similar GaAs photonic platform with InAs QDs, Schwartz et al. (2018) demonstrated HBT measurements entirely on chip. The device comprised InAs QDs in GaAs ridge WGs and a directional coupler beam splitter with integrated NbN SNSPDs, as represented in Fig. 13. Light emitted by a resonantly excited QD in a WG were split in a directional coupler beam splitter and then detected by the on-chip SNSPDs. To effectively suppress the excitation laser stray light, aluminum cover layers were produced above the SNSPDs and WG regions (buffered by an oxide layer), allowing operation without the need for spectral filtering or time gating. The single-photon nature of the QD emission was proven under continuous-wave

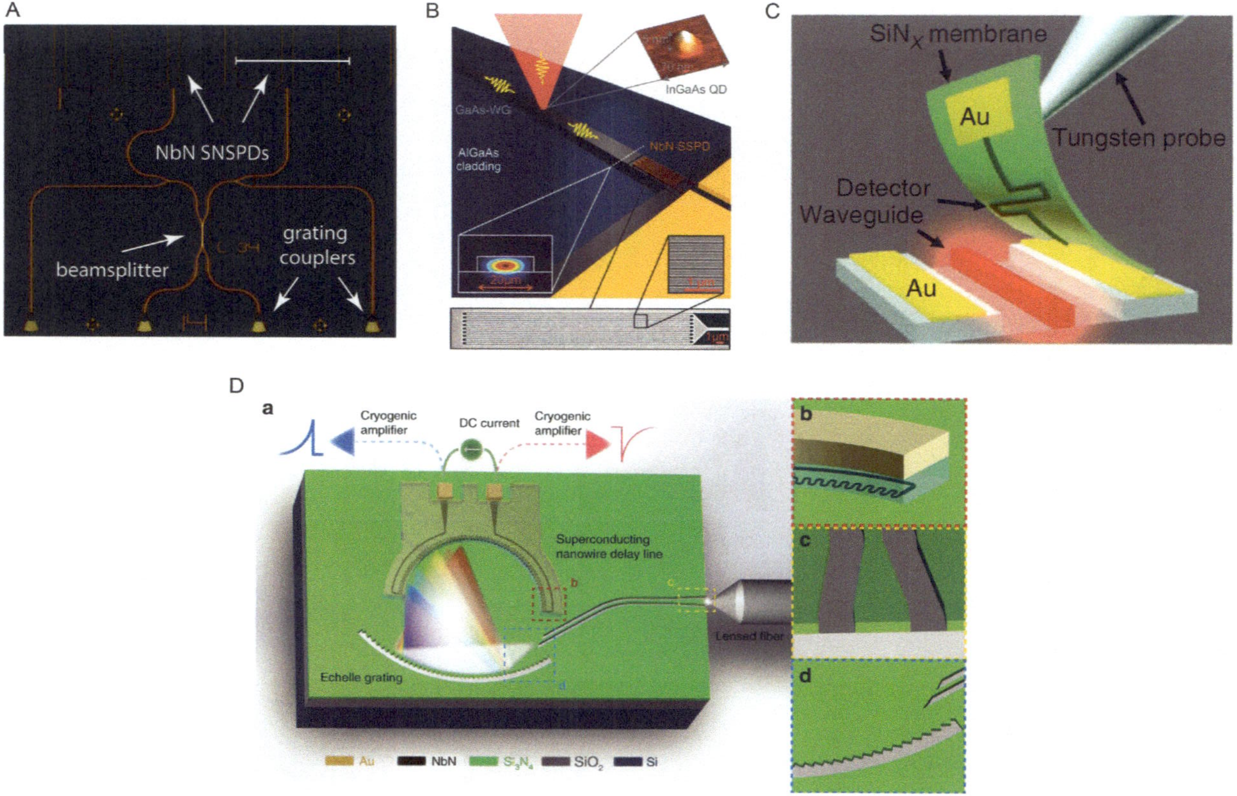

Fig. 21 See figure legend on opposite page.

and pulsed excitation, via the on-chip HBT circuit. The detection efficiencies for the two SNSPDs used in the experiment were significantly different, up to ≈47% and ≈16% close to the critical currents, and ≈1.8% and ≈21.8% for the bias level used for the HBT measurement.

While in all the work described above SNSPDs were produced directly onto the final photonic chip, the pick-and-place technique was used in the work of Najafi et al. (2015), to produce a hybrid SOI integrated photonic circuit chip for photon correlation measurements. The pick-and-place technique here was chosen to circumvent the typically low fabrication yield for SNSPDs, which is limited by nanoscale defects (Heeres et al., 2013; Sahin et al., 2013a). As suggested in Fig. 21C, the SNSPDs were produced on SiN_x membranes in a separate run than that of the SOI photonic circuit, and then transferred on top of the latter with a tungsten tip. The close proximity between the WG and the SNSPD allowed single-photon detection. This process allowed ten low-jitter detectors to be integrated on one circuit with 100% device yield. With an average system detection efficiency beyond 10%, and estimated on-chip detection efficiency of 14%–52% for four detectors operated simultaneously.

Fig. 21 (A) Integrated silicon photonic beamsplitter with integrated NbN SNSPDs from Schuck et al., (2016). (B) Representation of the GaAs-based photonic circuit demonstrated from Reithmaier et al. (2015b), which comprised InAs QD single photon sources and NbN SNSPDs. (C) Representation of the pick-and-place technique used by Najafi et al. (2015), to produce a hybrid SOI photonic circuit with NbN SNSPDs. (D) Representation of the integrated photonic spectrometer of Cheng et al. (2019). An on-chip focusing echelle grating disperses the incoming, WG-coupled optical signal across the SNSPD. The SNSPD functions as a single-photon detector and a slow microwave delay line, which allows it to map the dispersed photons. *Panel (A): Adapted from Schuck, C., Guo, X., Fan, L., Ma, X., Poot, M., Tang, H.X., 2016. Quantum interference in heterogeneous superconducting-photonic circuits on a silicon chip. Nat. Commun. 7 (1), 10352. doi:10.1038/ncomms10352 with permission from Springer Nature; Panel (B): Reprinted from Reithmaier, G., Kaniber, M., Flassig, F., Lichtmannecker, S., Müller, K., Andrejew, A., Vučković, J., Gross, R., Finley, J.J., 2015a. On-chip generation, routing, and detection of resonance fluorescence. Nano Lett. 15 (8), 5208–5213. https://doi.org/10.1021/acs.nanolett.5b01444. PMID: 26102603 with permission from the American Chemical Society; Panel (C): Reprinted from Najafi, F., Mower, J., Harris, N.C., Bellei, F., Dane, A., Lee, C., Hu, X., Kharel, P., Marsili, F., Assefa, S., et al., 2015. On-chip detection of non-classical light by scalable integration of single-photon detectors. Nat. Commun. 6, 5873. https://doi.org/10.1038/ncomms6873 with permission from Springer Nature; Panel (D): Reprinted from Cheng, R., Zou, C.-L., Guo, X., Wang, S., Han, X., Tang, H.X., 2019. Broadband on-chip single-photon spectrometer. Nat. Commun. 10 (1), 4104. https://doi.org/10.1038/s41467-019-12149-x with permission from Springer Nature.*

Photon-number resolving detectors were also integrated with on-chip waveguides. In the work of Gerrits et al. (2011) a tungsten transition-edge sensor (TES) operating in the 1550 nm telecom band, evanescently coupled a UV-written silica waveguide, was demonstrated. In the experiment, up to five photons were resolved in the guided optical mode—which was closely matched to that of an optical fiber—with a maximum detection efficiency of ≈7.2%. One important consideration regarding the fabricated chip was that the top surface was smooth to less than 1 nm, a critical need to preserve the tungsten superconducting transition temperature. A $T_c \approx 90$ mK was observed, for a 40 nm thick film, and the device was operated at 12 mK in a dilution refrigerator. In later work, Sahin et al. (2013b) demonstrated NbN photon-number resolving detectors on $GaAs/Al_{0.75}Ga_{0.25}As$ ridge waveguides. Operating at 2.1 K, detection of up to four photons was reported, with a maximum efficiency of 24% at 1310 nm. The detector featured a series connection of four nanowires, whereas each nanowire represented distinct detecting elements, sensing different parts of the same waveguide mode. The number of switching wires could furthermore be determined from the output voltage. An important factor contributing to lowering of detection efficiencies here was the quality of the sputtered NbN. While sputtering requires high temperature to promote surface diffusion, critical high film quality, As-oxide desorption at temperatures above 350°C led to rougher GaAs surfaces.

As an illustration of the potential for novel applications brought by heterogeneous integration of single-photon detectors, the work of Cheng et al. (2019) demonstrated a chip-scale integrated photonic single-photon spectrometer, covering a broad wavelength range, from spanning from 600 to 2000 nm. The spectrometer comprised an on-chip, WG-coupled dispersive echelle grating, and a single-element, NbN propagating superconducting nanowire detector, as shown in Fig. 21D. The detector also functioned as a slow-wave microwave transmission line, such that the arrival time-difference for generated signals at its two ends could lead to high-resolution spatial mapping of the dispersed photons. The echelle and the WG leading to it was composed of Si_3N_4.

6. Outlook

Progresses in the field of IQPCs have been considerable but very unequal, depending on the used platform. We propose a comparison between the different explored platforms in order to evaluate the advancement of device integration and provide an outlook on the different challenges inherent to each structure. Table 1 gives an overview on the advances of the field, based on the different components and operations to be implemented in an ideal IQPC.

So far, the Si-based platforms have seen the fastest and most impressive developments of the past decade. The number of optical components integrated into the IQPCs has expanded exponentially over the years (Wang et al., 2020). It has grown from a single multimode interferometer, used for demonstration of the on-chip Hong–Ou Mandel experiment, to a state-of-the-art large-scale IQPC comprising 671 optical components designed for the demonstration of multidimensional entanglement (Wang et al., 2018b). Recent demonstrations in integrated classical photonics, embedding up to 57,600 photonic switches on chip (Seok et al., 2019) seem to indicate that further development of very large-scale quantum photonics on chip, with prospects of million components IQPC can be expected over the next years. The IQPC will have to rely on the processing of very large number of single photons and will require further efforts in the realization of multiplexed parametric SPSs (Collins et al., 2013; Kaneda and Kwiat, 2019). Such a massive deployment of optical components will also require a significant improvement of the low-loss waveguides at hand (Bauters et al., 2011; Li et al., 2012), for critical application such as high rate boson sampling, efficient generation of cluster states and fault-tolerant quantum computation. Another major technical issue typically encountered, on all platforms, is the rejection of the pump laser addressing the integrated SPSs. Promising solutions involving cascaded microrings (Ong et al., 2013) and MZIs (Piekarek et al., 2017) have recently been proposed and already achieve up to 100 dB rejection. GaAs IQPCs emerged as the second most reliable platform for on-chip quantum operations. They present the advantage of relying on potentially deterministic SPS, which are relatively easy to integrate. The quasi-totality of the building blocks has been demonstrated but the challenge remains in the integration of all the parts, and their good functioning, into the same chip. The recent demonstration of fully integrated basic quantum experiment in a state of the art chip reveals that the technology reached a certain level of maturity (Schwartz et al., 2018). Even if antibunching is demonstrated, it is dramatically degraded by the remaining pump laser. The filtering of the latter is also a well identified issue for the GaAs platform. This might be tackled by AlGaAs/GaAs multi-quantum well waveguides (Sakata and Takeuchi, 1991). An equivalent scheme combining PhC cavities and ridge waveguides has been proposed

Table 1 Accomplished functionalities and the state of the art for the platforms reviewed in this section.

Platform	Waveguiding	Beamsplitting	Control of phase	On-chip detection	Boson sampling	Quantum logic	Deterministic quantum logic	Large-scale operations
Si-based chips	Yes	Yes	Yes	Yes	Yes	Yes	No	In progress
GaAs	Yes	Yes	In progress	Yes	No	No	No	No
Diamond-based	Yes	No	No	Yes	No	No	No	No

for the realization of on-chip HOM (Dietrich et al., 2016). This solution allows for an efficient filtering of the excitonic line thanks to the PhC cavity but suffers from typically very lossy interfaces between the photonic crystal structure and the ridge waveguides. Further control of the excitonic emission energy via Stark effect induced by the introduction p- and n-regions into the PhC. The tunability of the on-chip SPSs is a crucial prerequisite and should be in anyway urgently addressed. Even if semiconductor technology might come second to Si-based IQPC in terms of large-scale integration, very ambitious deterministic quantum logic schemes based on GaAs artificial 1D-atom seem to be reachable today, with the actual state of the art (Sollner et al., 2015). In order to increase the index contrast, hybrid structures are attracting increasing attention. Since GaAs/SiO_2 has a similar index contrast as Si/SiO_2, significant work in this direction has already been accomplished (see Section 5) and other III–V materials than GaAs are also considered in this context. InP, for example, might be an attractive candidate, based on its successes in classical photonics. However, the implementation of single-photon sources in this material is so far limited (Buckley et al., 2012), and concerns about decoherence in these structures have to be lifted.

As discussed in Section 5, heterogeneous integration offers great potential for circumventing many of the trade-offs that exist in single-material systems. While big strides have been made toward creating IQPCs with quantum emitters, reconfigurable linear optics and single-photon or photon number-resolving detectors, many challenges still remain. Because quantum photonic devices that may operate in the quantum advantage or supremacy regime are expected to require large resources—in terms of numbers of photonic qubits, logical channels, detectors, ancilla photons—methods to produce all such elements in large scale on chip is key. Large-scale reconfigurable waveguide-based linear optical networks, capable of implementing all possible linear quantum optical protocols, have been demonstrated in silicon (Carolan et al., 2015a), and can likewise be implemented in a number of other types of materials—such as silicon nitride (Taballione et al., 2019), silicon oxynitride (Ellis et al., 2018; Murray et al., 2015), aluminum nitride (Wan et al., 2019; Xiong et al., 2012a) and lithium niobate (Desiatov et al., 2019)—for which mature, top-down fabrication techniques already exist. Scalable incorporation of quantum emitters onto such on-chip networks however constitutes a crucial bottleneck, originating in the wide heterogeneity and reliability of quantum emitters properties. Generally, quantum emitters are highly sensitive to various physical factors, such as local fluctuating electric and magnetic fields, phonons, mechanical stress. All such factors taken together lead to significant variations in energy, quantum efficiency, spectral diffusion and coherence of the quantum emitter; their influence must be effectively controlled to ensure sufficient emitter performance for quantum photonic applications. In particular, the effects of fabrication methods upon the quantum emitter's "natural" characteristics must be carefully considered. In addition, as argued in Section 3, techniques that allow identification of single quantum emitters with the desirable characteristics within a population before fabrication are a necessity toward scalable fabrication of single-emitter devices. The ability to fully characterize a quantum emitter's location and orientation on a sample, as well as its spectrum, is furthermore critical to ensure efficient emitter coupling to desirable spatial optical modes supported by nanophotonic structures (Lodahl et al., 2015). This in turn has direct influence on the overall device efficiency. While deterministic fabrication of single QD devices in heterogeneous samples has been demonstrated by Schnauber et al. (2019), a detailed characterization of the selected quantum dots prefabrication (regarding dipole moment orientation, quantum efficiency, spectral purity, etc.) was not performed. Significant advances may take place, with the development of high-throughput single quantum emitter spectroscopic characterization techniques. Another promising approach toward scalable integration of quantum emitters was demonstrated by Wan et al. (2019), in which arrays of nanophotonic diamond waveguides containing deterministically implanted Si and Ge color centers were placed onto AlN photonic circuits through a pick-and-place technique. Development of high-throughput pick-and-place techniques could further increase the scalability of such technique.

7. Conclusion

In summary, this chapter gives an overview of the basics and current developments in the field of integrated quantum nanophotonics and related fields of modern photonic quantum technology. Important methods namely the FEM and FDTD approaches of the numerical modeling of integrated photonic structures were presented, which are used effectively for the design and understanding of such elements. With regard to the production of integrated photonic circuits, the focus was also on novel methods that enable the deterministic integration of quantum emitters in waveguide structures. An important method is the in situ EBL technology platform, which promises a precise and very reproducible integration of quantum emitters in complex circuits and will therefore play an important technological basis for scalable integrated quantum nanophotonics. Furthermore, current approaches and results for photonic circuits based on homogeneous materials such as Si and GaAs were presented. These more traditional approaches were and are still very successful in realizing important sub-functionalities such as linear quantum gates and circuits for boson sampling. They are comparatively easy to fabricate, which makes them particularly attractive for the study of basic physical properties such as the chiral light–matter interaction discussed. Additional functionalities and fully integrated photonic quantum circuits, however, require heterogeneous material compositions. In this context, current approaches for heterogeneous quantum circuits were presented, which, for example, include not only waveguide structures but also single-photon detectors. Furthermore, open questions and challenges were discussed and an outlook on possible future developments was given, which among other things aimed at modular quantum circuits for the implementation of large-scale quantum networks.

References

Aghaeimeibodi, S., Desiatov, B., Kim, J.-H., Lee, C.-M., Buyukkaya, M.A., Karasahin, A., Richardson, C.J.K., Leavitt, R.P., Lončar, M., Waks, E., 2018. Integration of quantum dots with lithium niobate photonics. Appl. Phys. Lett. 113 (22), 221102. https://doi.org/10.1063/1.5054865.

Aharonovich, I., Castelletto, S., Johnson, B.C., McCallum, J.C., Simpson, D.A., Greentree, A.D., Prawer, S., 2010. Chromium single-photon emitters in diamond fabricated by ion implantation. Phys. Rev. B 81, 121201. https://doi.org/10.1103/PhysRevB.81.121201.

Aharonovich, I., Englund, D., Toth, M., 2016. Solid-state single-photon emitters. Nat. Photon. 10 (10), 631–641. https://doi.org/10.1038/nphoton.2016.186.

Akopian, N., Lindner, N.H., Poem, E., Berlatzky, Y., Avron, J., Gershoni, D., Gerardot, B.D., Petroff, P.M., 2006. Entangled photon pairs from semiconductor quantum dots. Phys. Rev. Lett. 96 (13), 130501. https://doi.org/10.1103/PhysRevLett.96.130501.

Arcari, M., Söllner, I., Javadi, A., Lindskov Hansen, S., Mahmoodian, S., Liu, J., Thyrrestrup, H., Lee, E.H., Song, J.D., Stobbe, S., Lodahl, P., 2014. Near-unity coupling efficiency of a quantum emitter to a photonic crystal waveguide. Phys. Rev. Lett. 113, 093603. https://doi.org/10.1103/PhysRevLett.113.093603.

Arute, F., Arya, K., Babbush, R., Bacon, D., Bardin, J.C., Barends, R., Biswas, R., Boixo, S., Brandao, F.G.S.L., Buell, D.A., Burkett, B., Chen, Y., Chen, Z., Chiaro, B., Collins, R., Courtney, W., Dunsworth, A., Farhi, E., Foxen, B., Fowler, A., Gidney, C., Giustina, M., Graff, R., Guerin, K., Habegger, S., Harrigan, M.P., Hartmann, M.J., Ho, A., Hoffmann, M., Huang, T., Humble, T.S., Isakov, S.V., Jeffrey, E., Jiang, Z., Kafri, D., Kechedzhi, K., Kelly, J., Klimov, P.V., Knysh, S., Korotkov, A., Kostritsa, F., Landhuis, D., Lindmark, M., Lucero, E., Lyakh, D., Mandrà, S., McClean, J.R., McEwen, M., Megrant, A., Mi, X., Michielsen, K., Mohseni, M., Mutus, J., Naaman, O., Neeley, M., Neill, C., Niu, M.Y., Ostby, E., Petukhov, A., Platt, J.C., Quintana, C., Rieffel, E.G., Roushan, P., Rubin, N.C., Sank, D., Satzinger, K.J., Smelyanskiy, V., Sung, K.J., Trevithick, M.D., Vainsencher, A., Villalonga, B., White, T., Yao, Z.J., Yeh, P., Zalcman, A.,

Neven, H., Martinis, J.M., 2019. Quantum supremacy using a programmable superconducting processor. Nature 574 (7779), 505–510. https://doi.org/10.1038/s41586-019-1666-5.

Ba Hoang, T., Beetz, J., Midolo, L., Skacel, M., Lermer, M., Kamp, M., Höfling, S., Balet, L., Chauvin, N., Fiore, A., 2012. Enhanced spontaneous emission from quantum dots in short photonic crystal waveguides. Appl. Phys. Lett. 100 (6), 061122. https://doi.org/10.1063/1.3683541.

Bauters, J.F., Heck, M.J.R., John, D.D., Barton, J.S., Bruinink, C.M., Leinse, A., Heideman, R.G., Blumenthal, D.J., Bowers, J.E., 2011. Planar waveguides with less than 0.1 db/m propagation loss fabricated with wafer bonding. Opt. Express 19 (24), 24090–24101. https://doi.org/10.1364/OE.19.024090.

Berenger, J.-P., 1994. A perfectly matched layer for the absorption of electromagnetic waves. J. Comput. Phys. 114 (2), 185–200. https://doi.org/10.1006/jcph.1994.1159.

Bogucki, A., Zinkiewicz, Ł., Pacuski, W., Wasylczyk, P., Kossacki, P., 2018. Optical fiber micro-connector with nanometer positioning precision for rapid prototyping of photonic devices. Opt. Express 26 (9), 11513–11518. https://doi.org/10.1364/OE.26.011513.

Bremer, L., Weber, K., Fischbach, S., Thiele, S., Schmidt, M., Kaganskiy, A., Rodt, S., Herkommer, A., Sartison, M., Portalupi, S.L., Michler, P., Giessen, H., Reitzenstein, S., 2020. Quantum dot single-photon emission coupled into single-mode fibers with 3d printed micro-objectives. APL Photon. 5 (10), 106101. https://doi.org/10.1063/5.0014921.

Briegel, H.J., Raussendorf, R., 2001. Persistent entanglement in arrays of interacting particles. Phys. Rev. Lett. 86 (5), 910–913. https://doi.org/10.1103/PhysRevLett.86.910.

Broome, M.A., Fedrizzi, A., Rahimi-Keshari, S., Dove, J., Aaronson, S., Ralph, T.C., White, A.G., 2013. Photonic boson sampling in a tunable circuit. Science 339 (6121), 794–798. https://doi.org/10.1126/science.1231440.

Buckley, S., Rivoire, K., Vučković, J., 2012. Engineered quantum dot single-photon sources. Rep. Prog. Phys. 75 (12), 126503. https://doi.org/10.1088/0034-4885/75/12/126503.

Carolan, J., Meinecke, J.D.A., Shadbolt, P.J., Russell, N.J., Ismail, N., Wörhoff, K., Rudolph, T., Thompson, M.G., O'Brien, J.L., Matthews, J.C.F., Laing, A., 2014. On the experimental verification of quantum complexity in linear optics. Nat. Photon. 8 (8), 621–626. https://doi.org/10.1038/nphoton.2014.152.

Carolan, J., Harrold, C., Sparrow, C., Martín-López, E., Russell, N.J., Silverstone, J.W., Shadbolt, P.J., Matsuda, N., Oguma, M., Itoh, M., Marshall, G.D., Thompson, M.G., Matthews, J.C.F., Hashimoto, T., O'Brien, J.L., Laing, A., 2015a. Universal linear optics. Science 349 (6249), 711–716. https://doi.org/10.1126/science.aab3642.

Carolan, J., Harrold, C., Sparrow, C., Martín-López, E., Russell, N.J., Silverstone, J.W., Shadbolt, P.J., Matsuda, N., Oguma, M., Itoh, M., Marshall, G.D., Thompson, M.G., Matthews, J.C.F., Hashimoto, T., O'Brien, J.L., Laing, A., 2015b. Universal linear optics. Science 349 (6249), 711–716. https://doi.org/10.1126/science.aab3642.

Chang, L., Boes, A., Guo, X., Spencer, D.T., Kennedy, M.J., Peters, J.D., Volet, N., Chiles, J., Kowligy, A., Nader, N., Hickstein, D.D., Stanton, E.J., Diddams, S.A., Papp, S.B., Bowers, J.E., 2018. Heterogeneously integrated GaAs waveguides on insulator for efficient frequency conversion. Laser Photon. Rev. 12 (10), 1800149. https://doi.org/10.1002/lpor.201800149.

Cheng, R., Zou, C.-L., Guo, X., Wang, S., Han, X., Tang, H.X., 2019. Broadband on-chip single-photon spectrometer. Nat. Commun. 10 (1), 4104. https://doi.org/10.1038/s41467-019-12149-x.

Coles, R.J., Price, D.M., Dixon, J.E., Royall, B., Clarke, E., Kok, P., Skolnick, M.S., Fox, A.M., Makhonin, M.N., 2016. Chirality of nanophotonic waveguide with embedded quantum emitter for unidirectional spin transfer. Nat. Commun. 7 (1), 11183. https://doi.org/10.1038/ncomms11183.

Collins, M., Xiong, C., Rey, I., et al., 2013. Integrated spatial multiplexing of heralded single-photon sources. Nat. Commun. 4 (7198), 2582.

Crespi, A., Ramponi, R., Osellame, R., et al., 2015. Integrated photonic quantum gates for polarization qubits. Nat. Commun. 2 (7948), 566.

Davanco, M., Liu, J., Sapienza, L., Zhang, C.-Z., De Miranda Cardoso, J.V., Verma, V., Mirin, R., Nam, S.W., Liu, L., Srinivasan, K., 2017. Heterogeneous integration for on-chip quantum photonic circuits with single quantum dot devices. Nat. Commun. 8 (1), 889. https://doi.org/10.1038/s41467-017-00987-6.

Daveau, R.S., Balram, K.C., Pregnolato, T., Liu, J., Lee, E.H., Song, J.D., Verma, V., Mirin, R., Nam, S.W., Midolo, L., Stobbe, S., Srinivasan, K., Lodahl, P., 2017. Efficient fiber-coupled single-photon source based on quantum dots in a photonic-crystal waveguide. Optica 4 (2), 178–184. https://doi.org/10.1364/OPTICA.4.000178.

de Lasson, J.R., Frandsen, L.H., Gutsche, P., Burger, S., Kim, O.S., Breinbjerg, O., Ivinskaya, A., Wang, F., Sigmund, O., Häyrynen, T., Lavrinenko, A.V., Mørk, J., Gregersen, N., 2018. Benchmarking five numerical simulation techniques for computing resonance wavelengths and quality factors in photonic crystal membrane line defect cavities. Opt. Express 26 (9), 11366–11392. https://doi.org/10.1364/OE.26.011366.

Deotare, P.B., McCutcheon, M.W., Frank, I.W., Khan, M., Lončar, M., 2009. High quality factor photonic crystal nanobeam cavities. Appl. Phys. Lett. 94 (12), 121106. https://doi.org/10.1063/1.3107263.

Desiatov, B., Shams-Ansari, A., Zhang, M., Wang, C., Lončar, M., 2019. Ultra-low-loss integrated visible photonics using thin-film lithium niobate. Optica 6 (3), 380–384. https://doi.org/10.1364/OPTICA.6.000380.

Dietrich, C.P., Fiore, A., Thompson, M.G., Kamp, M., Höfling, S., 2016. GaAs integrated quantum photonics: Towards compact and multi-functional quantum photonic integrated circuits. Laser Photonics Rev. 10 (6), 870–894. https://doi.org/10.1002/lpor.201500321.

Ding, X., He, Y., Duan, Z.C., Gregersen, N., Chen, M.C., Unsleber, S., Maier, S., Schneider, C., Kamp, M., Höfling, S., Lu, C.-Y., Pan, J.-W., 2016. On-demand single photons with high extraction efficiency and near-unity indistinguishability from a resonantly driven quantum dot in a micropillar. Phys. Rev. Lett. 116 (2), 020401. https://doi.org/10.1103/physrevlett.116.020401.

Dousse, A., Lanco, L., Suffczyński, J., Semenova, E., Miard, A., Lemaître, A., Sagnes, I., Roblin, C., Bloch, J., Senellart, P., 2008. Controlled light-matter coupling for a single quantum dot embedded in a pillar microcavity using far-field optical lithography. Phys. Rev. Lett. 101 (26), 267404. https://doi.org/10.1103/PhysRevLett.101.267404.

Ellis, D.J.P., Bennett, A.J., Dangel, C., Lee, J.P., Griffiths, J.P., Mitchell, T.A., Paraiso, T.K., Spencer, P., Ritchie, D.A., Shields, A.J., 2018. Independent indistinguishable quantum light sources on a reconfigurable photonic integrated circuit. Appl. Phys. Lett. 112 (21), 211104. https://doi.org/10.1063/1.5028339.

Elshaari, A.W., Zadeh, I.E., Fognini, A., Reimer, M.E., Dalacu, D., Poole, P.J., Zwiller, V., Jöns, K.D., 2017. On-chip single photon filtering and multiplexing in hybrid quantum photonic circuits. Nat. Commun. 8 (1), 379. https://doi.org/10.1038/s41467-017-00486-8.

Elshaari, A.W., Büyüközer, E., Zadeh, I.E., Lettner, T., Zhao, P., Schöll, E., Gyger, S., Reimer, M.E., Dalacu, D., Poole, P.J., Jöns, K.D., Zwiller, V., 2018. Strain-tunable quantum integrated photonics. Nano Lett. 18 (12), 7969–7976. https://doi.org/10.1021/acs.nanolett.8b03937.

Engin, E., Bonneau, D., Natarajan, C.M., Clark, A.S., Tanner, M.G., Hadfield, R.H., Dorenbos, S.N., Zwiller, V., Ohira, K., Suzuki, N., Yoshida, H., Iizuka, N., Ezaki, M., O'Brien, J.L., Thompson, M.G., 2013. Photon pair generation in a silicon micro-ring resonator with reverse bias enhancement. Opt. Express 21 (23), 27826–27834. https://doi.org/10.1364/OE.21.027826.

Fang, A.W., Park, H., Cohen, O., Jones, R., Paniccia, M.J., Bowers, J.E., 2006. Electrically pumped hybrid AlGaInAs-silicon evanescent laser. Opt. Express 14 (20), 9203–9210. https://doi.org/10.1364/OE.14.009203.

Faraon, A., Fushman, I., Englund, D., Stoltz, N., Petroff, P., Vučković, J., 2008. Dipole induced transparency in waveguide coupled photonic crystal cavities. Opt. Express 16 (16), 12154–12162. https://doi.org/10.1364/OE.16.012154.

Faraon, A., Santori, C., Huang, Z., Fu, K.-M.C., Acosta, V.M., Fattal, D., Beausoleil, R.G., 2013. Quantum photonic devices in single-crystal diamond. New J. Phys. 15 (2), 025010. https://doi.org/10.1088/1367-2630/15/2/025010.

Fattah, S., Hoang, T.B., Midolo, L., Dietrich, C.P., Li, L.H., Linfield, E.H., Schouwenberg, J.F.P., Xia, T., Pagliano, F.M., van Otten, F.W.M., Fiore, A., 2013. Efficient coupling of single photons to ridge-waveguide photonic integrated circuits. Appl. Phys. Lett. 102 (13), 131105. https://doi.org/10.1063/1.4799669.

Fischbach, S., Schlehahn, A., Thoma, A., Srocka, N., Gissibl, T., Ristok, S., Thiele, S., Kaganskiy, A., Strittmatter, A., Heindel, T., et al., 2017. Single quantum dot with microlens and 3d-printed micro-objective as integrated bright single-photon source. ACS Photonics 4 (6), 1327–1332.

Flamini, F., Magrini, L., Rab, A.S., Spagnolo, N., D'Ambrosio, V., Mataloni, P., Sciarrino, F., Zandrini, T., Crespi, A., Ramponi, R., Osellame, R., 2015. Thermally reconfigurable quantum photonic circuits at telecom wavelength by femtosecond laser micromachining. Light Sci. Appl. 4 (11). https://doi.org/10.1038/lsa.2015.127. e354.

Gaebel, T., Popa, I., Gruber, A., Domhan, M., Jelezko, F., Wrachtrup, J., 2004. Stable single-photon source in the near infrared. New J. Phys. 6, 98. https://doi.org/10.1088/1367-2630/6/1/098.

Georgescu, I.M., Ashhab, S., Nori, F., 2014. Quantum simulation. Rev. Mod. Phys. 86 (1), 153–185. https://doi.org/10.1103/RevModPhys.86.153.

Gerrits, T., Thomas-Peter, N., Gates, J.C., Lita, A.E., Metcalf, B.J., Calkins, B., Tomlin, N.A., Fox, A.E., Linares, A.L., Spring, J.B., Langford, N.K., Mirin, R.P., Smith, P.G.R., Walmsley, I.A., Nam, S.W., 2011. On-chip, photon-number-resolving, telecommunication-band detectors for scalable photonic information processing. Phys. Rev. A 84 (6), 060301. https://doi.org/10.1103/PhysRevA.84.060301.

Grassani, D., Azzini, S., Liscidini, M., Galli, M., Strain, M.J., Sorel, M., Sipe, J.E., Bajoni, D., 2015. Micrometer-scale integrated silicon source of time-energy entangled photons. Optica 2 (2), 88–94. https://doi.org/10.1364/OPTICA.2.000088.

Gschrey, M., Gericke, F., Schüssler, A., Schmidt, R., Schulze, J.H., Heindel, T., Rodt, S., Strittmatter, A., Reitzenstein, S., 2013. In situ electron-beam lithography of deterministic single-quantum-dot mesa-structures using low-temperature cathodoluminescence spectroscopy. Appl. Phys. Lett. 102 (25), 251113 (pages 4).

Gschrey, M., Schmidt, R., Kaganskiy, A., Rodt, S., Reitzenstein, S., 2014. Study of high-resolution electron-beam resists for applications in low-temperature lithography. J. Vac. Sci. Technol. B 32 (6), 061601.

Gschrey, M., Schmidt, R., Schulze, J.-H., Strittmatter, A., Rodt, S., Reitzenstein, S., 2015a. Resolution and alignment accuracy of low-temperature in situ electron beam lithography for nanophotonic device fabrication. J. Vac. Sci. Technol. B 33 (2), 021603.

Gschrey, M., Schmidt, R., Schulze, J.-H., Strittmatter, A., Rodt, S., Reitzenstein, S., 2015b. Resolution and alignment accuracy of low-temperature in situ electron beam lithography for nanophotonic device fabrication. J. Vac. Sci. Technol. B 33 (2), 021603. https://doi.org/10.1116/1.4914914.

Gschrey, M., Thoma, A., Schnauber, P., Seifried, M., Schmidt, R., Wohlfeil, B., Krüger, L., Schulze, J.H., Heindel, T., Burger, S., Schmidt, F., Strittmatter, A., Rodt, S., Reitzenstein, S., 2015c. Highly indistinguishable photons from deterministic quantum-dot microlenses utilizing three-dimensional in situ electron-beam lithography. Nat. Commun. 6, 7662. https://doi.org/10.1038/ncomms8662.

Hamilton, C.S., Kruse, R., Sansoni, L., Barkhofen, S., Silberhorn, C., Jex, I., 2017. Gaussian boson sampling. Phys. Rev. Lett. 119 (17), 170501. https://doi.org/10.1103/PhysRevLett.119.170501.

Harris, N.C., Grassani, D., Simbula, A., Pant, M., Galli, M., Baehr-Jones, T., Hochberg, M., Englund, D., Bajoni, D., Galland, C., 2014. Integrated source of spectrally filtered correlated photons for large-scale quantum photonic systems. Phys. Rev. X 4 (4), 041047. https://doi.org/10.1103/PhysRevX.4.041047.

Harris, N.C., Steinbrecher, G.R., Prabhu, M., Lahini, Y., Mower, J., Bunandar, D., Chen, C., Wong, F.N.C., Baehr-Jones, T., Hochberg, M., Lloyd, S., Englund, D., 2017. Quantum transport simulations in a programmable nanophotonic processor. Nat. Photon. 11 (7), 447–452. https://doi.org/10.1038/nphoton.2017.95.

Hausmann, B.J.M., Shields, B., Quan, Q., Maletinsky, P., McCutcheon, M., Choy, J.T., Babinec, T.M., Kubanek, A., Yacoby, A., Lukin, M.D., Loncar, M., 2012. Integrated diamond networks for quantum nanophotonics. Nano Lett. 12 (3), 1578–1582. https://doi.org/10.1021/nl204449n.

Hausmann, B.J.M., Bulu, I.B., Deotare, P.B., McCutcheon, M., Venkataraman, V., Markham, M.L., Twitchen, D.J., Loncar, M., 2013. Integrated high-quality factor optical resonators in diamond. Nano Lett. 13 (5), 1898–1902. https://doi.org/10.1021/nl3037454.

Heeres, R., Kouwenhoven, L., Zwiller, V., 2013. On-chip generation of high-dimensional entangled quantum states and their coherent control. Nat. Nanotechnol. 8 (7948), 719–722.

Heindel, T., Thoma, A., von Helversen, M., Schmidt, M., Schlehahn, A., Gschrey, M., Schnauber, P., Schulze, J.H., Strittmatter, A., Beyer, J., Rodt, S., Carmele, A., Knorr, A., Reitzenstein, S., 2017. A bright triggered twin-photon source in the solid state. Nat. Commun. 8, 14870. https://doi.org/10.1038/ncomms14870.

Hensen, B., Bernien, H., Dréau, A.E., Reiserer, A., Kalb, N., Blok, M.S., Ruitenberg, J., Vermeulen, R.F.L., Schouten, R.N., Abellán, C., Amaya, W., Pruneri, V., Mitchell, M.W., Markham, M., Twitchen, D.J., Elkouss, D., Wehner, S., Taminiau, T.H., Hanson, R., 2015. Loophole-free bell inequality violation using electron spins separated by 1.3 kilometres. Nature 526 (7575), 682–686. https://doi.org/10.1038/nature15759.

Hepp, S., Bauer, S., Hornung, F., Schwartz, M., Portalupi, S.L., Jetter, M., Michler, P., 2018. Bragg grating cavities embedded into nano-photonic waveguides for Purcell enhanced quantum dot emission. Opt. Express 26 (23), 30614–30622.

Hiscocks, M.P., Ganesan, K., Gibson, B.C., Huntington, S.T., Ladouceur, F., Prawer, S., 2008. Diamond waveguides fabricated by reactive ion etching. Opt. Express 16 (24), 19512–19519.

Hochberg, M., Baehr-Jones, T., 2010. Towards fabless silicon photonics. Nat. Photon. 4 (8), 492.

Horn, R., Abolghasem, P., Bijlani, B.J., Kang, D., Helmy, A.S., Weihs, G., 2012. Monolithic source of photon pairs. Phys. Rev. Lett. 108 (15), 153605. https://doi.org/10.1103/PhysRevLett.108.153605.

Iwasaki, T., Ishibashi, F., Miyamoto, Y., Doi, Y., Kobayashi, S., Miyazaki, T., Tahara, K., Jahnke, K.D., Rogers, L.J., Naydenov, B., Jelezko, F., Yamasaki, S., Nagamachi, S., Inubushi, T., Mizuochi, N., Hatano, M., 2015. Germanium-vacancy single color centers in diamond. Sci. Rep. 5 (1), 12882. https://doi.org/10.1038/srep12882.

Javadi, A., Söllner, I., Arcari, M., Hansen, S.L., Midolo, L., Mahmoodian, S., Kirsanske, G., Pregnolato, T., Lee, E.H., Song, J.D., Stobbe, S., Lodahl, P., 2015. Single-photon non-linear optics with a quantum dot in a waveguide. Nat. Commun. 6, 8655. https://doi.org/10.1038/ncomms9655.

Javadi, A., et al., 2018. Spin-photon interface and spin-controlled photon switching in a nanobeam waveguide. Nat. Nanotechnol. 13, 398–403.

Jöns, K.D., Rengstl, U., Oster, M., Hargart, F., Heldmaier, M., Bounouar, S., Ulrich, S.M., Jetter, M., Michler, P., 2015. Monolithic on-chip integration of semiconductor waveguides, beamsplitters and single-photon sources. J. Phys. D Appl. Phys. 48 (8), 085101. https://doi.org/10.1088/0022-3727/48/8/085101.

Kaganskiy, A., Gschrey, M., Schlehahn, A., Schmidt, R., Schulze, J.-H., Heindel, T., Strittmatter, A., Rodt, S., Reitzenstein, S., 2015. Advanced in-situ electron-beam lithography for deterministic nanophotonic device processing. Rev. Sci. Instrum. 86 (7), 073903.

Kaganskiy, A., Heuser, T., Schmidt, R., Rodt, S., Reitzenstein, S., 2016. Csar 62 as negative-tone resist for high-contrast e-beam lithography at temperatures between 4 k and room temperature. J. Vac. Sci. Technol. B 34 (6), 061603.

Kahl, O., Ferrari, S., Rath, P., Vetter, A., Nebel, C., Pernice, W.H.P., 2016. High efficiency on-chip single-photon detection for diamond nanophotonic circuits. J. Lightwave Technol. 34 (2), 249–255.

Kaneda, F., Kwiat, P.G., 2019. High-efficiency single-photon generation via large-scale active time multiplexing. Sci. Adv. 5 (10), eaaw8586. https://doi.org/10.1126/sciadv.aaw8586.

Katsumi, R., Ota, Y., Kakuda, M., Iwamoto, S., Arakawa, Y., 2018. Transfer-printed single-photon sources coupled to wire waveguides. Optica 5 (6), 691–694. https://doi.org/10.1364/OPTICA.5.000691.

Katsumi, R., Ota, Y., Osada, A., Yamaguchi, T., Tajiri, T., Kakuda, M., Iwamoto, S., Akiyama, H., Arakawa, Y., 2019. Quantum-dot single-photon source on a CMOS silicon photonic chip integrated using transfer printing. APL Photon. 4 (3), 036105. https://doi.org/10.1063/1.5087263.

Katsumi, R., Ota, Y., Osada, A., Tajiri, T., Yamaguchi, T., Kakuda, M., Iwamoto, S., Akiyama, H., Arakawa, Y., 2020. In situ wavelength tuning of quantum-dot single-photon sources integrated on a CMOS-processed silicon waveguide. Appl. Phys. Lett. 116 (4), 041103. https://doi.org/10.1063/1.5129325.

Khasminskaya, S., Pyatkov, F., Słowik, K., Ferrari, S., Kahl, O., Kovalyuk, V., Rath, P., Vetter, A., Hennrich, F., Kappes, M.M., et al., 2016. Fully integrated quantum photonic circuit with an electrically driven light source. Nat. Photon. 10 (11), 727–732. https://doi.org/10.1038/nphoton.2016.178. https://doi.org/10.1038/nphoton.2016.178.

Kim, J.-H., Aghaeimeibodi, S., Richardson, C.J.K., Leavitt, R.P., Englund, D., Waks, E., 2017. Hybrid integration of solid-state quantum emitters on a silicon photonic chip. Nano Lett. 17 (12), 7394–7400. https://doi.org/10.1021/acs.nanolett.7b03220.

Kiršanskė, G., Thyrrestrup, H., Daveau, R.S., Dreeßen, C.L., Pregnolato, T., Midolo, L., Tighineanu, P., Javadi, A., Stobbe, S., Schott, R., Ludwig, A., Wieck, A.D., Park, S.I., Song, J.D., Kuhlmann, A.V., Söllner, I., Löbl, M.C., Warburton, R.J., Lodahl, P., 2017. Indistinguishable and efficient single photons from a quantum dot in a planar nanobeam waveguide. Phys. Rev. B 96 (16), 165306. https://doi.org/10.1103/PhysRevB.96.165306.

Knill, E., Laflamme, R., Milburn, G.J., 2001. A scheme for efficient quantum computation with linear optics. Nature 409 (6816), 46–52. https://doi.org/10.1038/35051009.

Komljenovic, T., Huang, D., Pintus, P., Tran, M.A., Davenport, M.L., Bowers, J.E., 2018. Photonic Integrated Circuits Using Heterogeneous Integration on Silicon. Proc. IEEE 106 (12), 2246–2257. https://doi.org/10.1109/JPROC.2018.2864668.

Kues, M., Reimer, C., Roztocki, P., et al., 2017. On-chip generation of high-dimensional entangled quantum states and their coherent control. Nature 546 (7948), 622–626.

Kuhlmann, A.V., Prechtel, J.H., Houel, J., Ludwig, A., Reuter, D., Wieck, A.D., Warburton, R.J., 2015. Transform-limited single photons from a single quantum dot. Nat. Commun. 6 (1), 8204. https://doi.org/10.1038/ncomms9204.

Laucht, A., Pütz, S., Günthner, T., Hauke, N., Saive, R., Frédérick, S., Bichler, M., Amann, M.C., Holleitner, A.W., Kaniber, M., Finley, J.J., 2012. A waveguide-coupled on-chip single-photon source. Phys. Rev. X 2 (1), 011014. https://doi.org/10.1103/PhysRevX.2.011014.

Lecamp, G., Lalanne, P., Hugonin, J.P., 2007. Very large spontaneous-emission β factors in photonic-crystal waveguides. Phys. Rev. Lett. 99 (2), 023902. https://doi.org/10.1103/PhysRevLett.99.023902.

Leclercq, J.L., Rojo-Romeo, P., Seassal, C., Mouette, J., Letartre, X., Viktorovitch, P., 2003. 3D structuring of multilayer suspended membranes including 2D photonic crystal structures. J. Vac. Sci. Technol. B 21 (6), 2903–2906. https://doi.org/10.1116/1.1627796.

Li, G., Yao, J., Thacker, H., Mekis, A., Zheng, X., Shubin, I., Luo, Y., hyoung Lee, J., Raj, K., Cunningham, J.E., Krishnamoorthy, A.V., 2012. Ultralow-loss, high-density SOI optical waveguide routing for macrochip interconnects. Opt. Express 20 (11), 12035–12039. https://doi.org/10.1364/OE.20.012035.

Li, Q., Davanco, M., Srinivasan, K., 2016. Efficient and low-noise single-photon-level frequency conversion interfaces using silicon nanophotonics. Nat. Photon. 10 (6), 406–414. https://doi.org/10.1038/nphoton.2016.64.

Liu, F., Brash, A.J., O'Hara, J., Martins, L.M.P.P., Phillips, C.L., Coles, R.J., Royall, B., Clarke, E., Bentham, C., Prtljaga, N., Itskevich, I.E., Wilson, L.R., Skolnick, M.S., Fox, A.M., 2018. High Purcell factor generation of indistinguishable on-chip single photons. Nat. Nanotechnol. 13 (9), 835–840. https://doi.org/10.1038/s41565-018-0188-x.

Liu, J., Su, R., Wei, Y., Yao, B., da Silva, S.F.C., Yu, Y., Iles-Smith, J., Srinivasan, K., Rastelli, A., Li, J., Wang, X., 2019. A solid-state source of strongly entangled photon pairs with high brightness and indistinguishability. Nat. Nanotechnol. 14 (6), 586–593. https://doi.org/10.1038/s41565-019-0435-9.

Lodahl, P., Mahmoodian, S., Stobbe, S., 2015. Interfacing single photons and single quantum dots with photonic nanostructures. Rev. Mod. Phys. 87 (2), 347–400. https://doi.org/10.1103/RevModPhys.87.347.

Lodahl, P., Mahmoodian, S., Stobbe, S., Rauschenbeutel, A., Schneeweiss, P., Volz, J., Pichler, H., Zoller, P., 2017. Chiral quantum optics. Nature 541 (7638), 473–480. https://doi.org/10.1038/nature21037.

Loredo, J.C., Broome, M.A., Hilaire, P., Gazzano, O., Sagnes, I., Lemaitre, A., Almeida, M.P., Senellart, P., White, A.G., 2017. Boson sampling with single-photon fock states from a bright solid-state source. Phys. Rev. Lett. 118 (13), 130503. https://doi.org/10.1103/PhysRevLett.118.130503.

Lu, X., Li, Q., Westly, D.A., et al., 2019. Chip-integrated visible-telecom entangled photon pair source for quantum communication. Nat. Phys. 15 (7198), 373–381.

Lund, A.P., Laing, A., Rahimi-Keshari, S., Rudolph, T., O'Brien, J.L., Ralph, T.C., 2014. Boson sampling from a Gaussian state. Phys. Rev. Lett. 113 (10), 100502. https://doi.org/10.1103/PhysRevLett.113.100502.

Lund-Hansen, T., Stobbe, S., Julsgaard, B., Thyrrestrup, H., Sünner, T., Kamp, M., Forchel, A., Lodahl, P., 2008. Experimental realization of highly efficient broadband coupling of single quantum dots to a photonic crystal waveguide. Phys. Rev. Lett. 101 (11), 113903. https://doi.org/10.1103/PhysRevLett.101.113903.

Luxmoore, I.J., Wasley, N.A., Ramsay, A.J., Thijssen, A.C.T., Oulton, R., Hugues, M., Kasture, S., Achanta, V.G., Fox, A.M., Skolnick, M.S., 2013a. Interfacing spins in an InGaAs quantum dot to a semiconductor waveguide circuit using emitted photons. Phys. Rev. Lett. 110 (3). https://doi.org/10.1103/physrevlett.110.037402.

Luxmoore, I.J., Wasley, N.A., Ramsay, A.J., Thijssen, A.C.T., Oulton, R., Hugues, M., Fox, A.M., Skolnick, M.S., 2013b. Optical control of the emission direction of a quantum dot. Appl. Phys. Lett. 103 (24), 241102. https://doi.org/10.1063/1.4845975.

Makhonin, M.N., Dixon, J.E., Coles, R.J., Royall, B., Luxmoore, I.J., Clarke, E., Hugues, M., Skolnick, M.S., Fox, A.M., 2014. Waveguide coupled resonance fluorescence from on-chip quantum emitter. Nano Lett. 14 (12), 6997–7002. https://doi.org/10.1021/nl5032937.

Manga Rao, V.S.C., Hughes, S., 2007. Single quantum-dot Purcell factor and β factor in a photonic crystal waveguide. Phys. Rev. B 75 (20), 205437. https://doi.org/10.1103/PhysRevB.75.205437.

Markov, I.L., Fatima, A., Isakov, S.V., Boixo, S., 2018. Quantum supremacy is both closer and farther than it appears. arxiv:1807.10749.

Marshall, G.D., Politi, A., Matthews, J.C.F., Dekker, P., Ams, M., Withford, M.J., O'Brien, J.L., 2009. Laser written waveguide photonic quantum circuits. Opt. Express 17 (15), 12546–12554. https://doi.org/10.1364/OE.17.012546.

Masada, G., Miyata, K., Politi, A., Hashimoto, T., O'Brien, J.L., Furusawa, A., 2015. Continuous-variable entanglement on a chip. Nat. Photon. 9 (7198), 319.

May, G.S., Sze, S.M., 2003. Fundamentals of Semiconductor Fabrication. vol. 1 Wiley, Weinheim, ISBN: 978-0-471-23279-7.

Metcalf, B., Spring, J., Humphreys, P., Thomas-Peter, N., Barbieri, M., Kolthammer, W.S., Jin, X.-M., Langford, N.K., Kundys, D., Gates, J.C., Smith, B.J., Smith, P.G.R., Walmsley, I.A., 2014. Quantum teleportation on a photonic chip. Nat. Photon. 8, 770–774.

Michler, P., Kiraz, A., Becher, C., Schoenfeld, W.V., Petroff, P.M., Zhang, L., Hu, E., Imamoglu, A., 2000. A quantum dot single-photon turnstile device. Science 290 (5500), 2282–2285. https://doi.org/10.1126/science.290.5500.2282.

Midolo, L., Pregnolato, T., Kirsanske, G., Stobbe, S., 2015. Soft-mask fabrication of gallium arsenide nanomembranes for integrated quantum photonics. Nanotechnology 26 (48), 484002. https://doi.org/10.1088/0957-4484/26/48/484002.

Mouradian, S.L., Schröder, T., Poitras, C.B., Li, L., Goldstein, J., Chen, E.H., Walsh, M., Cardenas, J., Markham, M.L., Twitchen, D.J., Lipson, M., Englund, D., 2015. Scalable Integration of Long-Lived Quantum Memories into a Photonic Circuit. Physical Review X 5 (3), 031009. https://doi.org/10.1103/PhysRevX.5.031009.

Mrowinski, P., Schnauber, P., Gutsche, P., Kaganskiy, A., Schall, J., Burger, S., Rodt, S., Reitzenstein, S., 2019. Directional emission of a deterministically fabricated quantum dot-bragg reflection multimode waveguide system. ACS Photon. 6 (9), 2231–2237.

Muller, A., Flagg, E.B., Bianucci, P., Wang, X.Y., Deppe, D.G., Ma, W., Zhang, J., Salamo, G.J., Xiao, M., Shih, C.K., 2007. Resonance fluorescence from a coherently driven semiconductor quantum dot in a cavity. Phys. Rev. Lett. 99 (18), 187402. https://doi.org/10.1103/PhysRevLett.99.187402.

Murray, E., Ellis, D.J.P., Meany, T., Floether, F.F., Lee, J.P., Griffiths, J.P., Jones, G.A.C., Farrer, I., Ritchie, D.A., Bennett, A.J., Shields, A.J., 2015. Quantum photonics hybrid integration platform. Appl. Phys. Lett. 107 (17), 171108. https://doi.org/10.1063/1.4935029.

Musial, A., Zolnacz, K., Srocka, N., Kravets, O., Große, J., Olszewski, J., Poturaj, K., Wojcik, G., Mergo, P., Dybka, K., et al., 2020. Plug&play fibre-coupled 73 kHz single-photon source operating in the telecom o-band. Adv. Quantum Technol., 202000018.

Najafi, F., Mower, J., Harris, N.C., Bellei, F., Dane, A., Lee, C., Hu, X., Kharel, P., Marsili, F., Assefa, S., et al., 2015. On-chip detection of non-classical light by scalable integration of single-photon detectors. Nat. Commun. 6, 5873. https://doi.org/10.1038/ncomms6873.

Nemoto, K., Trupke, M., Devitt, S.J., Stephens, A.M., Scharfenberger, B., Buczak, K., Nöbauer, T., Everitt, M.S., Schmiedmayer, J., Munro, W.J., 2014. Photonic architecture for scalable quantum information processing in diamond. Phys. Rev. X 4 (3), 031022. https://doi.org/10.1103/PhysRevX.4.031022.

Nielsen, M.A., 2004. Optical quantum computation using cluster states. Phys. Rev. Lett. 93 (4), 040503. https://doi.org/10.1103/PhysRevLett.93.040503.

Ong, J.R., Kumar, R., Mookherjea, S., 2013. Ultra-high-contrast and tunable-bandwidth filter using cascaded high-order silicon microring filters. IEEE Photon. Technology. Lett. 25 (16), 1543–1546.

Osada, A., Ota, Y., Katsumi, R., Kakuda, M., Iwamoto, S., Arakawa, Y., 2019. Strongly coupled single-quantum-dot-cavity system integrated on a CMOS-processed silicon photonic chip. Phys. Rev. Appl. 11 (2), 024071. https://doi.org/10.1103/PhysRevApplied.11.024071.

Oskooi, A.F., Roundy, D., Ibanescu, M., Bermel, P., Joannopoulos, J.D., Johnson, S.G., 2010. Meep: A flexible free-software package for electromagnetic simulations by the FDTD method. Comput. Phys. Commun. 181 (3), 687–702. https://doi.org/10.1016/j.cpc.2009.11.008.

Paesani, S., Ding, Y., Santagati, R., Chakhmakhchyan, L., Vigliar, C., Rottwitt, K., Oxenløwe, L.K., Wang, J., Thompson, M.G., Laing, A., 2019. Generation and sampling of quantum states of light in a silicon chip. Nat. Phys. 15 (9), 925–929. https://doi.org/10.1038/s41567-019-0567-8.

Patel, R.B., Bennett, A.J., Farrer, I., Nicoll, C.A., Ritchie, D.A., Shields, A.J., 2010. Two-photon interference of the emission from electrically tunable remote quantum dots. Nat. Photon. 4 (9), 632–635. https://doi.org/10.1038/nphoton.2010.161.

Pernice, W.H.P., Schuck, C., Minaeva, O., Li, M., Goltsman, G.N., Sergienko, A.V., Tang, H.X., 2012. High-speed and high-efficiency travelling wave single-photon detectors embedded in nanophotonic circuits. Nat. Commun. 3 (1), 1325. https://doi.org/10.1038/ncomms2307.

Pertsch, T., Peschel, U., Lederer, F., Burghoff, J., Will, M., Nolte, S., Tünnermann, A., 2004. Discrete diffraction in two-dimensional arrays of coupled waveguides in silica. Opt. Lett. 29 (5), 468–470. https://doi.org/10.1364/OL.29.000468.

Peruzzo, A., Laing, A., Politi, A., Rudolph, T., O'Brien, J.L., 2011. Multimode quantum interference of photons in multiport integrated devices. Nat. Commun. 2 (1), 224. https://doi.org/10.1038/ncomms1228.

Peruzzo, A., McClean, J., Shadbolt, P., Yung, M.-H., Zhou, X.-Q., Love, P.J., Aspuru-Guzik, A., O'Brien, J.L., 2014. A variational eigenvalue solver on a photonic quantum processor. Nat. Commun. 5 (1), 4213. https://doi.org/10.1038/ncomms5213.

Peyskens, F., Chakraborty, C., Muneeb, M., Van Thourhout, D., Englund, D., 2019. Integration of single photon emitters in 2D layered materials with a silicon nitride photonic chip. Nat. Commun. 10 (1), 4435. https://doi.org/10.1038/s41467-019-12421-0.

Piekarek, M., Bonneau, D., Miki, S., Yamashita, T., Fujiwara, M., Sasaki, M., Terai, H., Tanner, M.G., Natarajan, C.M., Hadfield, R.H., O'Brien, J.L., Thompson, M.G., 2017. High-extinction ratio integrated photonic filters for silicon quantum photonics. Opt. Lett. 42 (4), 815–818. https://doi.org/10.1364/OL.42.000815.

Piprek, J., 2018. Handbook of Optoelectronic Device Modeling and Simulation. vol. 2 CRC Press, Boca Raton, ISBN: 9781315152318.

Politi, A., Cryan, M.J., Rarity, J.G., Yu, S., O'Brien, J.L., 2008. Silica-on-silicon waveguide quantum circuits. Science 320 (5876), 646–649. https://doi.org/10.1126/science.1155441.

Politi, A., Matthews, J.C.F., O'Brien, J.L., 2009. Shor's quantum factoring algorithm on a photonic chip. Science 325 (5945), 1221. https://doi.org/10.1126/science.1173731.1221.

Politi, A., Matthews, J.C.F., Thompson, M.G., O'Brien, J.L., 2009. Integrated quantum photonics. IEEE J. Sel. Top. Quantum Electron. 15 (6), 1673–1684.

Prtljaga, N., Coles, R.J., O'Hara, J., Royall, B., Clarke, E., Fox, A.M., Skolnick, M.S., 2014. Monolithic integration of a quantum emitter with a compact on-chip beam-splitter. Appl. Phys. Lett. 104 (23), 231107. https://doi.org/10.1063/1.4883374.

Qiang, X., Zhou, X., Wang, J., Wilkes, C.M., Loke, T., O'Gara, S., Kling, L., Marshall, G.D., Santagati, R., Ralph, T.C., Wang, J.B., O'Brien, J.L., Thompson, M.G., Matthews, J.C.F., 2018. Large-scale silicon quantum photonics implementing arbitrary two-qubit processing. Nat. Photon. 12 (9), 534–539. https://doi.org/10.1038/s41566-018-0236-y.

Rakher, M.T., Srinivasan, K., 2011. Subnanosecond electro-optic modulation of triggered single photons from a quantum dot. Appl. Phys. Lett. 98 (21), 211103. https://doi.org/10.1063/1.3593007.

Rao, V.S.C.M., Hughes, S., 2007. Single quantum dot spontaneous emission in a finite-size photonic crystal waveguide: Proposal for an efficient "on chip" single photon gun. Phys. Rev. Lett. 99 (19), 193901. https://doi.org/10.1103/PhysRevLett.99.193901.

Rath, P., Kahl, O., Ferrari, S., Sproll, F., Lewes-Malandrakis, G., Brink, D., Ilin, K., Siegel, M., Nebel, C., Pernice, W., 2015a. Superconducting single-photon detectors integrated with diamond nanophotonic circuits. Light Sci. Appl. 4 (10). https://doi.org/10.1038/lsa.2015.111. e338.

Rath, P., Ummethala, S., Nebel, C., Pernice, W.H.P., 2015b. Diamond as a material for monolithically integrated optical and optomechanical devices. Phys. Status Solidi (a) 212 (11), 2385–2399. https://doi.org/10.1002/pssa.201532494.

Raussendorf, R., Briegel, H.J., 2001. A one-way quantum computer. Phys. Rev. Lett. 86 (22), 5188–5191. https://doi.org/10.1103/PhysRevLett.86.5188.

Reithmaier, G., Lichtmannecker, S., Reichert, T., Hasch, P., Muller, K., Bichler, M., Gross, R., Finley, J.J., 2013. On-chip time resolved detection of quantum dot emission using integrated superconducting single photon detectors. Sci. Rep. 3, 1901. https://doi.org/10.1038/srep01901.

Reithmaier, G., Kaniber, M., Flassig, F., Lichtmannecker, S., Müller, K., Andrejew, A., Vučković, J., Gross, R., Finley, J.J., 2015a. On-chip generation, routing, and detection of resonance fluorescence. Nano Lett. 15 (8), 5208–5213. https://doi.org/10.1021/acs.nanolett.5b01444. PMID: 26102603.

Reithmaier, G., Kaniber, M., Flassig, F., Lichtmannecker, S., Müller, K., Andrejew, A., Vučković, J., Gross, R., Finley, J.J., 2015b. On-Chip generation, routing, and detection of resonance fluorescence. Nano Lett. 15 (8), 5208–5213. https://doi.org/10.1021/acs.nanolett.5b01444.

Rengstl, U., Schwartz, M., Herzog, T., Hargart, F., Paul, M., Portalupi, S.L., Jetter, M., Michler, P., 2015. On-chip beamsplitter operation on single photons from quasi-resonantly excited quantum dots embedded in gaas rib waveguides. Appl. Phys. Lett. 107 (2), 021101. https://doi.org/10.1063/1.4926729.

Robledo, L., Childress, L., Bernien, H., Hensen, B., Alkemade, P.F.A., Hanson, R., 2011. High-fidelity projective read-out of a solid-state spin quantum register. Nature 477 (7366), 574–578. https://doi.org/10.1038/nature10401.

Rodt, S., Reitzenstein, S., Heindel, T., 2020. Deterministically fabricated solid-state quantum-light sources. J. Phys. Condens. Matter 32 (15), 153003.

Rose, B.C., Huang, D., Zhang, Z.-H., Stevenson, P., Tyryshkin, A.M., Sangtawesin, S., Srinivasan, S., Loudin, L., Markham, M.L., Edmonds, A.M., Twitchen, D.J., Lyon, S.A., de Leon, N.P., 2018. Observation of an environmentally insensitive solid-state spin defect in diamond. Science 361 (6397), 60–63. https://doi.org/10.1126/science.aao0290.

Sahin, D., Gaggero, A., Hoang, T.B., Frucci, G., Mattioli, F., Leoni, R., Beetz, J., Lermer, M., Kamp, M., Höfling, S., Fiore, A., 2013a. Integrated autocorrelator based on superconducting nanowires. Opt. Express 21 (9), 11162–11170. https://doi.org/10.1364/OE.21.011162.

Sahin, D., Gaggero, A., Zhou, Z., Jahanmirinejad, S., Mattioli, F., Leoni, R., Beetz, J., Lermer, M., Kamp, M., Höfling, S., et al., 2013b. Waveguide photon-number-resolving detectors for quantum photonic integrated circuits. Appl. Phys. Lett. 103 (11), 111116. https://doi.org/10.1063/1.4820842.

Sakata, H., Takeuchi, S., 1991. Grating-assisted directional coupler filters using AlGaAs/GaAs MQW waveguides. IEEE Photon. Technol. Lett. 3 (10), 899–901.

Santagati, R., Silverstone, J.W., Strain, M.J., Sorel, M., Miki, S., Yamashita, T., Fujiwara, M., Sasaki, M., Terai, H., Tanner, M.G., Natarajan, C.M., Hadfield, R.H., O'Brien, J.L., Thompson, M.G., 2017. Silicon photonic processor of two-qubit entangling quantum logic. J. Opt. 19 (11), 114006. https://doi.org/10.1088/2040-8986/aa8d56.

Sapienza, L., Davanço, M., Badolato, A., Srinivasan, K., 2015. Nanoscale optical positioning of single quantum dots for bright and pure single-photon emission. Nat. Commun. 6 (1), 7833. https://doi.org/10.1038/ncomms8833.

Schlehahn, A., Gaafar, M., Vaupel, M., Gschrey, M., Schnauber, P., Schulze, J.H., Rodt, S., Strittmatter, A., Stolz, W., Rahimi-Iman, A., et al., 2015. Single-photon emission at a rate of 143 MHz from a deterministic quantum-dot microlens triggered by a mode-locked vertical-external-cavity surface-emitting laser. Appl. Phys. Lett. 107 (4), 041105.

Schlehahn, A., Schmidt, R., Hopfmann, C., Schulze, J.-H., Strittmatter, A., Heindel, T., Gantz, L., Schmidgall, E.R., Gershoni, D., Reitzenstein, S., 2016. Generating single photons at gigahertz modulation-speed using electrically controlled quantum dot microlenses. Appl. Phys. Lett. 108 (2), 021104.

Schlehahn, A., Fischbach, S., Schmidt, R., Kaganskiy, A., Strittmatter, A., Rodt, S., Heindel, T., Reitzenstein, S., 2018. A stand-alone fiber-coupled single-photon source. Sci. Rep. 8 (1), 1–7.

Schnauber, P., Schmidt, R., Kaganskiy, A., Heuser, T., Gschrey, M., Rodt, S., Reitzenstein, S., 2016. Using low-contrast negative-tone PMMA at cryogenic temperatures for 3d electron beam lithography. Nanotechnology 27 (19), 195301.

Schnauber, P., Schall, J., Bounouar, S., Höhne, T., Park, S.-I., Ryu, G.-H., Heindel, T., Burger, S., Song, J.-D., Rodt, S., Reitzenstein, S., 2018. Deterministic integration of quantum dots into on-chip multimode interference beamsplitters using in situ electron beam lithography. Nano Lett. 18 (4), 2336–2342. https://doi.org/10.1021/acs.nanolett.7b05218.

Schnauber, P., Singh, A., Schall, J., Park, S.I., Song, J.D., Rodt, S., Srinivasan, K., Reitzenstein, S., Davanco, M., 2019. Indistinguishable photons from deterministically integrated single quantum dots in heterogeneous GaAs/Si3N4 quantum photonic circuits. Nano Lett. 19 (10), 7164–7172. https://doi.org/10.1021/acs.nanolett.9b02758.

Schneider, P.-I., Srocka, N., Rodt, S., Zschiedrich, L., Reitzenstein, S., Burger, S., 2018. Numerical optimization of the extraction efficiency of a quantum-dot based single-photon emitter into a single-mode fiber. Opt. Express 26 (7), 8479–8492.

Schuck, C., Guo, X., Fan, L., Ma, X., Poot, M., Tang, H.X., 2016. Quantum interference in heterogeneous superconducting-photonic circuits on a silicon chip. Nat. Commun. 7 (1), 10352. https://doi.org/10.1038/ncomms10352.

Schwagmann, A., Kalliakos, S., Farrer, I., Griffiths, J.P., Jones, G.A.C., Ritchie, D.A., Shields, A.J., 2011. On-chip single photon emission from an integrated semiconductor quantum dot into a photonic crystal waveguide. Appl. Phys. Lett. 99 (26), 261108. https://doi.org/10.1063/1.3672214.

Schwartz, M., Rengstl, U., Herzog, T., Paul, M., Kettler, J., Portalupi, S.L., Jetter, M., Michler, P., 2016. Generation, guiding and splitting of triggered single photons from a resonantly excited quantum dot in a photonic circuit. Opt. Express 24 (3), 3089–3094. https://doi.org/10.1364/OE.24.003089.

Schwartz, M., Schmidt, E., Rengstl, U., Hornung, F., Hepp, S., Portalupi, S.L., llin, K., Jetter, M., Siegel, M., Michler, P., 2018. Fully on-chip single-photon Hanbury-Brown and Twiss experiment on a monolithic semiconductor-superconductor platform. Nano Lett. 18 (11), 6892–6897. https://doi.org/10.1021/acs.nanolett.8b02794.

Schweickert, L., Jöns, K.D., Zeuner, K.D., Covre da Silva, S.F., Huang, H., Lettner, T., Reindl, M., Zichi, J., Trotta, R., Rastelli, A., Zwiller, V., 2018. On-demand generation of background-free single photons from a solid-state source. Appl. Phys. Lett. 112 (9), 093106. https://doi.org/10.1063/1.5020038.

Seok, T.J., Kwon, K., Henriksson, J., Luo, J., Wu, M.C., 2019. 240 ×240 wafer-scale silicon photonic switches. In: Optical Fiber Communication Conference (OFC) 2019, Optical Society of America. http://www.osapublishing.org/abstract.cfm?URI=OFC-2019-Th1E.5. paper Th1E.5.

Shadbolt, P., Verde, M., Peruzzo, A., et al., 2012. Generating, manipulating and measuring entanglement and mixture with a reconfigurable photonic circuit. Nat. Photon. 6, 45–49.

Silverstone, J., Santagati, R., Bonneau, D., et al., 2015. Qubit entanglement between ring-resonator photon-pair sources on a silicon chip. Nat. Commun. 6 (7948), 7948.

Silverstone, J.W., Bonneau, D., O'Brien, J.L., Thompson, M.G., 2016. Silicon quantum photonics. IEEE J. Sel. Top. Quantum Electron. 22 (6), 390–402.

Singh, A., Li, Q., Liu, S., Yu, Y., Lu, X., Schneider, C., Höfling, S., Lawall, J., Verma, V., Mirin, R., Nam, S.W., Liu, J., Srinivasan, K., 2019. Quantum frequency conversion of a quantum dot single-photon source on a nanophotonic chip. Optica 6 (5), 563–569. https://doi.org/10.1364/OPTICA.6.000563.

Sollner, I., et al., 2015. Deterministic photon-emitter coupling in chiral photonic circuits. Nat. Nanotechnol. 10 (9), 775.

Somaschi, N., Giesz, V., De Santis, L., Loredo, J.C., Almeida, M.P., Hornecker, G., Portalupi, S.L., Grange, T., Antón, C., Demory, J., Gómez, C., Sagnes, I., Lanzillotti-Kimura, N.D., Lemaítre, A., Auffeves, A., White, A.G., Lanco, L., Senellart, P., 2016. Near-optimal single-photon sources in the solid state. Nat. Photon. 10 (5), 340–345. https://doi.org/10.1038/nphoton.2016.23.

Sotillo, B., Bharadwaj, V., Hadden, J.P., Sakakura, M., Chiappini, A., Fernandez, T.T., Longhi, S., Jedrkiewicz, O., Shimotsuma, Y., Criante, L., Osellame, R., Galzerano, G., Ferrari, M., Miura, K., Ramponi, R., Barclay, P.E., Eaton, S.M., 2016. Diamond photonics platform enabled by femtosecond laser writing. Sci. Rep. 6 (1), 35566. https://doi.org/10.1038/srep35566.

Sparrow, C., Martín-López, E., Maraviglia, N., Neville, A., Harrold, C., Carolan, J., Joglekar, Y.N., Hashimoto, T., Matsuda, N., O'Brien, J.L., Tew, D.P., Laing, A., 2018. Simulating the vibrational quantum dynamics of molecules using photonics. Nature 557 (7707), 660–667. https://doi.org/10.1038/s41586-018-0152-9.

Spring, J.B., Metcalf, B.J., Humphreys, P.C., Kolthammer, W.S., Jin, X.-M., Barbieri, M., Datta, A., Thomas-Peter, N., Langford, N.K., Kundys, D., Gates, J.C., Smith, B.J., Smith, P.G.R., Walmsley, I.A., 2013. Boson sampling on a photonic chip. Science 339 (6121), 798–801. https://doi.org/10.1126/science.1231692.

Spring, J.B., Mennea, P.L., Metcalf, B.J., Humphreys, P.C., Gates, J.C., Rogers, H.L., Söller, C., Smith, B.J., Kolthammer, W.S., Smith, P.G.R., Walmsley, I.A., 2017. Chip-based array of near-identical, pure, heralded single-photon sources. Optica 4 (1), 90–96. https://doi.org/10.1364/OPTICA.4.000090.

Srocka, N., Musiał, A., Schneider, P.-I., Mrowiński, P., Holewa, P., Burger, S., Quandt, D., Strittmatter, A., Rodt, S., Reitzenstein, S., et al., 2018. Enhanced photon-extraction efficiency from InGaAs/GaAs quantum dots in deterministic photonic structures at 1.3 μm fabricated by in-situ electron-beam lithography. AIP Adv. 8 (8), 085205.

Srocka, N., Mrowiński, P., Große, J., von Helversen, M., Heindel, T., Rodt, S., Reitzenstein, S., 2020. Deterministically fabricated quantum dot single-photon source emitting indistinguishable photons in the telecom o-band. Appl. Phys. Lett. 116 (23), 231104. https://doi.org/10.1063/5.0010436.

Sugimoto, Y., Tanaka, Y., Ikeda, N., Nakamura, Y., Asakawa, K., Inoue, K., 2004. Low propagation loss of 0.76 db/mm in GaAs-based single-line-defect two-dimensional photonic crystal slab waveguides up to 1 cm in length. Opt. Express 12 (6), 1090–1096. https://doi.org/10.1364/OPEX.12.001090.

Sun, S., Kim, H., Luo, Z., Solomon, G.S., Waks, E., 2018. A single-photon switch and transistor enabled by a solid-state quantum memory. Science 361 (6397), 57–60. https://doi.org/10.1126/science.aat3581.

Szameit, A., Nolte, S., 2010. Discrete optics in femtosecond-laser-written photonic structures. J. Phys. B At. Mol. Opt. Phys. 43 (16), 163001. https://doi.org/10.1088/0953-4075/43/16/163001.

Taballione, C., Wolterink, T.A.W., Lugani, J., Eckstein, A., Bell, B.A., Grootjans, R., Visscher, I., Geskus, D., Roeloffzen, C.G.H., Renema, J.J., Walmsley, I.A., Pinkse, P.W.H., Boller, K.-J., 2019. Reconfigurable quantum photonic processor based on silicon nitride waveguides. Opt. Express 27 (19), 26842–26857. https://doi.org/10.1364/OE.27.026842.

Taflove, A., Hagness, S.C., 2005. Computational Electrodynamics: The Finite-Difference Time-Domain Method. In: Artech House antennas and propagation library, third ed. Artech House, Boston, MA. https://cds.cern.ch/record/1698084.

Tillmann, M., Dakić, B., Heilmann, R., Nolte, S., Szameit, A., Walther, P., 2013. Experimental boson sampling. Nat. Photon. 7 (7), 540–544. https://doi.org/10.1038/nphoton.2013.102.

Togan, E., Chu, Y., Trifonov, A., et al., 2010. Quantum entanglement between an optical photon and a solid-state spin qubit. Nature 466, 730–740.

Unsleber, S., He, Y.-M., Gerhardt, S., Maier, S., Lu, C.-Y., Pan, J.-W., Gregersen, N., Kamp, M., Schneider, C., Höfling, S., 2016. Highly indistinguishable on-demand resonance fluorescence photons from a deterministic quantum dot micropillar device with 74% extraction efficiency. Opt. Express 24 (8), 8539–8546. https://doi.org/10.1364/OE.24.008539.

Van den Nest, M., Miyake, A., Dür, W., Briegel, H.J., 2006. Universal resources for measurement-based quantum computation. Phys. Rev. Lett. 97 (15), 150504. https://doi.org/10.1103/PhysRevLett.97.150504.

Vuckovic, J., Painter, O., Xu, Y., Yariv, A., Scherer, A., 1999. Finite-difference time-domain calculation of the spontaneous emission coupling factor in optical microcavities. IEEE J. Quantum Electron. 35 (8), 1168–1175. https://doi.org/10.1109/3.777216.

Wan, N.H., Lu, T.-J., Chen, K.C., Walsh, M.P., Trusheim, M.E., De Santis, L., Bersin, E.A., Harris, I.B., Mouradian, S.L., Christen, I.R., Bielejec, E.S., Englund, D., 2019. Large-scale integration of near-indistinguishable artificial atoms in hybrid photonic circuits. arXiv:1911.05265 [quant-ph]. http://arxiv.org/abs/1911.05265.

Wang, J., Santamato, A., Jiang, P., Bonneau, D., Engin, E., Silverstone, J.W., Lermer, M., Beetz, J., Kamp, M., H S., 2014. Gallium arsenide (GaAs) quantum photonic waveguide circuits. Opt. Commun. 327, 49–55. https://doi.org/10.1016/j.optcom.2014.02.040 (special issue on Nonlinear Quantum Photonics).

Wang, H., He, Y., Li, Y.-H., Su, Z.-E., Li, B., Huang, H.-L., Ding, X., Chen, M.-C., Liu, C., Qin, J., Li, J.-P., He, Y.-M., Schneider, C., Kamp, M., Peng, C.-Z., Höfling, S., Lu, C.-Y., Pan, J.-W., 2017. High-efficiency multiphoton boson sampling. Nat. Photon. 11 (6), 361–365. https://doi.org/10.1038/nphoton.2017.63.

Wang, H., Li, W., Jiang, X., He, Y.M., Li, Y.H., Ding, X., Chen, M.C., Qin, J., Peng, C.Z., Schneider, C., Kamp, M., Zhang, W.J., Li, H., You, L.X., Wang, Z., Dowling, J.P., Höfling, S., Lu, C.-Y., Pan, J.-W., 2018a. Toward scalable boson sampling with photon loss. Phys. Rev. Lett. 120 (23), 230502. https://doi.org/10.1103/PhysRevLett.120.230502.

Wang, J., Paesani, S., Ding, Y., Santagati, R., Skrzypczyk, P., Salavrakos, A., Tura, J., Augusiak, R., Mančinska, L., Bacco, D., Bonneau, D., Silverstone, J.W., Gong, Q., Acín, A., Rottwitt, K., Oxenløwe, L.K., O'Brien, J.L., Laing, A., Thompson, M.G., 2018b. Multidimensional quantum entanglement with large-scale integrated optics. Science 360 (6386), 285–291. https://doi.org/10.1126/science.aar7053.

Wang, H., Qin, J., Ding, X., Chen, M.-C., Chen, S., You, X., He, Y.-M., Jiang, X., You, L., Wang, Z., Schneider, C., Renema, J.J., Höfling, S., Lu, C.-Y., Pan, J.-W., 2019. Boson sampling with 20 input photons and a 60-mode interferometer in a 10^{14}-dimensional Hilbert space. Phys. Rev. Lett. 123 (25), 250503. https://doi.org/10.1103/PhysRevLett.123.250503.

Wang, J., Sciarrino, F., Laing, A., Thompson, M.G., 2020. Integrated photonic quantum technologies. Nat. Photon. 14 (5), 273–284. https://doi.org/10.1038/s41566-019-0532-1.

Wei, Y.-J., He, Y.-M., Chen, M.-C., Hu, Y.-N., He, Y., Wu, D., Schneider, C., Kamp, M., Höfling, S., Lu, C.-Y., Pan, J.-W., 2014. Deterministic and robust generation of single photons from a single quantum dot with 99.5% indistinguishability using adiabatic rapid passage. Nano Lett. 14 (11), 6515–6519. https://doi.org/10.1021/nl503081n.

Weimann, S., Kremer, M., Plotnik, Y., Lumer, Y., Nolte, S., Makris, K.G., Segev, M., Rechtsman, M.C., Szameit, A., 2017. Topologically protected bound states in photonic parity-time-symmetric crystals. Nat. Mater. 16 (4), 433–438. https://doi.org/10.1038/nmat4811.

White, D., Branny, A., Chapman, R.J., Picard, R., Brotons-Gisbert, M., Boes, A., Peruzzo, A., Bonato, C., Gerardot, B.D., 2019. Atomically-thin quantum dots integrated with lithium niobate photonic chips [Invited]. Opt. Mater. Express 9 (2), 441–448. https://doi.org/10.1364/OME.9.000441.

Xiong, C., Pernice, W.H.P., Sun, X., Schuck, C., Fong, K.Y., Tang, H.X., 2012a. Aluminum nitride as a new material for chip-scale optomechanics and nonlinear optics. New J. Phys. 14 (9), 095014. https://doi.org/10.1088/1367-2630/14/9/095014.

Xiong, C., Pernice, W.H.P., Tang, H.X., 2012b. Low-loss, silicon integrated, aluminum nitride photonic circuits and their use for electro-optic signal processing. Nano Lett. 12 (7), 3562–3568. https://doi.org/10.1021/nl3011885.

Yuan, Z., Kardynal, B.E., Stevenson, R.M., Shields, A.J., Lobo, C.J., Cooper, K., Beattie, N.S., Ritchie, D.A., Pepper, M., 2002. Electrically driven single-photon source. Science 295 (5552), 102–105. https://doi.org/10.1126/science.1066790.

Zadeh, I.E., Elshaari, A.W., Jöns, K.D., Fognini, A., Dalacu, D., Poole, P.J., Reimer, M.E., Zwiller, V., 2016. Deterministic integration of single photon sources in silicon based photonic circuits. Nano Lett. 16 (4), 2289–2294. https://doi.org/10.1021/acs.nanolett.5b04709.

Zhang, H., Eaton, S.M., Herman, P.R., 2006. Low-loss Type II waveguide writing in fused silica with single picosecond laser pulses. Opt. Express 14 (11), 4826–4834. https://doi.org/10.1364/OE.14.004826.

Zhou, X., Kulkova, I., Lund-Hansen, T., Hansen, S.L., Lodahl, P., Midolo, L., 2018. High-efficiency shallow-etched grating on GaAs membranes for quantum photonic applications. Appl. Phys. Lett. 113 (25), 251103. https://doi.org/10.1063/1.5055622.

żołnacz, K., Musiał, A., Srocka, N., Große, J., Schlösinger, M.J., Schneider, P.-I., Kravets, O., Mikulicz, M., Olszewski, J., Poturaj, K., et al., 2019. Method for direct coupling of a semiconductor quantum dot to an optical fiber for single-photon source applications. Opt. Express 27 (19), 26772–26785.

CHAPTER FIVE

Quantum optics with quantum dot ensembles

Marc Aßmann and Manfred Bayer*
Experimentelle Physik 2 Technische Universität Dortmund, Dortmund, Germany
*Corresponding author: e-mail address: manfred.bayer@physik.uni-dortmund.de

Contents

1. Introduction

Quantum optics in solid state (Jahnke, 2012) has become an emerging field of condensed matter physics due to the development of two components:

- semiconductor nanostructures providing three-dimensional carrier confinement (quantum dots), so that the spectrum of carrier energy levels is discretized, mimicking that of atoms and
- optical resonators, in which light is confined three-dimensionally, so that narrow resonances appear in the otherwise continuous density of states for photons.

Generally speaking, these factors have the following impact: The light confinement in combination with high resonator finesse leads to an enhancement of the electric field strength of the optical modes and thus increases the strength of light–matter coupling. Further, the reduced scattering of quantum dot carriers due to suppression of free motion causes long dephasing times during which the coupling can act constructively.

Semiconductors and Semimetals, Volume 105
ISSN 0080-8784
https://doi.org/10.1016/bs.semsem.2020.10.003

This contribution describes certain aspects of light–matter interaction. In contrast to other chapters in this volume, we consider here quantum dot ensembles. This approach has the disadvantage of the unavoidable inhomogeneity of such an ensemble as reflected by the widely varying optical transition energies and oscillator strengths, resulting from variation of dot size, composition, strain, etc. On the other hand, a fraction of resonant quantum dots may contribute to collective interaction phenomena within the ensemble.

As outlined, we will shed light on different facets of the light field coupling to a quantum dot ensemble. We will start with the impact of the vacuum light field, addressing the emission of quantum dots after nonresonant excitation with a laser. Then we turn to situations where the statistics of the light field exciting the quantum dots becomes of importance, facilitating the study of quantum optical effects.

2. Enhancement of quantum dot spontaneous emission: Purcell effect

The standard consideration of light–matter interaction in basic quantum mechanics lectures is to quantize the matter part, while treating the light classically. Fermi's golden rule for calculating transition rates between electronic levels captures the processes of absorption and stimulated emission, missing the spontaneous emission. The latter has to be introduced by hand, in order to achieve a satisfactory description of the experiment, for example through the concept of Einstein's coefficients when considering the thermodynamic balance of excitation and deexcitation. In that way, the strength of spontaneous emission can be derived correctly.

Quantizing the light field, leads to a more consistent introduction of spontaneous emission at the expense of higher technical complexity. Therefore, we will refrain here first to the heuristic description using classical fields. The basics of the Purcell effect can be understood by considering the change of the vacuum electric field amplitude E_0 of a light field with a certain frequency when confined in a resonator. Setting the phonon number to zero, the total energy density $\propto E_0^2$ times the resonator volume V has to be equal to the energy of the vacuum fluctuations given by half of the photon energy, $\hbar\omega/2$. In that way one obtains

$$E_0 \propto \sqrt{\frac{\hbar\omega}{2V}}. \tag{1}$$

This shows that the vacuum field amplitude increases with decreasing resonator volume if the light field remains well-confined. Calculating the emission rate corresponding to a transition from an excited state (no matter how it was created) to the crystal vacuum ground state perturbatively in electric dipole approximation, the interaction operator is given by the product of the electric dipole moment with the field, $-e\mathbf{r}\cdot\mathbf{E}_0$, where e is the elementary charge of an electron at position $\mathbf{r}$. Thus the emission rate, which contains the matrix element of this operator squared, is inversely proportional to the cavity volume.

A more detailed analysis as first elaborated by Purcell et al. (1946) gives the following expression for the decay rate $1/\tau$ relative to the decay rate in free space $1/\tau_0$:

$$\frac{\tau_0}{\tau} = \frac{2}{3}F_P\frac{|E_0(r)|^2}{|E_{0,max}|^2}\frac{\Delta\lambda_C^2}{\Delta\lambda_C^2 + 4(\lambda_C - \lambda_E)^2} + \alpha. \tag{2}$$

Here $E_0(r)$ is the amplitude of the vacuum electric field at the position of the dipole $\mathbf{r}$, which is normalized by the maximum field strength $E_{0,max}$. The emitter sends out light at wavelength λ_E relative to the optical cavity mode located at wavelength λ_C, whose finesse Q determines the spectral width $\Delta\lambda_C/\lambda_C$. Most importantly, F_P is the Purcell factor given by

$$F_P = \frac{3}{4\pi}Q\frac{\lambda^3}{V}, \tag{3}$$

confirming the dependence of the decay rate on cavity volume V in case of resonance of emitter and optical mode, which needs to have also a sufficiently high finesse. In case of off-resonance, the Lorentzian in Eq. (2) leads to a corresponding reduction of the decay rate. Note that all wavelengths are taken inside the material. In the equation there is, however, the second term α on the right-hand side, which captures the emission into so-called leaky modes. This contribution would disappear for a resonator with perfect optical confinement in the resonator, leading to formation of standing waves between its edges with unity reflectivity. This is practically hard to achieve, as one wants to achieve photonic connection to the outside of the resonator. Therefore, the confined resonator modes typically appear as resonances on a considerable continuum arising from coupling to the modes outside of the resonator, which are termed "leaky modes."

During the last 2 decades, the Purcell effect has been studied for many different implementations of quantum dots in a resonator. For the resonators there is a huge variety of realizations such as micropillars, microdisks, photonic crystal defects, etc., which we cannot discuss in detail but rather have to refer to the literature (Vahala, 2003).

The first indications for the Purcell effect were observed for micropillars with embedded disordered quantum wells, in which excitons (electron-hole pairs) are localized for cryogenic conditions, similar to the situation in quantum dots. In these studies, an increase of the emission rate with decreasing micropillar size was reported as expected from Eqs. (2) and (3) (see Fig. 1)

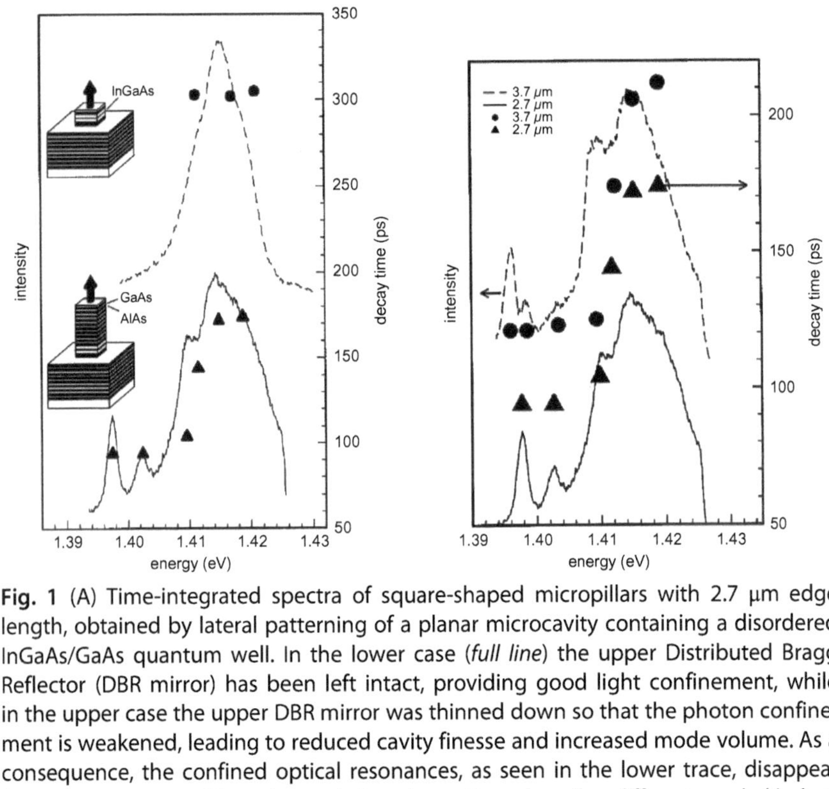

Fig. 1 (A) Time-integrated spectra of square-shaped micropillars with 2.7 μm edge length, obtained by lateral patterning of a planar microcavity containing a disordered InGaAs/GaAs quantum well. In the lower case (*full line*) the upper Distributed Bragg Reflector (DBR mirror) has been left intact, providing good light confinement, while in the upper case the upper DBR mirror was thinned down so that the photon confinement is weakened, leading to reduced cavity finesse and increased mode volume. As a consequence, the confined optical resonances, as seen in the lower trace, disappear from the spectrum. Then the emission decay time (see the *different symbols*) does not depend on photon energy, while it does in the high finesse cavity, where for the resonances with the optical modes a clear reduction of the decay time is seen. In the higher energy emission band, the emission into these modes overlaps with the emission into leaky modes. (B) Time-integrated spectra of square-shaped micropillars with 3.7 and 2.7 μm edge length, fabricated from the same microcavity as before. Here clearly the emission decay times of the mode-resonant emission are shorter in the cavity with smaller mode volume than in the larger cavity, as expected for the Purcell effect (Ohnesorge et al., 1997).

(Ohnesorge et al., 1997). These findings were strikingly confirmed by the implementation of "true," self-assembled quantum dots with strong carrier confinement into the micropillars (Gérard et al., 1998).

Since then the enhancement of the spontaneous emission rate has been studied comprehensively for many different optical resonator and quantum dot types, also on the single dot level (Kress et al., 2005; Lodahl et al., 2004). The highest Purcell factors so far reach 40 (Liu et al., 2018), while recent estimates show that a factor of 200 seems to be in reach by steady miniaturization of the cavity volume by which the reduction of the cavity finesse that is hard to avoid in this development becomes by far overcompensated.

Much harder to achieve is the avoidance of the spontaneous emission into unwanted, leaky modes. A sort of passive approach to this problem is to accelerate the wanted emission into specific modes as much as possible so that the impact on the other channels on the emission intensity becomes suppressed. The active approach is to suppress the leaky modes in the photon density of states. Often these modes have an emission pattern out of structured cavities that is strongly different from that of the desired modes. This opens up the possibility to surround the resonators with mirrors that suppress these modes as much as possible, for which designs with over time increasing complexity have been developed. From these activities a suppression by a factor of 10 seems feasible (see Fig. 2).

Overall it is worth to note that care has to be exercised to assess changes of the spontaneous emission rate, either an enhancement or a deenhancement. The measurement of only one quantity such as the time-integrated emission intensity, even resonator-size dependent, is not sufficient for a clear-cut assessment of the targeted change, but several tests need to be performed, for example, time-integrated and -resolved as well as angle-resolved photoluminescence studies. For example, a change in the intensity of the time-integrated emission from a resonator may be caused by a change of the emission pattern in space.

3. Superradiance

One of the most striking manifestations of quantum optical effects on ensembles of emitters is superradiance, which was proposed theoretically for the first time in a seminal paper by Dicke (1954). First experimental observations of superradiance were realized for atoms (Gibbs et al., 1977; Skribanowitz et al., 1973), but afterwards the range of systems, in which similar effects can be observed has been extended significantly and also includes

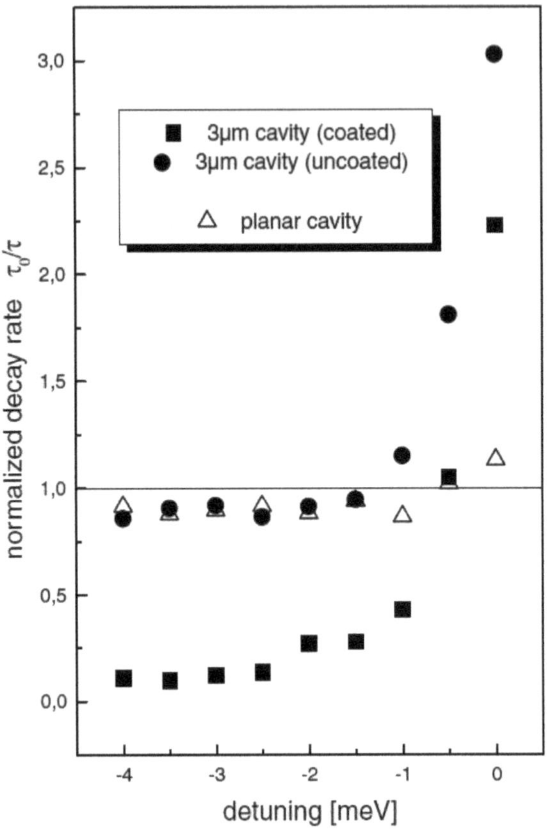

Fig. 2 Photoluminescence decay rate in a micropillar with 3 μm diameter as function of the relative energy distance between the emitter and the ground optical mode in the pillar. For a planar cavity one observes almost no dependence of the decay rate on the detuning. In the pillar with just etched sidewalls and thus limited reflectivity one observes the Purcell effect for resonance with an optical mode, but no change compared to the planar system for off-resonance. In contrast, for the pillar with gold-coated sidewalls, the decay rate is reduced by a factor of 8 as a consequence of the modification of the photon density of states (Bayer et al., 2001).

solid-state systems (Rainò et al., 2018; Scheibner et al., 2007). Several excellent review articles on the basics of superradiance exist (Andreev et al., 1980; Brandes, 2005; Gross and Haroche, 1982), so in the following, we will only sum up the basics of the effect and then discuss its observation in ensembles of quantum dots placed inside a micropillar cavity designed to achieve lasing.

Superradiance may be regarded as the generalization of spontaneous emission for ensembles consisting of more than one emitter. Very loosely speaking superradiance relates to spontaneous emission from a single emitter as lasing relates to stimulated emission from a single emitter. If one considers the simplest emitter possible, a two-level system in the excited state $|e\rangle$, it may decay toward the ground state $|g\rangle$ via spontaneous emission of a photon with a decay rate Γ given by the transition dipole moment $\vec{d}$. Considering instead the case of two emitters in the excited state, the interaction will now include a term proportional to the sum of the two transition dipole moments. As these are complex, the sum will be sensitive to the relative phase of both terms, which in turn may result in constructive or destructive interference of the emission pathways. Accordingly, the spontaneous decay splits into a fast superradiant decay channel corresponding to constructive interference and a slow subradiant decay channel corresponding to destructive interference. Assuming that one considers a small system, where all emitters are indistinguishable and the system is symmetric with respect to permutation of emitters, it becomes possible to form an insightful formal analogy to spin systems. Each two-level system may be considered as a fictitious spin 1/2 system, where $|e\rangle$ and $|g\rangle$ may be mapped onto the spin-up and spin-down states, respectively. Within this picture, the dipole operators are then fully analogous to Pauli spin matrices. Accordingly, such a state of N two-level systems may be mapped to a superposition of N spin 1/2 states. Accordingly, the algebra of spin addition and the basic insights already known from spin physics carry over to superradiance. For example, it becomes clear immediately that for $N=2$, the adequate states to describe the system will be the analogues of spin singlet and triplet states, which in turn gives an intuitive picture for the appearance of superradiant and subradiant states.

4. Superradiance from quantum dot lasers

As outlined above, quantum dots provide an appealing platform to study the interplay between several typical types of emission processes: spontaneous thermal emission for independent emitters, coherent radiation for stimulated emission, and superradiant spontaneous coherent emission due to the formation of correlations between emitters (Jahnke et al., 2016; Kreinberg et al., 2017). However, in terms of photon number statistics, the superradiant effects tend to diminish in the limit of a large number of quantum dots (Temnov and Woggon, 2005) and also in the limit of large inhomogeneous broadening. In order to facilitate the investigation of emission processes on the level of effectively only few emitters, we embed them inside an optical resonator. The device is designed as a micropillar structure with dimensions chosen such that they become comparable to the wavelength of the light emitted from the quantum dots. For such a reduced mode volume, only few quantum dots couple resonantly to the optical microcavity mode. While one may consider such micropillar lasers as nanolasers in first approximation, it is well known that the low mode volume results in significant deviations from standard lasers. For example the low mode volume also results in high values of the β-factor, which describes the amount of spontaneous emission directed toward the cavity mode as compared to the total spontaneous emission into all modes, including leaky modes. A large β-factor in turn implies that the lasing threshold is not that well defined and in the ideal case of $\beta = 1$ the laser will even become thresholdless (Rice and Carmichael, 1994). Also, lasing and strong coupling may coexist (Gies et al., 2011, 2017; Nomura et al., 2010) and generally speaking lasing thresholds may become very low and backaction of the emission of photons from the resonator on the dynamics inside the cavity may become non-negligible (Ulrich et al., 2007; Wiersig et al., 2009).

In order to study superradiant effects in the emission from nanolasers, we used a micropillar structure with a diameter of 6 μm containing a single layer of quantum dots at the antinode of the intracavity field inserted into a helium-flow cryostat and cooled to 10 K. The mirrors of the cavity consist of upper and lower distributed Bragg reflectors consisting of 20 and 23 alternating $\lambda/4$ pairs of AlAs (79 nm) and GaAs (67 nm) layers, respectively, which result in vertical photon confinement and emission. The cavity shows a quality factor of about 5250 and the quantum dot density is approximately 3×10^{10} cm^{-2}, so that within the cavity resonance window of about 1 meV approximately 200 quantum dots show a central emission wavelength within this spectral window and may couple resonantly to the cavity. However, additional nonresonant coupling between off-resonant quantum dots and the cavity is possible to some degree (Ates et al., 2009; Laucht et al., 2010; Winger et al., 2009). In order to recreate the incoherent excitation scenario discussed before, far-off-resonant optical excitation of the quantum dots is used. To this end, the excitation wavelength is set to 780 nm and the pulse repetition frequency amounts to 75.39 MHz. The excitation pulses are focused onto a single pillar using a microscope objective with a numerical aperture of 0.26, which is also used to collect the emission from the sample. The excitation spot at the sample position has a diameter of about 10 μm and thus covers the whole micropillar. The emission is filtered spectrally using an interference filter with a spectral width of approximately 1 nm, so that only the ground state emission is selected. The emission is then directed toward a Hamamatsu S-20 streak camera with a temporal resolution of less than 2 ps to perform time-resolved measurements of the emission dynamics and ultrafast photon correlation measurements (Aßmann et al., 2009, 2010a,b).

As it turns out, the emission from quantum dot nanolasers shows deviations from conventional laser emission and for pulsed nonresonant excitation inter-emitter coupling may play a very prominent role. However, it is highly nontrivial to identify such effects in experiments. In most cases, superradiance is a transient phenomenon and as such may appear during some parts of the emitted pulse, but be absent in others. Accordingly, experimental techniques with high temporal resolution, such as the ultrafast photon correlation measurements mentioned above, are required to reveal the different emission regimes. The emission regimes that one may expect in the emission from nanolasers are summarized in Fig. 3.

Below threshold and in the absence of correlations between emitters, the nanolaser is expected to emit thermal light, which corresponds to a second-order correlation function close to $g^{(2)}(0) = 2$. If photon exchange between emitters becomes reasonably efficient compared to emission of the photon from the cavity, this exchange results in the formation of phase-sensitive correlations between the emitters involved, which in turn may result in

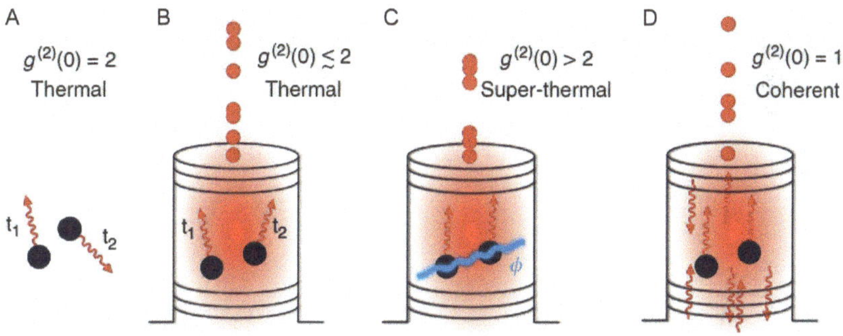

Fig. 3 Typical emission regimes: (A) Spontaneous emission from independent emitters results in thermal radiation. (B) Placing the emitters inside a cavity of small mode volume results in spontaneous emission into a single resonator mode. Below threshold the emission is close to thermal. (C) Inter-emitter coupling by photon exchange results in the formation of correlations between different emitters. A relative phase ϕ is established spontaneously. The system becomes effectively an entangled many-body system and the emission corresponds to a superradiant pulse showing enhanced photon bunching. (D) Above the lasing threshold stimulated emission becomes prominent and the emission becomes second-order coherent (Jahnke et al., 2016).

superradiant pulses showing giant photon bunching with $g^{(2)}(0) > 2$. Finally, above threshold stimulated emission becomes the dominant process, which results in $g^{(2)}(0) = 1$.

In the following, we will discuss three independent signatures of this inter-emitter coupling, which are especially prominent close to the lasing threshold region: the emission of superradiant pulses, which show a duration about one order of magnitude shorter compared to individual quantum dots in a cavity, excitation trapping, which suppresses the magnitude of the emitted intensity by nearly two orders of magnitude and enhanced photon bunching in the time-resolved second-order correlation function of the emission from the quantum dot laser. In the following, we will discuss these signatures.

To this end, we will first give a brief intuitive explanation showing why a tendency toward enhanced photon bunching is expected for superradiant emission from quantum dots. It is based on a more detailed account that may be found elsewhere (Temnov and Woggon, 2005). Let us assume a simplified system consisting only of two quantum dots, which are excited incoherently in a nonresonant manner. The level scheme of this system is shown in Fig. 4. It consists of the ground state, the doubly excited state and two semiexcited states, one of which is dark and one of which is bright. In the figure, all states are given both in angular momentum nomenclature and in the terminology of ground and excited states in order to emphasize the isomorphism of a superradiant system of two emitters and a system of 2 spin 1/2 particles. We consider the limit of low nonresonant pump rates, so a

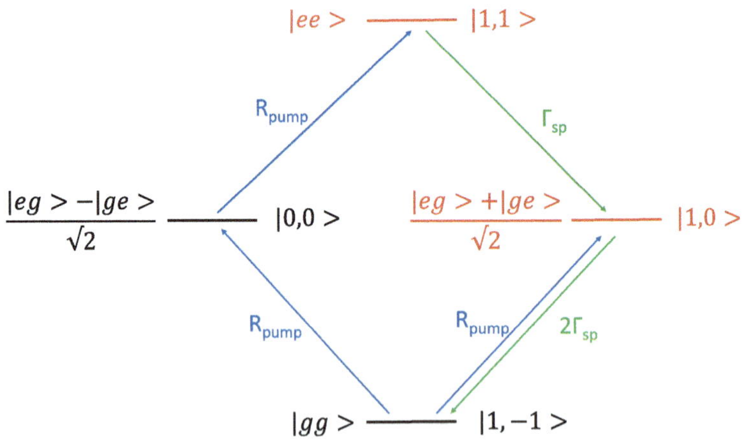

Fig. 4 Simplified level scheme of two indistinguishable quantum dots for nonresonant pumping. Optically active states are marked in *red*. Optically dark states are marked in *black*. Transitions that are possible via nonresonant excitation are marked in *blue*. Transitions that are possible due to emission processes are marked in *green*.

single excitation added to the system in its ground state may excite the system either to the dark or to the bright state with equal probability. If the system ends up in the bright state, the spontaneous decay rate from this state is much larger than the pump rate, so in virtually all cases a single photon will be emitted. If the system instead ends up in the dark state, the opposite is the case. Spontaneous emission is suppressed, so the next excitation added to the system will bring it to the doubly excited state, from which it may decay via a cascade, where the system first spontaneously emits a photon and relaxes toward the bright state, where a second photon will be emitted in quick succession due to the superradiantly enhanced emission rate. Due to the latter pathway, there is a significantly enhanced statistical probability for the emission of photon pairs, which translates to enhanced photon bunching in the limit of many emitters. However, the amount of photon bunching is the strongest for two emitters and becomes reduced with more emitters as the probability of photon pair emission becomes reduced.

Fig. 5 shows a detailed experimental characterization of the time-resolved emission from the nanolaser. The bottom panel displays the

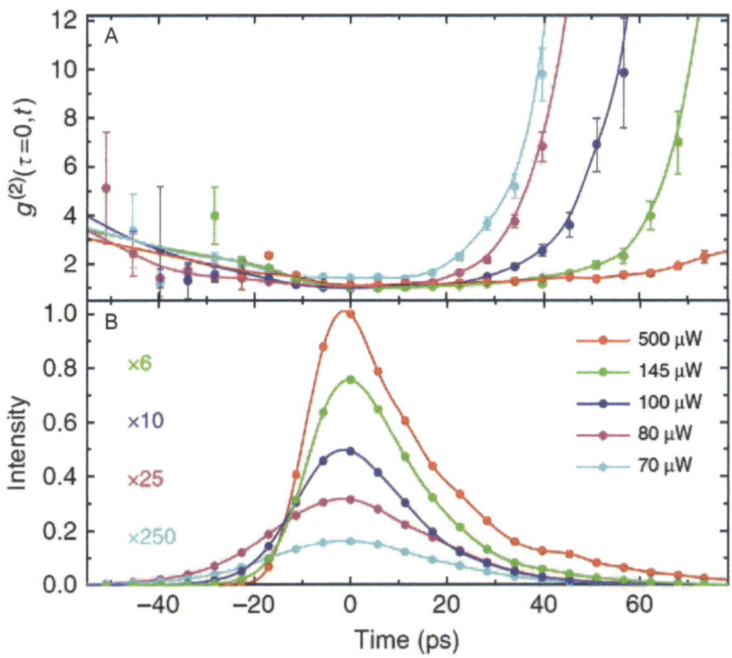

Fig. 5 Second-order correlation function (*top panel*) and time-resolved emission (*bottom panel*) after pulsed nonresonant excitation of a nanolaser with varying pump powers. *Solid lines* are a guide to the eye. *Error bars* correspond to standard counting errors (Jahnke et al., 2016).

time-resolved intensity from the cavity mode, while the upper panel shows the time-resolved photon statistics of the emission in terms of the time-resolved equal-time photon correlation function $g^{(2)}(0, t)$.

Interestingly, already at low pump powers of about 70 μW, the duration of the emitted pulse becomes considerably shorter than the spontaneous recombination time of individual quantum dots subject to the Purcell effect induced by the cavity. The latter amounts to about 200 ps, while the pulse duration amounts to approximately 20 ps and thus is approximately one order of magnitude shorter compared to the expected value for independent quantum dots. One prominent effect that might be responsible for shortening of pulses in nanolasers is stimulated emission. However, monitoring the second-order correlation function directly shows that stimulated emission is not the dominant process here. During the emission maximum of the pulse, we find $g^{(2)}(0) = 2$, which indicates thermal emission and shows that the nanolaser still operates below the lasing threshold. It is even more interesting that prior to the maximum and in an even more pronounced manner significantly after the peak $g^{(2)}(0) \gg 2$ is obtained and may take on values as large as 10.

For slightly larger pump powers of about 80 μW, stimulated emission may already be reached for a short time during the maximum of the emission pulse as indicated by $g^{(2)}(0) \approx 1$ close to the peak. However at earlier and later times the emission changes to thermal and in the wings of the pulse signatures of superradiance may be found again in the photon statistics. Only for high pump powers far above threshold, e.g., at 500 μW $g^{(2)}(0)$ stays close to unity during the whole pulse, so the emission becomes similar to the coherent emission expected from a regular laser. Interestingly, the pulse duration and general shape do not change too strongly across the different pump power regimes. However, the photon correlation measurements clearly show significant differences between the different emission regimes.

This giant photon bunching is already a strong indication for inter-emitter coupling and superradiance taking place. However, on its own this experimental finding does not identify the presence of inter-emitter coupling unambiguously. Other effects, such as mode competition between microcavity modes of different polarizations may in principle result in similar experimental signatures. Accordingly, microscopic theory is required to connect the photon correlations found in experiments to inter-emitter correlations. Microscopic theories based on the cluster expansion technique (Chow et al., 2014; Gies et al., 2007) allow one to formulate a hierarchy of coupled equations of motion for expectation values of carrier occupation numbers, carrier-photon correlations and photon correlation functions. The results for such a model that explicitly includes inter-emitter coupling effects that result in superradiance are shown in Fig. 6. In contrast to experiment, these inter-emitter coupling effects may be switched off while keeping all other parameters constant in theory, which gives a clear intuitive picture of their effects on the emitted pulse.

For weak pump powers, theory predicts thermal emission during the maximum of the pump pulse and giant photon bunching in connection with a considerable shortening of the output pulse after the maximum, which agrees well with the effects seen in the experiment. When the terms responsible for entanglement between different emitters are switched off in theory, the emission stays thermal for the whole duration of the decay and the decay itself becomes significantly longer and reaches the standard spontaneous emission lifetime of quantum dots modified solely by the Purcell effect. For stronger pump powers, stimulated emission may set in, which reduces the inter-emitter coupling significantly or may even destroy it completely if stimulated emission becomes the dominant factor that determines the dynamics of the system.

These findings nicely show that the giant photon bunching is a good indicator for the presence of superradiance. However, in order to gain a deeper understanding of its effects, it is worthwhile to also investigate the system dynamics and the emitted intensity in the presence and absence of inter-emitter coupling as well.

Fig. 7 shows such a comparison. For weak pumping, theory predicts that the emission dynamics follows the slow spontaneous lifetime for independent emitters, while the system dynamics is significantly faster if cooperative effects are considered, which in turn results in the formation of a superradiant emission pulse. Interestingly, the presence of superradiant coupling also reduces the mean photon number inside the cavity drastically below and around the lasing threshold, where radiative emitter coupling is prevailing and stimulated emission is not too significant. However, that does not necessarily mean that the total number of excitations present in the system is reduced. Instead, the excitation is rather stored in terms of the inter-emitter coupling as a collective coherence among the emitters. Very loosely speaking, one may consider the photons to be exchanged between emitters, which drives an inter-emitter polarization similar to the polarization that forms when light propagates inside a solid material. These excitations are then not present as photons inside the cavity mode. This excitation trapping strongly depends on the amount of spontaneous emission directed into

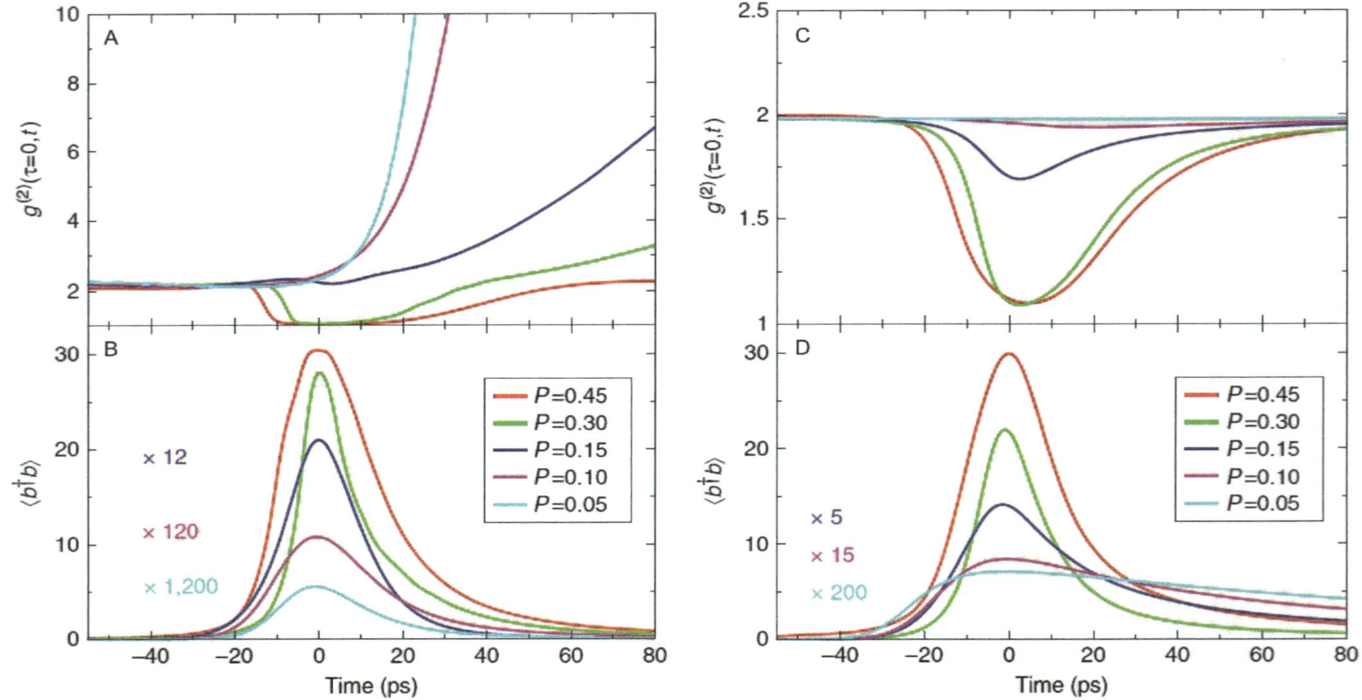

Fig. 6 Comparison of theory results for photon correlations (*top*) and time-resolved emission intensity (*bottom*) in the presence and absence of superradiant coupling for different excitation powers. Panels (A) and (B) include inter-emitter coupling, while panels (C) and (D) do not (Jahnke et al., 2016).

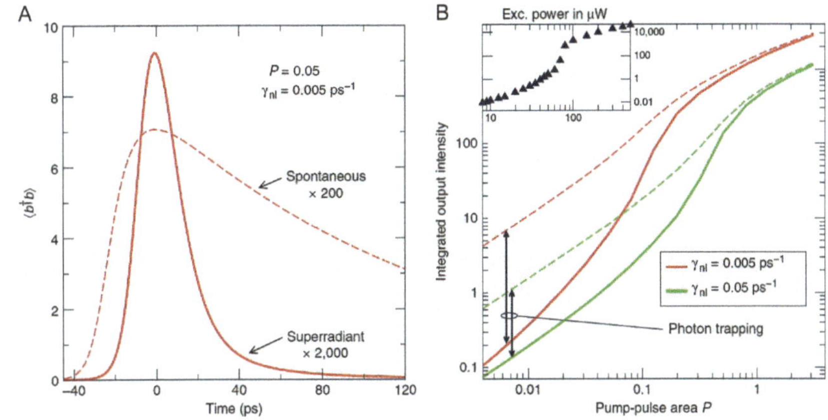

Fig. 7 Comparison of system dynamics in the presence and absence of inter-emitter coupling. Panel (A) shows the differences in the time evolution of the cavity photon number and the formation of a superradiant pulse, while panel (B) shows the integrated emission and indicates excitation trapping. The *inset* shows an experimental input–output curve for comparison (Jahnke et al., 2016).

nonlasing modes for the nanolaser in question. If spontaneous emission into nonlasing modes becomes enhanced, the lasing threshold shifts to higher pump powers and the excitation trapping effect is reduced. Accordingly, a low mode volume indeed is an important prerequisite to observe superradiant effects in microcavity lasers. Interestingly, excitation trapping counteracts another prominent effect frequently observed in nanolasers: almost thresholdless lasing.

Taking a basic rate equation laser model into account (Rice and Carmichael, 1994), the jump in intensity in the threshold region depends on the β-factor of the nanolaser in question and scales as β^{-1} in a double-logarithmic plot, which results in a not too well-defined threshold for high values of β. The magnitude of this jump is also often taken as an experimental signature used to deduce the β-factor of a laser structure. However, as shown in Fig. 7, excitation trapping may obscure this effect and result in significant deviations from the simple β^{-1}-scaling of the kink in the input–output curve. The jump may become much more pronounced if superradiant coupling is taken into account as excitation trapping is efficient for low pump powers, but becomes negligible as soon as stimulated emission becomes dominant, which results in an effectively enhanced jump in the input–output curve.

It is instructive to compare these predictions to experimental data as well. First, the inset of the same figure shows an experimentally determined input–output curve, which also demonstrates a significant jump of more than two orders of magnitude in the emitted intensity across the lasing threshold, although the nominal β-factor of the structure in question is on the order of 0.1. The enhanced kink observed in experiments also highlights the role of inter-emitter coupling for the laser structure studied here. Accordingly, it would also be highly desirable to directly compare experimental results in the presence and absence of superradiant coupling or to systematically vary the amount of inter-emitter coupling. For other emitters such as atoms one standard way to perform systematic investigations of superradiant coupling may be achieved by varying the number of emitters. For epitaxial semiconductor quantum dots, doing so is nontrivial and error-prone as changes in the quantum dot density during the growth process usually also introduce a multitude of additional changes including the confinement energies of the quantum dot energy levels, the strength of light–matter coupling and dephasing processes and many more. Therefore, it is highly important that there are several signatures for superradiance, which are fulfilled simultaneously. Still, it is worthwhile to compare different types of quantum dots, which are expected to show strong and no inter-emitter coupling, respectively. Although they are not completely comparable, the differences arising give a good grasp of the influence of inter-emitter coupling.

To this end, Fig. 8 shows a comparison of two different types of quantum dots. In the left panel, black dots represent the input–output curve for the

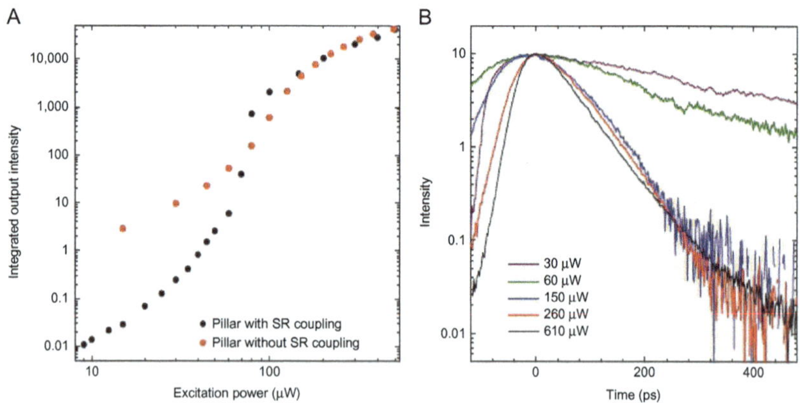

Fig. 8 Input–output curves for two samples with relevant and negligible inter-emitter coupling respectively (A). The dynamics of the emission are shown in comparison for the sample without inter-emitter coupling (B) (Jahnke et al., 2016).

sample investigated so far. These are InGaAs quantum dots with a diameter of 20–30 nm, which show a ground state emission energy of 1.38 eV and strong carrier confinement, which results in a sublevel splitting on the order of 40 meV. Red dots instead denote the input–output curve for a sample consisting of quaternary AlInGaAs quantum dots that have a larger band gap, but are also much larger, so that the emission energy coincides with that of the first sample. Due to the enhanced size, carrier quantization is much weaker in this kind of quantum dot, so that the sublevel splitting amounts only to about 5 meV. This reduced level splitting causes dephasing processes to be much more prominent in this sample, which should in turn reduce inter-emitter coupling drastically. The right panel of Fig. 8 also shows the time-resolved emission for this kind of sample. Here, the emission clearly shows a behavior that is contrary to the results presented for the other sample. For weak pumping, the sample shows an exponential decay with a characteristic time scale of about 450 ps, which is a factor of approximately 20 longer compared to the decay observed for the sample which shows inter-emitter coupling. For higher pump powers, a significant shortening of the output pulse occurs, which may be traced back to the onset of stimulated emission. Indeed, when micropillars of different diameters are compared for the latter kind of quantum dot, the exponential decay time at weak pump powers changes strongly as a result of the different mode volumes, which in turn result in different Purcell factors. Accordingly, the dynamics is clearly dominated by the properties of individual quantum dots and not by collective inter-emitter coupling effects for the latter kind of sample. Although several factors may contribute to these deviations, most probably the most important factor here is the modified electronic confinement, which results in enhanced dephasing rates. This difference is also clearly reflected in the input–output curves, which show an reduced jump by two orders of magnitude for the sample without inter-emitter coupling as compared to the sample where superradiant effects are prominent, which nicely matches the theoretical prediction for the two cases.

5. Photon statistics excitation spectroscopy

The aim of spectroscopy lies in getting information about a system by harnessing its light–matter interaction. To this end, a standard approach is to tailor the properties of a light field that is used to excite the system and investigate its response. Standard properties that may be tailored include the intensity, spectral profile, polarization, or temporal duration of the light fields used. Here, technological progress has steadily increased the range of possible degrees of freedom over time. While polarization-resolved studies are an essential tool since the early days of spectroscopy, the invention of the laser opened up the possibility to study nonlinear effects at high intensities or ultrafast dynamics by applying ultrashort pulses. Recent technological advances opened up the possibility to add another degree of freedom to the spectroscopy toolbox: photon statistics. For light fields that share the same wavelength, polarization, and mean intensity, the response of a system may still differ dramatically depending on whether the light field is in a Fock state, thermal state, or whether it is squeezed light. Accordingly, major efforts are currently undertaken, to apply the photon statistics degree of freedom to sensing, spectroscopy, imaging, and interferometry (Mukamel et al., 2020).

Still, designing, characterizing, and benchmarking optical devices and their interaction with light fields is a nontrivial task. Ideally, one should find a superoperator that describes the whole performance of a device and gives its response to any possible input light field. The process of determining all entries of this superoperator is called quantum process tomography. One might consider this an impossible task due to the vast amounts of possible light field states that might be used to excite any given system, which are hard to realize inside a single lab. However, fortunately it is known that the coherent states form an overcomplete basis of the underlying Hilbert space and that any given density matrix may be represented as well as a sum of coherent state density matrices (Glauber, 1963; Sudarshan, 1963). This representation is based on the Glauber-Sudarshan P-distribution of the state of interest, which unfortunately may become highly singular for nonclassical states (Leonhardt, 1997). Fortunately, it is possible to find a smooth approximation (Klauder, 1966) that may become arbitrarily close to the true Glauber-Sudarshan P-distribution and thus the whole superoperator may be reconstructed using only readily available coherent states as input states (Lobino et al., 2008; Rahimi-Keshari et al., 2011).

For processes, which do not show a dependence on the phase of the light field used for excitation, it may be sufficient to consider only the probability distribution of the output photon numbers as the quantity of interest. This is typically the case for nonresonant excitation, where any phase information present in the excited carriers dephases quickly. In the following, we will provide two examples of how photon statistics may be utilized in spectroscopy on quantum dot ensembles. First, we present an experiment, where the photon statistics of a nonresonant excitation beam is actually varied and investigate the changes in the input–output curve (Aßmann and Bayer, 2011; Kazimierczuk et al., 2015). Second, we present an experiment, where a version of quantum state tomography is used to identify quantum memory effects in a quantum dot laser (Berger et al., 2014). Although we focus on quantum dot ensembles, it is worthwhile to note that photon statistics are also relevant for the light–matter interaction of individual quantum dots (Strauß et al., 2016).

One of the most insightful experiments that shows the importance of photon statistics is the seemingly simple experimental determination of the input–output curve of some laser. While this is often considered as an intrinsic property of the laser structure under investigation, closer examination shows that it depends strongly on the input light field. Loosely speaking, the laser may be considered as an amplifier, which provides a strongly nonlinear response to the input light field. Along with the nonlinear response, the emission from the laser will typically become coherent, which may be clearly identified by a gradual decrease of $g^{(2)}(0)$ from 2 to 1 across the lasing threshold, which indicates the transition from thermal to coherent light. In a simplified description, the laser may thus be considered as an amplifier that maps some input photon number to some output photon number. Interestingly, this simple picture is already sufficient to show that the shape of the input–output curve will depend on the photon statistics used to excite the structure. By definition, the input–output curve gives the mean emitted photon number for the mean input photon number. For any classical light field, in any instant the actually present photon number will not necessarily be the mean photon number, but there will be fluctuations around it. While these fluctuations do not have any influence on the mean output photon number if the input–output curve is linear, any nonlinearity in the response will have a strong effect. If the curvature of the response function is positive, above-average photon numbers will result in a strong increase of the instantaneous output photon number, while below-average photon numbers result in a smaller decrease of the instantaneous output photon number, which taken together results in an enhanced output mean photon number. The enhancement grows with the variance of the input photon number. If the curvature of the input–output curve is negative, the results are exactly the other way round. An example for a typical calculated input–output curve is shown in Fig. 9. The curve typically encountered in standard experiments is the one for a Poissonian input photon number distribution

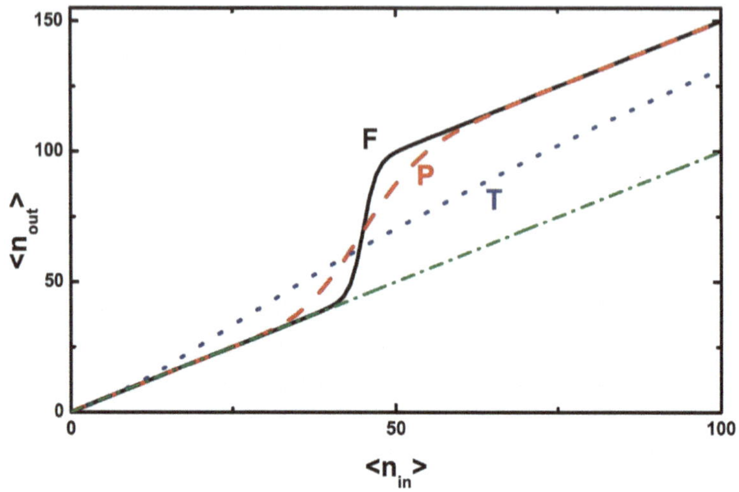

Fig. 9 Calculated input–output curve of a generic laser for excitation using coherent (C), thermal (T), or Fock (F) states. The *green line* shows the mean input photon number for comparison (Aßmann and Bayer, 2011).

$$p_c(n) = e^{-\langle n\rangle}\frac{\langle n\rangle^n}{n!}, \tag{4}$$

for which the mean photon number is the most probable photon number and the photon number variance also equals the mean photon number. This input photon number distribution is expected for exciting the system with a laser. Using a thermal light source, such as an incandescent lamp will result in excitation of the system with thermal light, which follows a Bose-Einstein photon number distribution:

$$p_t(n) = \frac{1}{(1+\langle n\rangle)(1+\frac{1}{\langle n\rangle})^n}. \tag{5}$$

Here, the most probable photon number is always 0 and the probability distribution decreases monotonically with n. Thus, the mean photon number has no distinguished role and both much larger and much smaller values occur with large probability. Accordingly, the variance of the photon number is quite large and amounts to $\langle n\rangle^2 + \langle n\rangle$. One clearly sees that the output photon number distribution for this kind of pump also shows a step-like increase around threshold, but it is broadened significantly compared to the case of coherent pumping. One also clearly sees that the output photon number may be larger or smaller compared to the case of coherent pumping, depending on whether the curvature of the input–output curve is predominantly positive or negative in the region of interest. Finally, the input–output curve for a pump source that emits Fock states is shown. For Fock states, the actual photon number always equals the mean photon number and the photon number variance is exactly 0. Accordingly, the step-like threshold in the input–output curve is defined better compared to the other pump scenarios. This again shows clearly, why it would be highly desirable to have tunable nonclassical light sources for precision spectroscopy.

The results discussed so far already show that the photon statistics degree of freedom of the excitation beam may be used to identify nonlinearities as only nonlinear systems will be sensitive to changes in the input photon number variance. However, it is even more interesting that also the photon number noise of the output light field is necessarily predicted to show a drastic enhancement as well.

In order to test these predictions experimentally, it is necessary to create a light source that may be switched from a thermal to a coherent light source without changing the spectral, temporal, or polarization properties of the excitation light field. Also, the coupling of the light field to the sample must necessarily be exactly the same in both cases. In order to realize a light source that offers these properties, we built a pseudothermal Martienssen lamp that is used to excite the sample nonresonantly (Martienssen and Spiller, 1964). It consists of a pulsed laser that delivers pulses with a pulse duration of approximately 100 fs at a repetition rate of 75.39 MHz and a wavelength of 790 nm. The laser light is then scattered from a rotating ground glass disk, which results in a typical time-varying speckle pattern (Aßmann and Bayer, 2012). If one now collects a fraction of the scattered light from an area that is smaller than the typical speckle size, this fraction of the light field will show thermal photon number statistics with a coherence time determined by the grit polish of the ground glass disk, its rotation velocity and the radial spot position on the disk. In our case, the ground glass disk rotates at a frequency of approximately 8 Hz and we collect the scattered light from the ground glass disk using a 35 mm lens and a single mode fiber placed at a distance of 30 cm from the disk. This way, we are able to realize a light field that shows almost perfect pseudothermal photon number statistics with $g^{(2)}(0) = 1.97 \pm 0.02$ and a coherence time of 13.7 μs. In order to switch the light source from pseudothermal to coherent, we simply stop the rotation of the disk, so the scattering process and the speckle pattern become static and the intensity of the scattered beam is not modulated in time, anymore. The sample we use is a micropillar laser similar to the sample containing quaternary quantum dots discussed before as an example for a quantum dot laser, where inter-emitter coupling is absent. The pillar diameter of the laser used for the following experiments amounts to 8 μm. For detection, the emitted light is guided toward a spectrometer to determine the input–output curve or to a Hanbury Brown-Twiss setup in order to determine $g^{(2)}(0)$ of the emission. For the latter, we deliberately used photodiodes with a temporal resolution of about 500 ps, which are unable to resolve the intrinsic photon bunching of the thermal light emitted by the micropillar laser below threshold. This kind of photon bunching is known to occur on a timescale of only few picoseconds for the samples investigated here (Aßmann et al., 2010a; Ulrich et al., 2007). By choosing a detection setup that is not sensitive to these fluctuations, we can be sure that all changes observed in the output photon number statistics may be traced back to the photon number statistics of the excitation light field and also influences of other short-term effects such as relaxation oscillations may be excluded (Ulrich et al., 2007; Wiersig et al., 2009).

A comparison of the experimental input–output curves for this pillar using coherent and pseudothermal light, respectively, is shown in Fig. 10. For excitation with thermal light, it can be seen clearly that the onset of

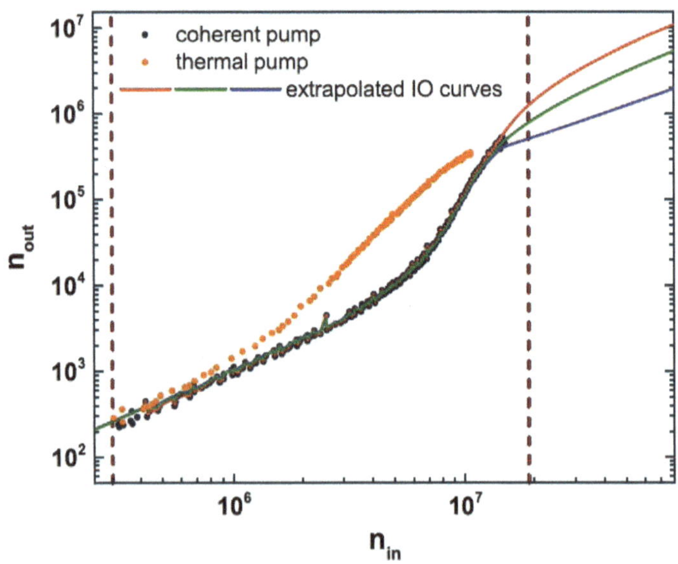

Fig. 10 Measured input–output curves (*dots*) for coherent (*right curve*) and thermal excitation (*left curve*). *Solid lines* represent three possible interpolations beyond the range of investigated pump powers. *Dashed lines* mark the range of excitation powers that will be investigated further in the following (Kazimierczuk et al., 2015).

the step in the input–output curve starts significantly earlier as compared to coherent excitation. The pump power, where the nonlinearity sets in is reduced by a factor of approximately 3 for the thermal pump beam. This difference is a consequence of the larger variance of the thermal input photon number distribution. Due to the large spread in the instantaneous photon numbers, a significant part of them will belong to the nonlinear region already for mean photon numbers below the nominal threshold value.

Fig. 11 shows the values of $g^{(2)}(0)$ of the emission from the micropillar both for coherent and thermal excitation. For coherent excitation, the emission shows an almost constant value of $g^{(2)}(0) = 1$. This is exactly the

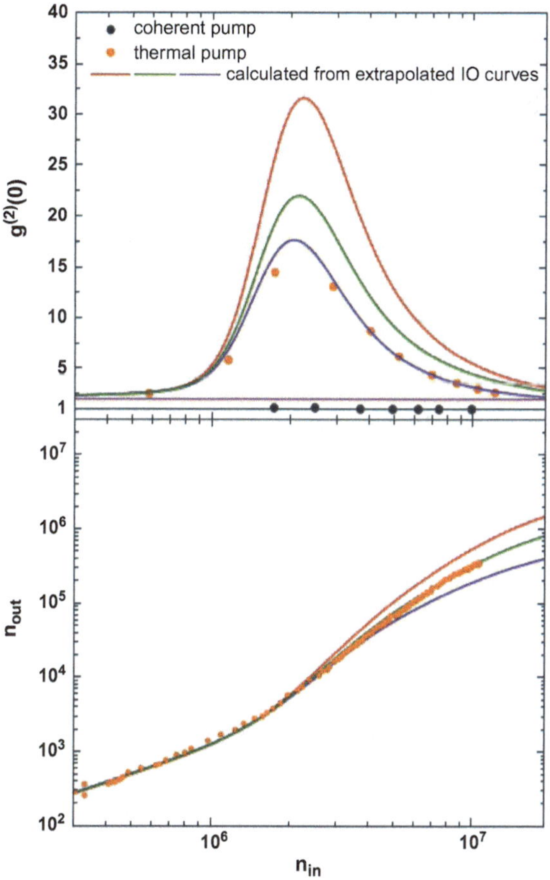

Fig. 11 *Upper panel*: Measured values of $g^{(2)}(0)$ for pseudothermal and coherent excitation. *Lower panel*: Thermal input–output curves. *Solid lines* in both panels represent the expected curves assuming the three extrapolations of the coherent input–output curve shown in Fig. 10 (Kazimierczuk et al., 2015).

expected behavior as the detectors used are slow and thus not overly sensitive to fast photon number fluctuations. For pseudothermal excitation, the results are drastically different. Far below and far above threshold, the photon number statistics follow the pseudothermal photon number distribution of the input light field, which results in $g^{(2)}(0) = 2$. Within the nonlinear region, however, there is a significant overshoot in the photon number noise of the emission and values as large as $g^{(2)}(0) = 15$ may be reached. This is again a significant consequence of the large variance of the thermal input photon number distribution. As it includes a wide range of input photon numbers, these will be subject to very different amplification ratios if they fall within the nonlinear region. Here, larger photon numbers will be subject to a stronger amplification as compared to smaller photon numbers. Accordingly, the photon number noise will necessarily become enhanced for pseudothermal excitation.

Interestingly, excitation using thermal light also yields information about the response of the system to mean photon numbers larger than those actually investigated in the experiment. Both for thermal and coherent pumping the full input–output curve could not be recorded due to the limited range of pump powers available. The three solid lines shown in Fig. 10 represent three different extrapolations showing how the curve might continue for coherent pumping. Judging solely from the data recorded for coherent pumping, all of them are equally plausible. However, the solid lines in Fig. 11 provide some further information on the correct shape of the input–output curve. Here, the solid lines represent the expected values for the output photon number and $g^{(2)}(0)$ for thermal pumping calculated from the three possible input–output curves for coherent pumping discussed above. Obviously, the differences are significant already within the threshold region and the results directly rule out the curve that shows the highest saturation photon number. For the two other extrapolations, the curve with the lowest saturation photon number reproduces the shape of $g^{(2)}(0)$ well, while the intermediate one reproduces the input–output curve well. This implies that the correct continuation of the input–output curve should lie between these shapes. Considering that in the experiment the detectors are slow, so that high-frequency photon number noise is not detected, it seems likely that the intermediate curve is the closest to the real one.

6. Quantum-optical spectroscopy

Finally, we would like to emphasize that it is not only possible to change the input photon number statistics to perform spectroscopy in order to get more detailed information about the system that is investigated, but also that this process works the other way round as well: Performing detailed spectroscopy with well-defined known classical states makes it possible to calculate the response of the system to any possible input state, even non-classical ones. This surprising result is based on the fact that coherent states form an overcomplete basis of the entire phase space. Accordingly, each density matrix may be represented as a sum of different coherent state density matrices. Although this representation may become highly nontrivial, it is always possible to find a well-behaved approximation that is arbitrarily close to the real representation. Accordingly, it becomes possible to determine the system response to arbitrary light fields, if the response to a full set of coherent states is known. This approach is known as quantum process tomography (Lobino et al., 2008; Rahimi-Keshari et al., 2011) or quantum spectroscopy (Kira et al., 2011).

To this end, we investigate a microring cavity with an outer diameter of 2.2 μm and an inner diameter of 1.5 μm containing self-assembled InP quantum dots (Berger et al., 2014). The quantum dots host several confined energy levels with a sublevel spacing of 3–5 meV. We used pulsed quasi-resonant excitation of the quantum dots at a wavelength of about 705 nm, which is 30.7 meV above the energy of the main cavity mode, but below the band gap of GaInP, of which the ring cavity consists. The emission is collected using a 100× microscope objective in a confocal geometry. The spectrally resolved emission is then detected using a 500 mm focal length spectrometer. For these pump conditions, even the highest pump power will generate less than 0.26 electrons per electronic level, so the system is below the lasing threshold. Also, the quantum dot density is on the order of 1.5×10^{9} cm^{-2} only, so on average only one quantum dot will couple resonantly to the cavity mode, while off-resonant dots contribute to a broad background.

The top panel of Fig. 12 shows the input–output curve for these excitation conditions. At first sight, it seems to be linear as expected for a laser way below threshold. However, a close examination reveals systematic small-amplitude oscillations around the mean value as shown in the middle panel of Fig. 12. Solving the full quantum dynamics for the light–matter coupling of the four relevant electronic levels of the quantum dots including dephasing makes it possible to identify the origin of these weak oscillations: The pump beam induces detuned Rabi oscillations, which result in enhanced emission at the peak of the quantum dot excitation probability amplitude and reduced emission at its dip. As the detuning is quite significant, the excitation of the dot remains small and oscillates very quickly.

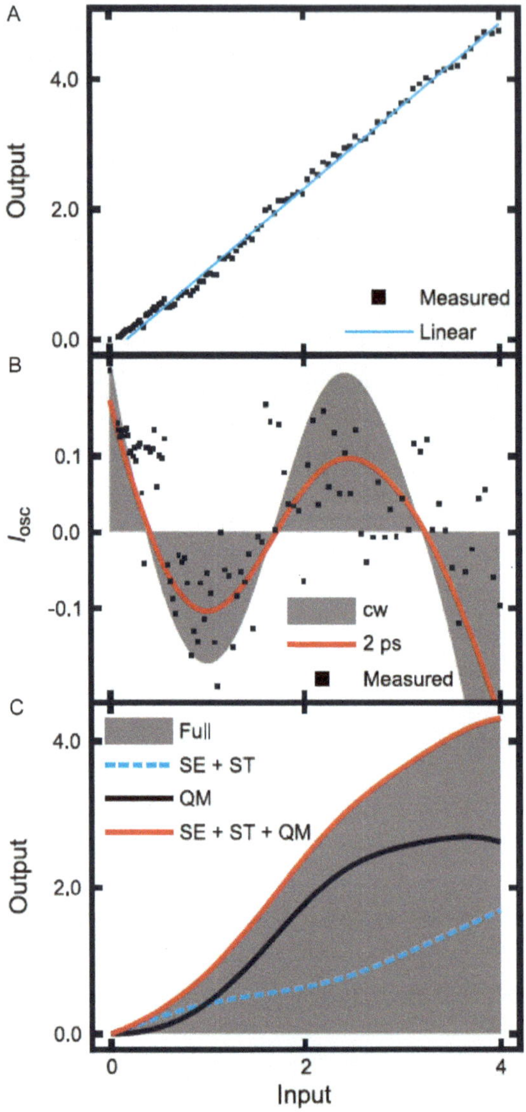

Fig. 12 *Top panel*: Input–output curve and best linear fit to it. *Middle panel*: Oscillating deviations from linearity as measured experimentally and as expected from a full theoretical model. *Bottom panel*: Computed input–output curves taking contributions due to stimulated emission (ST), spontaneous emission (SE), and quantum memory effects (QM) into account, respectively (Berger et al., 2014).

These oscillations do not only have an influence on the shape of the input–output curve, but also introduce memory effects that depend on excitations present prior to emission of a photon. These memory effects represent accumulated photon-density correlations, which are necessarily nonlinear with respect to the pump photon number, which in turn results in nonlinearities in the input–output curve. Taking all of these effects into account results in the theory curves also shown in the middle panel of Fig. 12, which match the experimental results reasonably well.

In fact, there are several contributions to the emitted light field and it is instructive to investigate in detail, in which manner they depend on the input photon number. Three main contributions to the emission need to be considered here. First, spontaneous emission will yield a contribution whenever the system is not in the ground state. Next, stimulated emission may contribute to the emitted signal. This contribution acts on the emission via photon correlations within the cavity. Finally, the quantum memory contribution depends both on the classical Rabi frequency and the correlations between photons and carrier densities. Contributions including higher-order correlations might in principle contribute to the signal as well, but turn out to be of negligible magnitude for the system at hand. Within the full theoretical model, the individual contributions of these three terms may be switched on and off individually, which gives a good impression of their importance. The magnitude of these contributions is shown in the lower panel of Fig. 12.

The full result is shown as the sum of all three terms and given by the line bordering the shaded area. Below that, the dashed line yields the summed contribution of spontaneous and stimulated emission, while the solid line in between corresponds to the contributions arising due to the quantum memory effect. It is interesting to note that the sum of spontaneous and stimulated emission increases monotonically with the input photon number. In contrast, the quantum memory contributions contain rather large oscillations, which remain visible partially in the total output intensity. Accordingly, it is mandatory to include the quantum memory effects in order to explain the shape of the input–output curve.

Finally, it would be interesting to see whether the oscillations in the input–output curve may be controlled experimentally. However, doing so requires light sources with tunable photon statistics. However, just as in the case of photon statistics excitation spectroscopy mentioned above, it is again possible to calculate the response to arbitrary input fields given that one knows the responses to all coherent input fields. A detailed analysis shows that the appearance of quantum memory effects depends on the simultaneous appearance of coherence and induced squeezing. Accordingly, a good test in order to determine whether the full model is a good description of the experimental data consists of calculating the response to input fields that lack coherence. If the model is correct, the response to such input fields should

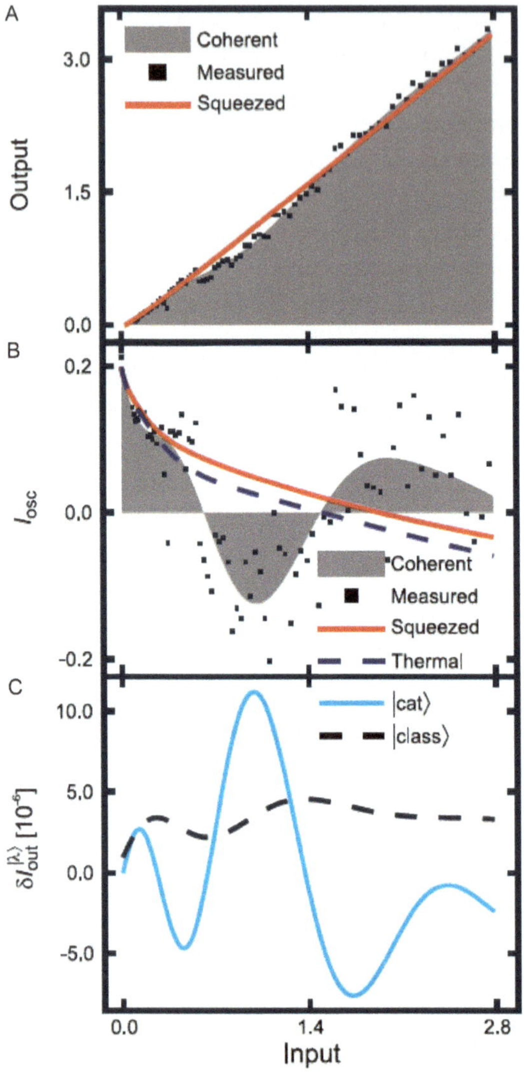

Fig. 13 *Top panel*: Measured input–output curve compared to model predictions for coherent and squeezed vacuum input states. *Middle panel*: Measured nonlinear behavior compared to model predictions for coherent, thermal, and squeezed vacuum input states. *Bottom panel*: Differential response of the system for a classical differential (*dashed line*) and a differential for a stretching cat state (*solid line*) (Berger et al., 2014).

not show any oscillations as the coherences driving the quantum memory terms will be absent. The top panel of Fig. 13 compares the calculated responses to coherent excitation and excitation via a squeezed vacuum state. The former includes coherence and is expected to generate squeezing via resonance fluorescence, while the latter includes squeezing, but no coherence and should thus not induce any oscillations. The results agree with this assumption and clearly show no oscillations for excitation using a squeezed vacuum state. The middle panel of Fig. 13 shows the same calculations, but only the deviations from linearity are plotted. In addition, the calculated response for thermal pumping is shown as well, which also does not include coherences and also does not result in any oscillations. The obvious question is now whether the oscillations may become enhanced by exciting the system with a light field that shows both coherence and stronger squeezing. A good candidate state is a stretching Schrödinger cat state, which consists of a superposition of two different coherent states. By comparing the response of the system to a standard coherent state to its response to a stretched cat, where both coherent states involved are displaced along the same axis as the classical coherent state, it becomes possible to isolate the influence of additional quantum fluctuations in the excitation light field. The results are shown in the bottom panel of Fig. 13 in terms of differentials. The dashed line gives the differential caused by the presence of only coherence. It represents the change in the output intensity if one considers some coherent excitation beam and then changes the amplitude of excitation field such, that its mean photon number increases by one photon. In contrast, the solid line represents the change in the output intensity, if one only changes the quantum fluctuations of the excitation light field. It represents the change in the output intensity if one again considers some coherent excitation beam and then changes it to a stretched cat state of the same mean photon number, so only the fluctuations change. In this case the oscillations become enhanced more strongly as compared to the case of changing just the amplitude of a coherent state. Moreover, one clearly sees that the response to the cat state shows a superlinear increase when the classical pump shows a sublinear output and vice versa. As these oscillations are out of phase, it is in principle possible to fully control the quantum memory effects via the quantum fluctuations of the excitation light field, which indeed opens up the possibility to use photon statistics as a novel degree of freedom in spectroscopy.

7. Summary

In contrast to the general trend to emphasize the role of individual emitters in quantum optical effects, we have discussed several exciting results that highlight the intriguing quantum effects that may arise for ensembles of emitters. It is important to emphasize that also the experimental tools required to identify these effects differ from the ones one would typically apply to investigate individual emitters. While single-photon counting is by far the most frequently applied experimental detection technique for quantum optics at the single photon level, it is expected that continuous variable detection techniques are most suitable for studies on ensembles (McAlister and Raymer, 1997; Roumpous and Cundiff, 2013; Smithey et al., 1993). As technological advancements have made it possible to refine these techniques to levels that match the demands of semiconductor spectroscopy (Lüders et al., 2018; Thewes et al., 2019, 2020), it will be interesting to see which genuine ensemble quantum optics effects will be identified in the future.

Acknowledgments

We gratefully acknowledge funding by the Deutsche Forschungsgemeinschaft (DFG) within the framework of TRR 142, project C1 (project number 231447078).

References

Andreev, A.V., Emel'yanov, V.I., Il'inskiĭ, Y.A., 1980. Collective spontaneous emission (dicke superradiance). Soviet Physics Uspekhi 23 (8), 493–514. https://doi.org/10.1070/pu1980v023n08abeh005024.

Aßmann, M., Bayer, M., 2011. Nonlinearity sensing via photon-statistics excitation spectroscopy. Phys. Rev. A 84 (5), 053806. https://doi.org/10.1103/PhysRevA.84.053806.

Aßmann, M., Veit, F., Bayer, M., van der Poel, M., Hvam, J.M., 2009. Higher-order photon bunching in a semiconductor microcavity. Science 325 (5938), 297. https://doi.org/10.1126/science.1174488.

Aßmann, M., Bayer, M., 2012. Coherence time measurements using a single detector with variable time resolution. Opt. Lett. 37 (14), 2811–2813. https://doi.org/10.1364/OL.37.002811. http://ol.osa.org/abstract.cfm?URI=ol-37-14-2811.

Aßmann, M., Veit, F., Bayer, M., Gies, C., Jahnke, F., Reitzenstein, S., Höfling, S., Worschech, L., Forchel, A., 2010a. Ultrafast tracking of second-order photon correlations in the emission of quantum-dot microresonator lasers. Phys. Rev. B 81 (16), 165314. https://doi.org/10.1103/PhysRevB.81.165314.

Aßmann, M., Veit, F., Tempel, J.-S., Berstermann, T., Stolz, H., van der Poel, M., Hvam, J.M., Bayer, M., 2010b. Measuring the dynamics of second-order photon correlation functions inside a pulse with picosecond time resolution. Opt. Express 18 (19), 20229–20241. https://doi.org/10.1364/OE.18.020229. http://www.opticsexpress.org/abstract.cfm?URI=oe-18-19-20229.

Ates, S., Ulrich, S.M., Ulhaq, A., Reitzenstein, S., Löffler, A., Höfling, S., Forchel, A., Michler, P., 2009. Non-resonant dot-cavity coupling and its potential for resonant single-quantum-dot spectroscopy. Nat. Photon. 3 (12), 724–728. https://doi.org/10.1038/nphoton.2009.215.

Bayer, M., Reinecke, T.L., Weidner, F., Larionov, A., McDonald, A., Forchel, A., 2001. Inhibition and enhancement of the spontaneous emission of quantum dots in structured microresonators. Phys. Rev. Lett. 86 (14), 3168–3171. https://doi.org/10.1103/PhysRevLett.86.3168.

Berger, C., Huttner, U., Mootz, M., Kira, M., Koch, S.W., Tempel, J.S., Aßmann, M., Bayer, M., Mintairov, A.M., Merz, J.L., 2014. Quantum-memory effects in the emission of quantum-dot microcavities. Phys. Rev. Lett. 113 (9), 093902. https://doi.org/10.1103/PhysRevLett.113.093902.

Brandes, T., 2005. Coherent and collective quantum optical effects in mesoscopic systems. Phys. Rep. 408 (5), 315–474. https://doi.org/10.1016/j.physrep.2004.12.002.

Chow, W.W., Jahnke, F., Gies, C., 2014. Emission properties of nanolasers during the transition to lasing. Light Sci. Appl. 3 (8), e201. https://doi.org/10.1038/lsa.2014.82.

Dicke, R.H., 1954. Coherence in spontaneous radiation processes. Phys. Rev. 93 (1), 99–110. https://doi.org/10.1103/PhysRev.93.99.

Gérard, J.M., Sermage, B., Gayral, B., Legrand, B., Costard, E., Thierry-Mieg, V., 1998. Enhanced spontaneous emission by quantum boxes in a monolithic optical cavity. Phys. Rev. Lett. 81 (5), 1110–1113. https://doi.org/10.1103/PhysRevLett.81.1110.

Gibbs, H.M., Vrehen, Q.H.F., Hikspoors, H.M.J., 1977. Single-pulse superfluorescence in cesium. Phys. Rev. Lett. 39 (9), 547–550. https://doi.org/10.1103/PhysRevLett.39.547.

Gies, C., Wiersig, J., Lorke, M., Jahnke, F., 2007. Semiconductor model for quantum-dot-based microcavity lasers. Phys. Rev. A 75 (1), 013803. https://doi.org/10.1103/PhysRevA.75.013803.

Gies, C., Florian, M., Gartner, P., Jahnke, F., 2011. The single quantum dot-laser: lasing and strong coupling in the high-excitation regime. Opt. Express 19 (15), 14370–14388. https://doi.org/10.1364/OE.19.014370. http://www.opticsexpress.org/abstract.cfm?URI=oe-19-15-14370.

Gies, C., Gericke, F., Gartner, P., Holzinger, S., Hopfmann, C., Heindel, T., Wolters, J., Schneider, C., Florian, M., Jahnke, F., Höfling, S., Kamp, M., Reitzenstein, S., 2017. Strong light-matter coupling in the presence of lasing. Phys. Rev. A 96 (2), 023806. https://doi.org/10.1103/PhysRevA.96.023806.

Glauber, R.J., 1963. Coherent and incoherent states of the radiation field. Phys. Rev. 131 (6), 2766–2788. https://doi.org/10.1103/PhysRev.131.2766.

Gross, M., Haroche, S., 1982. Superradiance: an essay on the theory of collective spontaneous emission. Phys. Rep. 93 (5), 301–396. https://doi.org/10.1016/0370-1573(82)90102-8.

Jahnke, F., 2012. Quantum Optics with Semiconductor Nanostructures. Woodhead Publishing.

Jahnke, F., Gies, C., Aßmann, M., Bayer, M., Leymann, H.A.M., Foerster, A., Wiersig, J., Schneider, C., Kamp, M., Höfling, S., 2016. Giant photon bunching, superradiant pulse emission and excitation trapping in quantum-dot nanolasers. Nat. Commun. 7 (1), 11540. https://doi.org/10.1038/ncomms11540.

Kazimierczuk, T., Schmutzler, J., Aßmann, M., Schneider, C., Kamp, M., Höfling, S., Bayer, M., 2015. Photon-statistics excitation spectroscopy of a quantum-dot micropillar laser. Phys. Rev. Lett. 115 (2), 027401. https://doi.org/10.1103/PhysRevLett.115.027401.

Kira, M., Koch, S.W., Smith, R.P., Hunter, A.E., Cundiff, S.T., 2011. Quantum spectroscopy with Schrödinger-cat states. Nat. Phys. 7 (10), 799–804. https://doi.org/10.1038/nphys2091.

Klauder, J.R., 1966. Improved version of optical equivalence theorem. Phys. Rev. Lett. 16 (12), 534–536. https://doi.org/10.1103/PhysRevLett.16.534.

Kreinberg, S., Chow, W.W., Wolters, J., Schneider, C., Gies, C., Jahnke, F., Höfling, S., Kamp, M., Reitzenstein, S., 2017. Emission from quantum-dot high-b microcavities: transition from spontaneous emission to lasing and the effects of superradiant emitter coupling. Light Sci. Appl. 6 (8), e17030. https://doi.org/10.1038/lsa.2017.30.

Kress, A., Hofbauer, F., Reinelt, N., Kaniber, M., Krenner, H.J., Meyer, R., Böhm, G., Finley, J.J., 2005. Manipulation of the spontaneous emission dynamics of quantum dots in two-dimensional photonic crystals. Phys. Rev. B 71, 241304(R). https://doi.org/10.1103/PhysRevB.71.241304.

Laucht, A., Kaniber, M., Mohtashami, A., Hauke, N., Bichler, M., Finley, J.J., 2010. Temporal monitoring of nonresonant feeding of semiconductor nanocavity modes by quantum dot multiexciton transitions. Phys. Rev. B 81 (24), 241302. https://doi.org/10.1103/PhysRevB.81.241302.

Leonhardt, U., 1997. Measuring the Quantum State of Light. Cambridge University Press, Cambridge.

Liu, F., Brash, A.J., O'Hara, J., Martins, L.M.P.P., Phillips, C.L., Coles, R.J., Royall, B., Clarke, E., Bentham, C., Prtljaga, N., Itskevich, I.E., Wilson, L.R., Skolnick, M.S., Fox, A.M., 2018. High purcell factor generation of indistinguishable on-chip single photons. Nat. Nanotechnol. 13 (9), 835–840. https://doi.org/10.1038/s41565-018-0188-x.

Lobino, M., Korystov, D., Kupchak, C., Figueroa, E., Sanders, B.C., Lvovsky, A.I., 2008. Complete characterization of quantum-optical processes. Science 322 (5901), 563–566. https://doi.org/10.1126/science.1162086. https://science.sciencemag.org/content/322/5901/563.

Lodahl, P., Floris van Driel, A., Nikolaev, I.S., Irman, A., Overgaag, K., Vanmaekelbergh, D., Vos, W.L., 2004. Controlling the dynamics of spontaneous emission from quantum dots by photonic crystals. Nature 430, 654–657. https://doi.org/10.1038/nature02772.

Lüders, C., Thewes, J., Aßmann, M., 2018. Real time $g^{(2)}$ monitoring with 100 kHz sampling rate. Opt. Express 26 (19), 24854–24863. https://doi.org/10.1364/OE.26.024854.

Martienssen, W., Spiller, E., 1964. Coherence and fluctuations in light beams. Am. J. Phys. 32 (12), 919–926. https://doi.org/10.1119/1.1970023.

McAlister, D.F., Raymer, M.G., 1997. Ultrafast photon-number correlations from dual-pulse, phase-averaged homodyne detection. Phys. Rev. A 55 (3), R1609(R). https://doi.org/10.1103/PhysRevA.55.R1609.

Mukamel, S., Freyberger, M., Schleich, W., Bellini, M., Zavatta, A., Leuchs, G., Silberhorn, C., Boyd, R.W., Sánchez-Soto, L.L., Stefanov, A., Barbieri, M., Paterova, A., Krivitsky, L., Shwartz, S., Tamasaku, K., Dorfman, K., Schlawin, F., Sandoghdar, V., Raymer, M., Marcus, A., Varnavski, O., Goodson, T., Zhou, Z.-Y., Shi, B.-S., Asban, S., Scully, M., Agarwal, G., Peng, T., Sokolov, A.V., Zhang, Z.-D., Zubairy, M.S., Vartanyants, I.A., del Valle, E., Laussy, F., 2020. Roadmap on quantum light spectroscopy. J. Phys. B At. Mol. Opt. Phys. 53 (7), 072002. https://doi.org/10.1088/1361-6455/ab69a8.

Nomura, M., Kumagai, N., Iwamoto, S., Ota, Y., Arakawa, Y., 2010. Laser oscillation in a strongly coupled single-quantum-dot-nanocavity system. Nat. Phys. 6 (4), 279–283. https://doi.org/10.1038/nphys1518.

Ohnesorge, B., Bayer, M., Forchel, A., Reithmaier, J.P., Gippius, N.A., Tikhodeev, S.G., 1997. Enhancement of spontaneous emission rates by three-dimensional photon confinement in Bragg microcavities. Phys. Rev. B 56 (8), R4367(R). https://doi.org/10.1103/PhysRevB.56.R4367.

Purcell, E.M., Torrey, H.C., Pound, R.V., 1946. Resonance absorption by nuclear magnetic moments in a solid. Phys. Rev. 69 (1-2), 37–38. https://doi.org/10.1103/PhysRev.69.37.

Rahimi-Keshari, S., Scherer, A., Mann, A., Rezakhani, A.T., Lvovsky, A.I., Sanders, B.C., 2011. Quantum process tomography with coherent states. New J. Phys. 13 (1), 013006. https://doi.org/10.1088/1367-2630/13/1/013006.

Rainò, G., Becker, M.A., Bodnarchuk, M.I., Mahrt, R.F., Kovalenko, M.V., Stöferle, T., 2018. Superfluorescence from lead halide perovskite quantum dot superlattices. Nature 563 (7733), 671–675. https://doi.org/10.1038/s41586-018-0683-0.

Rice, P.R., Carmichael, H.J., 1994. Photon statistics of a cavity-qed laser: a comment on the laser-phase-transition analogy. Phys. Rev. A 50 (5), 4318–4329. https://doi.org/10.1103/PhysRevA.50.4318.

Roumpous, G., Cundiff, S.T., 2013. Multichannel homodyne detection for quantum optical tomography. J. Opt. Soc. Am. B 30 (5), 1303–1316. https://doi.org/10.1364/JOSAB.30.001303.

Scheibner, M., Schmidt, T., Worschech, L., Forchel, A., Bacher, G., Passow, T., Hommel, D., 2007. Superradiance of quantum dots. Nat. Phys. 3 (2), 106–110. https://doi.org/10.1038/nphys494.

Skribanowitz, N., Herman, I.P., MacGillivray, J.C., Feld, M.S., 1973. Observation of dicke superradiance in optically pumped HF Gas. Phys. Rev. Lett. 30 (8), 309–312. https://doi.org/10.1103/PhysRevLett.30.309.

Smithey, D.T., Beck, M., Raymer, M.G., Faridani, A., 1993. Measurement of the Wigner distribution and the density matrix of a light mode using optical homodyne tomography: application to squeezed states and the vacuum. Phys. Rev. Lett. 70 (9), 1244–1247. https://doi.org/10.1103/PhysRevLett.70.1244.

Strauß, M., Placke, M., Kreinberg, S., Schneider, C., Kamp, M., Höfling, S., Wolters, J., Reitzenstein, S., 2016. Photon-statistics excitation spectroscopy of a single two-level system. Phys. Rev. B 93 (24), 241306. https://doi.org/10.1103/PhysRevB.93.241306.

Sudarshan, E.C.G., 1963. Equivalence of semiclassical and quantum mechanical descriptions of statistical light beams. Phys. Rev. Lett. 10 (7), 277–279. https://doi.org/10.1103/PhysRevLett.10.277.

Temnov, V.V., Woggon, U., 2005. Superradiance and subradiance in an inhomogeneously broadened ensemble of two-level systems coupled to a low-*q* cavity. Phys. Rev. Lett. 95 (24), 243602. https://doi.org/10.1103/PhysRevLett.95.243602.

Thewes, J., Lüders, C., Aßmann, M., 2019. Eavesdropping attack on a trusted continuous-variable quantum random-number generator. Phys. Rev. A 100 (5), 052318. https://doi.org/10.1103/PhysRevA.100.052318.

Thewes, J., Lüders, C., Aßmann, M., 2020. Conditional spectroscopy via nonstationary optical homodyne quantum state tomography. Phys. Rev. A 101 (2), 023824. https://doi.org/10.1103/PhysRevA.101.023824.

Ulrich, S.M., Gies, C., Ates, S., Wiersig, J., Reitzenstein, S., Hofmann, C., Löffler, A., Forchel, A., Jahnke, F., Michler, P., 2007. Photon statistics of semiconductor microcavity lasers. Phys. Rev. Lett. 98 (4), 043906. https://doi.org/10.1103/PhysRevLett.98.043906.

Vahala, K.J., 2003. Optical microcavities. Nature 424, 839–846. https://doi.org/10.1038/nature01939.

Wiersig, J., Gies, C., Jahnke, F., Aßmann, M., Berstermann, T., Bayer, M., Kistner, C., Reitzenstein, S., Schneider, C., Höfling, S., Forchel, A., Kruse, C., Kalden, J., Hommel, D., 2009. Direct observation of correlations between individual photon emission events of a microcavity laser. Nature 460 (7252), 245–249. https://doi.org/10.1038/nature08126.

Winger, M., Volz, T., Tarel, G., Portolan, S., Badolato, A., Hennessy, K.J., Hu, E.L., Beveratos, A., Finley, J., Savona, V., Imamoğlu, A., 2009. Explanation of photon correlations in the far-off-resonance optical emission from a quantum-dot-cavity system. Phys. Rev. Lett. 103 (20), 207403. https://doi.org/10.1103/PhysRevLett.103.207403.

CHAPTER SIX

Valley excitons: From monolayer semiconductors to moiré superlattices

Hongyi Yu[a,*], Xiaodong Xu[b,c], and Wang Yao[d,e,*]

[a]Guangdong Provincial Key Laboratory of Quantum Metrology and Sensing & School of Physics and Astronomy, Sun Yat-Sen University (Zhuhai Campus), Zhuhai, China
[b]Department of Physics, University of Washington, Seattle, WA, United States
[c]Department of Materials Science and Engineering, University of Washington, Seattle, WA, United States
[d]Department of Physics, The University of Hong Kong, Pok Fu Lam, Hong Kong, China
[e]HKU-UCAS Joint Institute of Theoretical and Computational Physics at Hong Kong, Pok Fu Lam, Hong Kong, China
*Corresponding authors: e-mail address: yuhy33@mail.sysu.edu.cn; wangyao@hku.hk

Contents

1. Introduction

Monolayer and bilayer structures of group-VIB transition metal dichalcogenides (TMDs) have been the focus of vast research activities on semiconductor optics in the atomically thin two-dimensional (2D) limit. This family of TMDs has the chemical formula MX_2 (M=Mo or W, X=S, Se or Te). The bulk crystal has a layered structure, with adjacent layers

Semiconductors and Semimetals, Volume 105
ISSN 0080-8784
https://doi.org/10.1016/bs.semsem.2020.09.004

loosely bound together by the weak van der Waals force. A monolayer unit consists of *X-M-X* tri-atomic planes covalently bonded to form a 2D hexagonal lattice (Fig. 1A). While bulk and few-layer TMDs have indirect band gaps, the monolayers exhibit direct band gaps in the visible frequency range (Mak et al., 2010; Splendiani et al., 2010), with conduction and valence band edges located at the degenerate but inequivalent **K** and $-\mathbf{K}$ corners of the hexagonal Brillouin zone (Fig. 1A). The two degenerate band extrema constitute an extra quantum degree of freedom for low energy carriers, known as valley, which can be exploited as an information carrier, in addition to the spin, in future electronics and optoelectronics (Xiao et al., 2012; Xu et al., 2014).

Optical properties of monolayer TMDs are dominated by the neutral exciton and charged exciton (trion), which are hydrogen-like bound states formed by band-edge electron and hole at the valleys. The reduced screening of Coulomb interaction in 2D geometry leads to exceptionally large exciton binding energies (hundreds of meV) and trion charging energies (tens of meV) (Jones et al., 2013; Mak et al., 2013; Ross et al., 2013), implying their room temperature stability. Valley excitons have a variety of bright and dark configurations due to the valley and spin degrees of freedom of the electron and hole constituents (Yu et al., 2015a). Bright excitons' interconversion with photons is subject to a valley optical selection rule (Xiao et al., 2012), making possible the optical generation and detection of excitonic valley polarization and coherence. The magnetic moments associated with the valley pseudospin and spin give rise to distinct Zeeman shift of the various exciton configurations in external magnetic field (Aivazian et al., 2015; Li et al., 2014; MacNeill et al., 2015; Srivastava et al., 2015b). The strong Coulomb binding also results in a sizable electron-hole exchange that effectively couples exciton's valley index with the center-of-mass momentum, making possible valley dependent Hall transport of excitons (Yu et al., 2014). The rich valley physics of excitons underlies the versatile control possibilities of valley pseudospin in monolayer semiconductors.

Monolayer TMDs can be stacked to form van der Waals heterostructures without the need of lattice matching. Because of the difference in bandgap and work function between the TMDs compounds, a heterobilayer typically has a type-II band alignment, i.e., electron and hole energetically favor opposite layers. The low energy excitons in such bilayers then have an interlayer configuration, where the separation of electron and hole leads to a substantially longer exciton lifetime while still exhibiting bright luminescence, as well as a permanent electric dipole that allows electric control of exciton (Rivera et al., 2015, 2018). The lattice-mismatch and twisting between the

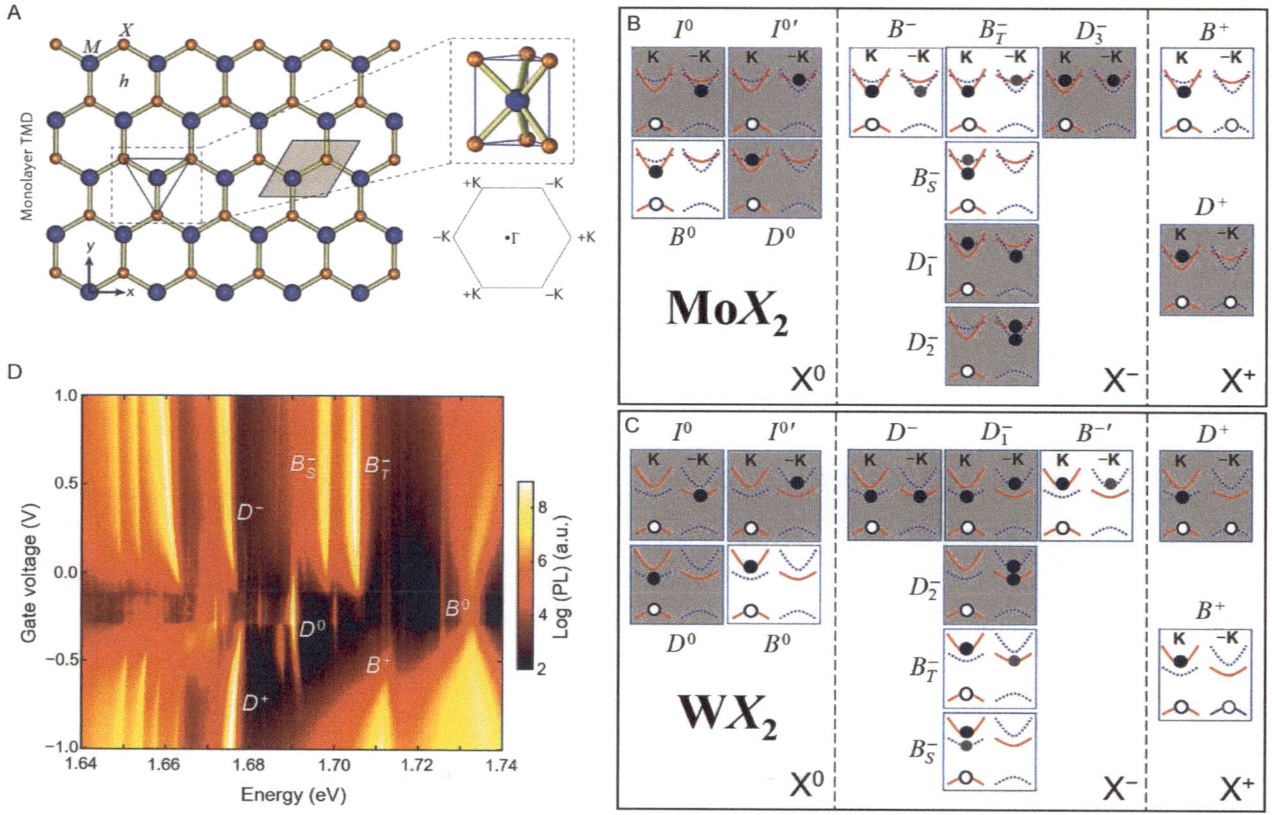

Fig. 1 See figure legend on next page.

layers inevitably lead to the formation of moiré pattern in the heterobilayers. At small twist angles, the moiré pattern realizes a lateral 2D superlattice with a period that ranges from a few to several tens nm. The local-to-local variation of interlayer atomic registry spatially modulate the energy, electric dipole, and valley optical selection rules of interlayer excitons (Yu et al., 2017). This has offered unprecedented opportunities to engineer the valley exciton properties.

In this chapter, we give an overview of the theoretical foundation and experimental progresses of the valley-related phenomena for excitons in 2D TMDs. Section 2 focuses on monolayer valley excitons, and Section 3 introduces the more recent development of interlayer valley excitons in moiré superlattices.

2. Valley excitons in monolayers

2.1 Binding energies and Bohr radii

The neutral exciton (X^0) is an elementary excitation consisting of a conduction band electron and a valence band hole bound together by the Coulomb interaction. The electron-hole relative motion is in the discrete

Fig. 1—Cont'd (A) Lattice structure of a monolayer TMDs. *M*, *X* and *h* correspond to the transition metal atom, chalcogen atom and hollow center of the hexagon, respectively. The diamond indicates a monolayer unit cell. The upper right is a side view of the trigonal prismatic coordination. The lower right shows the hexagonal Brillouin zone. (B) Valley and spin configurations of X^0, X^- and X^+ in monolayer MoX_2 (X=S/Se/Te). Solid (empty) dots denote electrons (holes), and red solid (blue dashed) curves denote spin-up (spin-down) bands. Only half of the configurations are shown, and the rest are their time-reversal counterparts. The white background denotes bright $X^0/X^\pm$, where the recombining electron-hole pair is in black color. The others with gray backgrounds are dark $X^0/X^\pm$. (C) The valley and spin configurations in WX_2. (D) PL spectrum as a function of the back-gate voltage tuning the carrier density. $V>0$ (<0) corresponds to electron (hole) doping. The peaks marked with symbols are identified as the direct recombination of the corresponding ones in (C), whereas the rest are the various phonon replicas from the dark $X^0/X^\pm$. *Panel (A): Adapted from Liu, G.-B., Xiao, D., Yao, Y., Xu, X., Yao, W., 2015. Electronic structures and theoretical modelling of two-dimensional group-Vib transition metal dichalcogenides. Chem. Soc. Rev. 44, 2643–2663 by permission of the Royal Society of Chemistry. Panels (B, C): Adapted from Yu, H., Cui, X., Xu, X., Yao, W., 2015a. Valley excitons in two-dimensional semiconductors. Natl. Sci. Rev. 2, 57–70. Panel (D): Adapted from He, M., Rivera, P., Tuan, D. V., Wilson, N. P., Yang, M., Taniguchi, T., Watanabe, K., Yan, J., Mandrus, D. G., Yu, H., Dery, H., Yao, W., Xu, X., 2020. Valley phonons and exciton complexes in a monolayer semiconductor. Nat. Commun. 11, 618, both under the terms of the Creative Commons CC BY license (https://creativecommons.org/licenses/by/4.0/).*

hydrogen-like orbitals, and we focus here on the ground-state (1*s*) orbital. The energy to create X^0 is lower than the quasiparticle band-gap E_g, by the X^0 binding energy E_b (an analog of the hydrogen Rydberg energy). This binding energy is exceptionally large in monolayer TMDs compared to that in III-V semiconductor quantum wells, due to the joint effects of: (1) the reduced dielectric screening in the 2D geometry, (2) large effective masses for the electron and hole (both in the order of the free electron mass). Numerical calculations find E_b in the order of hundreds of meV in monolayers of different TMD compounds (Berkelbach et al., 2013; Cheiwchanchamnangij and Lambrecht, 2012; Feng et al., 2012; Qiu et al., 2013; Shi et al., 2013), which can be further tuned by varying the dielectric environment (e.g., the substrate and capping layer) (Raja et al., 2017). The calculation results are in agreement with the experimentally determined values using a combination of measurements including the X^0 ground state resonance in photoluminescence (PL), excited state resonances measured in absorption/reflection spectra and two-photon absorption, and quasi-particle gap determined by scanning tunneling spectroscopy (He et al., 2014; Ugeda et al., 2014; Wang et al., 2015; Ye et al., 2014; Zhang et al., 2014; Zhu et al., 2015). While having such a large binding energy, the X^0 wavefunction in monolayer TMDs is still the Wannier type, with the electron-hole relative motion extending over tens of unit cells, and the momentum space distributions of electron and hole are therefore well localized at the $\pm\mathbf{K}$ valleys (Qiu et al., 2013; Ye et al., 2014). Numerical calculations give an X^0 Bohr radius a_B about 1–2 nm (Qiu et al., 2013, Ye et al., 2014), which is in agreement with the values extracted from the X^0 diamagnetic shift measured at high magnetic field (Stier et al., 2016). This value of a_B is by an order of magnitude smaller than that in GaAs quantum wells, leading to a sizable oscillator strength (inversely proportional to a_B^2). The experimentally measured resonant excitonic absorption of a single TMD monolayer can reach 10%–20% (Mak et al., 2010).

The negative/positive trion (X^-/X^+) can be viewed as a weakly bound state of an X^0 and an excess electron/hole. Trion emits at the energy $E_g - E_b - E_c$, where E_c is the energy required for its dissociation into an X^0 plus a free electron/hole, known as the charging energy. E_c can be directly measure from the PL spectrum, and is found in the order of several tens meV in monolayer TMDs (Jones et al., 2013; Mak et al., 2013; Mitioglu et al., 2013; Ross et al., 2013), which gives the first experimental indication of the large X^0 binding energy E_b (the ratio of E_b/E_c roughly $\sim$10). For average electron-hole distances in the trion, a variational calculation gives a values of ~1 nm for the strongly bound electron-hole pair, and ~2.5 nm for the weakly bound electron-hole pair (Berkelbach et al., 2013).

It is worth pointing out that, the dielectric screening of the Coulomb interaction in the 2D geometry is not a constant one, but becomes weaker at larger electron-hole separations. Thus the excited states (2*s*, 2*p*, 3*s* … orbitals) with larger electron-hole separations experience smaller dielectric screening than that of the ground state, and the resultant energies of the excitonic Rydberg series in monolayer TMDs show significant deviation from the 2D hydrogen model (Chernikov et al., 2014; Wang et al., 2015; Ye et al., 2014).

2.2 Valley and spin configurations of monolayer excitons

With the Wannier type wavefunctions well localized in the momentum space valleys, X^0 and trions are characterized by the spin and valley configurations of its electron and hole constituents (Yu et al., 2015a). Strong spin-orbit-coupling (SOC) from the transition metal *d*-orbitals results in pronounced spin splitting at the valleys. The mirror symmetry about the *M* atom plane, together with the time reversal relation between **K** and −**K**, dictates that the spin splitting is of the form $\hat{H}_{SOC}=\lambda\tau\hat{S}_z$, i.e., an effective coupling between the out-of-plane spin $\hat{S}_z$ with the valley index $\tau=\pm1$. The valence (conduction) band edges at ±**K** valley mainly consist of the metal $d_{x^2-y^2}\pm id_{xy}$ (d_{z^2}) orbital (Kormányos et al., 2015; Liu et al., 2015; Xiao et al., 2012). For the $d_{x^2-y^2}\pm id_{xy}$ orbitals with magnetic quantum number $m=\pm2$, the strong on-site SOC introduces a giant spin splitting for the valence band edge, with $\lambda_v\sim0.15$ eV (~0.45 eV) in MoX_2 (WX_2). This effectively leads to the spin-valley locking at the valence band edge, i.e., valley **K** (−**K**) only have spin-down (up) holes. For the conduction band near ±**K**, the d_{z^2} orbital corresponds to a magnetic quantum number $m=0$. It is the small components of chalcogen $p_x\mp ip_y$ orbitals, as well as the perturbative coupling to remote $d_{xz}\mp id_{yz}$ bands (Liu et al., 2013), that give rise to a finite spin splitting λ_c at the conduction band edge. λ_c ranges from a few meV for MoS_2 to tens of meV for $MoSe_2$, $MoTe_2$, WS_2 and WSe_2 according to first-principle calculations (Kośmider et al., 2013; Liu et al., 2013), supported by recent experiments (Marinov et al., 2017; Wang et al., 2016). It is important to note that λ_c has opposite signs in MoX_2 and WX_2, due to the competition between the two contributions involving metal $d_{xz}\mp id_{yz}$ and chalcogen $p_x\mp ip_y$ orbitals (Liu et al., 2013). In MoX_2 (WX_2)

the lowest conduction sub-band has the same (opposite) spin as the valence band edge in a given valley, see Fig. 1B, C.

The small λ_c suggests that both spin sub-bands of the electrons are relevant in forming the low energy X^0 and trions with the spin-valley locked band edge holes. Such X^0 can have eight distinct spin and valley configurations (see Fig. 1B–C). Those with antiparallel (parallel) spins for its electron and hole constituents are referred as the spin-singlet (spin-triplet) X^0. Those with electron and hole constituents in the same (opposite) valley are denoted as intra-valley (inter-valley) X^0. For X^+, the two holes must be in opposite valleys due to the Pauli exclusion principle so X^+ only has 4 spin and valley configurations. X^- can have 12 configurations. Due to the conduction band spin-splitting and the effective mass difference between the two spin sub-bands, X^0 and trions with different spin and valley configurations generally have different energies (see Fig. 1B–C) (He et al., 2020; Li et al., 2019b).

Optical interband transitions conserve momentum (and therefore valley index), and spin-flip transitions are orders weaker compared to the spin-conserved ones. So the species that can efficiently interconvert with photon is the spin-singlet intra-valley X^0, referred as bright X^0, whereas X^0 in other spin-valley configurations are the dark ones. Correspondingly, bright $X^{\pm}$ consists of a bright X^0 capturing an excess hole/electron. Bright X^0 and $X^{\pm}$ dominate the absorption of the monolayer TMDs. The radiative recombination of a dark $X^0/X^{\pm}$ requires a spin-flip or valley-flip or both, which can be facilitated by a phonon or impurity scattering. In monolayer WX_2 where the dark $X^0/X^{\pm}$ are the ground state configuration with long lifetimes, their luminescence in the form of various phonon replicas can be observed in clean samples under a low temperature (Dery and Song, 2015; He et al., 2020; Li et al., 2019a,b; Liu et al., 2019). The various spin and valley configurations of the bright and dark $X^0/X^{\pm}$ result in a remarkably rich PL spectrum for monolayer WX_2 (see Fig. 1D).

2.3 Symmetry dictated valley- and spin-dependent polarization selection rules

The X^0-photon interconversion is governed by a valley- and spin-dependent polarization selection rule, dictated by the crystal symmetry. The monolayer lattice has an in-plane $2\pi/3$-rotational ($\hat{C}_3$) symmetry around M (metal site), X (chalcogen site) or h (hollow center of the hexagon formed by M and X) (Fig. 1A), as well as an out-of-plane mirror ($\widehat{\sigma}_h$) symmetry about the metal atom plane. At the high-symmetry point **K**, the conduction (c) and valence (v) band Bloch states $\psi_{n,\mathbf{K},Sz}$ ($n=c$, v) are eigenfunctions of $\hat{C}_3$ and $\widehat{\sigma}_h$ operations (Yu et al., 2018), i.e.,

$$\begin{aligned} \widehat{C}_3\psi_{n,\mathbf{K},S_z} &= e^{-i\frac{2\pi}{3}(m_n+S_z)}\psi_{n,\mathbf{K},S_z} \\ \widehat{\sigma}_h\psi_{n,\mathbf{K},S_z} &= e^{-i\pi S_z}\psi_{n,\mathbf{K},S_z} \end{aligned} \quad (1)$$

Here m_n is the $\hat{C}_3$ quantum number which depends on both the band index n and the rotation center M, X or h, as summarized in Table 1 below (Liu et al., 2015). Besides, the spin part of the wavefunction also contributes a phase factor to the $\hat{C}_3$ and $\widehat{\sigma}_h$ eigenvalues depending on the spin quantum number S_z.

A **K** valley X^0 in s-orbital and zero center-of-mass momentum has the same symmetry as an electron-hole pair at the **K** point. The optical dipole of the exciton then has a $\hat{C}_3$ quantum number $m_c - m_v + S_z - S'_z$ (mod 3) and a $\widehat{\sigma}_h$ quantum number $S_z - S'_z$. As given in Table 1, though m_c and m_v depend on the rotation center, their difference $m_c - m_v$ equals a constant +1 under modulo 3. This leads to the optical selection rule: a spin-singlet X^0, i.e., $S_z = S'_z$, in **K** (–**K**) valley couples to a photon with an in-plane $\sigma+$ ($\sigma-$) circular polarization (Xiao et al., 2012). The spin-triplet X^0 can couple to a photon of out-of-plane (z) linear polarization, while first-principles calculations indicate that its oscillator strength is two to three orders of magnitude weaker compared to the spin-singlet ones (Yu et al., 2018). Such polarization selection rules are schematically shown in Fig. 2A–B. For p-type or other excited states, the angular momentum of the electron-hole relative motion also contributes to the $\hat{C}_3$ quantum number (Glazov et al., 2017; Gong et al., 2017). For example, **K** valley spin-singlet X^0 in the $p_+ \equiv p_x + ip_y$ orbital has a $\hat{C}_3$ quantum number $m_c - m_v + 1$ (mod 3) $= -1$, which couples to a $\sigma-$polarized photon. However, the oscillator strength of the p orbital X^0 is also small, only 1% of that of the $1s$ X^0, thus hard to observe in one-photon process (Ye et al., 2014).

Table 1 The $\hat{C}_3$ quantum numbers of the conduction and valence band Bloch states at K point, for three rotation centers M, X and h, respectively (see Fig. 1A).

	M	X	h
m_c	0	−1	+1
m_v	−1	+1	0

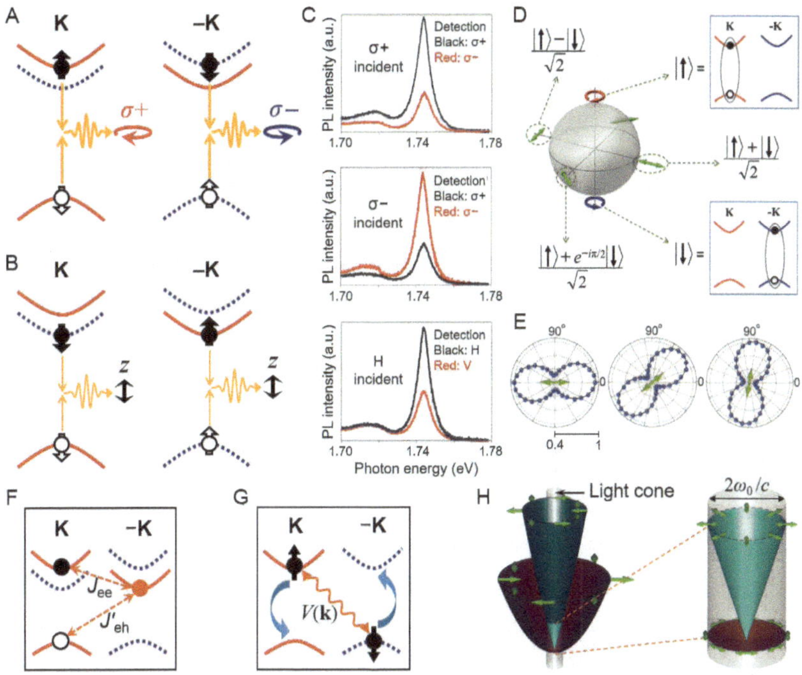

Fig. 2 (A) The polarization selection rules for the spin-conserved interband transitions in $\pm\mathbf{K}$ valleys. (B) Selection rules for the spin-flip interband transitions. (C) Polarization-resolved PL in monolayer WSe_2 by $\sigma+$ (upper), $\sigma-$ (middle) and linearly (lower) polarized excitations. (D) Bloch vector representations of the excitonic valley pseudospin. The double arrow denotes a linear polarization. (E) Normalized X^0 PL peak intensity (ranged from 0.4 to 1) as a function of the linear-polarization detection angle for a given incident laser polarization (double arrows). (F) The diagonal exchange interaction (J_{ee} and J'_{eh}) between the recombining electron-hole pair and the excess electron in the inter-valley X^- in WX_2. (G) The off-diagonal electron-hole exchange interaction depicted in terms of the conduction and valence states. (H) The bright X^0 dispersion under the exchange-induced valley-orbit coupling, whose valley pseudospin (single arrows) is locked with the wave vector direction. Inside the light cone where the interconversion of the X^0 with a photon is directly allowed, the emitted photon has a linear polarization longitudinal (transverse) to the wave vector in the upper (lower) X^0 branch. *Panel (C): Reproduced from Xu, X., Yao, W., Xiao, D., Heinz, T. F., 2014. Spin and pseudospins in layered transition metal dichalcogenides. Nat. Phys. 10, 343–350, Springer Nature. Panels: (D, H): Adapted from Yu, H., Cui, X., Xu, X., Yao, W., 2015a. Valley excitons in two-dimensional semiconductors. Natl. Sci. Rev. 2, 57–70 under the terms of the Creative Commons CC BY license (https://creativecommons.org/licenses/by/4.0/). Panels (E): Reproduced from Jones, A. M., Yu, H., Ghimire, N. J., Wu, S., Aivazian, G., Ross, J. S., Zhao, B., Yan, J., Mandrus, D. G., Xiao, D., Yao, W., Xu, X., 2013. Optical generation of excitonic valley coherence in monolayer WSe2. Nat. Nanotechnol., 8 634–638, Springer Nature. Panels (F–G): Adapted from Yu, H., Liu, G.-B., Gong, P., Xu, X., Yao, W., 2014. Dirac cones and Dirac saddle points of bright excitons in monolayer transition metal dichalcogenides. Nat. Commun. 5, 3876, Springer Nature.*

Polarization-resolved PL measurements have demonstrated the generation of valley-polarized spin-singlet X^0 by circularly polarized excitation in monolayer MoS_2 and WX_2, whose radiative recombination leads to circularly polarized PL (Fig. 2C) (Cao et al., 2012; Jones et al., 2013; Mak et al., 2012; Sallen et al., 2012; Zeng et al., 2012). Meanwhile, a linearly polarized photon corresponds to the coherent superposition of the $\sigma+$ and $\sigma-$ polarized ones, which can excite an X^0 in a coherent superposition of its two valley configurations. Experiments in monolayer WX_2 have shown that, the X^0 PL emission carries a significant linear polarization (30%–40%) whose direction always coincides with that of the linearly polarized excitation (Jones et al., 2013). It is intuitive to represent the X^0 valley pseudospin using the Bloch sphere (Xu et al., 2014; Yu et al., 2015a), as shown in Fig. 2D. The north (south) pole corresponds to the **K** (–**K**) valley or valley pseudospin up (down), which couples to a $\sigma+$ ($\sigma-$) polarized photon. The equator corresponds to an equal superposition of the two valleys, i.e., in-plane valley pseudospin, which couples to a linearly polarized photon (Fig. 2E). An arbitrary point on this Bloch sphere of X^0 valley pseudospin can be selectively accessed through the choice of excitation polarization.

The photon emission from the dark spin-triplet X^0 has also been observed in WX_2. With the out-of-plane optical dipole of the spin-triplet X^0, these experiments either detect photons propagating in the in-plane direction (Wang et al., 2017a), or use surface plasmon polariton or nano-optical antenna-tip to enhance the photon emission (Park et al., 2018; Zhou et al., 2017). The obtained energy of the spin-triplet X^0 ($\sim$40 meV below the spin-singlet) verifies its ground state nature in WX_2. Due to the weak optical dipole strength, the intrinsic radiative lifetime ($\sim$100 ps) of the spin-triplet X^0 is much longer than that of the spin-singlet ($\sim$0.1–1 ps (Moody et al., 2015)).

For the trion, its emission polarization is determined by the valley configuration of the recombining electron-hole pair. Under a circularly polarized excitation, bright trion's PL is also polarized, showing the optically generated valley polarization. However, under a linearly polarized excitation, the trion PL does not show linear polarization (Jones et al., 2013). This is in contrast to the X^0, where the observed linearly polarized PL is a signature of the optically generated valley coherence.

We note that, for X^+, its two valley configurations have the extra hole in opposite valleys respectively (Fig. 1B–C). Consider the photon emission by a linear superposition of these two trion valley configurations: the $\sigma+$ ($\sigma-$) photon emission leaves behind an excess hole in the –**K** (**K**) valley, resulting in a spin-photon entanglement, which destroys the coherence between the

$\sigma+$ and $\sigma-$ components. So linear polarized PL is not relevant even if there is coherence created between the X^+ valley configurations. The same picture applies to low-energy bright X^- in MoX_2.

In WX_2, because of its reversed sign of spin-splitting in the conduction band, X^- has four low-energy configurations characterized by the valley index τ of the recombining electron-hole pair, and the spin S_z ($= \pm 1/2$) of the excess electron which lies in a separate spin sub-band, see Fig. 1C. The configurations with $\tau S_z > 0$ (<0) corresponds to the inter-valley (intra-valley) X^- (Yu et al., 2014). Apparently, a coherent superposition of an intra-valley and an inter-valley X^-, e.g., the superposition of ($\tau = 1$, $S_z = 1/2$) and ($\tau = -1$, $S_z = 1/2$), can couple to linearly polarized photon. However, the diagonal exchange interaction (c.f. Section 2.4 below) between the recombining electron-hole pair and the excess electron in the inter-valley X^- leads to an energy splitting ($\sim$7 meV (Jones et al., 2016)) from the intra-valley one, see Fig. 2F, which quenches the linear photon emission.

2.4 Valley-orbit coupling and valley Hall effect from electron-hole exchange

The Coulomb interaction between an electron and a hole can be separated into a direct part and an exchange part. The direct Coulomb is responsible for the X^0 binding. The electron-hole exchange interaction can be viewed as a virtual annihilation-creation process of the electron-hole pair (Fig. 2G), which is significant in spin-singlet X^0 but not in spin-triplet ones, and has a strength proportional to the probability for the electron-hole to overlap spatially. Bright X^0 is affected by both the intra-valley and inter-valley electron-hole exchange. The intra-valley exchange is a diagonal process where the initial and final states of the virtual annihilation-creation process are in the same valley, which raises X^0 energy by $\sim$10 meV, contributing to the energy splitting between the bright and spin-triplet dark X^0 (Fig. 1B–C) (He et al., 2020). The inter-valley exchange is an off-diagonal process annihilating a bright X^0 in one valley while creating one in the other, i.e., a valley pseudospin flip process. For bright X^0 with small center-of-mass wave vectors $\mathbf{k}$, this pseudospin flip term is proportional to $e^{-2i\theta}V(\mathbf{k})k^2$, with $V(\mathbf{k})$ the momentum-space form of Coulomb potential and θ the angle of $\mathbf{k}$ (Yu et al., 2014). In the absence of screening by free carriers, $V(\mathbf{k}) \propto 1/k$, the bright X^0 dispersion is described by a valley-orbit coupled Hamiltonian (Yu et al., 2015a),

$$\hat{H}_X = E_0 + Jk + Jk\left(e^{-2i\theta}\,\hat{\tau}_+ + e^{2i\theta}\,\hat{\tau}_-\right). \tag{2}$$

Here $\hat{\tau}_{\pm}\equiv\hat{\tau}_x \pm i\hat{\tau}_y$ is the 2×2 Pauli matrices of the valley pseudospin, $J\sim1\,\text{eV}\cdot\text{Å}$ is a constant characterizing the strength of electron-hole exchange. The bright X^0 dispersion is split into two branches, both with **k** dependent in-plane valley pseudospins, coupled to photons of linear polarization longitudinal and transverse to **k** respectively. Remarkably, the transverse branch has a parabolic dispersion with the regular X^0 mass, while the longitudinal branch is massless, see Fig. 2H (Qiu et al., 2015; Wu et al., 2015; Yu et al., 2015a). In III-V semiconductors, such splitting of X^0 dispersion by electron-hole exchange is known as the longitudinal-transverse (LT) splitting.

The LT splitting combined with momentum scattering can result in an efficient valley depolarization mechanism for the bright X^0 (Maialle et al., 1993). In the strong scattering limit ($\overline{\Omega}\tau_p \ll 1$) the valley depolarization rate is given by $\sim\overline{\Omega}^2\tau_p$, with $\overline{\Omega}$ the ensemble-averaged LT splitting and τ_p the X^0 momentum relaxation time. The strong Coulomb interaction in monolayer TMDs gives rise to a large LT splitting (J inversely proportional to a_B^2), which leads to a short valley lifetime (<1 ps).

For X^+, and X^- in MoX_2, their two valley configurations are not coupled by the Coulomb exchange, because of the extra electron/hole in opposite valleys. For the bright X^- in WX_2, the electron-hole exchange can couple the intra-valley and inter-valley configurations, e.g., ($\tau=1$, $S_z=1/2$) with ($\tau=-1$, $S_z=1/2$) (Yu et al., 2014). For the inter-valley X^-, the excess electron has the same spin as the recombining electron-hole pair, and their exchange (Fig. 2F) further contributes a blue shift that results in an energy splitting $\delta\sim7\,\text{meV}$ from the intra-valley X^- (Jones et al., 2016). The bright X^- dispersion in WX_2 is then described by,

$$\hat{H}_{X^-} = E_- + J'_{\mathbf{k}} + \delta S_z\hat{\tau}_z + J_{\mathbf{k}}\left(e^{-2i\theta}\,\hat{\tau}_+ + e^{2i\theta}\,\hat{\tau}_-\right). \tag{3}$$

$J_{\mathbf{k}}$ ($J'_{\mathbf{k}}$) is the magnitude of the inter-valley (intra-valley) exchange from the recombining electron-hole pair, whose form can in general be different from that in a bright X^0. The splitting δ suppresses the valley depolarization due to the exchange-induced valley flip term, in agreement with the measured longer valley lifetime of X^- (~10 ps) compared to that of the X^0 (Wang et al., 2014). Moreover, as the joint effect of these exchange couplings (last three terms in Eq. (3)), both the inter- and intra-valley X^- at small **k** acquire a sizable Berry curvature with valley-dependent sign, giving rise to a pronounced trion valley Hall effect in an in-plane electric field (Yu et al., 2014).

3. Valley excitons in moiré superlattices

3.1 Interlayer excitons in type-II heterobilayer

When two different monolayer TMDs are vertically stacked to form a heterostructure, typically the conduction and valence band edges are located in opposite layers, referred as the type-II band alignment (see Fig. 3A) (Özçelik et al., 2016). Taking MoX_2/WX_2 as an example, the lowest energy electron is in the MoX_2 layer whereas the lowest energy hole is in the WX_2 layer. The band offset between the two layers is generally in the order of several hundred meV, depending on the compounds. The valence band offset is found to be ~0.3 eV in $MoSe_2/WSe_2$ (Wilson et al., 2017), whereas a valence (conduction) band offset as large as 0.83 eV (0.76 eV) has been measured in MoS_2/WSe_2 (Chiu et al., 2015). Given the small layer separation (vertical distance between the two M-atom planes $d \sim 0.6$ nm), the interlayer Coulomb interaction between the electron and hole leads to the formation of a tightly bound interlayer exciton (IX) (Rivera et al., 2015). The measured binding energy for IX is ~0.2 eV, smaller than that of the monolayer X^0, but in the same order (Wilson et al., 2017). The calculated Bohr radius is ~2 nm (Donck and Peeters, 2018), also comparable to that of X^0. With the large conduction and valence band offset, the energy of IX is then significantly lower than X^0 in individual layers. For example, IX in heterobilayer $MoSe_2/WSe_2$ (MoS_2/WSe_2) has an energy ~1.4 eV (~1 eV), where X^0 have the energies ~1.6, 1.7, 1.9 eV, respectively, in $MoSe_2$, WSe_2, MoS_2 (Karni et al., 2019; Rivera et al., 2015).

In contrast to the intralayer X^0 which has a large optical dipole but zero electric dipole, the vertical separation between the electron and hole in IX results in its permanent electric dipole (ed) and substantially reduced optical dipole. Despite its weak optical oscillator strength, luminescence from IX has been observed in various heterobilayers (Fig. 3B), and can have dominant spectral weight at low temperatures in high quality heterobilayers (Rivera et al., 2015). In contrast, luminescence from the intralayer X^0 is quenched by the rapid energy relaxation to form IX through the ultrafast interlayer charge transfer process (<0.1 ps (Hong et al., 2014)). IX can be generated through the excitation of intralayer X^0 followed by the spin/valley-conserved interlayer charge transfer (Schaibley et al., 2016). Resonant excitation of IX is inefficient due to its weak oscillator strength, about two to three orders of magnitude weaker than that of X^0, as measured from the photocurrent responses (Ross et al., 2017).

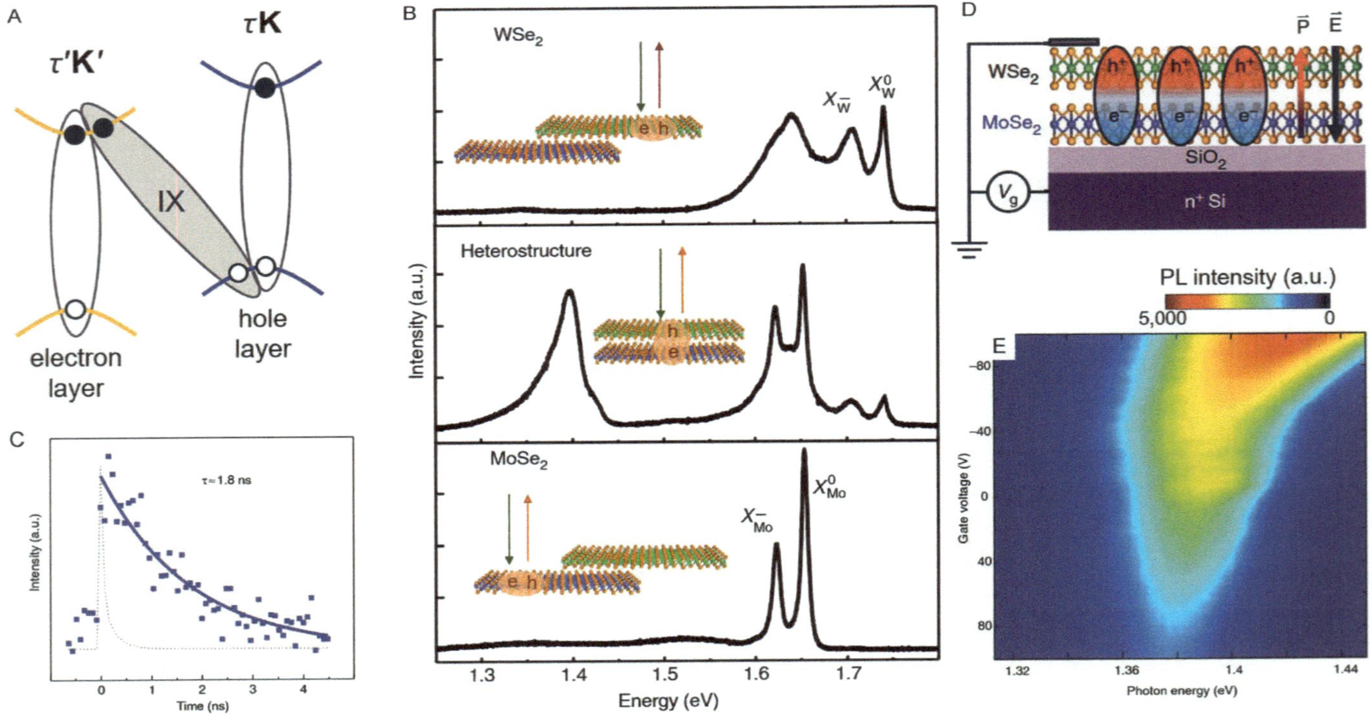

Fig. 3 (A) Schematic of a type-II band alignment, where the IX has a lower energy than the intralayer X^0. (B) PL from monolayer WSe_2 region (top), monolayer $MoSe_2$ region (bottom) and heterobilayer region (middle) in a $MoSe_2/WSe_2$ heterostructure. The low energy peak near 1.4 eV is from the IX emission. (C) Time-resolved PL shows an IX lifetime of $\tau \approx 1.8$ ns. (D) Device illustration for the gate tunability of the IX energy. (E) The IX PL spectrum as a function of the gate voltage. *Panels (B–E): Reproduced from Rivera, P., Schaibley, J. R., Jones, A. M., Ross, J. S., Wu, S., Aivazian, G., Klement, P., Seyler, K., Clark, G., Ghimire, N. J., Yan, J., Mandrus, D. G., Yao, W., Xu, X., 2015. Observation of long-lived interlayer excitons in monolayer MoSe2-WSe2 heterostructures. Nat. Commun. 6, 6242, Nature Publishing Group.*

IX's weak oscillator strength leads to a long radiative lifetime, see Fig. 3C. The spatial separation of electron and hole also quenches their Coulomb exchange (as the main cause of X^0 valley depolarization). PL lifetime from 1 to 100 ns (Jauregui et al., 2019; Miller et al., 2017; Rivera et al., 2015) and valley lifetime ranged from 40 ns to more than 100 ns (Baranowski et al., 2017; Rivera et al., 2016) have been reported. The permanent electric dipole of IX enables the electrostatic tuning of its energy by a perpendicular electric field (Fig. 3D–E), and leads to stronger exciton-exciton interactions (Rivera et al., 2016). The directional flow of IX current controlled by an inhomogeneous electric field has been demonstrated (Unuchek et al., 2018, 2019).

IX luminescence can be observed only when the two layers have nearly aligned crystalline directions, i.e., with an angle close to either 0 or 60 degrees. The small layer twisting and difference in lattice constants result in the displacement between the electron and hole valleys in the momentum space, and the formation of a long-period moiré pattern in the real space. These can have a profound impact on the properties of IX. Below we establish the description of IX from the momentum space and real space perspectives, respectively.

3.2 Momentum eigenstates of interlayer exciton and light cones

In most heterobilayers, the band offsets (several hundred meV) of the type-II band alignment are much larger than the interlayer hopping matrix element at the $\pm\mathbf{K}$ valleys (Wang et al., 2017b). It is then convenient to start with the IX momentum eigenstates in the limit of vanishing interlayer coupling, which are also eigenstates of the electron-hole interlayer Coulomb interaction (Yu et al., 2015b). Under the effective mass approximation, the electron-hole Hamiltonian is given by

$$\widehat{H}_{IX} = \frac{\hbar^2}{2m_e}\left(-i\nabla_{\mathbf{r}_e} - \tau'\mathbf{K}'\right)^2 + \frac{\hbar^2}{2m_h}\left(-i\nabla_{\mathbf{r}_h} + \tau\mathbf{K}\right)^2 + V_{\text{Coulomb}}(\mathbf{r}_e - \mathbf{r}_h). \quad (4)$$

Here $\mathbf{K}$ ($\mathbf{K}'$) refers to the Brillouin zone corner from the hole (electron) layer, where the spin-singlet intralayer X^0 emit $\sigma+$ photon, and τ' (τ) is the electron (hole) valley index. $-i\nabla_{\mathbf{re}} - \tau'\mathbf{K}'$ ($-i\nabla_{\mathbf{rh}} + \tau\mathbf{K}$) is the operator for electron (hole) kinematic momentum, corresponding to the displacement from $\tau'\mathbf{K}'$ ($\tau\mathbf{K}$) point. Analysis shows that the electron-hole interlayer

Coulomb interaction V_{Coulomb} conserves the sum of their kinematic momenta $\mathbf{Q}$, i.e., eigenvalue of $(-i\nabla_{\mathbf{r}e}-\tau'\mathbf{K}')+(-i\nabla_{\mathbf{r}h}+\tau\mathbf{K})$, as well as the electron/hole valley index τ'/τ (Yu et al., 2015b). The eigenstates of the Coulomb interaction can then be characterized by the quantum numbers $\mathbf{Q}$ and (τ', τ),

$$X_{\tau'\tau,\mathbf{Q}} = \sum_{\Delta\mathbf{Q}} \Phi(\Delta\mathbf{Q}) \psi_{c,\tau'\mathbf{K}'+\frac{m_e}{M}\mathbf{Q}+\Delta\mathbf{Q}}(\mathbf{r}_e) \psi^{*}_{v,\tau\mathbf{K}-\frac{m_h}{M}\mathbf{Q}+\Delta\mathbf{Q}}(\mathbf{r}_h). \tag{5}$$

Here $\psi_{c,\tau'\mathbf{K}'+\mathbf{k}'}(\mathbf{r}_e)$ and $\psi_{v,\tau\mathbf{K}+\mathbf{k}}(\mathbf{r}_h)$ are the conduction and valence band Bloch states, respectively, $M\equiv m_e+m_h$. $X_{\tau'\tau,\mathbf{Q}}$ is also IX's momentum eigenstate, with a kinematic momentum $\mathbf{Q}$ and kinetic energy $\hbar^2Q^2/2M$.

The term *light cone* here refers to the momentum-space regions where interconversion of IX with photon is directly allowed (without phonon or impurity assistance). Radiative recombination of an electron at $\tau'\mathbf{K}'+\mathbf{k}'$ and a hole at $\tau\mathbf{K}+\mathbf{k}$ then requires $\tau\mathbf{K}+\mathbf{k}+\mathbf{G}=\tau'\mathbf{K}'+\mathbf{k}'+\mathbf{G}'$, with $\mathbf{G}$ ($\mathbf{G}'$) being a reciprocal lattice vector of the hole (electron) layer. The light cones are thus located at $\mathbf{Q}=(\tau\mathbf{K}+\mathbf{G})-(\tau'\mathbf{K}'+\mathbf{G}')$. The bright IX has the valley configurations $(\tau', \tau)=(\pm,\pm)$ for small twist from 0 degree, and $(\tau', \tau)=(\pm,\mp)$ for small twist from 60 degree. Below we focus on the configurations $(\tau', \tau)=(\pm,+)$, and those with $(\tau', \tau)=(\mp,-)$ are simply their time reversal.

When $\tau'\mathbf{K}'$ is displaced from $\mathbf{K}$, IX with a zero kinetic energy ($\mathbf{Q}=0$) is optically dark (Fig. 4A). The light cones are at finite $\mathbf{Q}$ values, forming an ordered hexagonal array, see Fig. 4B. The optical dipole strengths at these light cones decay exponentially with Q, thus only the three light cones with the smallest Q values - termed as the main light cones - are important, which are located at $\mathbf{Q}_0\equiv\mathbf{K}-\tau'\mathbf{K}'$ and its $\hat{C}_3$ rotations (Fig. 4C). $Q_0\approx K(\delta^2+\delta\theta^2)^{1/2}$, where δ is the lattice mismatch and $\delta\theta$ is the twisting angle. Other light cones at larger Q values have orders of magnitude weaker optical dipoles, where the radiative recombination is through the Umklapp-type process (Fig. 4C). The main light cone has a finite kinetic energy $\hbar^2Q_0^2/2M$, which increases rapidly with $\delta\theta$. This explains why the IX luminescence is observable in heterobilayers at small $\delta\theta$ only, where $Q_0\ll K$. With a low symmetry at the finite kinematic momentum, the optical dipole of each main light cone has all the $\sigma+$, $\sigma-$ and z components,

$$\begin{aligned}
\mathbf{D}_{\mathbf{Q}_0} &= D_+(d)\boldsymbol{\sigma}_+ + D_-(d)\boldsymbol{\sigma}_- + D_z(d)\mathbf{z}\ , \\
\mathbf{D}_{\hat{C}_3\mathbf{Q}_0} &= e^{-i\frac{2\pi}{3}(S_z-S_z')}D_+(d)\boldsymbol{\sigma}_+ + e^{-i\frac{2\pi}{3}(1+S_z-S_z')}D_-(d)\boldsymbol{\sigma}_- + e^{-i\frac{2\pi}{3}(2+S_z-S_z')}D_z(d)\mathbf{z}, \\
\mathbf{D}_{\hat{C}_3^2\mathbf{Q}_0} &= e^{-i\frac{4\pi}{3}(S_z-S_z')}D_+(d)\boldsymbol{\sigma}_+ + e^{-i\frac{4\pi}{3}(1+S_z-S_z')}D_-(d)\boldsymbol{\sigma}_- + e^{-i\frac{4\pi}{3}(2+S_z-S_z')}D_z(d)\mathbf{z},
\end{aligned} \tag{6}$$

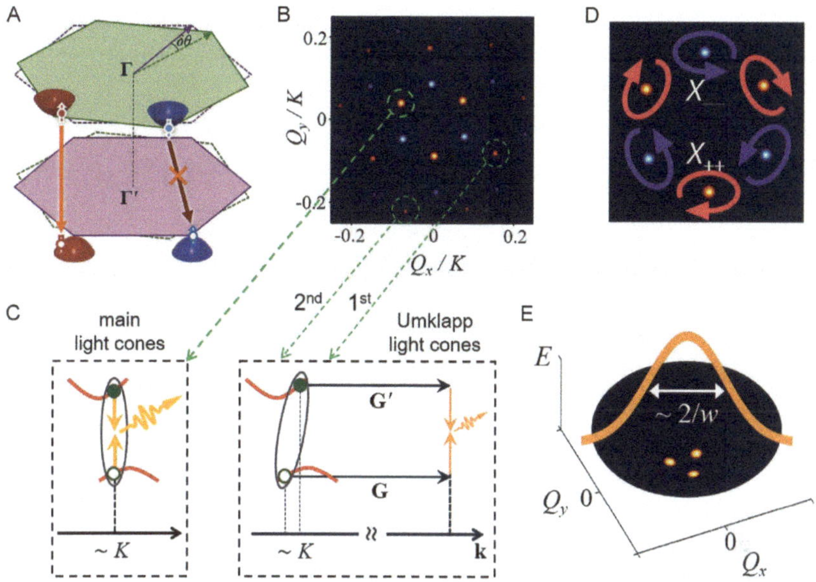

Fig. 4 (A) The BZ corners of the hole layer (lower hexagon) is displaced from those of the electron layer (upper hexagon) by a relative twist angle $\delta\theta$. A lowest-energy electron-hole pair cannot radiatively recombine due to the momentum mismatch (long tilted arrow). Nevertheless, an electron-hole pair with finite kinetic energy can have the momentum mismatch compensated, allowing their radiative recombination (long vertical arrow). (B) Bright dots denote the IX states that can directly couple to photons, i.e., the light cones, shown in IX's kinematic momentum space. (C) The six largest spots in (B) are the main light cones (three from each valley configuration), and the others are the Umklapp-type light cones. (D) The six main light cones have elliptically polarized optical selection rules. (E) The yellow curve denotes an IX wavepacket with real space width *w*. For small *w*, the momentum distribution covers all three main light cones (bright spots), and the wavepacket's optical dipole is determined by their interference. *Panel (A): Adapted from: Rivera, P., Yu, H., Seyler, K. L., Wilson, N. P., Yao, W., Xu, X., 2018. Interlayer valley excitons in heterobilayers of transition metal dichalcogenides. Nat. Nanotechnol., 13, 1004–1015, Springer Nature. Panels (B–D): Reprinted figure with permission from Yu, H., Wang, Y., Tong, Q., Xu, X., Yao, W., 2015b. Anomalous light cones and valley optical selection rules of interlayer excitons in twisted heterobilayers. Phys. Rev. Lett., 115, 187002 (https://doi.org/10.1103/PhysRevLett.115.187002), Copyright 2015 by the American Physical Society. Panel (E): From Yu, H., Liu, G.-B., Tang, J., Xu, X., Yao, W., 2017. Moiré excitons: from programmable quantum emitter arrays to spin-orbit-coupled artificial lattices. Sci. Adv. 3, e1701696, reprinted with permission from AAAS.*

and the corresponding photon emission is elliptically polarized in general (Fig. 4D) (Yu et al., 2015b, 2018). In the above equations, $S_z'-S_z$ is the total spin of IX, which equals 0 for spin-singlet and -1 for spin-triplet. $\boldsymbol{\sigma}_{\pm} \equiv (\mathbf{x} \pm i\mathbf{y})^*/\sqrt{2}$ are the polarization vectors for the $\sigma\pm$ polarized photon, with **x**, **y** and **z** the three Cartesian unit vectors. The three main light cones are related by the $\hat{C}_3$ rotations. Both photocurrent measurements and first-principles calculations indicate that IX's optical dipole at the main light cone is about ~5% of that of the intralayer X^0 (Jin et al., 2019a; Ross et al., 2017; Wu et al., 2018; Yu et al., 2015b, 2017). The IX optical dipole strength $D_{+/-/z}(d)$ is an exponential function of the interlayer distance d, and can be enhanced by reducing d through applying pressure.

With the above basis of momentum eigenstates, a lowest energy wave packet for IX centered at position $\mathbf{R}_c$ can be constructed (Yu et al., 2018)

$$\mathcal{X}_{\tau'\tau,\mathbf{R}_c} = \sum_{\mathbf{Q}} e^{-i\mathbf{Q}\cdot\mathbf{R}_c} W(\mathbf{Q}) X_{\tau'\tau,\mathbf{Q}}. \tag{7}$$

Here $W(\mathbf{Q}) = W(Q)$ is the s-wave momentum distribution centered at $\mathbf{Q}=0$, with a width ~$2/w$ (Fig. 4E).

3.3 Exciton superlattice in the Moiré pattern

In the limit of $Q_0 \ll K$, corresponding to δ, $\delta\theta \ll 1$, the spatial variation of interlayer atomic registries leads to a periodic moiré pattern, with a period $b = a/(\delta^2 + \delta\theta^2)^{1/2}$ much larger than the monolayer lattice constants a (Fig. 5A). Writing $\mathbf{g}'_{1,2}$ ($\mathbf{g}_{1,2}$) as the electron (hole) layer primitive reciprocal lattice vectors, the primitive reciprocal lattice vectors of the moiré superlattice are given by $\mathbf{g}'_1 - \mathbf{g}_1$ and $\mathbf{g}'_2 - \mathbf{g}_2$, with $|\mathbf{g}'_{1,2} - \mathbf{g}_{1,2}| \ll g'_1, g_1$.

In a long-period moiré pattern, it is convenient to examine the IX properties from a local perspective. Within a region with length-scale small compared to moiré period b but large compared to lattice constant a, the interlayer atomic registry is nearly indistinguishable from a lattice-matched stacking (see Fig. 5A–B). Thus the local electronic structure can be approximated by that of the R- (0-degree aligned) or H-type (60-degree aligned) lattice-matched bilayer of the corresponding registry, characterized by the in-plane displacement vector $\mathbf{r}_0$ pointing from an M site in the hole layer to a nearby M site in the electron layer. Different local regions in the moiré correspond to different $\mathbf{r}_0$. Special attention should be paid to the $\hat{C}_3$-symmetric local regions with interlayer registries denoted as $R_h{}^h$, $R_h{}^X$, $R_h{}^M$, $H_h{}^X$, $H_h{}^h$ and $H_h{}^M$ (Fig. 5B). Here $R_h{}^\mu$ ($H_h{}^\mu$) denotes the R-type (H-type) configuration with the electron-layer μ site sitting on top of the hole-layer h site. $\mathbf{r}_0$ is a smooth and periodic function of the position variable $\mathbf{R}$ in the moiré. In the rigid lattice limit (in the absence of structural relaxation), $\mathbf{r}_0$ is a linear function of $\mathbf{R}$. Choosing the spatial coordinate origin at a locale with $\mathbf{r}_0 = 0$, we find the linear mapping function $\mathbf{K}\cdot\mathbf{r}_0(\mathbf{R}) = \mathbf{Q}_0\cdot\mathbf{R}$ (mod 2π) (Yu et al., 2017). On the other hand, when the moiré period gets

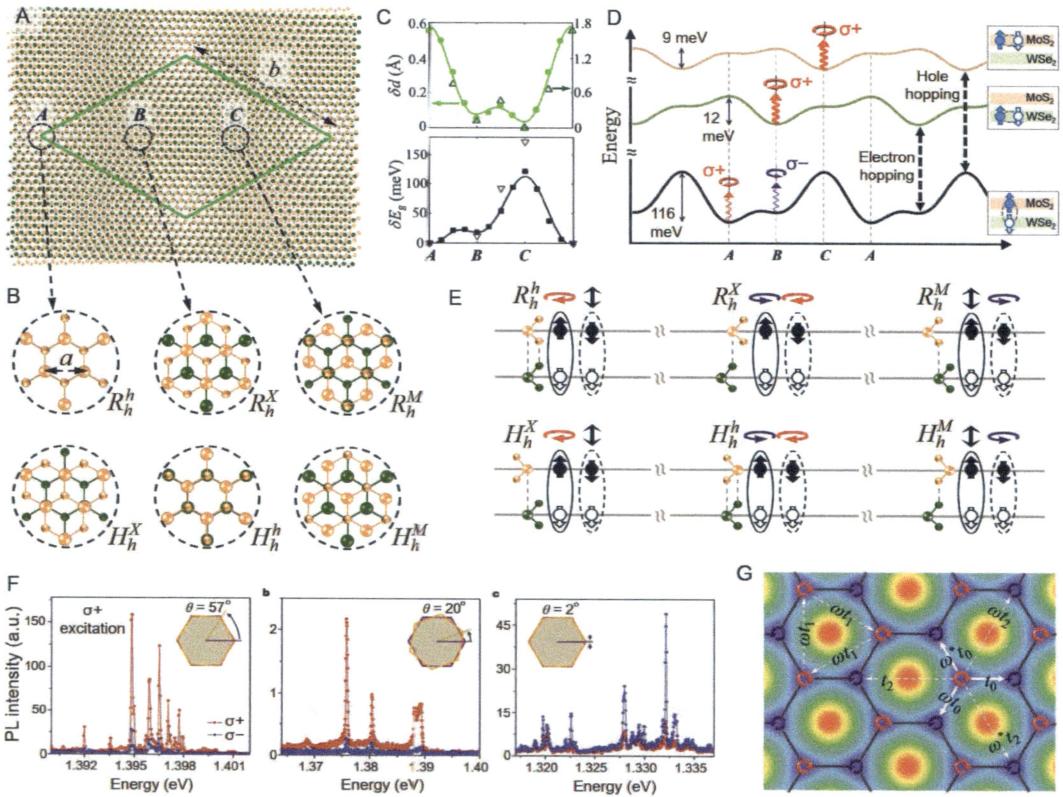

Fig. 5 See figure legend on next page.

large, the bilayer moiré can undergo structural relaxation to shrink the area of the energetically unfavorable $R_h{}^h$ and $H_h{}^M$ stacking regions while enlarging the area of the $R_h{}^X$, $R_h{}^M$, $H_h{}^X$, and $H_h{}^h$ stacking regions (Rosenberger et al., 2020), which changes the mapping function $\mathbf{r}_0(\mathbf{R})$.

The equilibrium interlayer distance d is determined by the van der Waals interaction, which depends sensitively on the interlayer atomic registry $\mathbf{r}_0$, now a function of $\mathbf{R}$ in the moiré pattern. This introduces a periodic modulation of the interlayer distance $d(\mathbf{R})$ (see Fig. 5C) (Yu et al., 2017; Zhang et al., 2017). The local regions where d reaches its maximum have the X sites of one layer sit on top of the X sites of the other layer, i.e., $R_h{}^h$ and $H_h{}^M$ configurations (Liu et al., 2014; van der Zande et al., 2014). On the other hand, d reaches its minimum at local regions where the X sites of one layer sit on top of the M sites or h center of the other layer, i.e., $R_h{}^X$, $R_h{}^M$, $H_h{}^X$, or $H_h{}^h$ configurations. The modulation range of the interlayer distance is $\delta d \approx$ 0.6 Å from first-principles calculations, while the scanning tunneling microscopy measurement gives an upper bound 1.3–1.7 Å (Pan et al., 2018; Zhang et al., 2017). The electrical dipole of IX is $ed(\mathbf{R})$. In a uniform perpendicular electric field, IX then experience different Stark shifts at different local regions.

Fig. 5—Cont'd (A) Schematic of a moiré pattern generated by a small twisting from a R-stacking heterobilayer (0° aligned). Green diamond is a supercell. ***A***, ***B*** and ***C*** are regions of three high symmetry local stacking, where the interlayer atomic registries are shown in (B). (B) The six high symmetry stackings of R-type bilayers (0° aligned, upper part) and H-type bilayers (60° aligned, lower part). (C) The interlayer distance d (top) and variation δE_g of the local bandgap (bottom) as functions of the atomic registries in the MoS_2/WSe_2 bilayer. Solid symbols are from the first-principles calculations, and empty symbols are from the scanning tunneling microscopy/spectroscopy measurement in the MoS_2/WSe_2 moiré pattern. (D) The moiré potential landscapes for the intralayer and interlayer excitons in MoS_2/WSe_2, with the optical selection rule shown at the potential minima. (E) The polarization selection rules for spin-singlet and spin-triplet IX at the six high symmetry stackings in (B). (F) Polarization-resolved PL spectra of trapped IX in $MoSe_2/WSe_2$ heterobilayers with three different alignment angles. Under a $\sigma+$ polarized excitation, the PL from the samples of 57° and 20° angles is mainly $\sigma-$ polarized (red), whereas that of 2° twist angle is mainly $\sigma+$ polarized (blue). (G) Under a critical electric field value, the two degenerate potential minima have opposite photon emission polarizations, and have complex hopping between them for the spin-up IX. $\omega \equiv \exp(-2i\pi/3)$. *Panels (A–D, G): From Yu, H., Liu, G.-B., Tang, J., Xu, X., Yao, W. 2017. Moiré excitons: from programmable quantum emitter arrays to spin-orbit-coupled artificial lattices. Sci. Adv. 3, e1701696, adapted with permission from AAAS. Panel (F): From Seyler, K. L., Rivera, P., Yu, H., Wilson, N. P., Ray, E. L., Mandrus, D. G., Yan, J., Yao, W., Xu, X. 2019. Signatures of moiré-trapped valley excitons in MoSe2/WSe2 heterobilayers. Nature 567, 66–70, reprinted with permission from Springer Nature.*

The more pronounced moiré effects come from the sensitive dependence of interlayer hopping and interlayer optical transition on the stacking registry $\mathbf{r}_0$. At the high symmetry stacking $R_h{}^h$, $R_h{}^X$, $R_h{}^M$, $H_h{}^X$, $H_h{}^h$ or $H_h{}^M$ where the $\hat{C}_3$-symmetry is retained, the interlayer hopping is dictated by the $\hat{C}_3$ eigenvalues or quantum numbers (c.f. Table 1) of the band edge Bloch functions (Rivera et al., 2018). Consider the band n in the hole-layer and the band n' in the electron-layer, the interlayer hopping between the Bloch states $\psi_{n,\mathbf{K}}$ and $\psi_{n',\mathbf{K}}$ is nonzero only when their $\hat{C}_3$ quantum numbers are the same under the choice of a common rotation center. Note that the rotation center in the electron-layer can be different from that in the hole-layer depending on the bilayer stacking. If we set the hole-layer rotation center as h, then those of the electron-layers are h, X and M for $R_h{}^h$, $R_h{}^X$ and $R_h{}^M$ stackings respectively. The corresponding $\hat{C}_3$ quantum numbers for $\psi_{n,\mathbf{K}}$ and $\psi_{n',\mathbf{K}}$ around the rotation centers M, X and h are given in Table 1. So the interlayer hopping between the two conduction (valence) bands at $\mathbf{K}$ is finite only for $R_h{}^h$ and $H_h{}^X$ ($R_h{}^h$ and $H_h{}^h$), and vanishes for the other four registries. Clearly, such sensitive dependence on stacking registry comes from the dependence of $\hat{C}_3$ quantum numbers on the rotation center, unique to the band edge carriers at the high symmetry Brillouin zone $\mathbf{K}$ corners. A general interlayer stacking $\mathbf{r}_0$ breaks the $\hat{C}_3$ rotational symmetry, where the interlayer hopping strength is finite.

Under the two-center-approximation and keeping only the leading Fourier components, the $\mathbf{r}_0$ dependence of the hopping matrix element between $\psi_{n,\mathbf{K}}$ and $\psi_{n',\mathbf{K}}$ can be written as (Wang et al., 2017b)

$$h_{nn'}(d,\mathbf{r}_0) \approx t_{nn'}(d)\left(e^{-i\mathbf{K}\cdot\mathbf{r}_0} + e^{-i\hat{C}_3\mathbf{K}\cdot\mathbf{r}_0}e^{-i\frac{2\pi}{3}(m_n-m_{n'})} + e^{-i\hat{C}_3^2\mathbf{K}\cdot\mathbf{r}_0}e^{-i\frac{4\pi}{3}(m_n-m_{n'})}\right). \tag{8}$$

Here $t_{nn'}(d)$ is an $\mathbf{r}_0$-independent factor that decays exponentially with the increase of the interlayer distance d, ranging from several meV to several tens meV at the equilibrium d value. m_n and $m_{n'}$ are the $\hat{C}_3$ quantum numbers of the Bloch states $\psi_{n,\mathbf{K}}$ and $\psi_{n',\mathbf{K}}$, respectively. In most heterobilayer combinations, the band offset between n and n' is much larger than $h_{nn'}$, so the effect of interlayer hopping can be treated perturbatively. The overall effect can be well accounted by an energy shift $\delta E_{n/n',\mathbf{K}}(d,\mathbf{r}_0)$ to the Bloch state $\psi_{n/n',\mathbf{K}}$, which originates from the combination of a second-order energy correction due to $h_{nn'}(d,\mathbf{r}_0)$ (Wang et al., 2017b) and a static electric potential from the hopping-induced charge transfer (Li and Wu, 2017).

In the moiré pattern, the above dependence of interlayer hopping on stacking registry $\mathbf{r}_0$, combined with the mapping between $\mathbf{r}_0$ and location in moiré $\mathbf{R}$, leads to the location dependent local hopping amplitude. Ignoring the atomic reconstruction, $\mathbf{K}\cdot\mathbf{r}_0(\mathbf{R})=\mathbf{Q}_0\cdot\mathbf{R}$ (mod 2π), and also keeping the $\mathbf{R}$-dependence in interlayer distance, we obtain function form of the local interlayer hopping

$$h_{nn'}(\mathbf{R}) \approx t_{nn'}(d(\mathbf{R}))\left(e^{-i\mathbf{Q}_0\cdot\mathbf{R}} + e^{-i\hat{C}_3\mathbf{Q}_0\cdot\mathbf{R}}e^{-i\frac{2\pi}{3}(m_n-m_{n'})} + e^{-i\hat{C}_3^2\mathbf{Q}_0\cdot\mathbf{R}}e^{-i\frac{4\pi}{3}(m_n-m_{n'})}\right). \tag{9}$$

As discussed above, the overall effect of $h_{nn'}(\mathbf{R})$ is an energy shift $\delta E_{n/n',\mathbf{K}}(\mathbf{R})$ to the band edges in the two layers, which now depends on location. This leads to a spatially-varying periodic moiré potential on the band edge carriers, i.e., a superlattice for electrons and holes. In an R-type heterobilayer MoS_2/WSe_2 moiré pattern with a period $b=8.7\,\text{nm}$, scanning tunneling spectroscopy has revealed a variation range of $\sim$150 meV on the local interlayer band gap (Fig. 5C) (Zhang et al., 2017).

The interlayer hopping $h_{nn'}(\mathbf{R})$ can no longer be treated perturbatively when its magnitude is comparable to or larger than the band offset, e.g., in a homobilayer or certain heterobilayers like $MoSe_2/WS_2$ (Alexeev et al., 2019). $h_{nn'}(\mathbf{R})$ can then lead to position dependent layer-hybridization of the band edge carriers (Wu et al., 2019; Yu et al., 2020).

For the IX, the moiré potential $\delta E_{n/n',\mathbf{K}}(\mathbf{R})$ can be considered as a perturbation added to the electron-hole Coulomb binding. As the length scale of $\delta E_{n/n'}(\mathbf{R})$ is much larger than the IX Bohr radius ($a_B \sim 2\,\text{nm}$), its major effect is a superlattice potential $V(\mathbf{R})$ on the IX center-of-mass motion. In contrast, the electron-hole relative motion is barely affected. The excitonic superlattice potential is.

$$V(\mathbf{R}) \approx \delta E_{c',\mathbf{K}}(\mathbf{R}) - \delta E_{v,\mathbf{K}}(\mathbf{R}) + \mathscr{E}d(\mathbf{R}). \tag{10}$$

The last term accounts for the dipole energy in a perpendicular electric field $\mathscr{E}$, and the $\mathbf{R}$-dependence of interlayer distance can be exploited to engineer the potential (Yu et al., 2017). In principle the modulation of the interlayer distance $d(\mathbf{R})$ can affect the IX binding energy which also contributes to $V(\mathbf{R})$. We expect such contribution to be small though, as $\delta d \sim 0.1\,\text{nm}$ is much smaller than a_B.

By the $\hat{C}_3$ symmetry, a local region with the registry R_h^h, R_h^X or R_h^M (H_h^X, H_h^h, or H_h^M) in a near R-type (H-type) moiré pattern should correspond to local extrema of $V(\mathbf{R})$. For the example of MoS_2/WSe_2 heterobilayer shown in Fig. 5C–D, both the scanning tunneling spectroscopy and first-principles calculations show that the global minimum (maximum) of $V(\mathbf{R})$ are at the local registry $R_h{}^h$ ($R_h{}^M$) at zero ε, whereas $R_h{}^X$ corresponds to a local minimum (Yu et al., 2017, Zhang et al., 2017). For other heterobilayers like $MoSe_2/WSe_2$, first-principles calculations show a similar moiré potential.

With $V(\mathbf{R})$ varying in a range ~150 meV, the depth of the confinement potential at its energy minima is several tens meV (Fig. 5D), which realizes a highly ordered array of nanodots where low energy IXs are tightly confined. Hopping between the neighboring nanodots drops exponentially with b. At relatively large b where the hopping is negligible, a nanodot loaded with a single IX or two IX can be exploited as emitter of single photon or entangled photon pair respectively (Yu et al., 2017). In a small moiré, inter-dot hopping can couple the nanodots, leading to mini-bands of IX dispersions.

The electric field $\mathscr{E}$ can switch the global minimum from $R_h{}^h$ to $R_h{}^X$, utilizing the distinct values of d at these two stacking regions (Fig. 5D). When energies at $R_h{}^h$ and $R_h{}^X$ are nearly degenerate, $V(\mathbf{R})$ defines a honeycomb lattice where IX acquires a sub-lattice pseudospin in addition to the spin and valley pseudospin (Yu et al., 2017). Below we show that this sub-lattice pseudospin can be addressed through the distinct optical selection rules at the two trapping sites.

3.4 Location-dependent polarization selection rules

The low energy IX trapped in the moiré nanodots are in wave packet form, with a spatial extension w smaller than the moiré period b. The polarization of the photon emission is determined by the symmetry of the atomic registry at the region where the wave packet localizes. Note that the out-of-plane mirror symmetry $\widehat{\sigma}_h$ is absent in the heterobilayer, thus the conservation of the $\widehat{\sigma}_h$ quantum number is no longer a constraint. For the high symmetry regions with stacking registries $R_h{}^h$, $R_h{}^X$, $R_h{}^M$, $H_h{}^X$, $H_h{}^h$ and $H_h{}^M$, the emission polarization is solely determined by the $\hat{C}_3$ quantum number of IX. Choosing the rotation center to be h in the hole-layer, a $\mathbf{K}$-valley valence band edge with spin-up has a $\hat{C}_3$ quantum number $m_v + S_z = 1/2$. This rotation center corresponds to h, X and M respectively in the electron-layer under the $R_h{}^h$, $R_h{}^X$ and $R_h{}^M$ ($H_h{}^h$, $H_h{}^X$ and $H_h{}^M$) stacking, which determines the $\hat{C}_3$ quantum number of the corresponding conduction band edge $\tau' m_{c'} + S'_z$. The $\hat{C}_3$ quantum number of IX is then $\tau' m_{c'} + S'_z - 1/2$ (mod 3). Reading the values of $\tau' m_{c'} + S'_z$ from Table 1, the polarization selection rule of IX at the various interlayer stacking registries and spin-valley configurations can be obtained, as summarized in Table 2 and shown in Fig. 5E.

At a given stacking registry spin-singlet and triplet IX from the same valley emit photons of distinct polarizations, while for a given spin and valley configuration, IX also have different emission polarizations at different registries. Considering the tens of meV splitting between the singlet and triplet, as well as the energy difference in the moiré potential $V(\mathbf{R})$, it is possible to distinguish the spin, valley and sub-lattice configurations of IX through their emission energies and polarizations (Ciarrocchi et al., 2019; Hsu et al., 2018; Jin et al., 2019a; Tran et al., 2019).

For an IX wave packet at a general location where the local registry does not have $\hat{C}_3$ symmetry, the optical dipole can have all $\sigma+$, $\sigma-$ and z components. From the wave function form Eq. (7), the wave packet's optical dipole is the coherent superposition of those at the three light cones

$$\mathcal{D}_{\tau'+}(\mathbf{R}_c) = W(Q_0)\left(e^{-i\mathbf{Q}_0\cdot\mathbf{R}_c}\mathbf{D}_{\mathbf{Q}_0} + e^{-i\hat{C}_3\mathbf{Q}_0\cdot\mathbf{R}_c}\mathbf{D}_{\hat{C}_3\mathbf{Q}_0} + e^{-i\hat{C}_3^2\mathbf{Q}_0\cdot\mathbf{R}_c}\mathbf{D}_{\hat{C}_3^2\mathbf{Q}_0}\right). \tag{11}$$

Clearly the wave packet dipole is inversely proportional to the wave packet extension w in the real space. For w small compared to moiré period b, $1/w > 1/b \sim Q_0$ (Fig. 4E), the wave packet has a finite optical dipole with the polarization vector dependent on its central location $\mathbf{R}_c$ in the moiré, determined by the interference of the three main light cones. Using $\mathbf{K}\cdot\mathbf{r}_0(\mathbf{R}) = \mathbf{Q}_0\cdot\mathbf{R}$ (mod 2π), the optical dipole can be rewritten as

$$\mathcal{D}_{\tau'+}(\mathbf{R}_c) = W(Q_0)\left(e^{-i\mathbf{K}\cdot\mathbf{r}_0(\mathbf{R}_c)}\mathbf{D}_{\mathbf{Q}_0} + e^{-i\hat{C}_3\mathbf{K}\cdot\mathbf{r}_0(\mathbf{R}_c)}\mathbf{D}_{\hat{C}_3\mathbf{Q}_0} + e^{-i\hat{C}_3^2\mathbf{K}\cdot\mathbf{r}_0(\mathbf{R}_c)}\mathbf{D}_{\hat{C}_3^2\mathbf{Q}_0}\right). \tag{12}$$

This implies that the emitted photon polarizations are fully determined by the interlayer registry $\mathbf{r}_0(\mathbf{R}_c)$ at the wave packet center $\mathbf{R}_c$.

IX optical dipole polarization in lattice-matched $MoSe_2/WSe_2$ heterobilayers of various stacking registries $\mathbf{r}_0$ can be obtained from first-principles

Table 2 The polarization selection rules for the spin-singlet and spin-triplet bright IX at R_h^h, R_h^X, R_h^M, H_h^X, H_h^h, and H_h^M stacking configurations. The hole is assumed at **K** valley.

	R_h^h	R_h^X	R_h^M	H_h^X	H_h^h	H_h^M
Spin-singlet	$\sigma+$	$\sigma-$	z	$\sigma+$	$\sigma-$	z
Spin-triplet	z	$\sigma+$	$\sigma-$	z	$\sigma+$	$\sigma-$

band structure calculations. It is found that the analytical form given by Eq. (12) can well fit the first-principles results (Yu et al., 2018). Interestingly, the out-of-plane optical dipole of spin-singlet and in-plane optical dipole of spin-triplet IX, both symmetry-forbidden in the monolayer, now become as significant as other dipole components. As a result, both the spin-singlet and spin-triplet IX can contribute significantly to the heterobilayer PL signals, in both the out-of-plane and in-plane emission directions (Fig. 5E).

The first generation of experiments observing the IX PL peaks show a rather broad linewidths (Fig. 3B) (Rivera et al., 2015, 2016). These experiments are conducted on samples without hBN encapsulation at temperatures above 10 K and relatively high excitation power. In high quality $MoSe_2/WSe_2$ heterobilayer moiré at lower temperature (~1 K), by ramping down the excitation power, the broad IX peak evolves into a series of well separated ultranarrow lines with linewidth ~0.1 meV (Seyler et al., 2019). The pronounced power saturation behavior and power broadening suggest they are from three-dimensionally confined IX. These emitters exhibit a series of remarkable properties distinct from the excitonic emitters bound to defects, impurities or strain-induced confining potentials in monolayer TMDs (Chakraborty et al., 2015; He et al., 2015; Koperski et al., 2015; Kumar et al., 2015; Srivastava et al., 2015a; Tonndorf et al., 2015):

(1) In contrast to the monolayer emitters which show doublets with cross-linearly polarized emissions, the sharp lines from IX emitters exhibit strong circular polarizations (Fig. 5F), implying that the trapping potential retains the $\hat{C}_3$ symmetry.

(2) The helicity can be either co-polarized or cross-polarized with the excitation at the intralayer X^0 energy, consistent with the stacking registry control of IX spin-valley optical selection rules, in contrast to the fixed selection rule of the intralayer X^0. The exact sign of the emitter polarization is determined by the twist angle. For example, all emitters in samples near 0 (60) degree twist angle exhibit cross-circular (co-circular) polarization. This twist angle determined PL polarization precludes IX bound to extrinsic traps by impurities etc., whose random locations in the moiré and the associated random local stacking registries could result in distributed polarizations (Fig. 5E).

(3) Magneto-optical spectroscopies (discussed in detail in Section 3.5) find uniform g-factors not only among emitters of the same sample, but also only taking two values in various samples (one in R-type stacking, one in H-type stacking) determined by the twist angle. In contrast, monolayer emitters in a single sample have a wide distribution of g-factors.

(4) For 20-degree twist samples, the emitters properties (such as the sign of polarization and g-factors) are the same as those in 60-degree twist ones, but with an order magnitude weaker intensity. This is due to the formation of a moiré superlattice at a near commensurate angle (21.8-degree), which gives rise to the electron and hole recombination assisted by Umklapp scattering (Fig. 4C).

All the above evidences suggest that the observed sharp lines are from IX emitters trapped at the $\hat{C}_3$-symmetric moiré potential minima. Recent photon correlation measurement has also revealed the anti-bunching behavior of the narrow line emission, an unambiguous evidence of the quantum nature of moiré trapped IXs (Baek et al., 2020). Combining the twisted angle determined emission polarization and the g-factors, these sharp IX resonances can be identified as spin-singlet and spin-triplet ones located at certain moiré potential minima (Brotons-Gisbert et al., 2020; Seyler et al., 2019; Wang et al., 2019).

For small moiré period, the inter-dot hopping becomes non-negligible, which leads to dispersive mini bands for low energy IX well described by a tight-binding model (Yu et al., 2017). When the two local minima in a moiré supercell (e.g., $R_h{}^h$ and $R_h{}^X$) are tuned to nearly degenerate by a vertical electric field, it corresponds to a honeycomb lattice with a nearest-neighbor hopping strength that can reach several meV for moiré period $b < 10\,\text{nm}$. Due to the distinct $\hat{C}_3$ quantum numbers of the two sub-lattice sites, the hopping between them has a complex phase depending on both the direction of their displacement vector and the IX spin/valley index (Fig. 5G). This is a unique feature endowed by the moiré pattern, absent in graphene and other artificial honeycomb lattices. Such complex hopping realizes a sizable spin-orbit splitting of IX bands. When the IX energies at $R_h{}^h$ and $R_h{}^X$ are degenerate, the mini-bands feature a Dirac node (four-fold degenerate) at $\mathbf{\Gamma}$, and two Weyl nodes (two-fold degenerate) with opposite spin/valley indices at the mini Brillouin zone corners (Yu et al., 2017).

We note that intralayer X^0 in a heterobilayer moiré pattern also experience a relatively weak superlattice potential (see Fig. 5D) (Wu et al., 2017; Yu et al., 2017), which leads to multiple emergent peaks observed in the absorption spectra near the intralayer X^0 resonance due to the formation of minibands (Jin et al., 2019b).

3.5 g-factors of the inter- and intra-layer excitons in comparison with intralayer ones

For each exciton species, its two degenerate valley configurations are related by a time reversal, which have opposite Zeeman shifts in an perpendicular magnetic field B (Fig. 6A–C) (Aivazian et al., 2015; Li et al., 2014;

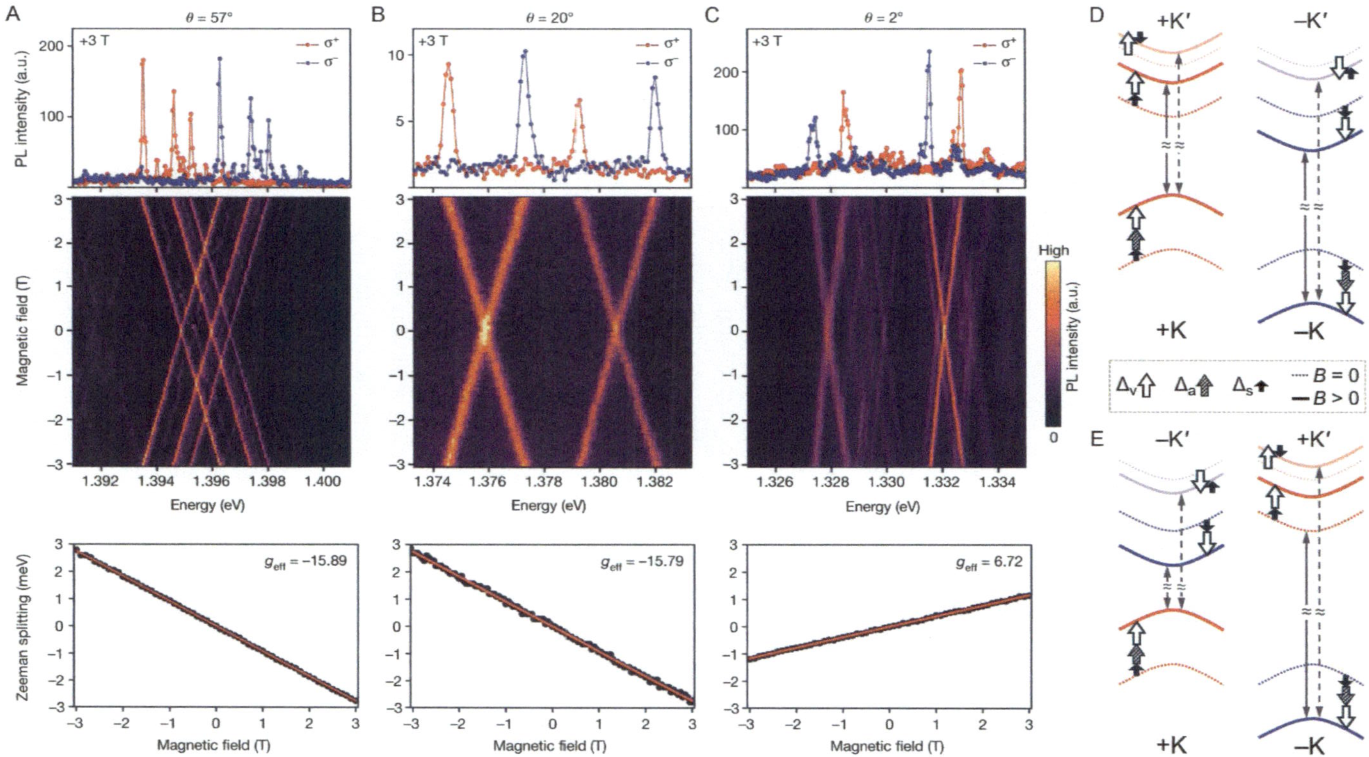

Fig. 6 Magnetic field dependent PL from trapped IX in $MoSe_2/WSe_2$ heterobilayers with alignment angles (A) 57°, (B) 20° and (C) 2°. Top, polarization-resolved PL spectra at 3 T under a linearly polarized excitation. Middle, the total PL intensity as a function of the magnetic field. Bottom, the extracted polarization g-factors from the Zeeman splitting. (D) The spin (Δ_s), atomic orbital (Δ_a), and lattice (Δ_v) contributions to the IX Zeeman shift in a R-type TMD heterobilayer. The solid (dotted) curves denote the band positions under a finite (zero) magnetic field. (E) Those in a H-type heterobilayer. *Panels (A–C): Reprinted from Seyler, K. L., Rivera, P., Yu, H., Wilson, N. P., Ray, E. L., Mandrus, D. G., Yan, J., Yao, W., Xu, X., 2019. Signatures of moiré-trapped valley excitons in MoSe2/WSe2 heterobilayers. Nature 567, 66–70, Nature Publishing Group.*

MacNeill et al., 2015; Srivastava et al., 2015b). The resultant valley Zeeman splitting scales linearly with B, which defines the exciton g-factor,

$$2\Delta = E_{\tau=+} - E_{\tau=-} = g\mu_B B. \tag{13}$$

The convention adopted here is $E_{\tau=+}$ ($E_{\tau=-}$) denotes the energy of the configuration with the hole in $\mathbf{K}$ ($-\mathbf{K}$) valley. For either monolayer X^0 or IX at the moiré trapping sites, their two valley configurations also emit photons of opposite circular polarization. This allows the determination of the g-factor from polarization resolved PL spectrum, even when the linewidth is much broader than Δ.

For IX in the moiré traps in $MoSe_2/WSe_2$ heterobilayers, their ultranarrow linewidth allows resolution of the Zeeman splitting, and in the PL map plotted as a function of energy and the magnetic field, the IX resonances exhibit "X" pattern crossing at $B=0$ (Fig. 6A–C). Remarkably, the g-factors are found to be uniform for the different emission lines across the same sample, and exhibit limited values in samples of varying twist angle from the 0-degree and 60-degree alignments (Fig. 6A–C). What the experiments measure is in fact the polarization Zeeman splitting, i.e. the energy difference between $\sigma+$ and $\sigma-$ polarized PL peaks, which defines a "polarization" g-factor, $g_p\mu_B B = E_{\sigma+} - E_{\sigma-}$. This quantity is the same as the valley g-factor defined in Eq. (13), except for a possible sign difference reflecting the valley optical selection rules. While g defined in Eq. (13) is always negative for various exciton species, g_p can be either positive or negative for IX in heterobilayer moiré patterns due to the position-dependent valley selection rules (Brotons-Gisbert et al., 2020; Seyler et al., 2019).

From Table 2, we can tell that spin-singlet IX at R_h^h, H_h^X, and spin-triplet IX at R_h^X, H_h^h stackings have the same valley selection rule as the intralayer exciton, where the valley configuration with hole index $\tau=+$ emits $\sigma+$ photon (i.e., $g_p=g$). For spin-singlet IX at R_h^X, H_h^h and spin-triplet IX at R_h^M, H_h^M stackings, the valley configuration with hole index $\tau=+$ emits $\sigma-$ photon instead, so $g_p=-g$. The sign of g_p is an important information to determine the trapping sites of IX emitters in a moiré pattern.

The different magnitudes of the g-factors are due to the different spin and valley pairing of the electron and hole constituents in near 0-degree and 60-degree alignments. The Zeeman shifts of the electron and hole have three contributions (Aivazian et al., 2015). (1) The spin contribution, $\Delta_s = 2S_z\mu_B B$, with $S_z = \pm 1/2$ the spin of the conduction or valence electron. (2) The lattice contribution related to the Berry phase effect, $\Delta_v = \tau\alpha_{v/c}\mu_B B$, which has a valley contrasted sign. Here

$$\alpha_n = \frac{-i}{m_0}\mathbf{z} \cdot \sum_{i \neq n} \frac{\langle u_{n,\mathbf{K}}|\widehat{\mathbf{p}}|u_{i,\mathbf{K}}\rangle \times \langle u_{i,\mathbf{K}}|\widehat{\mathbf{p}}|u_{n,\mathbf{K}}\rangle}{E_{n,\mathbf{K}} - E_{i,\mathbf{K}}}, \tag{14}$$

with m_0 the free electron mass and $u_{n,\mathbf{K}}$ the periodic part of the Bloch function. A simplified 2-band **k·p** model gives $\alpha_c \sim m_0/m_e$ and $\alpha_v \sim m_0/m_h$, where the electron and hole effective masses (m_e, m_h) are comparable (Xiao et al., 2012). (3) The atomic orbital contribution, $\Delta_a = l\mu_B B$. With $l_v = 2\tau$ ($l_c = 0$) for the valence (conduction) band $d_{x^2-y^2} + i\tau d_{xy}$ (d_{z^2}) orbital in $\tau\mathbf{K}$ valley, this contribution is for the hole only, and also has a valley dependent sign. Δ_s, Δ_v and Δ_a for electrons and holes on different spin and valley states are schematically shown in Fig. 6D–E. For an exciton in the valley configuration (τ', +), i.e., electron in $\tau'\mathbf{K}'$ and hole in $\mathbf{K}$ valley, the Zeeman shift is then $2\Delta = -2(1 - 2S'_z + \alpha_v - \tau'\alpha_c + 2)\mu_B B$, given by the difference between the conduction and valence band shifts. We note that bright IX (i.e. momentum direct) can have two types of electron-hole valley pairing depending on the heterobilayer stacking: $\tau' = +$ for near 0-degree alignment and $\tau' = -$ for near 60-degree alignment. The forms of g-factors for different spin configurations in monolayer and heterobilayers are compared in Table 3 below.

Combining the g-factors of intra- and interlayer exciton, the coefficients $\alpha_{v/c}$ for the lattice (Berry phase) contribution can be determined. Table 4 summarizes the g-factors of the various exciton species in monolayer WSe_2 and $MoSe_2/WSe_2$ heterobilayers determined from magneto PL spectra. Together with the forms listed in Table 3, the three monolayer exciton g-factors give $\alpha_v = 2.4$, $\alpha_c = 2.05$ and $\alpha'_c = 1.85$ in WSe_2, while the three heterobilayer g-factors give $\alpha_v = 2.47$ in WSe_2, and $\alpha_c = 1.13$, $\alpha'_c = 1.23$ in $MoSe_2$. Here α_c (α'_c) is for the conduction sub-band denoted by the red solid (dashed) curve. Besides, monolayer bright X^0 in $MoSe_2$ has a g-factor ~ -4 (Li et al., 2014; MacNeill et al., 2015), suggesting that $\alpha_v \approx \alpha_c$ in $MoSe_2$.

Table 3 The g-factors for bright excitons with different spin configurations in monolayer and heterobilayers.

	Monolayer	IX in near R-type bilayer (0-degree)	IX in near H-type bilayer (60-degree)
Spin-singlet	$-2(\alpha_v - \alpha_c + 2)$	$-2(\alpha_v - \alpha_c + 2)$	$-2(\alpha_v + \alpha_c + 2)$
Spin-triplet	$-2(\alpha_v - \alpha_c + 4)$	$-2(\alpha_v - \alpha_c + 4)$	$-2(\alpha_v + \alpha_c + 4)$

Table 4 The experimentally measured g-factors of bright and dark X^0 in monolayer WSe_2, and IX in $MoSe_2/WSe_2$ heterobilayer.

	Monolayer WSe_2 (He et al., 2020)			Near R-type $MoSe_2/WSe_2$ (Seyler et al., 2019)	Near H-type $MoSe_2/WSe_2$ (Wang et al., 2019)	
	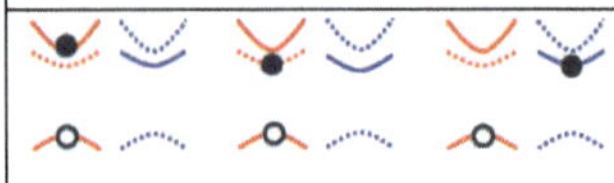			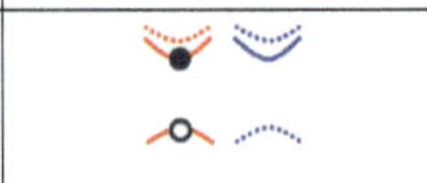	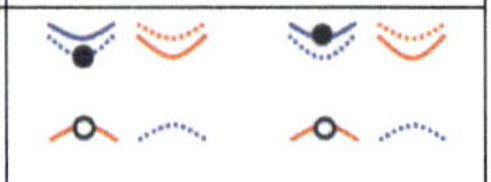	
g	−4.7	−9.1	−12.5	−6.7	−15.2	−10.7

References

Aivazian, G., Gong, Z., Jones, A.M., Chu, R.-L., Yan, J., Mandrus, D.G., Zhang, C., Cobden, D., Yao, W., Xu, X., 2015. Magnetic control of valley pseudospin in monolayer WSe2. Nat. Phys. 11, 148–152.

Alexeev, E.M., Ruiz-Tijerina, D.A., Danovich, M., Hamer, M.J., Terry, D.J., Nayak, P.K., Ahn, S., Pak, S., Lee, J., Sohn, J.I., Molas, M.R., Koperski, M., Watanabe, K., Taniguchi, T., Novoselov, K.S., Gorbachev, R.V., Shin, H.S., Fal'ko, V.I., Tartakovskii, A.I., 2019. Resonantly hybridized excitons in moiré superlattices in van der Waals heterostructures. Nature 567, 81–86.

Baek, H., Brotons-Gisbert, M., Koong, Z.X., Campbell, A., Rambach, M., Watanabe, K., Taniguchi, T., Gerardot, B.D., 2020. Highly tunable quantum light from moiré trapped excitons. arXiv. 2001.04305.

Baranowski, M., Surrente, A., Klopotowski, L., Urban, J.M., Zhang, N., Maude, D.K., Wiwatowski, K., Mackowski, S., Kung, Y.C., Dumcenco, D., Kis, A., Plochocka, P., 2017. Probing the interlayer exciton physics in a MoS2/MoSe2/MoS2 van der Waals heterostructure. Nano Lett. 17, 6360–6365.

Berkelbach, T.C., Hybertsen, M.S., Reichman, D.R., 2013. Theory of neutral and charged excitons in monolayer transition metal dichalcogenides. Phys. Rev. B 88, 045318.

Brotons-Gisbert, M., Baek, H., Molina-Sánchez, A., Campbell, A., Scerri, E., White, D., Watanabe, K., Taniguchi, T., Bonato, C., Gerardot, B.D., 2020. Spin–layer locking of interlayer excitons trapped in moiré potentials. Nat. Mater. 19, 630–636.

Cao, T., Wang, G., Han, W., Ye, H., Zhu, C., Shi, J., Niu, Q., Tan, P., Wang, E., Liu, B., Feng, J., 2012. Valley-selective circular dichroism of monolayer molybdenum disulphide. Nat. Commun. 3, 887.

Chakraborty, C., Kinnischtzke, L., Goodfellow, K.M., Beams, R., Vamivakas, A.N., 2015. Voltage-controlled quantum light from an atomically thin semiconductor. Nat. Nanotechnol. 10, 507–511.

Cheiwchanchamnangij, T., Lambrecht, W.R.L., 2012. Quasiparticle band structure calculation of monolayer, bilayer, and bulk MoS2. Phys. Rev. B 85, 205302.

Chernikov, A., Berkelbach, T.C., Hill, H.M., Rigosi, A., Li, Y., Aslan, O.B., Reichman, D.R., Hybertsen, M.S., Heinz, T.F., 2014. Exciton binding energy and non-hydrogenic Rydberg series in monolayer WS2. Phys. Rev. Lett. 113, 076802.

Chiu, M.-H., Zhang, C., Shiu, H.W., Chuu, C.-P., Chen, C.-H., Chang, C.-Y.S., Chen, C.-H., Chou, M.-Y., Shih, C.-K., Li, L.-J., 2015. Determination of band alignment in the single-layer MoS2/WSe2 heterojunction. Nat. Commun. 6, 7666.

Ciarrocchi, A., Unuchek, D., Avsar, A., Watanabe, K., Taniguchi, T., Kis, A., 2019. Polarization switching and electrical control of interlayer excitons in two-dimensional van der Waals heterostructures. Nat. Photon. 13, 131–136.

Dery, H., Song, Y., 2015. Polarization analysis of excitons in monolayer and bilayer transition-metal dichalcogenides. Phys. Rev. B 92, 125431.

Donck, M.V.D., Peeters, F.M., 2018. Interlayer excitons in transition metal dichalcogenide heterostructures. Phys. Rev. B 98, 115104.

Feng, J., Qian, X., Huang, C.-W., Li, J., 2012. Strain-engineered artificial atom as a broad-spectrum solar energy funnel. Nat. Photon. 6, 866–872.

Glazov, M.M., Golub, L.E., Wang, G., Marie, X., Amand, T., Urbaszek, B., 2017. Intrinsic exciton-state mixing and nonlinear optical properties in transition metal dichalcogenide monolayers. Phys. Rev. B 95, 035311.

Gong, P., Yu, H., Wang, Y., Yao, W., 2017. Optical selection rules for excitonic Rydberg series in the massive Dirac cones of hexagonal two-dimensional materials. Phys. Rev. B 95, 125420.

He, K., Kumar, N., Zhao, L., Wang, Z., Mak, K.F., Zhao, H., Shan, J., 2014. Tightly bound excitons in monolayer WSe2. Phys. Rev. Lett. 113, 026803.

He, Y.-M., Clark, G., Schaibley, J.R., He, Y., Chen, M.-C., Wei, Y.-J., Ding, X., Zhang, Q., Yao, W., Xu, X., Lu, C.-Y., Pan, J.-W., 2015. Single quantum emitters in monolayer semiconductors. Nat. Nanotechnol. 10, 497–502.

He, M., Rivera, P., Tuan, D.V., Wilson, N.P., Yang, M., Taniguchi, T., Watanabe, K., Yan, J., Mandrus, D.G., Yu, H., Dery, H., Yao, W., Xu, X., 2020. Valley phonons and exciton complexes in a monolayer semiconductor. Nat. Commun. 11, 618.

Hong, X., Kim, J., Shi, S.-F., Zhang, Y., Jin, C., Sun, Y., Tongay, S., Wu, J., Zhang, Y., Wang, F., 2014. Ultrafast charge transfer in atomically thin MoS2/WS2 heterostructures. Nat. Nanotechnol. 9, 682–686.

Hsu, W.-T., Lu, L.-S., Wu, P.-H., Lee, M.-H., Chen, P.-J., Wu, P.-Y., Chou, Y.-C., Jeng, H.-T., Li, L.-J., Chu, M.-W., Chang, W.-H., 2018. Negative circular polarization emissions from WSe2/MoSe2 commensurate heterobilayers. Nat. Commun. 9, 1356.

Jauregui, L.A., Joe, A.Y., Pistunova, K., Wild, D.S., High, A.A., Zhou, Y., Scuri, G., Greve, K.D., Sushko, A., Yu, C.-H., Taniguchi, T., Watanabe, K., Needleman, D.J., Lukin, M.D., Park, H., Kim, P., 2019. Electrical control of interlayer exciton dynamics in atomically thin heterostructures. Science 366, 870–875.

Jin, C., Regan, E.C., Wang, D., Utama, M.I.B., Yang, C.-S., Cain, J., Qin, Y., Shen, Y., Zheng, Z., Watanabe, K., Taniguchi, T., Tongay, S., Zettl, A., Wang, F., 2019a. Identification of spin, valley and moiré quasi-angular momentum of interlayer excitons. Nat. Phys. 15, 1140–1144.

Jin, C., Regan, E.C., Yan, A., Utama, M.I.B., Wang, D., Zhao, S., Qin, Y., Yang, S., Zheng, Z., Shi, S., Watanabe, K., Taniguchi, T., Tongay, S., Zettl, A., Wang, F., 2019b. Observation of moiré excitons in WSe2/WS2 heterostructure superlattices. Nature 567, 76–80.

Jones, A.M., Yu, H., Ghimire, N.J., Wu, S., Aivazian, G., Ross, J.S., Zhao, B., Yan, J., Mandrus, D.G., Xiao, D., Yao, W., Xu, X., 2013. Optical generation of excitonic valley coherence in monolayer WSe2. Nat. Nanotechnol. 8, 634–638.

Jones, A.M., Yu, H., Schaibley, J.R., Yan, J., Mandrus, D.G., Taniguchi, T., Watanabe, K., Dery, H., Yao, W., Xu, X., 2016. Excitonic luminescence upconversion in a two-dimensional semiconductor. Nat. Phys. 12, 323–327.

Karni, O., Barr, E., Lau, S.C., Gillen, R., Ma, E.Y., Kim, B., Watanabe, K., Taniguchi, T., Maultzsch, J., Barmak, K., Page, R.H., Heinz, T.F., 2019. Infrared interlayer exciton emission in MoS2/WSe2 heterostructures. Phys. Rev. Lett. 123, 247402.

Koperski, M., Nogajewski, K., Arora, A., Cherkez, V., Mallet, P., Veuillen, J.-Y., Marcus, J., Kossacki, P., Potemski, M., 2015. Single photon emitters in exfoliated WSe2 structures. Nat. Nanotechnol. 10, 503–506.

Kormányos, A., Burkard, G., Gmitra, M., Fabian, J., ZoLyomi, V., Drummond, N.D., Fal'ko, V., 2015. k·p theory for two-dimensional transition metal dichalcogenide semiconductors. 2D Mater. 2, 022001.

Kośmider, K., González, J.W., Fernández-Rossier, J., 2013. Large spin splitting in the conduction band of transition metal dichalcogenide monolayers. Phys. Rev. B 88, 245436.
Kumar, S., Kaczmarczyk, A., Gerardot, B.D., 2015. Strain-induced spatial and spectral isolation of quantum emitters in mono- and bilayer WSe2. Nano Lett. 15, 7567–7573.
Li, L., Wu, M., 2017. Binary compound bilayer and multilayer with vertical polarizations: two-dimensional ferroelectrics, multiferroics, and nanogenerators. ACS Nano 11, 6382–6388.
Li, Y., Ludwig, J., Low, T., Chernikov, A., Cui, X., Arefe, G., Kim, Y.D., van der Zande, A.M., Rigosi, A., Hill, H.M., Kim, S.H., Hone, J., Li, Z., Smirnov, D., Heinz, T.F., 2014. Valley splitting and polarization by the Zeeman effect in monolayer MoSe2. Phys. Rev. Lett. 113, 266804.
Li, Z., Wang, T., Jin, C., Lu, Z., Lian, Z., Meng, Y., Blei, M., Gao, M., Taniguchi, T., Watanabe, K., Ren, T., Cao, T., Tongay, S., Smirnov, D., Zhang, L., Shi, S.-F., 2019a. Momentum-dark intervalley exciton in monolayer tungsten diselenide brightened via chiral phonon. ACS Nano 13, 14107–14113.
Li, Z., Wang, T., Jin, C., Lu, Z., Lian, Z., Meng, Y., Blei, M., Gao, S., Taniguchi, T., Watanabe, K., Ren, T., Tongay, S., Yang, L., Smirnov, D., Cao, T., Shi, S.-F., 2019b. Emerging photoluminescence from the dark-exciton phonon replica in monolayer WSe2. Nat. Commun. 10, 2469.
Liu, G.-B., Shan, W.-Y., Yao, Y., Yao, W., Xiao, D., 2013. Three-band tight-binding model for monolayers of group-VIB transition metal dichalcogenides. Phys. Rev. B 88, 085433.
Liu, K., Zhang, L., Cao, T., Jin, C., Qiu, D., Zhou, Q., Zettl, A., Yang, P., Louie, S.G., Wang, F., 2014. Evolution of interlayer coupling in twisted molybdenum disulfide bilayers. Nat. Commun. 5, 4966.
Liu, G.-B., Xiao, D., Yao, Y., Xu, X., Yao, W., 2015. Electronic structures and theoretical modelling of two-dimensional group-Vib transition metal dichalcogenides. Chem. Soc. Rev. 44, 2643–2663.
Liu, E., Baren, J.V., Taniguchi, T., Watanabe, K., Chang, Y.-C., Lui, C.H., 2019. Valley-selective chiral phonon replicas of dark excitons and trions in monolayer WSe2. Phys. Rev. Res. 1, 032007.
MacNeill, D., Heikes, C., Mak, K.F., Anderson, Z., Kormányos, A., Zólyomi, V., Park, J., Ralph, D.C., 2015. Breaking of valley degeneracy by magnetic field in monolayer MoSe2. Phys. Rev. Lett. 114, 037401.
Maialle, M.Z., de Andrada e Silva, E.A., Sham, L.J., 1993. Exciton spin dynamics in quantum wells. Phys. Rev. B 47, 15776.
Mak, K.F., Lee, C., Hone, J., Shan, J., Heinz, T.F., 2010. Atomically thin MoS2: a new direct-gap semiconductor. Phys. Rev. Lett. 105, 136805.
Mak, K.F., He, K., Shan, J., Heinz, T.F., 2012. Control of valley polarization in monolayer MoS2 by optical helicity. Nat. Nanotechnol. 7, 494–498.
Mak, K.F., He, K., Lee, C., Lee, G.H., Hone, J., Heinz, T.F., Shan, J., 2013. Tightly bound trions in monolayer MoS2. Nat. Mater. 12, 207–211.
Marinov, K., Avsar, A., Watanabe, K., Taniguchi, T., Kis, A., 2017. Resolving the spin splitting in the conduction band of monolayer MoS2. Nat. Commun. 8, 1938.
Miller, B., Steinhoff, A., Pano, B., Klein, J., Jahnke, F., Holleitner, A., Wurstbauer, U., 2017. Long-lived direct and indirect interlayer excitons in van der Waals heterostructures. Nano Lett. 17, 5229–5237.
Mitioglu, A.A., Plochocka, P., Jadczak, J.N., Escoffier, W., Rikken, G.L.J.A., Kulyuk, L., Maude, D.K., 2013. Optical manipulation of the exciton charge state in single-layer tungsten disulfide. Phys. Rev. B 88, 245403.
Moody, G., Dass, C.K., Hao, K., Chen, C.-H., Li, L.-J., Singh, A., Tran, K., Clark, G., Xu, X., Berghäuser, G., Malic, E., Knorr, A., Li, X., 2015. Intrinsic homogeneous linewidth and broadening mechanisms of excitons in monolayer transition metal dichalcogenides. Nat. Commun. 6, 8315.

Özçelik, V.O., Azadani, J.G., Yang, C., Koester, S.J., Low, T., 2016. Band alignment of two-dimensional semiconductors for designing heterostructures with momentum space matching. Phys. Rev. B 94, 035125.
Pan, Y., Fölsch, S., Nie, Y., Waters, D., Lin, Y.-C., Jariwala, B., Zhang, K., Cho, K., Robinson, J.A., Feenstra, R.M., 2018. Quantum-confined electronic states arising from the Moiré Pattern of MoS2-WSe2 heterobilayers. Nano Lett. 18, 1849–1855.
Park, K.-D., Jiang, T., Clark, G., Xu, X., Raschke, M.B., 2018. Radiative control of dark excitons at room temperature by nano-optical antenna-tip Purcell effect. Nat. Nanotechnol. 13, 59–64.
Qiu, D.Y., Jornada, F.H.D., Louie, S.G., 2013. Optical spectrum of MoS2: many-body effects and diversity of exciton states. Phys. Rev. Lett. 111, 216805.
Qiu, D.Y., Cao, T., Louie, S.G., 2015. Nonanalyticity, valley quantum phases, and lightlike exciton dispersion in monolayer transition metal dichalcogenides: theory and first-principles calculations. Phys. Rev. Lett. 115, 176801.
Raja, A., Chaves, A., Yu, J., Arefe, G., Hill, H.M., Rigosi, A.F., Berkelbach, T.C., Nagler, P., Schüller, C., Korn, T., Nuckolls, C., Hone, J., Brus, L.E., Heinz, T.F., Reichman, D.R., Chernikov, A., 2017. Coulomb engineering of the bandgap and excitons in two-dimensional materials. Nat. Commun. 8, 15251.
Rivera, P., Schaibley, J.R., Jones, A.M., Ross, J.S., Wu, S., Aivazian, G., Klement, P., Seyler, K., Clark, G., Ghimire, N.J., Yan, J., Mandrus, D.G., Yao, W., Xu, X., 2015. Observation of long-lived interlayer excitons in monolayer MoSe2-WSe2 heterostructures. Nat. Commun. 6, 6242.
Rivera, P., Seyler, K.L., Yu, H., Schaibley, J.R., Yan, J., Mandrus, D.G., Yao, W., Xu, X., 2016. Valley-polarized exciton dynamics in a 2D semiconductor heterostructure. Science 351, 688–691.
Rivera, P., Yu, H., Seyler, K.L., Wilson, N.P., Yao, W., Xu, X., 2018. Interlayer valley excitons in heterobilayers of transition metal dichalcogenides. Nat. Nanotechnol. 13, 1004–1015.
Rosenberger, M.R., Chuang, H.-J., Phillips, M., Oleshko, V.P., Mccreary, K.M., Sivaram, S.V., Hellberg, C.S., Jonker, B.T., 2020. Twist angle-dependent atomic reconstruction and Moiré patterns in transition metal dichalcogenide heterostructures. ACS Nano 14, 4550–4558.
Ross, J.S., Wu, S., Yu, H., Ghimire, N.J., Jones, A.M., Aivazian, G., Yan, J., Mandrus, D.G., Xiao, D., Yao, W., Xu, X., 2013. Electrical control of neutral and charged excitons in a monolayer semiconductor. Nat. Commun. 4, 1474.
Ross, J.S., Rivera, P., Schaibley, J., Lee-Wong, E., Yu, H., Taniguchi, T., Watanabe, K., Yan, J., Mandrus, D., Cobden, D., Yao, W., Xu, X., 2017. Interlayer exciton optoelectronics in a 2D heterostructure p–n junction. Nano Lett. 17, 638–643.
Sallen, G., Bouet, L., Marie, X., Wang, G., Zhu, C.R., Han, W.P., Lu, Y., Tan, P.H., Amand, T., Liu, B.L., Urbaszek, B., 2012. Robust optical emission polarization in MoS2 monolayers through selective valley excitation. Phys. Rev. B 86. 081301(R).
Schaibley, J.R., Rivera, P., Yu, H., Seyler, K.L., Yan, J., Mandrus, D.G., Taniguchi, T., Watanabe, K., Yao, W., Xu, X., 2016. Directional interlayer spin-valley transfer in two-dimensional heterostructures. Nat. Commun. 7, 13747.
Seyler, K.L., Rivera, P., Yu, H., Wilson, N.P., Ray, E.L., Mandrus, D.G., Yan, J., Yao, W., Xu, X., 2019. Signatures of moiré-trapped valley excitons in MoSe2/WSe2 heterobilayers. Nature 567, 66–70.
Shi, H., Pan, H., Zhang, Y.-W., Yakobson, B.I., 2013. Quasiparticle band structures and optical properties of strained monolayer MoS2 and WS2. Phys. Rev. B 87, 155304.
Splendiani, A., Sun, L., Zhang, Y., Li, T., Kim, J., Chim, C.-Y., Galli, G., Wang, F., 2010. Emerging photoluminescence in monolayer MoS2. Nano Lett. 10, 1271–1275.
Srivastava, A., Sidler, M., Allain, A.V., Lembke, D.S., Kis, A., Imamoğlu, A., 2015a. Optically active quantum dots in monolayer WSe2. Nat. Nanotechnol. 10, 491–496.

Srivastava, A., Sidler, M., Allain, A.V., Lembke, D.S., Kis, A., Imamoğlu, A., 2015b. Valley zeeman effect in elementary optical excitations of a monolayer WSe2. Nat. Phys. 11, 141–147.

Stier, A.V., McCreary, K.M., Jonker, B.T., Kono, J., Crooker, S.A., 2016. Exciton diamagnetic shifts and valley Zeeman effects in monolayer WS2 and MoS2 to 65 Telsa. Nat. Commun. 7, 10643.

Tonndorf, P., Schmidt, R., Schneider, R., Kern, J., Buscema, M., Steele, G.A., Castellanos-Gomez, A., van der Zant, H.S.J., de Vasconcellos, S.M., Bratschitsch, R., 2015. Single-photon emission from localized excitons in an atomically thin semiconductor. Optica 2, 347–352.

Tran, K., Moody, G., Wu, F., Lu, X., Choi, J., Kim, K., Rai, A., Sanchez, D.A., Quan, J., Singh, A., Embley, J., Zepeda, A., Campbell, M., Autry, T., Taniguchi, T., Watanabe, K., Lu, N., Banerjee, S.K., Silverman, K.L., Kim, S., Tutuc, E., Yang, L., Macdonald, A.H., Li, X., 2019. Evidence for moiré excitons in van der Waals heterostructures. Nature 567, 71–75.

Ugeda, M.M., Bradley, A.J., Shi, S.-F., Jornada, F.H.D., Zhang, Y., Qiu, D.Y., Ruan, W., Mo, S.-K., Hussain, Z., Shen, Z.-X., Wang, F., Louie, S.G., Crommie, M.F., 2014. Giant bandgap renormalization and excitonic effects in a monolayer transition metal dichalcogenide semiconductor. Nat. Mater. 13, 1091–1095.

Unuchek, D., Ciarrocchi, A., Avsar, A., Watanabe, K., Taniguchi, T., Kis, A., 2018. Room-temperature electrical control of exciton flux in a van der Waals heterostructure. Nature 560, 340–344.

Unuchek, D., Ciarrocchi, A., Avsar, A., Sun, Z., Watanabe, K., Taniguchi, T., Kis, A., 2019. Valley-polarized exciton currents in a van der Waals heterostructure. Nat. Nanotechnol. 14, 1104–1109.

van der Zande, A.M., Kunstmann, J., Chernikov, A., Chenet, D.A., You, Y., Zhang, X., Huang, P.Y., Berkelbach, T.C., Wang, L., Zhang, F., Hybertsen, M.S., Muller, D.A., Reichman, D.R., Heinz, T.F., Hone, J.C., 2014. Tailoring the electronic structure in bilayer molybdenum disulfide via interlayer twist. Nano Lett. 14, 3869–3875.

Wang, G., Bouet, L., Lagarde, D., Vidal, M., Balocchi, A., Amand, T., Marie, X., Urbaszek, B., 2014. Valley dynamics probed through charged and neutral exciton emission in monolayer WSe2. Phys. Rev. B 90, 075413.

Wang, G., Marie, X., Gerber, I., Amand, T., Lagarde, D., Bouet, L., Vidal, M., Balocchi, A., Urbaszek, B., 2015. Giant enhancement of the optical second-harmonic emission of WSe2 monolayers by laser excitation at exciton resonances. Phys. Rev. Lett. 114, 097403.

Wang, Z., Shan, J., Mak, K.F., 2016. Valley- and spin-polarized Landau levels in monolayer WSe2. Nat. Nanotechnol. 12, 144–149.

Wang, G., Robert, C., Glazov, M.M., Cadiz, F., Courtade, E., Amand, T., Lagarde, D., Taniguchi, T., Watanabe, K., Urbaszek, B., Marie, X., 2017a. In-plane propagation of light in transition metal dichalcogenide monolayers: optical selection rules. Phys. Rev. Lett. 119, 047401.

Wang, Y., Wang, Z., Yao, W., Liu, G.-B., Yu, H., 2017b. Interlayer coupling in commensurate and incommensurate bilayer structures of transition metal dichalcogenides. Phys. Rev. B 95, 115429.

Wang, T., Miao, S., Li, Z., Meng, Y., Lu, Z., Lian, Z., Blei, M., Taniguchi, T., Watanabe, K., Tongay, S., Smirnov, D., Shi, S.-F., 2019. Giant Valley-Zeeman splitting from spin-singlet and spin-triplet interlayer excitons in WSe2/MoSe2 heterostructure. Nano Lett. 20, 694–700.

Wilson, N.R., Nguyen, P.V., Seyler, K.L., Rivera, P., Marsden, A.J., Laker, Z.P.L., Constantinescu, G.C., Kandyba, V., Barinov, A., Hine, N.D.M., Xu, X., Cobden, D.H., 2017. Determination of band offsets, hybridization, and exciton binding in 2D semiconductor heterostructures. Sci. Adv. 3, e1601832.

Wu, F., Qu, F., Macdonald, A.H., 2015. Exciton band structure of monolayer MoS2. Phys. Rev. B 91, 075310.
Wu, F., Lovorn, T., Macdonald, A.H., 2017. Topological exciton bands in Moiré heterojunctions. Phys. Rev. Lett. 118, 147401.
Wu, F., Lovorn, T., Macdonald, A.H., 2018. Theory of optical absorption by interlayer excitons in transition metal dichalcogenide heterobilayers. Phys. Rev. B 97, 035306.
Wu, F., Lovorn, T., Tutuc, E., Martin, I., Macdonald, A.H., 2019. Topological insulators in twisted transition metal dichalcogenide homobilayers. Phys. Rev. Lett. 122, 086402.
Xiao, D., Liu, G.-B., Feng, W., Xu, X., Yao, W., 2012. Coupled spin and valley physics in monolayers of MoS2 and other Group-VI dichalcogenides. Phys. Rev. Lett. 108, 196802.
Xu, X., Yao, W., Xiao, D., Heinz, T.F., 2014. Spin and pseudospins in layered transition metal dichalcogenides. Nat. Phys. 10, 343–350.
Ye, Z., Cao, T., O'brien, K., Zhu, H., Yin, X., Wang, Y., Louie, S.G., Zhang, X., 2014. Probing excitonic dark states in single-layer tungsten disulphide. Nature 513, 214–218.
Yu, H., Liu, G.-B., Gong, P., Xu, X., Yao, W., 2014. Dirac cones and Dirac saddle points of bright excitons in monolayer transition metal dichalcogenides. Nat. Commun. 5, 3876.
Yu, H., Cui, X., Xu, X., Yao, W., 2015a. Valley excitons in two-dimensional semiconductors. Natl. Sci. Rev. 2, 57–70.
Yu, H., Wang, Y., Tong, Q., Xu, X., Yao, W., 2015b. Anomalous light cones and valley optical selection rules of interlayer excitons in twisted heterobilayers. Phys. Rev. Lett. 115, 187002.
Yu, H., Liu, G.-B., Tang, J., Xu, X., Yao, W., 2017. Moiré excitons: from programmable quantum emitter arrays to spin-orbit-coupled artificial lattices. Sci. Adv. 3, e1701696.
Yu, H., Liu, G.-B., Yao, W., 2018. Brightened spin-triplet interlayer excitons and optical selection rules in van der Waals heterobilayers. 2D Mater. 5, 035021.
Yu, H., Chen, M., Yao, W., 2020. Giant magnetic field from moiré induced Berry phase in homobilayer semiconductors. Natl. Sci. Rev. 7, 12–20.
Zeng, H., Dai, J., Yao, W., Xiao, D., Cui, X., 2012. Valley polarization in MoS2 monolayers by optical pumping. Nat. Nanotechnol. 7, 490–493.
Zhang, C., Johnson, A., Hsu, C.-L., Li, L.-J., Shih, C.-K., 2014. Direct imaging of band profile in single layer MoS2 on graphite: quasiparticle energy gap, metallic edge states, and edge band bending. Nano Lett. 14, 2443–2447.
Zhang, C., Chuu, C.-P., Ren, X., Li, M.-Y., Li, L.-J., Jin, C., Chou, M.-Y., Shih, C.-K., 2017. Interlayer couplings, Moiré patterns, and 2D electronic superlattices in MoS2/WSe2 hetero-bilayers. Sci. Adv. 3, e1601459.
Zhou, Y., Scuri, G., Wild, D.S., High, A.A., Dibos, A., Jauregui, L.A., Shu, C., Greve, K.D., Pistunova, K., Joe, A.Y., Taniguchi, T., Watanabe, K., Kim, P., Lukin, M.D., Park, H., 2017. Probing dark excitons in atomically thin semiconductors via near-field coupling to surface plasmon polaritons. Nat. Nanotechnol. 12, 856–860.
Zhu, B., Chen, X., Cui, X., 2015. Exciton binding energy of monolayer WS2. Sci. Rep. 5, 9218.

CHAPTER SEVEN

Quantum light from optically dressed quantum dot states in microcavities

Glenn S. Solomon[a,*], Andreas Muller[b], and Edward Flagg[c]
[a]Joint Quantum Institute, National Institute of Standards and Technology & University of Maryland, College Park, MD, United States
[b]Department of Physics, University of South Florida, Tampa, FL, United States
[c]Department of Physics and Astronomy, West Virginia University, Morgantown, WV, United States
[*]Corresponding author: e-mail address: gsolomon@umd.edu

Contents

1. Introduction

Resonant or near-resonant excitation of an atomic transition is a bedrock concept in atomic, molecular, and optical (AMO) physics. It is fundamentally important in the engineering of quantum states in AMO and in many applications in quantum information technologies. Any physical realization of a quantum bit (qubit) requires at minimum a series of single-qubit operations, including quantum-state initialization, manipulation, and read-out (DiVincenzo, 2000). In AMO systems such as atoms and ions, this is most often achieved through resonant or near-resonant laser

Semiconductors and Semimetals, Volume 105
ISSN 0080-8784
https://doi.org/10.1016/bs.semsem.2020.10.004

operations such as state initialization, stimulated Raman adiabatic passage (STIRAP) and adiabatic population transfer, and single-shot read-out.

In resonant or near-resonant excitation, the physics of the light–matter interaction can be simplistically broken into several regimes based on the excitation (pump) power and the transition dipole strength. When the two-level system is excited by low-intensity (monochromatic) laser light, the semiclassical theory (Heitler, 1954) predicts that the fluorescent light is also monochromatic and coherent with the driving laser field. This is often called the Rayleigh or Heitler regime. As the pump power is further increased, the two-level system upper state population becomes nonzero, and the emission includes contributions from population decay. This is resonance fluorescence, and the light is no longer entirely coherent with the driving field; becoming less so as the driving field increases.

As the resonant or near-resonant monochromatic field intensity increases still further, new physics results and the fluorescence spectrum associated with the two-level transition splits from a single-line into a spectrum consisting of a doublet or triplet. In 1955 Autler and Townes showed experimentally, and provided approximate theoretical solutions with a classical drive field, that a molecular microwave transition splits into a doublet when one of the two levels in the transition is coupled to a third level through a strong and resonant microwave field (Autler and Townes, 1955). Mollow showed theoretically in 1969 using optical Bloch equations that at high resonant field intensity the fluorescence spectrum of the two-level system consists of a triplet (Mollow, 1969). The strong laser field creates mixed, light–matter states, often called "dressed states" in the parlance (Cohen-Tannoudji and Reynaud, 1977). When the dressing field is far detuned, the mixed-state doublet become approximately matter-like and light-like, yet some energy shift remains in the transition energy, called the AC Stark shift. A schematic energy diagram is shown in Fig. 1.

In this chapter, we investigate the strong-drive field processes in a semiconductor quantum dot (QD) in a microcavity. A microcavity creates enhanced coupling between the QD optical transitions and the collected modes of the electromagnetic field into which emission occurs. We are generally interested here in using the dressing of QD states by the strong coherent field to manipulate the energy spectrum of the QD emission. Thus, we do not collect the resonance fluorescence from this process, but apply an additional laser to measure photoluminescence (PL). This is similar to the Aulter-Townes 1955 experiment except they used, by necessity, a weak

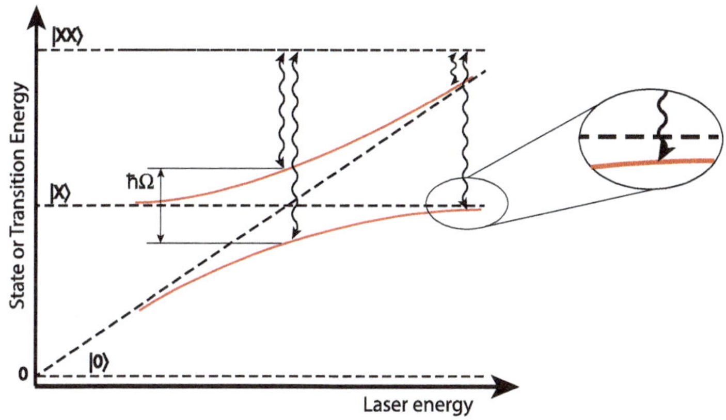

Fig. 1 Schematic of the transition energies in a three-level system with a strong driving laser field. The three uncoupled states, $|0\rangle$, $|X\rangle$, and $|XX\rangle$, are indicated by horizontal dashed lines, where the indices are the matter state quantum number, and here $|0\rangle$ is the vacuum state. The diagonal dashed line represents an additional photon from the laser field. The dressed states are in red. When $|XX\rangle$ radiatively decays to each of the dressed states an emission doublet results (wavy vertical lines). $\hbar\Omega$, where Ω is the Rabi frequency, is the dressed state energy splitting when the laser is on-resonance with the bare-state transition. The far detuned, AC-Stark shift is the small residual energy shift as the dressed states asymptotically approach the bare state, $|X\rangle$ at large laser detuning from the $|0\rangle-|X\rangle$ transition energy (an example is highlighted on right).

resonant probe field between the upper molecular level of interest and a state not coupled to the strong microwave driving field. Because our QD is a solid-state system, there is a continuous distribution of available conduction and valence states in the GaAs host crystal allowing for nonresonant probe excitation, and fast relaxation processes to the QD. However, as we discovered in these experiments, such relaxation processes lead to a time distribution in both the occupation of the state of interest, due to inherent relaxation rates and to different relaxation paths. This leads to distinguishability in photon emission (Flagg et al., 2012).

We begin with some background on semiconductor QDs and the microcavity device. Then we discuss three experiments. In the first, a laser dresses one QD optical transition in a set (manifold) of neutral quasiparticle states: the lowest energy, optically active neutral manifold containing a single biexciton, two excitons, and the vacuum state. We observe the detuning features on dressed transitions, and other transitions in the manifold using weak, above-band light. Despite the strong dressing field, the radiative decay of a higher energy transition connected to the dressed transition still produces single-photon (antibunched) emission. In a second experiment we use a far-detuned laser field in the AC Stark regime to restore the degeneracy of the two intermediate exciton transitions in biexciton–exciton manifold, producing entangled emission. Finally, in a third experiment we measure the degree of indistinguishability of single photons produced in two different QD microcavity samples. Here we use a strain field instead of an optical field to modify the transition energies. The chapter concludes with a theoretical proposal to use the AC Stark effect (Flagg and Solomon, 2015) to create a cycling transition in the QD system, an important component in the single-shot read-out of a qubit.

1.1 The quantum dot microcavity and optical pumping schemes

The quantum dot (QD) used here is made using epitaxial crystal growth by a strain-driven process due to the 7% lattice mismatch strain between InAs and the GaAs substrate. Typically, molecular-beam epitaxy is used for the epitaxy process, but metal-organic vapor-phase epitaxy can also be used. The InAs QD is a flattened dome-like structure with in-plane diameter on the order of 30 nm and vertical height of 3–5 nm, resulting in a significantly larger vertical confinement than in-plane confinement. The QD contains tens of thousands of atoms with an electronic structure of discrete states. This discrete-state physics is markedly different from energy- and crystal momentum-dependent bandstructures typical of a bulk semiconductor, or even 1D and 2D quantum-confined semiconductors which are characterized by a continuous density of states within an energy band. A charge-neutral, lowest-energy manifold contains a set of, optically allowed (dipole-allowed) transitions. This ground-state manifold contains one or two pairs of electrons ($\uparrow$, $\downarrow$) and holes ($\Uparrow$, $\Downarrow$) and is schematically shown in Fig. 2. Here, $|0\rangle$ is the vacuum state, and $\uparrow$, $\downarrow$; $\Uparrow$, $\Downarrow$ represent the two possible orthogonal spins of electrons in the lowest-level conduction states and pseudo-spins of holes in the highest-level valence states in the single particle picture. Here, $|X\rangle$ is the optically active or bright exciton state, a Coulomb-bound electron and hole with total spin ± 1. In this ground-state manifold there is one biexciton state, $|XX\rangle$ containing each spin configuration of electrons and holes. There are Coulomb interactions between all particles of the biexciton, which result in this system in a lower energy state than one formed from the two individual single exciton states. Two charge-neutral, dark exciton states ($|\Uparrow\uparrow\rangle$, $|\Downarrow\downarrow\rangle$) are also present at lower energies than the bright exciton states but are not

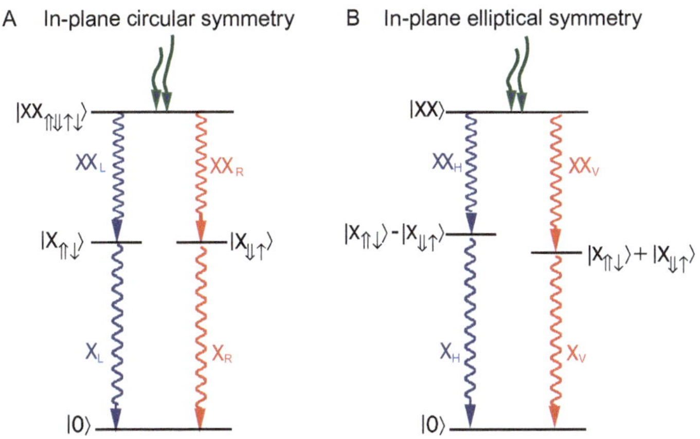

Fig. 2 Schematic of biexciton, $|XX\rangle$—exciton, $|X\rangle$ cascade when the quantum dot has circular in-plane symmetry (A) and elliptical in-plane symmetry (B). $\Downarrow$, $\Uparrow$ represent single hole valence states of opposite pseudo spin, while $\downarrow$, $\uparrow$ represent single electron conduction states of opposite spin. $|XX\rangle = |XX_{\Uparrow\Downarrow\uparrow\downarrow}\rangle$. L, R are left and right optical polarizations, and H, V are horizontal and vertical polarizations. Green downward arrows at the figure's top represent nonresonant pumping of $|XX\rangle$.

shown in Fig. 2. Transitions from $|XX\rangle$ to these dark states are dipole forbidden; they cannot be directly populated from the $|XX\rangle$ without a spin flip. Finally, charge exciton states are also present and are discussed in the final section of the chapter. However, their transitions are independent of those in the neutral ground-state manifold if charge conservation is maintained. See Poem and Gershoni (2011) for more details.

With circular in-plane symmetry the two bright exciton states are degenerate as in Fig. 2A, and the emitted photons are circularly polarized. However, because of the crystal structure and epitaxial growth process the in-plane symmetry is almost always elliptical, which mixes the exciton states, lifting their energy degeneracy, and making the optical transitions linearly polarized (Fig. 2B). Nonresonant excitation into conduction band states results in a variety of ground-state manifold populations, including charged states not discussed here, and depends on many variables. If the pump power is low with respect to the exciton decay rate, $|X\rangle$ and electrons and holes decaying to the ground-state manifold have populations at or below unity, the $|X\rangle$ states will be populated without decay through $|XX\rangle$. However, if the pump power is increased to a point where on average two electrons and holes accumulate in the QD, $|XX\rangle$ will be populated and decay through $|X\rangle$ to $|0\rangle$ in the cascade represented in Fig. 2. See Santori et al. (2002a) for details.

The samples used here contains a dilute ensemble of self-assembled InAs QDs (few per μm^2), unintentionally alloyed with GaAs through diffusion, and are grown at the center of a planar optical microcavity, all through molecular-beam epitaxy. The microcavity consists of two distributed Bragg reflectors (15.5 lower and 10 upper AlAs/GaAs quarter-wave pairs) and a full-wave GaAs spacer (see Fig. 3A). While forming pillar structures of a few micrometers in diameter creates discrete cavity modes and enhanced emission (Solomon et al., 2001), planar microcavities and ridge waveguides formed from these microcavities also have advantages. In particular, both allow for resonant or near-resonant side excitation via the waveguide mode, while the much weaker quantum light at the single photon level is

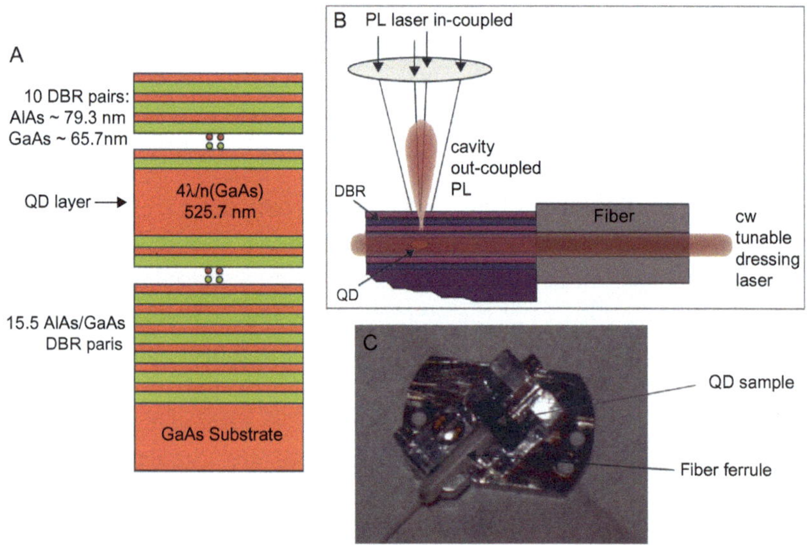

Fig. 3 (A) The quantum dot (QD) microcavity structure is schematically shown. An upper set of distributed-Bragg reflectors (DBR) of $\frac{1}{4}\lambda$ thick alternating layers of GaAs and AlAs is epitaxially deposited on a cavity region of GaAs. A bottom DBR layer completes the microcavity. λ is the cavity resonance, n is the index of refraction. A dilute layer of QDs is epitaxially deposited in the center of the cavity region. (B) Light can be coupled in and out of the microcavity from either above into the well-formed Gaussian microcavity mode or laterally through a planar waveguide mode. QD emission is predominately vertical due to stronger optical transitions and an enhancement due to the confined electric field of the cavity (Purcell enhancement). Here, the dressing laser is fiber coupled to the lateral waveguide mode and single QD emission is predominately coupled vertically, off-chip. (C) An image of the fiber-coupled QD microcavity device. Here, the fiber is glued to the side of the chip (black). The metal base mount is ≈2.5 cm across.

preferentially emitted vertically into the microcavity mode. This was first demonstrated by the Shih/Deppe group (Muller et al., 2007) in 2007, and we use it here in various configurations. Fig. 3B is a schematic of the device, where a tunable cw laser is used for resonant fluorescence or optical dressing of a quantum dot state. Here, an optical fiber is fixed to the side of the device coupling light into the waveguide mode of the microcavity. Fig. 3C is image of a sample used, where a fiber is glued to the side the microcavity sample.

All the experiments described in this chapter were made in a liquid He-bath cryostat at a constant base temperature near 4.2 K. The QD ensemble emission spectrum spans ≈880 to 990 nm. The cavity mode of interest is located at ≈920 nm.

2. Controlling the emission properties using optical dressing

Here we show the full dressing of an exciton–biexciton neutral complex in a single semiconductor QD in an experiment revealing the unique features of a dressed four-level system in emission. These include Mollow triplets and Autler–Townes doublets in both exciton and biexciton transitions and their avoided crossing spectra as a function of detuning. A pump laser at a frequency above the GaAs band-edge is used to generate photoluminescence (PL) from a single QD that is simultaneously dressed by a near-resonant control laser. This results in a unique, and to our knowledge unexplored approach, in which the above-band laser serves primarily to populate the levels that have been tailored by the control laser. We show that, unlike in pump–probe techniques, the two lasers are largely independent in the sense that the pump laser frequency or amplitude (that can be large) does not directly affect the shape of the emission spectrum. In particular, when applying a pulsed pump laser, we can obtain a pulsed spectrum that remains dressed by the control laser (Muller et al., 2008).

Some quick details about this particular experiment. The cavity mode of interest couples the QD emission to a 0.6-numerical aperture fiber-coupled microscope objective (probe area ≈1 μm^2). This objective was also used for above-band excitation with either a 633 nm HeNe laser or a ps-pulsed mode-locked Ti:sapphire laser at 730 nm. In contrast, the near-resonant control laser, a tunable cw Ti:sapphire ring laser, was introduced laterally in the waveguide mode of the cavity so as to minimize stray laser scattering into the collection mode (Muller et al., 2007) in a geometry schematically shown in Fig. 3B. This was achieved by bonding a single mode fiber directly to the cleaved edge ($[\bar{1}10]$) of the sample resulting in a compact and durable package. Optical measurements relied on a scanning Fabry-Pérot (FP) interferometer with a free spectral range of 17 GHz and resolution of ≈150 MHz, in series with a grating spectrometer. The PL signal from the sample was first resolved with the spectrometer without the FP, after which a selected spectral window was directed toward a single-photon counting detector with the FP inserted into the collection path.

With such a combination we resolve the full spectrum of a single QD, particularly its fine structure. For the neutral ground-state excitons, the well-known fine structure consists of a linear polarization doublet ($|X_H\rangle$, $|X_V\rangle$) split by δ_{HV} due to the anisotropic exchange interaction (Bayer et al., 2002; Gammon et al., 1996). "H" and "V" denote the two orthogonal in-plane polarization directions of these states. While we measure the relative orientation between polarizations we cannot determine their absolute angle to the sample because of our fiber-based system. The neutral bound exciton pair ($|XX\rangle$), the biexciton, has a two-path radiative cascade to the exciton vacuum ($|0\rangle$) through the two possible $|X_{H/V}\rangle$ states (Fig. 4A). Each path

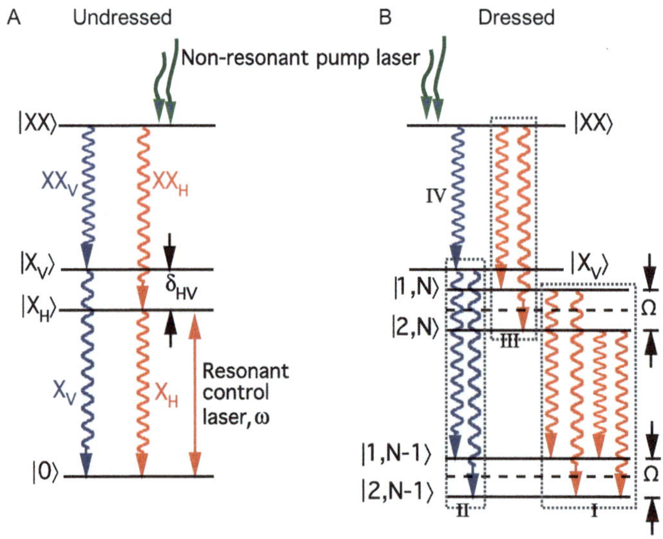

Fig. 4 Energy schematic and the allowed transitions in the bare (A), and X_H dressed (B) exciton–biexciton complex. Dashed box I contains the triplet emission while II and III are Autler–Townes doublets. IV is unchanged. Several repeated levels in the ladder have been omitted for clarity.

produces photon pairs of identical polarization XX_H/X_H and XX_V/X_V with an additional energy difference between $XX_{H/V}$ and $X_{H/V}$ due to the biexciton binding energy, here $E_B/\hbar \approx$ 650 GHz (2.6 meV) (Akopian et al., 2006). For the QD studied, we measured excitonic linewidths of $\approx$1 GHz (4 μeV) and a splitting of δ_{HV}=2.75 GHz (11 μeV). The exciton and biexciton emission was identified on the basis of intensity-dependent PL measurements in which a square law dependence was observed as anticipated for the biexcitonic emission.

We now apply the resonant control laser nominally polarized along $[\bar{1}10]$. We initially assume that the laser polarization is along $[\bar{1}10]$ and H so that it couples only to X_H. Then, the unperturbed states $|0\rangle \otimes |N+1\rangle$ and $|X_H\rangle \otimes |N\rangle$ are dressed into pairs of hybrid QD/field states $|1(N)\rangle$ and $|2(N)\rangle$ (Cohen-Tannoudji et al., 1992). N represents the photon number of the applied field of frequency ω. The splitting between the two states equals $\Omega' = \sqrt{\Omega^2 + \delta^2}$ in the presence of laser detuning $\delta = \omega - \omega_0$; ω_0 is the resonance frequency of the unperturbed transition, and Ω is the Rabi coupling of the resonant laser to X_H. Transitions between adjacent photon number manifolds give rise to a resonant emission spectrum consisting of a central line at ω and side-bands at $\omega \pm \Omega'$. This is illustrated in Fig. 4B, where the control laser, originally resonant with X_H creates four H-polarized transitions (two of which are degenerate), represented in the box labeled "I." In addition, Autler–Townes type splittings are expected from all transitions that share a common state with the states to which the laser couples. In Fig. 1B these are labeled "II" (V-polarized) and "III" (H-polarized).

Experimental results as a function of Rabi frequency are shown in Fig. 5. Detailed exciton emission lineshapes are displayed in Fig. 2A–C at $\Omega =$ (0, 1.26, 2.35) GHz, respectively, for both H and V polarizations. Note that the FP frequency (abscissas) is relative to the frequency of X_H in the excitonic transitions and relative to the frequency of XX_H in the biexcitonic transitions. As the Rabi frequency increases (by increasing the control laser intensity), X_H indeed splits into a triplet (Fig. 2D) according to the illustration in Fig. 1B. In Fig. 5E and F X_V and XX_H split into doublets. The remaining biexciton transition, XX_V, is unaffected at low Rabi frequency (Fig. 5G) as anticipated (IV in Fig. 4B).

At larger Rabi frequencies the observations deviate from the simple picture given in Fig. 4B. Additional features appear, that are, nonetheless, well described by including some coupling of the laser to X_V. This was not

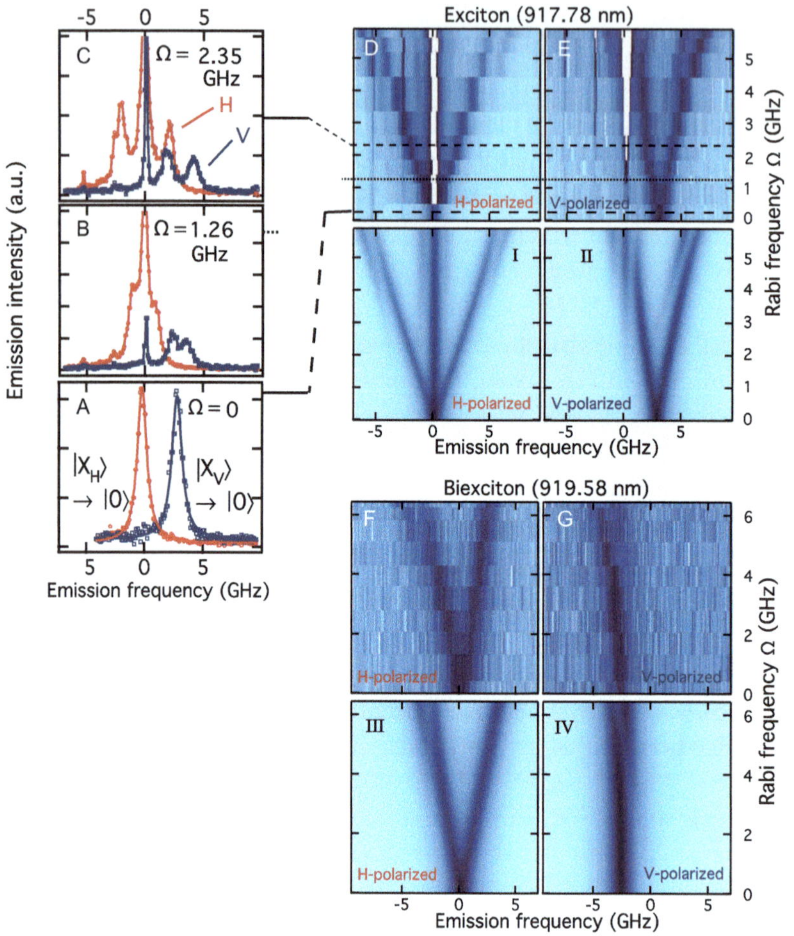

Fig. 5 Splittings of exciton and biexciton emission lines with variable Rabi frequency. Panels (A)–(C) show exciton emission spectra at specific Rabi frequencies, where for (A) at bottom left, $\Omega=0$ GHz and for (C) at top left, $\Omega=2.35$ GHz. Panels (D)–(G) show full color-scale maps of emission intensity vs emission (abscissas) and Rabi (ordinates) frequency. Panels below represent simulated data. Roman numerals refer to the boxed transitions in Fig. 4. In the simulation panels, transitions are numbered, where primes refer to the biexciton transitions. In (D) and (E), the clear regions around the laser frequency correspond to signal truncation due to excessive background laser scattering.

anticipated because the polarization doublets are usually oriented along [110] and $[\bar{1}10]$ and so the laser, polarized along $[\bar{1}10]$, would not couple to X_V. A full simulation based on a Hamiltonian with an interaction term written as

$$H_{int} = -\frac{\hbar}{2} \sum_{j=\mathrm{H,V}} \Omega_j |X_j, N\rangle\langle 0, N+1| + h.c.$$

can describe the results very well. For this particular QD, Ω_V/Ω_H=0.28. Since the coupling is related to the inner product of the field and optical dipole, either H is not completely aligned with $[\bar{1}10]$, the control field is significantly altered in the waveguide, or both. In this case, we obtain nine transitions (three dressed states in the N manifold connecting the three states in the $N-1$ manifold), and out of those, seven are nondegenerate (See Muller et al. (2008) for details).

The dressed states undergo avoided crossings with laser detuning, shown in Fig. 6 for a fixed Rabi frequency of 2.4 GHz. Fig. 6A shows the direct transitions (I), in which the avoided crossing separates the dressed states by an amount ($2\Omega'$), with a minimum equal to 2Ω at resonance. The Autler–Townes transitions (II and III) are shown in Fig. 6B and C and have splittings equal to Ω'. At large detuning, the optically active transitions asymptotically approach the original excitonic transitions, where the remaining finite energy offset is the AC Stark or light shift. The panels below each section in Fig. 6 show the result of simulations in excellent agreement with the data. Here, the linewidths were held constant and light shifts of $|XX\rangle$ due to the laser near $X_{H/V}$ were neglected. This is very reasonable because the biexcitonic binding energy is much larger than $\hbar\delta_{HV}$ and $\hbar\Omega$. As in the resonant case (Fig. 5) there are additional features due to the partial

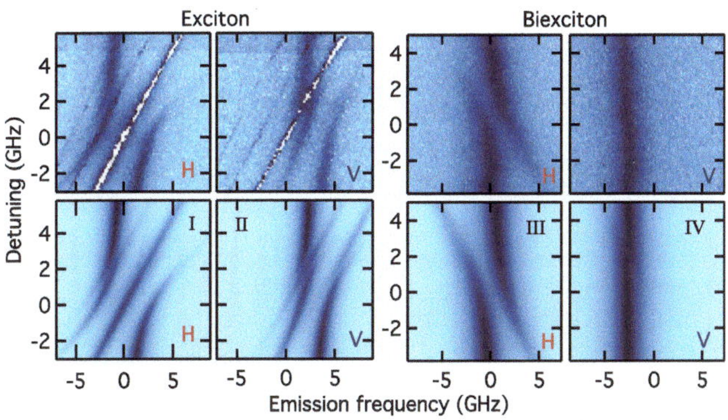

Fig. 6 Emission spectrum of exciton H and V transition, left and biexciton H and V transitions, right under detuning for a fixed Rabi frequency of 2.4 GHz. Panels below represent simulated data.

X_V coupling. For instance, when the laser is resonant with X_V then we observe a small triplet on X_V and an Autler–Townes doublet on X_H.

For convenience the measurements above were performed with a HeNe laser as an excitation source. However, as shown in Fig. 7, the above-band excitation can be pulsed as well, giving rise to triggered single photon emission for correlation measurements and on-demand sources. To simultaneously illustrate the capability to drive the biexciton rather than the exciton, the resonant laser is now applied to the XX_H transition (with the pump still on). The H-polarized emission from transitions to the ground state is then sent to a Hanbury-Brown-Twiss setup for second-order correlation. The FP-resolved spectrum appears as expected as an Autler–Townes doublet but since the pump is now pulsed, the second order correlation function $g^2(\tau)$ exhibits peaks spaced by the laser repetition period (13 ns) with a nearly missing peak at $\tau = 0$ due to photon antibunching (Michler et al., 2000). This remnant peak is unaffected by the presence of the control laser within the uncertainty of our measurements. We have also verified that the dressing did not depend on the pump laser. But as expected, when the pump laser intensity increases, the emission increases until saturation occurs.

In the absence of the above-band pump laser, we would expect the resonant control laser to generate resonance fluorescence from the resonantly dressed transitions in region I (Fig. 4). However, without the pump laser, no such emission could be clearly identified. In our current setup, the emission at the laser frequency (within our FP bandwidth) was still dominated by indirect laser scattering so that in the elastic (coherent) regime the resonance

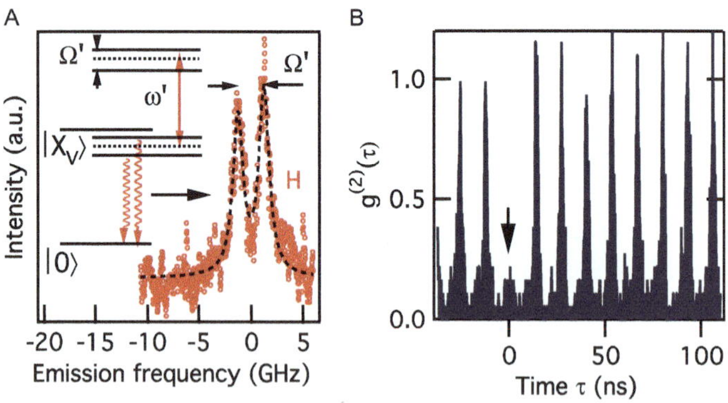

Fig. 7 Autler–Townes doublet in the H-polarized exciton emission spectrum (A) and the second-order correlation function (B) of this emission under resonant dressing of XX_H. The arrow at time difference, τ=0 indicates antibunching.

fluorescence would go unnoticed. On the other hand we cannot exclude the presence of spectral shifts observed in previous absorption experiments (Jundt et al., 2008; Xu et al., 2007b), due apparently to the strong resonant laser, that may detune and thus strongly weaken the resonance fluorescence.

The drawback of the current approach mainly lies in the lack of control over the excitation polarization due to the fixed side excitation, which can be overcome with additional technical complexity. To excite an arbitrary superposition of X_V and X_H one could for example angle-polish the sample or introduce a second phase-locked laser into the wave guide, rotated by 90 degrees to the first one. Additionally, fully utilizing the emission from both exciton and biexciton states, for instance, for deterministic generation of entangled photon pairs, will require filtering the residual background laser scattering. This could be done in slightly detuned excitation conditions by using a FP of appropriate band width, centered at the QD emission frequency.

3. Creating entangled photon states with optical dressing of the biexciton–exciton cascade

Entangled photons are an important element in quantum information science (Aspect et al., 1982), for example in quantum repeaters (Briegel et al., 1998) and quantum cryptography (Ekert, 1991), and enable tests of the intriguing properties of quantum mechanics (Einstein et al., 1935). They are frequently generated by nonlinear optical processes in bulk media (Kiess et al., 1993). Parametric down conversion, for example, yields a large photon-pair flux (Kwiat et al., 1995), but is typically broadband and subject to multi-photon emission. In contrast, isolated single atoms can provide entangled photons discretely, but require careful trapping and preparation.

Semiconductor QDs combine advantages of both. Despite comprising tens of thousands of atoms, they feature atom-like coherent optical properties (Bayer et al., 2000) with near transform-limited linewidths in the GHz range at cryogenic temperatures (Muller et al., 2008). Furthermore, they can be electrically gated, and integrated into microcavities for enhanced speed and efficiency, and could lead to chip-based devices delivering entangled photon pairs on demand. The InAs QDs in GaAs discussed in this chapter are good prospects for this application. In the radiative cascade from $|XX\rangle$, a first photon is emitted during a transition to $|X\rangle$, followed by a second photon emitted during a transition to the ground state, $|0\rangle$. Due to the additional binding energy, E_B, in $|XX\rangle$, the two photons have different wavelengths and can be spectrally separated. Ideally, $|X\rangle$ is twofold degenerate, resulting in two indistinguishable decay paths, as shown in Fig. 2A. Therefore, an entangled state such as the Bell state $|\Psi\rangle = (|H_X H_{XX}\rangle + |V_X V_{XX}\rangle)/\sqrt{2}$ can be produced, where H and V again denote horizontal and vertical linear polarizations, as proposed by Benson et al. (2000) and recently demonstrated with high fidelity (Young et al., 2009). The biexciton radiative cascade could thus lead to a semiconductor source of triggered entangled photon pairs.

As discussed earlier, the anisotropic electron–hole exchange interaction causes $|X\rangle$ to split into two states, $|X_H\rangle$ and $|X_V\rangle$, separated by the fine-structure splitting, Δ_{FS}, schematically shown in Fig. 2B. This is shown experimentally in Fig. 8, where a total of four linear polarized transitions now contribute to the emission spectrum. For this QD, $\Delta_{FS}/2\pi = 5.5$ GHz, and the typical full width at half-maximum of the individual transitions is $\gamma/2\pi \approx 1$ GHz. Since $\Delta_{FS} \gg \gamma$, the two decay paths can be distinguished, and entanglement vanishes (Santori et al., 2002a). The two-photon state is now an incoherent mixture described by the density operator $\hat{\rho} = (|HH\rangle\langle HH| + |VV\rangle\langle VV|)/2$ (omitting X and XX indices). Consequently, with the exception of rare coincidental QD symmetry

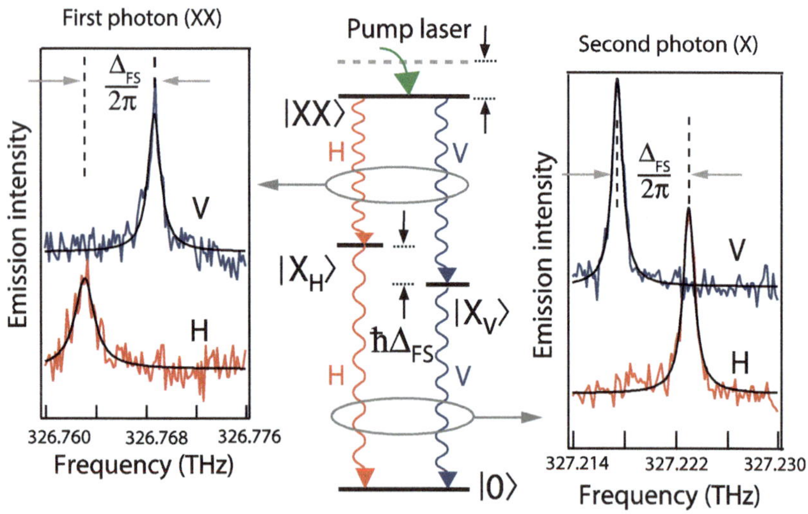

Fig. 8 Single QD emission spectrum. The biexcitonic (*excitonic*) transitions give rise to the left (*right*) spectrum. The two doublets are spectrally separated by $E_B/2\pi\hbar \approx$ 460 GHz ($\Delta\lambda \approx 1.3$ nm).

(Hafenbrak et al., 2007; Young et al., 2006), which can be improved by annealing (Langbein et al., 2004; Young et al., 2005), or by applying spectral (Akopian et al., 2006), or temporal (Stevenson et al., 2008) filtering, entanglement generation is precluded.

Here we demonstrate that polarization entanglement can be retrieved by externally and selectively tuning the fine structure with a far-detuned optical field (Muller et al., 2009a). Other proposed approaches in which external fields manipulate the fine structure splitting have used magnetic fields (Stevenson et al., 2006), DC electric fields (Gerardot et al., 2007; Kowalik et al., 2005; Vogel et al., 2007), or stress (Seidl et al., 2006), but have not yet shown successful creation of entanglement.

A strong continuous-wave (CW) laser near resonance with an optical transition "dresses" the electronic states, forming pairs of states separated by $\sqrt{\Omega^2 + \delta^2}$, where Ω is the Rabi frequency, and δ the laser detuning from resonance (Cohen-Tannoudji et al., 1992). Because Ω is proportional to the electric field amplitude of the laser, the dressed states separate in frequency with increasing intensity. If Ω is larger than the transition linewidth, the single QD dressed state splitting can be observed via the Autler–Townes effect as a doublet, or resonantly as a triplet (Flagg et al., 2009; Jundt et al., 2008; Muller et al., 2007, 2008; Vamivakas et al., 2009; Xu et al., 2007b). Alternatively, by fixing the laser intensity and detuning the laser, the system undergoes an anticrossing, and the spectrum asymptotically approaches the unperturbed spectrum for $|\delta| \rightarrow \infty$. In the asymptotic limit the residual finite offset, the AC Stark shift, represents a small correction to the natural resonance frequency.

The selectivity of dressing specific states and the ability to manipulate these states with the laser intensity (Ω) and detuning (δ) provides a technique to tailor QD energy states precisely. With this approach we can always find a value for δ (given Ω) at which the fine-structure splitting has vanished and where the residual transition in the other dressed state branch is small (Jundt et al., 2008). We make use of this shift to precisely tailor QD energy states.

The CW laser is applied to the $|X_H\rangle \rightarrow |XX\rangle$ transition of Fig. 8. The optical field dresses the $|XX\rangle$ and $|X_H\rangle$ states, resulting in an anticrossing in three transitions: $|XX\rangle \rightarrow |X_H\rangle$, $|XX\rangle \rightarrow |X_V\rangle$, and $|X_H\rangle \rightarrow |0\rangle$. Because the laser does not couple to V-polarized transitions, the $|X_V\rangle \rightarrow |0\rangle$ transition remains unchanged. As depicted in Fig. 9A, the laser is coupled into the waveguide mode of a planar optical microcavity. The structure serves the dual purpose of allowing improved sample-normal photon extraction from

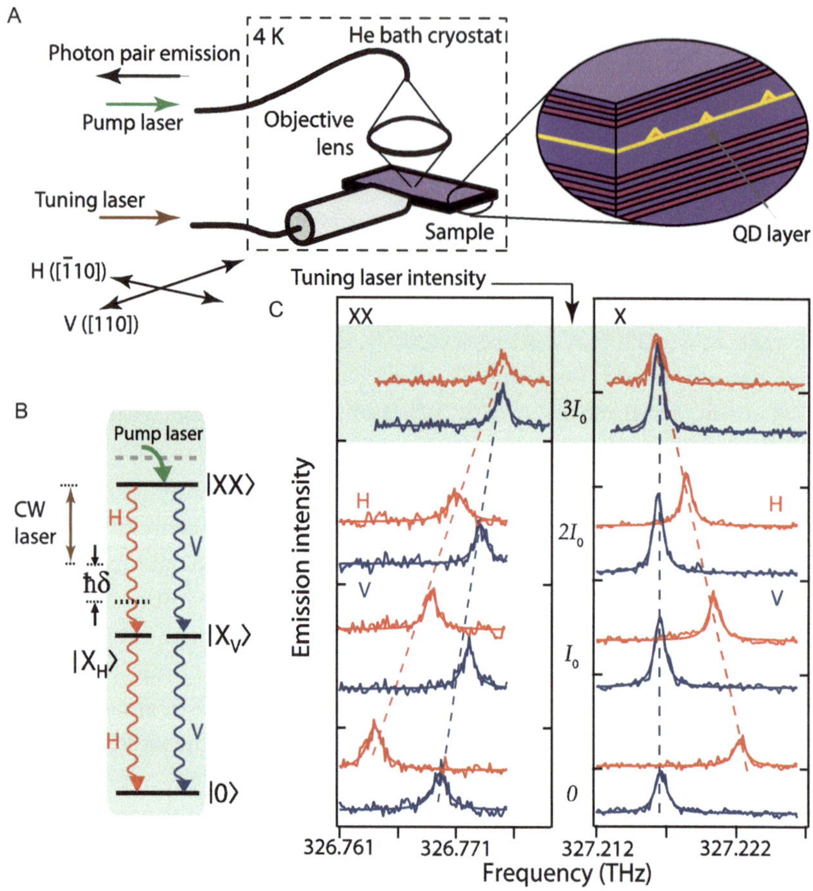

Fig. 9 (A) Experimental setup. (B) Energy diagram in the presence of the near-resonant tuning laser. (C) Spectra of the evolution of the exciton (left) and biexciton (right) transitions with increasing tuning laser intensity, I. When $I = 3I_0$ (uppermost spectra), the overlapped configuration sketched in (B) is achieved. $I_0 = \Omega/2\pi \approx 8.3$ GHz.

the QD via the microcavity, and reduced background laser scattering via good orthogonal (in-plane) laser coupling. It contains two distributed Bragg reflectors of alternate quarter-wave layers of GaAs and AlAs grown above (10 pairs) and beneath (15.5 pairs). The 4λ-thick center spacer defined this cavity, where $\lambda = 915$ nm is the resonant wavelength. A QD centered in the cavity interacts with a near-resonant CW laser launched into a lateral waveguide mode. The near-resonant tuning laser, a Ti-sapphire ring oscillator with a linewidth <1 MHz, is fiber coupled and the fiber rigidly bonded to the cleaved sample edge (Muller et al., 2008). The laser is H-polarized with frequency ω, and detuned from the $|X_H\rangle \Leftrightarrow |XX\rangle$ transition frequency, ω_0 by $\delta/2\pi = \omega_0 - \omega = 25.5$ GHz (Fig. 9B). To pump the biexcitonic state, we apply a second, picosecond pulsed laser (76 MHz repetition period) at 790 nm. It nonresonantly generates carriers in the semiconductor matrix which are subsequently captured in the QD. This is probabilistic state initialization since there is a probability that no carriers are captured or that captured carriers leave the QD charged. This results in decreased photon emission rates from $|XX\rangle$ and $|X\rangle$. The laser intensity was chosen such that biexciton states were significantly populated. All measurements were taken at 4.2 K in a He-bath cryostat using a fiber-coupled microscope objective.

The evolution of the QD emission spectrum with varying tuning laser intensity is shown in Fig. 9C for biexciton (left) and exciton (right) transitions at a fixed detuning of $\delta/2\pi = 25.5$ GHz. The laser intensity, and thus Ω, was increased to tune the fine structure until overlap was achieved (uppermost spectra) for $\Omega/2\pi = 25$ GHz. With increasing laser intensity, the $|XX\rangle \rightarrow |X_V\rangle$ and $|X_H\rangle \rightarrow |0\rangle$ transitions shift with near equal rates, while $|XX\rangle \rightarrow |X_H\rangle$ shifts faster and $|X_V\rangle \rightarrow |0\rangle$ is stationary. With the degeneracy restored, which-path information appears to be erased and photon entanglement from the $|XX\rangle$ and $|X\rangle$ decay created.

Conclusively establishing the formation of an entangled state requires reconstructing the two-photon density matrix, $\hat{\rho}$ (James et al., 2001); it contains all the information about the two-photon state. For this purpose, correlation measurements between the biexciton and exciton photons were performed. In the correlation measurements, the collected emission was first divided equally by a nonpolarizing beam splitter into two independent measurement branches. The photons in each branch were then passed through a set of variable retarders and a polarizer for projection into any desired polarization state. Monochromators spectrally selected the exciton emission on one branch and the biexciton emission on the other branch, before single-photon counting avalanche photodiodes detected the signal. Time correlated single photon counting and histogramming yielded cross-correlation functions $g_{ij}^{(2)}(\tau)$ between polarizations i and j. Using these correlation measurements, quantum-state tomography was applied to reconstruct the two-photon density matrix $\hat{\rho}$ after (James et al., 2001). The two-photon density matrix $\hat{\rho}$ was obtained through the relation $\hat{\rho} = \left(\sum_{\nu=1}^{16} \hat{M}_\nu n_\nu\right) / \left(\sum_{\nu=1}^{4} n_\nu\right)$ derived in James et al. (2001), where

$\hat{M}_\nu$ denote 16 measurement matrices that depend on a set of 16 tomographic states, and where n_ν is the number of coincidence events. We use the same set as in James et al. (2001), chosen to be tomographically complete.

The real and imaginary parts of the density matrix, $\hat{\rho}$, are shown in Fig. 10 for two different conditions. In the absence of any tuning laser (Fig. 10A), it has only diagonal elements and represents a state in which the polarization of the two photons are classically correlated, but not entangled. However, when degeneracy has been restored (Fig. 10B), $\hat{\rho}$ acquires off-diagonal elements (arrows) associated with an entangled state like $|\Psi\rangle$ above. $\hat{\rho}$ in Fig. 10B differs from the ideal matrix $\hat{\rho} = |\Psi\rangle\langle\Psi|$ with corner diagonal elements equal to 1/2. If it is entangled, it is not maximally entangled. It must be further evaluated to determine whether it is inseparable (entangled). We find that well-known tests, such as the concurrence, tangle, and entanglement of formation tests (James et al., 2001), as well as the Peres criterium (Peres, 1996), are all satisfied for $\hat{\rho}$ of Fig. 10B and thus the two-photon state is polarization-entangled. They are not satisfied for $\hat{\rho}$ of Fig. 10A. In Fig. 10B, $\hat{\rho}_{HHVV}$ and $\hat{\rho}_{VVHH}$ have imaginary components, implying that the entangled state obtained is not exactly the state Ψ, as defined above, but rather $|\Psi'\rangle = (|\mathrm{H_X H_{XX}}\rangle + \exp[i\theta]|\mathrm{V_X V_{XX}}\rangle)/\sqrt{2}$ which differs from Ψ by a phase delay between H and V. This phase delay is partially due to a reflection at our beamsplitter, and partially due to imperfect overlap conditions (Stevenson et al., 2008).

Fig. 10E–G shows the measured correlation contrast in three different polarization bases, H/V, D/$\bar{\mathrm{D}}$, and R/L. The contrast is defined as the difference between co-polarized and counter-polarized correlations divided by their sum, and is indexed by repetition period ($\tau_{rep} = 13.2$ ns). Fig. 10E corresponds to tuned conditions, i.e., as in Fig. 10B. In Fig. 10F and G the same contrast is shown when there is no fine-structure overlap, i.e., when the resonant laser is not strong enough to create overlap (Fig. 10F) or when it is so strong that is exceeds the overlap condition (Fig. 10G). This verifies that the R/L contrast, which is needed for entanglement, is only present in optimal overlap conditions.

There are several intrinsic and extrinsic contributions which reduce the photon entanglement. The Peres criterion says that the state is inseparable if the partial transpose of $\hat{\rho}$ has a nonpositive eigenvalue; it is satisfied in our systems if $\rho_{HH,VV}\rho_{VV,HH} > \rho_{HV,HV}\rho_{VH,VH}$ (Peres, 1996). It is satisfied for the density matrix in Fig. 10B. However, since the unmodified $\hat{\rho}$ already contains nonzero $\rho_{HV,\,HV}$ and $\rho_{VH,\,VH}$ elements, a maximally entangled state

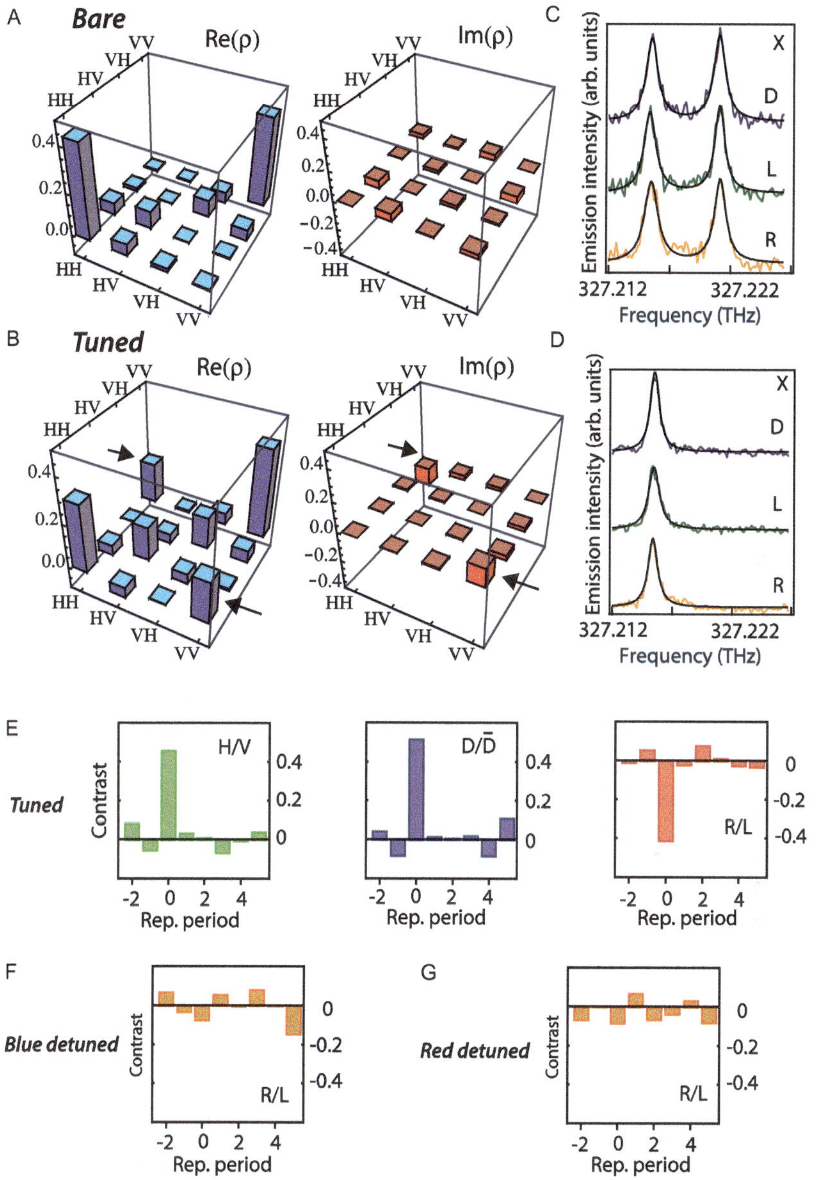

Fig. 10 (A) Real and imaginary parts of the two-photon density matrix, tomographically reconstructed from 16 coincidence measurements. No tuning laser is applied. (B) Same as in (A), but with a near-resonant laser that creates fine structure overlap as in Fig. 9C. Off-diagonal matrix elements appear and the state is now inseparable (entangled). (C) Spectra of exciton emission with detection set to D (diagonal), L (left-circular), and R (right-circular) polarizations, without tuning laser. (D) Same as in (C), but with the tuning laser adjusted to create fine structure overlap, as in (B). (E) Correlation contrast in different polarization bases under tuned condition as in (B) and (D). Correlation contrast in R/L basis under blue (F) and red (G) detuned conditions, i.e., when the laser does not create fine-structure overlap.

cannot be created from it by simply removing Δ_{FS} (Hudson et al., 2007). To emphasize that coherent control can be successfully used in a wide variety of QDs, we have chosen a QD with the largest splitting encountered in our sample (5.5 GHz). To remove this large splitting, the required Rabi frequencies are large enough so that the condition $\delta \gg \Omega$ breaks down. This results in a residual transition in the second branch of the $|X\rangle$ avoided cross spectrum that may not be negligible, and would be another cause of diminished entanglement. Finally, residual background light contributes noise, for example in the off-diagonal elements in Fig. 10A, which reduces the measured entanglement.

These conclusions are supported by data from another QD (QD2) with smaller initial $\Delta_{FS} = 2$ GHz for which entanglement qualifiers are much larger. Spectra and the corresponding density matrix for this QD are shown in Fig. 11 for various tuning conditions. In the absence of any near-resonant external laser, the density matrix already measures a very small amount of entanglement because of partial overlap between fine-structure states. This entanglement significantly increases when tuning is applied (Fig. 11B). Here $\Omega/2\pi = 16$ GHz and $\delta/2\pi = -30$ GHz. In Fig. 11C the fine structure splitting is purposefully increased by detuning the near-resonant laser in the opposite direction from resonance from that used in Fig. 11A ($\delta/2\pi = +57$ GHz). As anticipated, the off-diagonal matrix elements of ρ disappear completely. Table 1 summarizes the result of entanglement test for both QD1 and QD2 in the conditions shown. In general, we note that the degree of entanglement could be improved by further increasing δ and Ω. Besides increasing the laser power, improved coupling efficiency into the cavity waveguide will also increase Ω. Furthermore, increasing the transition dipole moment through crystal growth or changing the field strength by cavity effects are additional routes.

In this section, we have shown how a near-resonant optical field can be used to selectively tailor QD states. In particular, we have successfully canceled the exciton fine-structure splitting in a single QD. With which-path information removed, photon entanglement from the QD biexciton—exciton radiative decay was created. Our experiments establish a deterministic scheme that does not require QD preselection, overcoming a main drawback of strain-induced QDs—those in general use—for entangled photon generation. Our approach, which is completely noninvasive, uses a rigid fiber-coupled device and could lead to a compact, on-demand semiconductor source of entangled photons. It also opens new experimental possibilities in tailoring of quantum states and symmetries in QDs, and perhaps between different QDs.

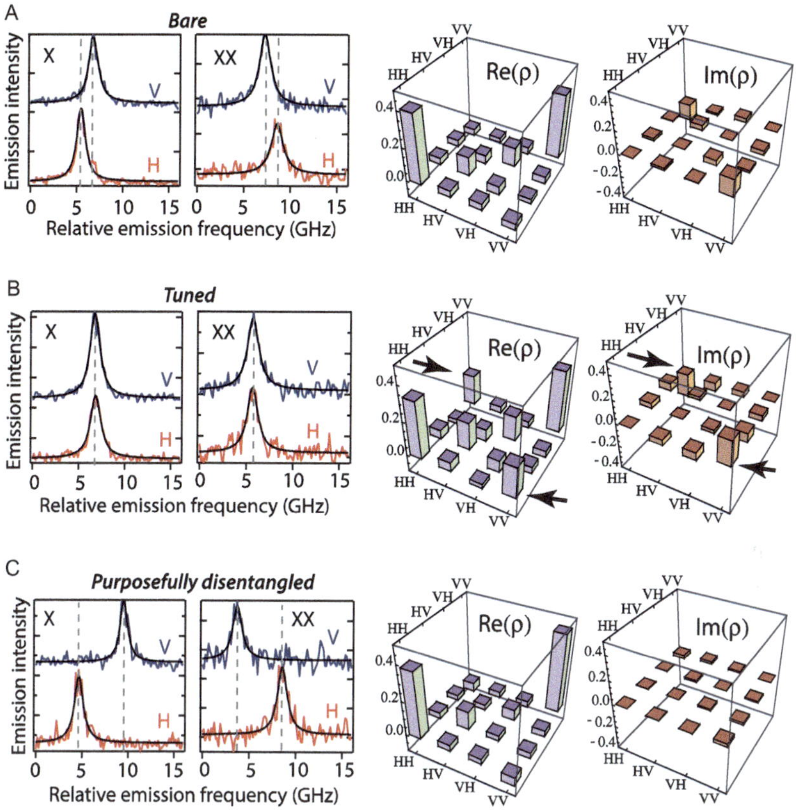

Fig. 11 QD spectra and density matrix for QD2 under different tuning conditions. (A) Without any applied near-resonant laser. (B) With a laser that creates conditions of spectral overlap. (C) With a laser that purposefully increases the fine-structure splitting by reversing the polarity of the laser detuning.

Table 1 Summary of entanglement tests.

$\Delta_{FS}/2\pi$	Concurrence[a,b]	Peres crit[c]	Entangle form[d]
QD1 bare	0	+0.03	0
5.5 GHz tuned	0.16±0.05	−0.08	0.06
Bare	0.09±0.03	−0.04	0.02
GHz tuned	0.32±0.03	−0.16	0.17
Detuned	0	+0.04	0

[a]See James et al. (2001)
[b]Error margins refer to one standard deviation.
[c]Smallest eigenvalue of the partial transpose of ρ (Peres, 1996).
[d]Entanglement of formation (James et al., 2001).

4. Creating indistinguishable photons from the emission from quantum dots in separate samples

When two classical optical fields interfere on a beamsplitter, the joint probability of detection at the two outputs can be as low as 50% (Loudon, 2000). In seminal work, Hong et al. (1987) showed that pairs of photons produced by parametric down conversion (PDC) and interfered can have a reduction in coincidence detection well below 50%, reaching zero for an ideal source of indistinguishable photon pairs. Such interference, where two single-photon Fock states coalesce into a two-photon Fock state, has become one of the central measurements in quantum optics. The coalescence probability hinges on the indistinguishability of the photons involved.

Photon pairs produced by PDC are highly indistinguishable, but the number of pairs produced per pulse is super-Poissonian (Walls and Milburn, 1995). A sub-Poissonian source of indistinguishable single photons, however, is a fundamental component of emerging concepts in quantum information. For example, such sources could be used to realize a quantum C-NOT gate using linear optical elements (Knill et al., 2001). Quantum emitters, such as trapped atoms (Darquie et al., 2005; McKeever et al., 2004), ions (Maunz et al., 2007), and QDs (Michler et al., 2000; Santori et al., 2001), are good sources of single photons. Single photons with a high degree of indistinguishability have been produced by a single QD in an optical microcavity (Santori et al., 2002b) while separate sources of mutually indistinguishable photons have been produced by pairs of trapped atoms (Beugnon et al., 2006) and ions (Maunz et al., 2007). Lifetime-limited, spectrally identical photons have been produced by two separate molecules (Lettow et al., 2007) and indistinguishability (Lettow et al., 2010). Interference between separate solid-state photon sources, like QDs or impurities in crystals, has been performed, but either does not beat the classical limit (Sanaka et al., 2009) or shows no interference due to dephasing (Benyoucef et al., 2009). Continuous-wave (CW) excited QDs have shown interference which exceeds the classical limit in a narrow time window (Ates et al., 2009; Patel et al., 2008), but a CW measurement does not give the whole picture of indistinguishability.

Here, we demonstrate the interference of photons emitted by two semiconductor QD states, each in a different sample (Flagg et al., 2010). We observe a clear signature of coalescence in the second-order correlation and the coincidence detection probability is below the classical limit. We tune the QD states into resonance using externally applied strain.

Using pulsed excitation allows us to measure the total coalescence probability which would be obscured with CW excitation. The probability of coalescence is reduced from unity mainly because of dephasing of the QD states. The data are fit well by a model of two-photon interference using measured values from the emission of each QD.

The two QD single-photon sources are in separate samples. One sample is a normal 4-λ planar distributed Bragg reflector (DBR) microcavity discussed earlier. The other sample is an open cavity comprising a lower DBR (35.5 pairs) and an upper external mirror attached to an optical fiber which we recently described in Muller et al. (2009b). Fig. 12 shows a schematic of the two QD samples and the interferometer. The open cavity sample is glued to a piezoelectric transducer (PZT) such that changing the voltage applied to the PZT strains the sample and tunes the emission energy of the QDs (Seidl et al., 2006). Both samples are maintained at 8 K in a cryostat. The QDs are excited by a mode-locked Ti:sapphire laser with a wavelength of 800 nm and a repetition rate of 76.1 MHz (period = 13.14 ns).

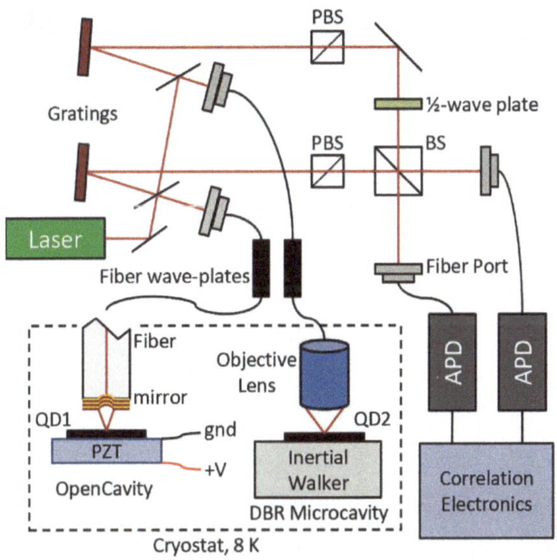

Fig. 12 Schematic of QD samples and interferometer. Two separate samples are used. One (*left*) uses an ablated optical fiber with deposited DBR layers as the top cavity, where light is directly coupled from a single QD into the fiber. The other (*right*) is a planar microcavity sample where light is coupled into a fiber through a fiber-coupled objective. Quantum light in the fibers is focused on to a beam splitter (BS), detections are made with two avalanche photodiodes (APDs) and correlation electronics store detector clicks in the time bins.

After finding a QD in the open cavity which demonstrated high-quality antibunching and a narrow linewidth, denoted QD1, we scan the DBR microcavity for a second QD, denoted QD2, whose emission energy is within the $\approx$10 GHz tuning range of QD1's energy. The emission from both QDs is coupled into optical fibers and variable fiber waveplates ensure proper polarization matching. The light exiting the fibers is filtered by diffraction gratings and sent to interfere at a nonpolarizing beamsplitter. The light from the beamsplitter outputs is coupled into fibers and guided to a pair of avalanche photodiodes (APDs) with a time resolution of 640 ps.

We confirm the spectral overlap of the two QDs and measure their emission linewidths using a scanning Fabry-Perot cavity, monochromator and photodetector with an overall resolution of 150 MHz. Fig. 13A shows the emission data for each QD fitted with a Lorentzian. The excitation laser powers for each QD are adjusted such that their emission intensities are the same; thus the areas under both curves are equal. The full-widths at half-maximum are $\Delta\nu_{\mathrm{QD1}} = 0.55$ GHz and $\Delta\nu_{\mathrm{QD2}} = 0.81$ GHz from which we extract the coherence times $T_2^{(\mathrm{QD1})} = 580$ ps and $T_2^{(\mathrm{QD2})} = 390$ ps. From polarization-dependent measurements (not shown) we determine that the

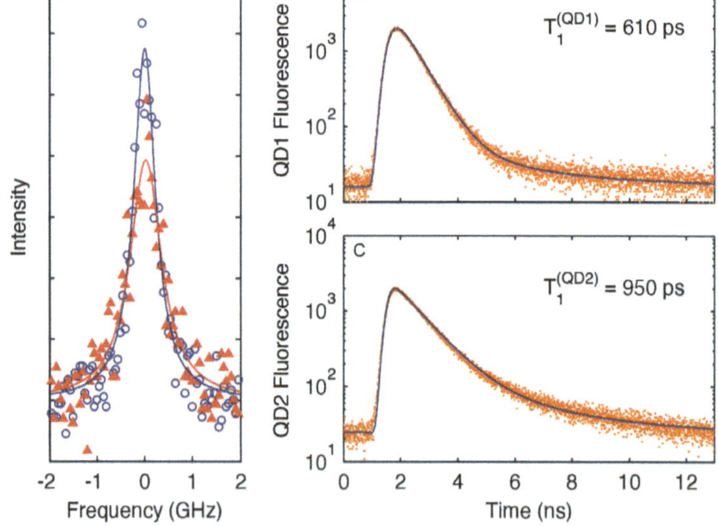

Fig. 13 (A) Area-normalized emission lines for QD1 (open bule circle) and QD2 (filled red triangle) each from different samples. (B) Time-dependent fluorescence from QD1. (C) Time-dependent fluorescence from QD2.

emission line from QD1 is a charged exciton and that from QD2 is one of the fine-structure split lines of an exciton.

To measure the QD lifetimes, each QD's emission is individually sent through a monochromator to an APD and the resulting population decay curves are shown in Fig. 13B and C on a log scale. The curve for QD1 is fit with a single exponential decay, while that for QD2 is fit with a bi-exponential because it is a neutral exciton and we must account for spin-flip transitions from dark states (Dalgarno et al., 2005). Both curves include the effect of the detector time resolution. The lifetimes are $T_1^{(QD1)} = 610$ ps and $T_1^{(QD2)} = 950$ ps; the dark state spin-flip time for QD2 is 4.0 ns. For both QDs, $T_2 < 2T_1$, i.e., the coherence times are not lifetime-limited.

The beamsplitter and APDs can be used as a Hanbury Brown & Twiss correlation measurement (Hanbury Brown and Twiss, 1956) if we input the emission from only one of the QDs. Fig. 14A and B shows the second-order auto-correlation measurement, $g^{(2)}(\tau)$, for each QD individually. The residual counts in the $\tau = 0$ peaks are 9% and 7% of the average amount in the other peaks for QD1 and QD2, respectively. No background subtraction has been applied and the lack of coincidences shows that the light from each QD is over 90% pure single-photon emission.

When photons from each QD interfere, the quality of two-photon interference depends on many experimental parameters, not only spectral overlap.

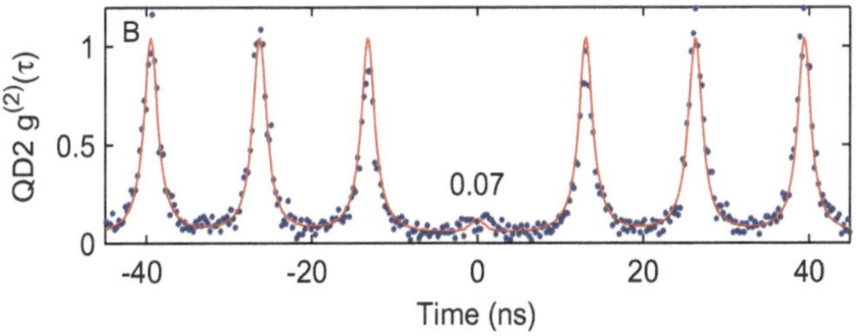

Fig. 14 (A) Auto-correlation for QD1; residual counts in the center peak are 9% of those in the other peaks. (B) Auto-correlation for QD2; residual counts in the center peak are 7%.

To maximize the spatial overlap of the interferometer input modes at the beamsplitter, we send CW laser light tuned to the wavelength of the QDs into both input ports. The field intensity at the output ports can be easily visualized with a CCD camera and the spatial mode overlap optimized. The spatial mode overlap is 98% as determined from the interference fringes of the laser light. By sending the emission from each QD separately through the interferometer we also measure and eliminate the time delay difference between the two light collection paths to ensure optimal temporal overlap of the photons. We attain optimal polarization alignment using the variable fiber waveplates on each input port of the interferometer and polarizing beamsplitters to highly attenuate the remaining undesired polarization. Photons in the two beam paths can be made distinguishable by rotating a 1/2-wave plate to make their polarizations orthogonal. At orthogonal polarization the two photons do not interfere but since the inputs are single photons, coincidence detection is still be below that of Poissonian light.

Despite the nonnegligible differences between the QDs in coherence time, lifetime, and charge state, their photons still interfere. Fig. 15A and B shows the second-order correlation of the light exiting the two output ports of the interferometer for orthogonal and parallel polarizations, respectively. No background or dark-count subtraction is performed on the data. For parallel polarizations, the height of the $\tau = 0$ peak is lower than that for orthogonal polarizations, indicating that photons from the two different QDs have a nonzero coalescence probability.

An interesting feature of pulsed correlation, which is present in CW but whose significance is obscured, is the reduction in coincidences in the center of the $\tau = 0$ peak for parallel polarizations. Fig. 15C shows a close-up of the peak for both relative polarizations. While the total coincidence counts in the $\tau = 0$ peak is not influenced by the time response of the detectors, the depth of the dip is affected. The dash-dotted curve in Fig. 15C is the result of a simulation based on the work in Kiraz et al. (2004) using the T_1 and T_2 values measured above. It shows the shape expected if the detectors were infinitely fast and the single QD $g^{(2)}(\tau)$ went to zero at $\tau = 0$. The solid curve is the same simulation including the effects of the detector response. The residual difference between the data and the solid curve can be explained by the fact that the QDs are not 100% pure single-photon sources which adds counts to the $\tau = 0$ peak not accounted for in the simulation. The data for orthogonal polarizations are fit very well by the dashed curve which is another simulation including the detector response.

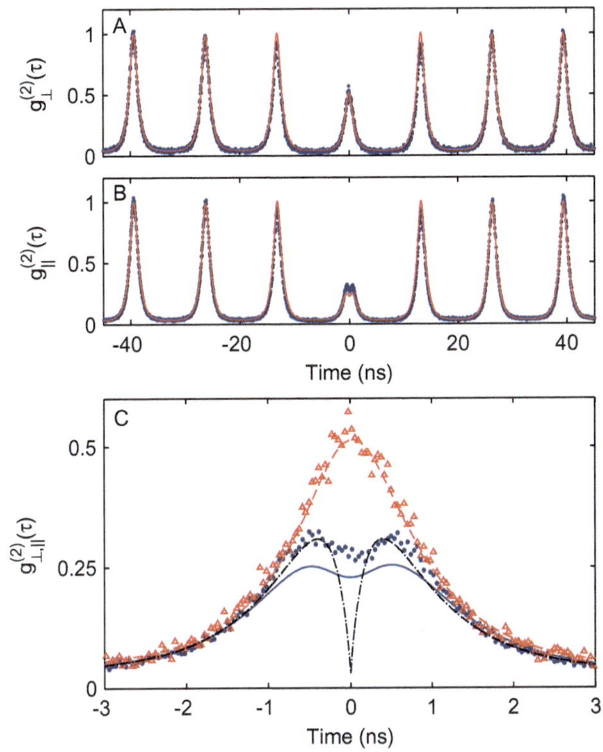

Fig. 15 (A) Correlation of the interference for orthogonal polarizations with simulated curve. (B) Correlation of the interference for parallel polarizations with simulated curve. (C) Close-up of $\tau = 0$ peak for orthogonal (open red triangle) and parallel (filled blue circle) polarizations. The solid and dashed curves are simulations including the detectors' time response, while the dash-dotted curve is the expected curve for infinitely fast detectors.

The probability of coalescence is given by

$$P_c = \frac{A_{\perp} - A_{\parallel}}{A_{\perp}} \tag{1}$$

where $A_{\perp,\parallel}$ is the integrated number of counts in $g_{\perp,\parallel}^{(2)}(\tau)$ during one repetition period around $\tau = 0$. From the data in Fig. 15 we obtain $P_c = 18\%$.[a] This value is reduced from unity mainly due to the presence of dephasing

[a] Removing the effect of dark-counts on the data raises Pc to 19.7%. Dark counts are uncorrelated with other counts, so their effect can determined from the correlation of detector signals (I1 + D1) ⊗ (I2 + D2) where Ii is the signal from photons, and Di is the dark counts.

(Bylander et al., 2003) as explained below. It would be erroneous to calculate P_c using the values of $g^{(2)}_{\perp,\parallel}(\tau)$ right at $\tau = 0$ because any photon emitted by a QD will have a nonzero temporal extent. Using the values at $\tau = 0$ results in a postselective measurement of coalescence,

$$P'_c = \frac{g^{(2)}_{\perp}(0) - g^{(2)}_{\parallel}(0)}{g^{(2)}_{\perp}(0)} \tag{2}$$

which represents the probability of coalescence conditional on detecting both photons within a narrow time window. A CW interference measurement can obtain the postselective value (Patel et al., 2008), P'_c, which has been shown to be quite high for a QD in a microcavity (Ates et al., 2009). The data in Fig. 15C lead to the value $P'_c = 47\%$. With infinitely fast detectors the value of $g^{(2)}_{\parallel}(0)$ would go nearly to zero as does the dash-dotted curve in Fig. 15C. The postselective probability would then be $P'_c = 96\%$ despite the significant counts remaining in the peak at nonzero τ. The coincidence detection rate is given by $A_{\perp,\parallel}/B$ where B is the average number of integrated counts in the peaks not centered at $\tau = 0$. The classical limit is $A_{\parallel}/B = 0.5$, which is the lowest coincidence rate for two classical fields (Loudon, 2000; Ou et al., 1989). From the data in Fig. 15 we extract the value $A_{\parallel}/B = 0.481 \pm 0.002$, which is below the classical limit within experimental error.

The central dip in Fig. 15C is caused by coalescence of the photons. It does not completely eliminate the $\tau = 0$ peak mainly because the QDs' coherence time is not lifetime-limited. Though the photons' temporal extent is given by the QD lifetimes, T_1, the time delay over which they can interfere is given by the coherence times, T_2. Thus the width of the peaks are determined by T_1, and the width of the dip is determined by T_2. If the coherence times were lifetime-limited we would have $T_2 = 2T_1$ and the dip would be wide enough to nearly eliminate the $\tau = 0$ peak. Some residual counts would remain because the two QD lifetimes are different. When $T_2 < 2T_1$ as in this case, the dip is narrow enough to leave significant counts in the peak and gets almost smoothed away by the finite time response of the detectors. Thus for a postselective measurement, both the time window and the detector response time must be less than T_2 (Patel et al., 2008). It is important to note that the imperfect coalescence in the present work is not caused mainly by the differences in the two QDs' lifetimes and coherence times, but by the fact that the coherence times are not lifetime-limited.

For photons from two QDs with the parameters we measure, the maximum theoretical coalescence probability is $P_{c,\max} = 29\%$, as determined by the simulations described above. We attribute the difference between $P_{c,\max}$ and our measured value of 18% to dark-counts and background scattering, which show up as a constant offset in the correlation functions, and to the nonzero $\tau = 0$ peaks of the individual QD auto-correlations in Fig. 14A and B.

The indistinguishability of two photons is a property independent of the measuring device. Therefore, while a CW photon source could be utilized in an application requiring indistinguishable photons as long as post-selection can be used, to accurately characterize the indistinguishability an unconditional measurement must be performed. This means pulsed excitation is required to fully evaluate future sources of indistinguishable photons.

5. An optical spin read-out method for a quantum dot using the AC Stark effect

There are many candidate systems that may act as qubits, including the spin degree of freedom of a single electron or hole trapped in a QD (Warburton, 2013). While previously we have focused on neutral exciton states, optically active transitions to charged exciton states are also possible and allowing the spin of the single particle ground state (electron or hole) to be influenced by a laser field. Spin initialization has been accomplished in a number of experimental situations involving a magnetic field in either the Faraday configuration, where the magnetic field is aligned parallel to the optical axis, (Atatüre et al., 2006; Kroutvar2004 et al., 2004) or the Voigt configuration, where the magnetic field is aligned orthogonal to the optical axis (Bracker et al., 2005; Carter et al., 2013; Gerardot et al., 2008; Press et al., 2008; Xu et al., 2007a). A magnetic field in the Voigt configuration allows both spin initialization and coherent manipulation because the field modifies the polarization selection rules of the optical transitions (Berezovsky et al., 2008; Imamoğlu et al., 1999; Press et al., 2008). Statistically significant spin read-out has been achieved in both Voigt and

Faraday configurations (Atatüre et al., 2007; Berezovsky et al., 2006; Heiss et al., 2008; Lu et al., 2010; Mikkelsen et al., 2007; Press et al., 2008; Vamivakas et al., 2009; Xu et al., 2007a), but the lack of a cycling transition in the Voigt configuration makes a single-shot read-out of the spin-state very difficult.

A single-shot measurement determines the state of the qubit faster than the back-action of the measurement disturbs the state. In a charged QD in a Voigt magnetic field, there is no optical transition that would leave the electron state unchanged with high fidelity (Xu et al., 2007a). In contrast, the Faraday magnetic field configuration results in cycling transitions (Bayer et al., 2002; Vamivakas et al., 2009), which produce photons but leave the electron spin-state largely unchanged after emission. These cycling transitions allowed the demonstration of single-shot spin-state read-out in the Faraday configuration (Delteil et al., 2014), but the optical selection rules preclude arbitrary coherent spin manipulation beyond initialization to an eigenstate. Therefore, in order to realize the three essential single-qubit operations of initialization, manipulation, and read-out, there is a need to combine the capabilities of the Voigt and Faraday configurations.

Here, we propose a scheme to read-out the spin-state of a single electron trapped in an optically active quantum dot using a cycling transition induced by the AC Stark effect of a strong optical field far detuned from resonance. In this scheme, a constant Voigt configuration magnetic field allows rapid spin initialization and picosecond spin rotation via STIRAP (Berezovsky et al., 2008; Carter et al., 2013; Imamoğlu et al., 1999; Press et al., 2008). We show that the AC Stark effect is capable of modifying the allowed optical transitions resulting in spin-selective cycling transitions, which could be used for a single-shot measurement.

As in Section 3, here the AC Stark effect is induced by a laser that is far detuned from the optical transitions but now the laser is circularly polarized instead of linearly polarized. In the limit of large detuning, the interaction between the laser and the QD does not significantly populate the excited states, but still shifts the energy levels coupled by that transition (Muller et al., 2008, 2009a; Unold et al., 2004; Xu et al., 2007a). When this AC Stark shift is much larger than the Zeeman splitting from the Voigt magnetic field, then the polarization selection rules are more similar to the Faraday configuration than the Voigt configuration. Therefore, we call the combined Voigt field with AC Stark shift the "pseudo-Faraday" configuration. Because the pseudo-Faraday configuration is induced by an optical field rather than a DC magnetic field, the system may be rapidly switched between Voigt and pseudo-Faraday configurations. This versatility will allow not only spin initialization and manipulation, but also single-shot spin read-out. Here, we summarize a detailed model proposal by Flagg and Solomon (2015).

While previously we considered the neutral exciton and biexciton ($|X\rangle$, $|XX\rangle$) with the ground state being the vacuum state ($|0\rangle$), we are interested in the charged exciton system. Now, the ground state can contain a single charge, either an electron ($|e\rangle$) or hole ($|h\rangle$). If the QD is in a diode structure, we do not rely on probabilities for the correct occupation of the ground-state charge; it can be stabilized and adjusted with a bias voltage applied to the diode (Warburton et al., 2000). He we assume the QD is negatively charged, the situation is the same for a positively charged QD.

The single electron z-projection spin states ($\uparrow$, $\downarrow$) are optically coupled to charged exciton ($|X^-\rangle$) spin states comprising a pair of electrons in the conduction band and a single heavy-hole in the valence band (Warburton et al., 1997). Since $|X^-\rangle$ contains two electrons of opposite spin, the spin of the $|X^-\rangle$ is determined solely by the spin of the hole. One of the electrons may recombine with the hole, emitting a photon and returning the QD to the single electron ground state. Due to conservation of angular momentum, each charged exciton state has an allowed transition to only the electron state of matching spin, and the transition to the opposite-spin electron state is only weakly allowed due to slight light-hole/heavy-hole mixing (Atatüre et al., 2006; Calarco et al., 2003; Dreiser et al., 2008). The energy level structure of the charged QD is shown schematically in Fig. 16A.

The Hamiltonian of $|X^-\rangle$ in both a magnetic field and an electric field is:

$$H = \hbar\omega_0(\sigma_+^\dagger\sigma_+ + \sigma_-^\dagger\sigma_-) - \vec{\mu}\cdot\vec{B} - \vec{d}\cdot\vec{E} \quad (3)$$

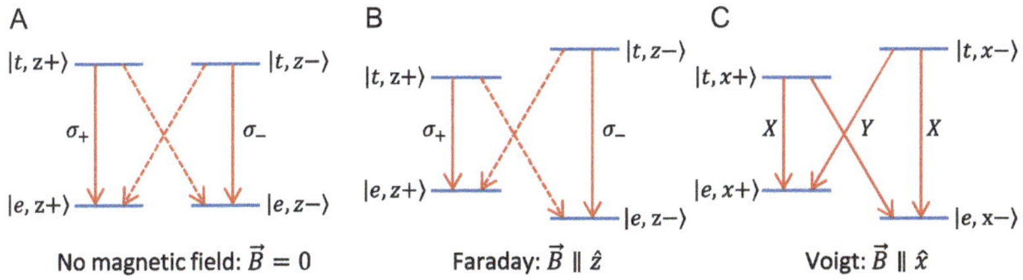

Fig. 16 Energy level structure of a charged quantum dot with (A) no magnetic field, (B) a Faraday magnetic field, and (C) a Voigt magnetic field. Allowed transitions are shown with solid lines; weakly allowed transitions are shown with dashed lines.

where ω_0 is the transition frequency, σ_+ and σ_- are the lowering operators for the z-projection spin-up and spin-down manifolds, respectively, $\vec{\mu}$ and $\vec{d}$ are the magnetic and electric dipole operators, and $\vec{B}$ and $\vec{E}$ are the magnetic and electric field amplitudes.

To describe the AC Stark shift, we assume an oscillatory form for the electric field, as in a single-frequency laser beam, and perform the standard rotating wave approximation. We further simplify the notation by introducing the complex Rabi frequencies associated with the two circularly polarized components: $\Omega_{1+} = dE_{1+}/\hbar$ and $\Omega_{1-} = dE_{1-}/\hbar$. We are primarily interested in circularly polarized light because it will produce an AC Stark shift that reduces the effect of the magnetic coupling between the spin states, as we will demonstrate below. Therefore, we assume only left-circularly polarized light ($\Omega_{1-} = 0$), and without loss of generality we can treat Ω_{1+} as purely real. We can rewrite the Hamiltonian from Eq. (3) of the charged QD system in the rotating frame using operators for the magnetic and electric fields, and including the AC Stark shift as

$$
\begin{aligned}
H_0 = &\frac{1}{2}\hbar\Delta_1(\sigma_+^\dagger\sigma_+ - \sigma_+\sigma_+^\dagger + \sigma_-^\dagger\sigma_- - \sigma_-\sigma_-^\dagger) \\
&+ \mu_B B_x(g_{e,x}(s_e^\dagger + s_e) + g_{h,x}(s_h^\dagger + s_h)) \\
&+ \frac{1}{2}\hbar\Omega_{1+}(\sigma_+^\dagger + \sigma_+)
\end{aligned} \tag{4}
$$

where $\Delta_1 = \omega_0 - \omega_1$ is the detuning of the laser frequency ω_1 from the QD resonance ω_0; μ_B is the Bohr magneton, $g_{e,x}$ and $g_{h,x}$ are the electron and hole g-factors for a magnetic field in the x-direction, and s_e and s_h are the electron and hole spin-flip operators, respectively.

If we make the assumption that the detuning Δ_1 is much larger than the Rabi frequency Ω_{1+}, then to first order in the ratio Ω_{1+}/Δ_1 the generalized Rabi frequency is

$$
\mathcal{W}_1 \approx \Delta_1 + \frac{\Omega_{1+}^2}{2\Delta_1} \tag{5}
$$

and we can approximate the eigenvalues and eigenvectors of the Hamiltonian as follows.

$$\lambda_1 = -\frac{\hbar}{2}\Delta_1 - \frac{\hbar\Omega_{1+}^2}{4\Delta_1} + \mathcal{O}\left(\frac{\Omega_{1+}^2}{\Delta_1^2}\right)$$

$$\lambda_3 = \frac{\hbar}{2}\Delta_1 + \frac{\hbar\Omega_{1+}^2}{4\Delta_1} + \mathcal{O}\left(\frac{\Omega_{1+}^2}{\Delta_1^2}\right)$$

$$|\nu_1\rangle = \begin{bmatrix} 1 \\ 0 \\ -\Omega_{1+}/2\Delta_1 \\ 0 \end{bmatrix} + \mathcal{O}(\Omega_{1+}^2/\Delta_1^2) \tag{6}$$

$$|\nu_3\rangle = \begin{bmatrix} \Omega_{1+}/2\Delta_1 \\ 0 \\ 1 \\ 0 \end{bmatrix} + \mathcal{O}(\Omega_{1+}^2/\Delta_1^2)$$

Note that these eigenvectors are normalized only to first order in Ω_{1+}/Δ_1. In the large detuning approximation, one state, $|\nu_1\rangle$, is more electron-like and the other, $|\nu_3\rangle$, is more $|X^-\rangle$-like. The AC Stark shifts of the energy levels are the second terms in the eigenvalues, which are quadratic in the Rabi frequency and thus linear in the laser power.

5.1 Pseudo-Faraday configuration

The Voigt geometry magnetic field couples the $z+$ spin states to the $z-$ spin states, while the AC Stark effect of a σ_+ polarized laser shifts only the $z+$ spin states. In the presence of both a magnetic field and a strong, far red-detuned σ_+ polarized laser, the spin eigenstates, ψ_i and the polarization selection rules result from a competition between the magnetic coupling between the $z+$ and $z-$ states and the AC Stark shift of the $z+$ states. When the AC Stark shift is very large compared to the magnetic coupling, then the magnetic field has little effect on the system. We call this situation the pseudo-Faraday configuration because the energy structure, eigenstates, and polarization selection rules are similar to the Faraday magnetic field configuration where $\vec{B} = \hat{z}B_z$.

The branching ratio quantifies the measurement back-action of the transition and will determine the number of cycles that can be used for a single-shot measurement. To produce effective cycling transitions suitable for spin read-out, the branching ratio must be much less than unity, meaning that the spin-preserving transition is far more likely than the spin-flipping transition. To obtain the branching ratio for one of the $|X^-\rangle$-like states, $|\psi_4\rangle$, we take the ratio of the rates for the spin-flipping transition, $|\psi_4\rangle \rightarrow |\psi_1\rangle$, and the spin-preserving transition, $|\psi_4\rangle \rightarrow |\psi_2\rangle$. Following the derivation in Flagg and Solomon (2015), this ratio reduces to

$$r_B = \frac{|\langle\psi_1||\mathrm{d}||\psi_4\rangle|^2}{|\langle\psi_2||\mathrm{d}||\psi_4\rangle|^2} \tag{7}$$

In Fig. 17, we plot the branching ratio as a function of Rabi frequency for fixed magnetic field and typical g-factors (Kroner et al., 2008): $g_{e,x} = 0.47$, $g_{h,x} = 0.24$. At zero Rabi frequency the system is in the Voigt configuration and the branching ratio is unity. As the Rabi frequency increases, the branching ratio reduces significantly, becoming similar to that of the Faraday configuration. For a magnetic field of 0.1 T, detuning of $\Delta_1/2\pi = 2$ THz, and Rabi frequency of $\Omega_{1+}/2\pi = 200$ GHz, we calculate the branching ratio to be 0.02. Measured values of branching ratio for solid-state systems range from 0.001 (Fernandez et al., 2009) to 0.04 (Müller et al., 2014). Our prediction of the branching ratio for a QD in the pseudo-Faraday configuration is within this range, suggesting that it may be possible to perform a single-shot read-out of the electron spin state via resonantly excited fluorescence.

5.2 Spin read-out operation

We have shown that applying a strong, far red-detuned, circularly polarized laser to a charged QD in a Voigt configuration magnetic field results in a situation similar to a Faraday magnetic field. We now discuss how a practical

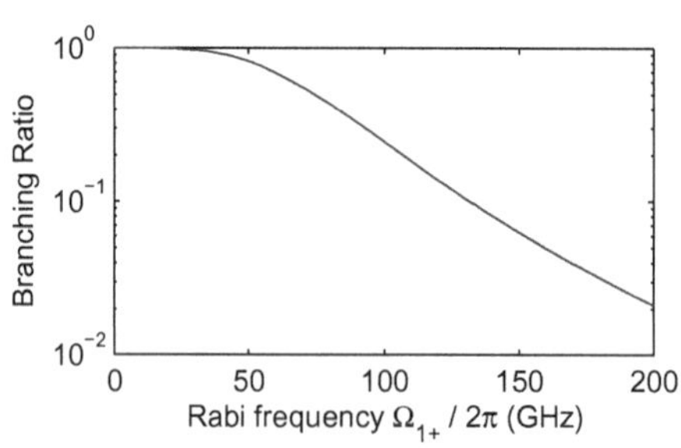

Fig. 17 Evolution of the branching ratio from the Voigt configuration ($B_x = 0.1$ T) to the pseudo-Faraday configuration. In the pseudo-Faraday configuration, the branching ratio is 0.02.

spin read-out scheme would work in the pseudo-Faraday configuration. To perform a spin read-out operation, we need to excite the system in a spin-selective manner and detect the fluorescence. Detection of a photon would correspond to the electron being in a certain spin state, and the fidelity of the measurement depends on the branching ratio and the spin-selectivity of the excitation.

In the pseudo-Faraday configuration, there are two spin-preserving transitions that are nondegenerate and two weakly allowed spin flipping transitions that are nearly degenerate. The spin-preserving transitions are circularly polarized, σ_+ and σ_-, and can both be excited by linearly polarized light. The energy difference between the two spin-preserving transitions, however, makes resonant excitation of one of them a spin-selective excitation even when linear polarization is used. Thus, spin-selective excitation may be accomplished by a linearly polarized laser tuned to resonance with one of the spin-preserving transitions. The fluorescence will be circularly polarized and may, therefore, be distinguished from the linearly polarized laser scattering by cross-polarized detection (Vamivakas et al., 2009). Alternatively, we can use a modal discrimination method, like shown earlier in the chapter where the resonant laser is introduced into the waveguide mode of a planar microcavity that confines the laser scattering, while the QD fluorescence is emitted into the orthogonal cavity mode.

After the electron has undergone some operations (e.g., initialization and/or manipulation) in the Voigt configuration, the AC Stark laser can be applied to transform to the pseudo-Faraday configuration where read-out will occur. The application of the AC Stark laser can be rapid compared to the switching time of a magnetic field, but if the laser field is applied too fast then the transition from Voigt to pseudo-Faraday will be nonadiabatic. Prior to the application of the AC Stark laser, the electron may be in any arbitrary superposition of the two Voigt eigenstates, which are the $x\pm$ projection states of the spin. The transition from Voigt to pseudo-Faraday configurations will occur adiabatically if the AC Stark laser is turned on slowly relative to the Larmor period of spin precession caused by the Voigt magnetic field. For a 0.1 T field this is about 120 ps. If the adiabatic condition is satisfied, the population of the $x\pm$ state in the Voigt configuration transitions without change to the population of the $z\mp$ state in the pseudo-Faraday configuration. This is for the case of a σ_+ polarized AC Stark laser; for a σ_- polarized laser, the mapping from x-basis to z-basis would be the opposite. Due to the adiabatic mapping of the Voigt x-basis to the pseudo-Faraday eigenbasis, detection of the spin-selectively excited fluorescence in the pseudo-Faraday configuration is equivalent to a projective measurement of the spin onto the $x\pm$ states in the Voigt configuration. Furthermore, a measurement in any arbitrary spin projection basis can be performed by preceding the AC Stark laser with a rotation of the electron spin Bloch sphere via stimulated Raman adiabatic passage by additional laser pulses (Berezovsky et al., 2008; Press et al., 2008). The capability of using multiple measurement bases for identically prepared states would allow full quantum state tomography (James et al., 2001) to be performed on the electron spin state.

We numerically calculate the density matrix evolution under experimental conditions. A spin read-out operation will consist of resonant excitation in the pseudo-Faraday configuration where the QD is in one of the two electron-like eigenstates, $z+$ or $z-$. Each electron-like eigenstate has one strong and one weakly allowed optical transition. To perform a projective measurement on the $z-$ eigenstate, for example, the resonant laser is tuned to the allowed σ_- polarized transition and the fluorescence is detected. Detection of a photon is interpreted as confirmation that the electron was in the $z-$ eigenstate. The fidelity of the spin read-out depends on the specificity of the excitation and fluorescence. If the resonant laser is tuned to the σ_- polarized transition, but the electron is in the $z+$ spin state, photon emission may still be detected, giving an erroneous signal.

To predict the fidelity of the spin read-out operation, we simulate it using typical QD parameters and a linearly polarized resonant laser tuned to the σ_- cycling transition. The experimentally controllable parameters used are: $B_x = 0.1$ T, $\Omega_{1+}/2\pi = 200$ GHz, $\Delta_1/2\pi = 2$ THz, and $\Omega_{2+}/2\pi = \Omega_{2-}/2\pi = 0.5$ GHz. Note that the AC Stark mixing parameter is kept small ($\Omega_{1+}/2\Delta_1 = 0.05$), but the AC Stark shift is still large: $\Omega_{1+}^2/2\Delta_1 = 2\pi(10\text{GHz})$. The probe laser is resonant with the σ_- cycling transition between the $z-$ electron-like and $|X^-\rangle$-like states. The QD parameters would not be controllable in practice, except by the choice of QD, but typical values are used here. For details of the model and the parameters used, see Flagg and Solomon (2015).

Fig. 18A shows the photon emission rate $R(t)$ as a function of time after resonant excitation begins. When the initial state is the $z-$ electron state (solid red curve) the emission starts strong and decays as the population is pumped into the $z+$ state, which is not being resonantly excited. When the initial state is the $z+$ electron state (dotted blue curve) the emission rate starts very low but rises slightly as a small amount of population is pumped into the $z-$ electron state. Fig. 18B shows the average number of photons detected $D(T)$ as a function of the detection window duration for the two

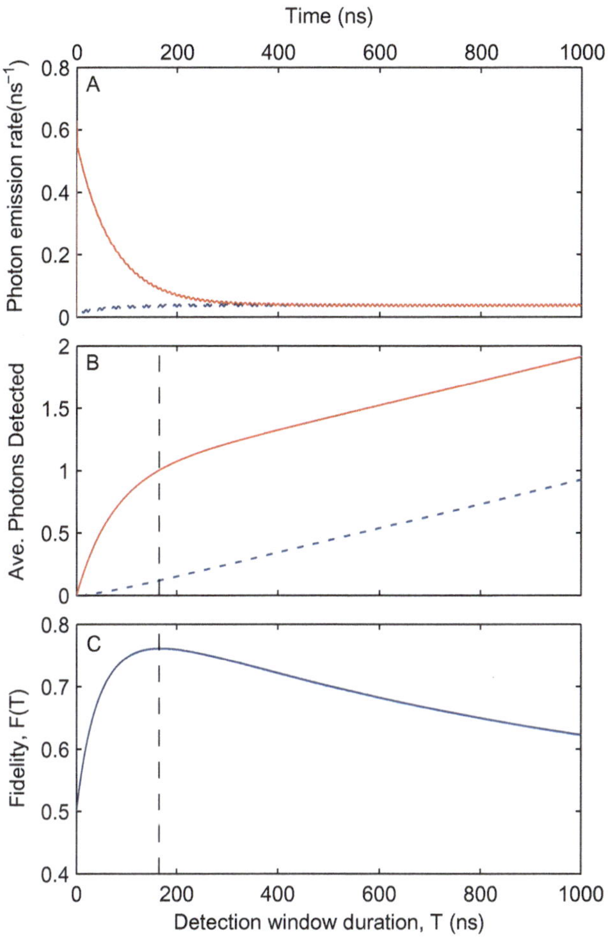

Fig. 18 Spin read-out operation under σ_- resonant excitation. In (A) and (B) the two initial conditions are the $z-$ electron state (solid red curve) and the $z+$ electron state (dashed blue curve). (A) Photon emission rate after σ_- resonant excitation begins. (B) Average number of photons detected as a function of detection window duration. (C) Fidelity of the spin measurement as a function of detection window duration. The vertical dashed line indicates the optimum detection window.

initial conditions. When the initial state is the $z+$ electron state (dotted blue curve), photon emission is relatively unlikely because the excitation laser is not resonant with the allowed σ_+ transition; emission is still possible, however, due to weak remaining magnetic spin mixing. When the initial state is the $z-$ electron state (solid red curve), photon emission is relatively likely. The fidelity of the spin measurement is $F = (1 - p_{z+} + p_{z-})/2$ where $p_{z\pm}$ is the probability of detecting at least one photon when the initial state is $z\pm$. The fidelity is plotted in Fig. 18C as a function of detection window duration. The maximum fidelity occurs for a detection window of 165 ns, indicated by a vertical dashed line, beyond which the value of p_{z-} saturates but p_{z+} keeps increasing. For the optimum detection window duration the calculated fidelity is 76.2%, which is slightly lower than the 82.3% measured in the true Faraday configuration in Delteil et al. (2014). The fidelity would be improved with larger AC Stark shift, smaller Voigt magnetic field, smaller electron spin dephasing and decay rates, or smaller $|X^-\rangle$ dephasing rate. The values chosen here are either typical or feasible for real experiments.

When the average number of photons detected is greater than unity, we can say that a single-shot measurement of the electron spin state is possible (Delteil et al., 2014). For these calculations, we have chosen an overall detection efficiency $\epsilon = 2.5\%$ that results in an average number of detected photons, $D(T) > 1$ for the optimum detection window duration: for $T = 165$ ns, $D(T) = 1.01$.

Acknowledgments

We acknowledge partial support from the NSF Physics Frontier Center at the Joint Quantum Institute (PFC@JQI).

References

Akopian, N., et al., 2006. Entangled photon pairs from semiconductor quantum dots. Phys. Rev. Lett. 96, 130501.

Aspect, A., Grangier, P., Roger, G., 1982. Experimental realization of Einstein-Podolsky-Rosen-Bohm *gedankenexperiment*: a new violation of Bell's inequalities. Phys. Rev. Lett. 49, 91.

Atatüre, M., Dreiser, J., Badolato, A., Högele, A., Karrai, K., Imamoğlu, A., 2006. Quantum-dot spin-state preparation with near-unity fidelity. Science 432, 551.

Atatüre, M., Dreiser, J., Badolato, A., Imamoğlu, A., 2007. Observation of faraday rotation from a single confined spin. Nat. Phys. 3, 101.

Ates, S., Ulrich, S.M., Reitzenstein, S., Loeffler, A., Forchel, A., Michler, P., 2009. Post-selected indistinguishable photons from the resonance fluorescence of a single quantum dot in a microcavity. Phys. Rev. Lett. 103 (16), 167402.

Autler, S.H., Townes, C.H., 1955. Stark effect in rapidly varying fields. Phys. Rev. 100, 703.

Bayer, M., et al., 2000. Hidden symmetries in the energy levels of excitonic 'artificial atoms'. Nature 405, 923.

Bayer, M., et al., 2002. Fine structure of neutral and charged excitons in self-assembled In(Ga)As/(Al)GaAs quantum dots. Phys. Rev. B 65, 195315.

Benson, O., Santori, C., Pelton, M., Yamamoto, Y., 2000. Regulated and entangled photons from a single quantum dot. Phys. Rev. Lett. 84, 2513.

Benyoucef, M., Wang, L., Rastelli, A., Schmidt, O.G., 2009. Toward quantum interference of photons from independent quantum dots. Appl. Phys. Lett. 95 (26), 261908.

Berezovsky, J., Mikkelsen, M.H., Gywat, O., Stoltz, N.G., Coldren, L.A., Awschalom, D.D., 2006. Nondestructive optical measurements of a single electron spin in a quantum dot. Science 314, 1916.
Berezovsky, J., Mikkelsen, M.H., Stoltz, N.G., Coldren, L.A., Awschalom, D.D., 2008. Picosecond coherent optical manipulation of a single electron spin in a quantum dot. Science 320, 349.
Beugnon, J., Jones, M.P.A., Dingjan, J., Darquie, B., Messin, G., Browaeys, A., Grangier, P., 2006. Quantum interference between two single photons emitted by independently trapped atoms. Nature 440 (7085), 779–782.
Bracker, A.S., Stinaff, E.A., Gammon, D., Ware, M.E., Tischler, J.G., Shabaev, A., Efros, A.L., Park, D., Gershoni, D., Korenev, V.L., Merkulov, I.A., 2005. Optical pumping of the electronic and nuclear spin of single charge-tunable quantum dots. Phys. Rev. Lett. 94, 047402.
Briegel, H.J., Dür, W., Cirac, J.I., Zoller, P., 1998. Quantum repeaters: The role of imperfect local operations in quantum communication. Phys. Rev. Lett. 81, 5932.
Bylander, J., Robert-Philip, I., Abram, I., 2003. Interference and correlation of two independent photons. Eur. Phys. J. D 22, 295–301.
Calarco, T., Datta, A., Fedichev, P., Pazy, E., Zoller, P., 2003. Spin-based all-optical quantum computation with quantum dots: understanding and suppressing decoherence. Phys. Rev. A 68, 012310.
Carter, S.G., Sweeney, T.M., Kim, M., Kim, C.S., Solenov, D., Economou, S.E., Reinecke, T.L., Yang, L., Bracker, A.S., Gammon, D., 2013. Quantum control of a spin qubit coupled to a photonic crystal cavity. Nat. Photon. 7, 329.
Cohen-Tannoudji, C., Reynaud, S., 1977. Dressed-atom description of resonance fluorescence and absorption spectra of a multi-level atom in an intense laser beam. J. Phys. B At. Mol. Phys. 10, 345.
Cohen-Tannoudji, C., Dupont-Roc, J., Grynberg, G., 1992. Atom-Photon Interactions. John Wiley and Sons, New York.
Dalgarno, P.A., Smith, J.M., Gerardot, B.D., Govorov, A.O., Karrai, K., Petroff, P.M., Warburton, R.J., 2005. Dark exciton decay dynamics of a semiconductor quantum dot. Phys. Status Solidi A 202 (14), 2591–2597.
Darquie, B., Jones, M.P.A., Dingjan, J., Beugnon, J., Bergamini, S., Sortais, Y., Messin, G., Browaeys, A., Grangier, P., 2005. Controlled single-photon emission from a single trapped two-level atom. Science 309, 454–456.
Delteil, A., Gao, W.B., Fallahi, P., Miguel-Sanchez, J., Imamoğlu, A., 2014. Observation of quantum jumps of a single quantum dot spin using submicrosecond single-shot optical readout. Phys. Rev. Lett. 112, 116802.
DiVincenzo, D.P., 2000. The physical implementation of quantum computation. Fortschr. Phys. 48, 771.
Dreiser, J., Atatüre, M., Galland, C., Müller, T., Badolato, A., Imamoğlu, A., 2008. Optical investigations of quantum dot spin dynamics as a function of external electric and magnetic fields. Phys. Rev. B 77, 075317.
Einstein, A., Podolsky, B., Rosen, N., 1935. Can quantum-mechanical description of physical reality be considered complete? Phys. Rev. 47, 777.
Ekert, A.K., 1991. Quantum cryptography based on Bell's theorem. Phys. Rev. Lett. 67, 661.
Fernandez, G., Volz, T., Desbuquois, R., Badolato, A., Imamoğlu, A., 2009. Optically tunable spontaneous Raman fluorescence from a single self-assembled InGaAs quantum dot. Phys. Rev, Lett. 103, 087406.
Flagg, E.B., Solomon, G.S., 2015. An optical spin read-out method for a quantum dot using the AC Stark effect. Phys. Rev. B 92, 245309.
Flagg, E.B., et al., 2009. Resonantly driven coherent oscillations in a solid-state quantum emitter. Nat. Phys. 5, 203.

Flagg, E.B., Muller, A., Polyakov, S.V., Ling, A., Migdall, A., Solomon, G.S., 2010. Interference of single photons from two separate semiconductor quantum dots. Phys. Rev. Lett. 104, 137401.

Flagg, E.B., Polyakov, S.V., Thomay, T., Solomon, G.S., 2012. Dynamics of nonclassical light from a single solid-state quantum emitter. Phys. Rev. Lett. 109, 163601.

Gammon, D., Snow, E.S., Shanabrook, B.V., Katzer, D.S., Park, D., 1996. Fine structure splitting in the optical spectra of single GaAs quantum dots. Phys. Rev. Lett. 76, 3005.

Gerardot, B.D., et al., 2007. Manipulating exciton fine structure in quantum dots with a lateral electric field. Appl. Phys. Lett. 90, 041101.

Gerardot, B.D., Brunner, D., Dalgarno, P.A., Öhberg, P., Seidl, S., Kroner, M., Karrai, K., Stoltz, N.G., Petroff, P.M., Warburton, R.J., 2008. Optical pumping of a single hole spin in a quantum dot. Nature 451, 441.

Hafenbrak, R., et al., 2007. Triggered polarization-entangled photon pairs from a single quantum dot up to 30 k. New J. Phys. 9, 315.

Hanbury Brown, R., Twiss, R.Q., 1956. Correlation between photons in two coherent beams of light. Nature 177 (4497), 27–29.

Heiss, D., Jovanov, V., Bichler, M., Abstreiter, G., Finley, J.J., 2008. Charge and spin readout scheme for single self-assembled quantum dots. Phys. Rev. B 77, 235442.

Heitler, W., 1954. The Quantum Theory of Radiation, third ed. Oxford University Press.

Hong, C.K., Ou, Z.Y., Mandel, L., 1987. Measurement of subpicosecond time intervals between two photons by interference. Phys. Rev. Lett. 59 (18), 2044–2046.

Hudson, A.J., et al., 2007. Coherence of an entangled exciton-photon state. Phys. Rev. Lett. 99, 266802.

Imamoğlu, A., Awschalom, D.D., Burkard, G., DiVincenzo, D.P., Loss, D., Sherwin, M., Small, A., 1999. Quantum information processing using quantum dot spins and cavity QED. Phys. Rev. Lett. 83, 4204.

James, D.F.V., Kwiat, P.G., Munro, W.J., White, A.G., 2001. Measurement of qubits. Phys. Rev. A 64, 052312.

Jundt, G., Robledo, L., Högele, A., Fält, S., Imamoğlu, A., 2008. Observation of dressed excitonic states in a single quantum dot. Phys. Rev. Lett. 100, 177401.

Kiess, T.E., Shih, Y.H., Sergienko, A.V., Alley, C.O., 1993. Einstein-Podolsky-Rosen-Bohm experiment using pairs of light quanta produced by type-II parametric down-conversion. Phys. Rev. Lett. 71, 3893.

Kiraz, A., Atature, M., Imamoğlu, A., 2004. Quantum-dot single-photon sources: Prospects for applications in linear optics quantum-information processing. Phys. Rev. A 69 (3), 032305.

Knill, E., Laflamme, R., Milburn, G.J., 2001. A scheme for efficient quantum computation with linear optics. Nature 409 (6816), 46–52.

Kowalik, K., et al., 2005. Influence of an in-plane electric field on exciton fine structure in InAs-GaAs self-assembled quantum dots. Appl. Phys. Lett. 86, 041907.

Kroner, M., Weiss, K.M., Biedermann, B., Seidl, S., Holleitner, A.W., Badolato, A., Petroff, P.M., Öhberg, P., Warburton, R.J., Karrai, K., 2008. Resonant two-color high-resolution spectroscopy of a negatively charged exciton in a self-assembled quantum dot. Phys. Rev, B 78, 075429.

Kroutvar2004, M., Ducommun, Y., Heiss, D., Bichler, M., Schuh, D., Abstreiter, G., Finley, J.J., 2004. Optically programmable electron spin memory using semiconductor quantum dots. Nature 432, 81.

Kwiat, P.G., et al., 1995. New high-intensity source of polarization-entangled photon Pairs. Phys. Rev. Lett. 75, 4337.

Langbein, W., et al., 2004. Control of fine-structure splitting and biexciton binding in $In_xGa_{1-x}As$ quantum dots by annealing. Phys. Rev. B 69, 161301(R).

Lettow, R., Ahtee, V., Pfab, R., Renn, A., Ikonen, E., Goetzinger, S., Sandoghdar, V., 2007. Realization of two Fourier-limited solid-state single-photon sources. Opt. Express 15 (24), 15842–15847.
Lettow, R., Rezus, Y.L.A., Renn, A., Zumofen, G., Ikonen, E., Götzinger, S., Sandoghdar, V., 2010. Quantum Interference of Tunably Indistinguishable Photons from Remote Organic Molecules. Phys. Rev. Lett. 104, 123605.
Loudon, R., 2000. The Quantum Theory of Light, third ed. Oxford University Press.
Lu, C.Y., Zhao, Y., Vamivakas, A.N., Matthiesen, C., Fält, S., Badolato, A., Atatüre, M., 2010. Direct measurement of spin dynamics in InAs/GaAs quantum dots using time-resolved resonance fluorescence. Phys. Rev. B 81, 035332.
Maunz, P., Moehring, D.L., Olmschenk, S., Younge, K.C., Matsukevich, D.N., Monroe, C., 2007. Quantum interference of photon pairs from two remote trapped atomic ions. Nat. Phys. 3 (8), 538–541.
McKeever, J., Boca, A., Boozer, A.D., Miller, R., Buck, J.R., Kuzmich, A., Kimble, H.J., 2004. Deterministic generation of single photons from one atom trapped in a cavity. Science 303 (5666), 1992–1994.
Michler, P., Kiraz, A., Becher, C., Schoenfeld, W.V., Petroff, P.M., Zhang, L.D., Hu, E., Imamoğlu, A., 2000. A quantum dot single-photon turnstile device. Science 290 (5500), 2282.
Mikkelsen, M.H., Berezovsky, J., Stoltz, N.G., Coldren, L.A., Awschalom, D.D., 2007. Optically detected coherent spin dynamics of a single electron in a quantum dot. Nat. Phys. 3, 770.
Mollow, B.R., 1969. Power spectrum of light scattered by two-level systems. Phys. Rev. 188, 1969.
Muller, A., et al., 2007. Resonance fluorescence from a coherently driven semiconductor quantum dot in a cavity. Phys. Rev. Lett. 99, 187402.
Muller, A., Fang, W., Lawall, J., Solomon, G.S., 2008. Emission spectrum of a dressed exciton-biexciton complex in a semiconductor quantum dot. Phys. Rev. Lett. 101, 027401.
Muller, A., Fang, W., Lawall, J., Solomon, G.S., 2009a. Creating polarization-entangled photon pairs from a semiconductor quantum dot using the optical Stark effect. Phys. Rev. Lett. 103, 217402.
Muller, A., Flagg, E.B., Metcalfe, M., Lawall, J., Solomon, G.S., 2009b. Coupling an epitaxial quantum dot to a fiber-based external-mirror microcavity. Appl. Phys. Lett. 95 (17), 173101.
Müller, T., Hepp, C., Pingault, B., Neu, E., Gsell, S., Schreck, M., Sternschulte, H., Steinmüller-Nethl, D., Becher, C., Atatüre, M., 2014. Optical signatures of silicon-vacancy spins in diamond. Nat. Commun. 5, 3328.
Ou, Z.Y., Gage, E.C., Magill, B.E., Mandel, L., 1989. 4th-order interference technique for determining the coherence time of a light-beam. J. Opt. Soc. Am. B 6 (1), 100–103.
Patel, R.B., Bennett, A.J., Cooper, K., Atkinson, P., Nicoll, C.A., Ritchie, D.A., Shields, A.J., 2008. Postselective two-photon interference from a continuous non-classical stream of photons emitted by a quantum dot. Phys. Rev. Lett. 100 (20), 207405.
Peres, A., 1996. Separability criterion for density matrices. Phys. Rev. Lett. 77, 1413.
Poem, E., Gershoni, D., 2011. Radiative cascades in semiconductor quantum dots. In: Bergman, L., McHalei, J.L. (Eds.), Handbook of Luminescent Semiconductor Materials. CRC Press, Taylor & Francis Group, Boca Raton, pp. 321–364 (Chapter 12).
Press, D., Ladd, T.D., Zhang, B., Yamamoto, Y., 2008. Complete quantum control of a single quantum dot spin using ultrafast optical pulses. Nature 456, 218.
Sanaka, K., Pawlis, A., Ladd, T.D., Lischka, K., Yamamoto, Y., 2009. Indistinguishable photons from independent semiconductor nanostructures. Phys. Rev. Lett. 103 (5), 053601.

Santori, C., Pelton, M., Solomon, G., Dale, Y., Yamamoto, E., 2001. Triggered single photons from a quantum dot. Phys. Rev. Lett. 86 (8), 1502–1505.
Santori, C., Fattal, D., Pelton, M., Solomon, G.S., Yamamoto, Y., 2002a. Polarization-correlated photon pairs from a single quantum dot. Phys. Rev. B 66, 045308.
Santori, C., Fattal, D., Vuckovic, J., Solomon, G.S., Yamamoto, Y., 2002b. Indistinguishable photons from a single-photon device. Nature 419 (6907), 594–597.
Seidl, S., Kroner, M., Hogele, A., Karrai, K., 2006. Effect of uniaxial stress on excitons in a self-assembled quantum dot. Appl. Phys. Lett. 88, 203113.
Solomon, G.S., Pelton, M., Yamamoto, Y., 2001. Single-mode spontaneous emission from a single quantum dot in a three-dimensional microcavity. Phys. Rev. Lett. 86, 3903.
Stevenson, R.M., et al., 2006. A semiconductor source of triggered entangled photon pairs. Nature 439, 179.
Stevenson, R.M., et al., 2008. Evolution of entanglement between distinguishable light states. Phys. Rev. Lett. 101, 170501.
Unold, T., et al., 2004. Optical Stark effect in a quantum dot: ultrafast control of single exciton polarizations. Phys. Rev. Lett. 92, 157401.
Vamivakas, N.A., Zhao, Y., Lu, C.-Y., Atatüre, M., 2009. Spin-resolved quantum-dot resonance fluorescence. Nat. Phys. 5, 198.
Vogel, M.M., et al., 2007. Influence of lateral electric fields on multiexcitonic transitions and fine structure of single quantum dots. Appl. Phys. Lett. 91, 051904.
Walls, D.F., Milburn, G.J., 1995. Quantum Optics. Springer.
Warburton, R.J., 2013. Single spins in self-assembled quantum dots. Nat. Mater. 12, 483.
Warburton, R.J., Durr, C.S., Karrai, K., Kotthaus, J.P., Medeiros-Ribeiro, G., Petroff, P.M., 1997. Charged excitons in self-assembled semiconductor quantum dots. Phys. Rev. Lett. 79, 5282–5285.
Warburton, R.J., Schäflein, C., Haft, D., Bickel, F., Lorke, A., Karrai, K., Garcia, J.M., Schoenfeld, W., Petroff, P.M., 2000. Optical emission from a charge-tunable quantum ring. Nature 405, 926.
Xu, X., Wu, Y., Sun, B., Huang, Q., Cheng, J., Steel, D.G., Bracker, A.S., Gammon, D., Emary, C., Sham, L.J., 2007a. Fast spin state initialization in a singly charged InAs-GaAs quantum dot by optical cooling. Phys. Rev. Lett. 99, 097401.
Xu, X., et al., 2007b. Coherent optical spectroscopy of a strongly driven quantum dot. Science 317, 929.
Young, R.J., et al., 2005. Inversion of exciton level splitting in quantum dots. Phys. Rev. B 72, 113305.
Young, R.J., et al., 2006. Improved fidelity of triggered entangled photons from single quantum dots. New J. Phys. 8, 29.
Young, R.J., et al., 2009. Bell-inequality violation with a triggered photon-pair source. Phys. Rev. Lett. 102, 030406.

CHAPTER EIGHT

Coherent control of a semiconductor quantum dot ensemble

Takeshi Suzuki[a,*] **and Rohan Singh**[b]
[a]ISSP, The University of Tokyo, Kashiwa, Japan
[b]Department of Physics, Indian Institute of Science Education and Research Bhopal, Bhopal, India
[*]Corresponding author: e-mail address: takeshi.suzuki@issp.u-tokyo.ac.jp

Contents

1. Introduction

Coherent control uses phase-sensitive light-matter interactions to optically manipulate carriers in a desired material. A fundamental operation is Rabi oscillation, where populations of the upper and lower states in a two-level system have been inverted or "flopped" depending on the intensity and

Semiconductors and Semimetals, Volume 105
ISSN 0080-8784
https://doi.org/10.1016/bs.semsem.2020.09.005

duration of the light-matter interaction. The initial demonstration of Rabi oscillation used an atomic system because of the sharp optical transition lines, which results in easily observable coherent effects (Gibbs, 1973). Beyond Rabi oscillation, many experiments such as Ramsay fringes and photon echo have been demonstrated and utilized in device applications (Allen and Eberly, 1987). Understanding such coherent interactions is important from the perspective of applications in quantum information where the excitations can act as a quantum bit (qubit). An ideal qubit candidate requires long coherence times along with possible control of inter-qubit interactions. While atomic systems are ideal with respect to showing long coherence times, it can be difficult to efficiently couple multiple qubits.

The advent of semiconductor quantum dots (QDs), which are often regarded as artificial atoms, provides a promising route to combine atomic properties in semiconductors, which form the backbone of electronic devices. Single QDs have been used to form logic gates (Atatüre et al., 2006; Li et al., 2003). These QDs are embedded in a solid-state material, which can be used to optically and electrically tune coupling between QDs (Martin and Cundiff, 2018; Nilsson et al., 2018). However, the interactions between electrons, phonons, and impurities in semiconductor QDs lead to fast relaxation, which has marked substantial differences from a real atomic system (Takagahara, 1999). Furthermore, fluctuation of size, compound, or shape in semiconductor QDs induce inhomogeneous broadening that limits efficient coherent control in an ensemble. It is imperative to improve our understanding of coherent light-matter interactions in QD ensembles to develop practical quantum information applications.

The complications due to inhomogeneous broadening are ubiquitous in semiconductor nanostructures. Over the years, several implementations of coherent nonlinear techniques such as transient four-wave mixing (TFWM) (Shah, 1999) have been utilized to study homogeneous optical properties in semiconductor nanostructures. TFWM has been frequently employed to reveal many-body correlations and coupling between carriers (Chemla and Shah, 2001). However, the detection of the signal intensity and spectral integration of the measured signal can lead to ambiguity in interpretation of the experimental observations (Koch et al., 1992).

Two-dimensional coherent spectroscopy (2DCS) has been developed as an extension of TFWM to overcome the aforementioned limitations through frequency-resolved detection of the nonlinear-signal field. The main advantages of 2DCS are (i) unambiguous separation between homogeneous and inhomogeneous linewidths (Siemens et al., 2010) and (ii) identification of the origin of coupling between multiple transitions (Dai et al., 2010). 2DCS is especially suitable for studying semiconductor QD ensembles because we can measure the intrinsic homogenous properties despite large inhomogeneous broadening. This advantage has been used to reveal relaxation dynamics of excitons (Moody et al., 2011b), exciton-phonon interactions (Moody et al., 2011a; Suzuki et al., 2018b), and binding energy of the biexciton (Moody et al., 2013a). These experiments can all be understood through perturbative light-matter interactions up to third order. At higher excitation density higher-order terms are needed to explain the experimental observations (Mermillod et al., 2016; Moody et al., 2013c; Stone et al., 2009). This approach, however, is inapplicable to a strictly non-perturbative phenomenon such as Rabi oscillation. Experimentally too, it is not possible to indefinitely increase pulse energies to reach the non-perturbative interaction with each of the three excitation pulses. To circumvent this experimental limitation, we have developed a prepulse-2DCS experiment, in which we use an initial, intense "prepulse" to prepare the QD system in a population state which is perturbatively "probed" using the usual 2DCS experiment (Suzuki et al., 2016, 2018a). We have demonstrated coherent control of a semiconductor QD ensemble using this novel technique. Coherent evolution of the inhomogeneous distribution has been clearly elucidated. Furthermore, these studies revealed nontrivial coherent evolution of population in inhomogeneously broadened multilevel system. Our results also demonstrate that the prepulse 2DCS can be a sensitive tool of coherent readout in any inhomogeneous broadened system.

In this chapter, we first describe the basic theoretical formalisms to describe coherent light-matter interactions using the optical Bloch equations. We will follow this with a brief explanation of the theoretical and experimental concepts of 2DCS along with some discussion of previous studies of semiconductor nanostructures. We will then discuss our recent study of coherent control for semiconductor QDs ensemble by using the prepulse 2DCS; we will show how we extracted coherent evolution of exciton population in the presence of inhomogeneous broadening. Finally, we conclude this chapter with an outlook on the future progress in this field.

2. Optical Bloch equations and coherent control

Incoherent light-matter interaction relies only on light intensity and relative populations of the two states in an optical transition which inhibits population inversion of a two-level system. Coherent control, on the other

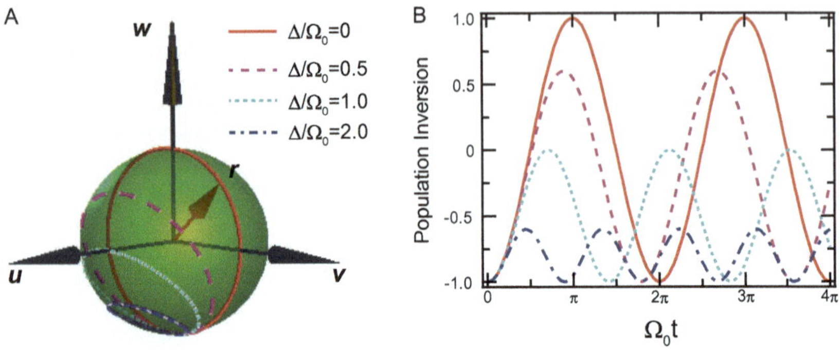

Fig. 1 (A) Illustration of the evolution of the Bloch vector representing a two-level system and (B) corresponding population evolution for different values of detuning between the excitation light and the transition energy. The red-solid, magenta-dashed, cyan-dotted, and blue dashed-dotted lines represent detuning given by $\Delta/\Omega_0 = 0, 0.5, 1,$ and 2, respectively.

hand, involves phase-sensitive manipulation of an optical transition which allows forming population inversion in a two-level system as illustrated in Rabi oscillation experiments.

The optical Bloch equations (OBEs) provide the theoretical formalism to understand and model coherent control of an optical transition. These equations are inspired from the Bloch equations which were originally developed to study the response of a spin to an external magnetic field. The magnetic field "rotates" the spin, which is treated as an arrow moving on a sphere, known as Bloch sphere, with one end fixed at its center; one such arrow is shown in Fig. 1A. The OBEs are readily obtained by substituting the spin-one-half particle with a two-level system and the external magnetic field by the electric field of the incident light (Allen and Eberly, 1987). If required, it is straightforward to extend this formalism to multiple states. Next we present the basic concepts of OBEs and coherent control, as will be relevant for the rest of the chapter.

2.1 Principles of optical Bloch equation

Let us consider a two-level system composed of the ground $|0\rangle$ and excited states $|1\rangle$. Any coherent state $|\psi\rangle$ can be written as a superposition of these two states

$$|\psi\rangle = C_0|0\rangle + C_1|1\rangle, \tag{1}$$

where C_0 and C_1 are complex numbers, satisfying $|C_0|^2 + |C_1|^2 = 1$.

In a statistical ensemble the system is often in a mixed state. Consequently it is useful to define the density matrix $\rho = |\psi\rangle\langle\psi|$ of the ensemble, explicitly written as

$$\rho = \begin{pmatrix} \rho_{00} & \rho_{01} \\ \rho_{10} & \rho_{11} \end{pmatrix} = \begin{pmatrix} |C_0|^2 & C_0^* C_1 \\ C_0 C_1^* & |C_1|^2 \end{pmatrix}. \tag{2}$$

The diagonal elements of ρ represent the populations in different states while the off-diagonal elements are the coherences between these states. OBEs are essentially used to study the evolution of ρ in the presence of an external electromagnetic radiation.

We introduce the Bloch representation before proceeding to a discussion of the dynamics of the density matrix; Bloch representation is very intuitive to see how the system develops with the electromagnetic field. The Bloch vector $\boldsymbol{r} = (u \ \ v \ \ w)$ is defined, where

$$u = 2\,\mathrm{Re}\,(\rho_{01}) = C_0^* C_1 + C_0 C_1^*, \tag{3}$$

$$v = 2\mathrm{Im}(\rho_{01}) = -i\left(C_0^* C_1 - C_0 C_1^*\right), \tag{4}$$

$$w = \rho_{11} - \rho_{00} = |C_1|^2 - |C_0|^2. \tag{5}$$

Here, u and v are the real and imaginary part of the coherence ρ_{01}, respectively, and w stands for the population inversion, which is defined as the difference in population of the excited and ground states. The sphere in Fig. 1A shows an illustration of a Bloch sphere with the arrow indicating a Bloch vector. The change in the orientation of the Bloch vector with time describes the evolution of the system. The lines on the Bloch sphere in Fig. 1A give some examples of such evolutions and will be discussed in more detail later. It is easily confirmed that $|u|^2 + |v|^2 + |w|^2 = 1$, which means the Bloch sphere is a unit sphere in $\boldsymbol{r}$ space. In the absence of decay mechanisms, head of the Bloch vector is attached on the Bloch sphere.

The Heisenberg equation for the density matrix is written as

$$\frac{d}{dt}\rho = -\frac{i}{\hbar}[H, \rho], \tag{6}$$

where H is the Hamiltonian of the system. The total Hamiltonian H for the two-level system interacting with the electromagnetic radiation, assuming dipole-like interaction is,

$$H = \begin{pmatrix} \varepsilon_0 & -\mu_{01}E(t) \\ -\mu_{01}E(t) & \varepsilon_1 \end{pmatrix}, \tag{7}$$

where $\varepsilon_i (i=0,1)$ is the energy of state i, μ_{01} is the transition dipole moment between states 0 and 1, and $E(t)$ is the time-dependent electric field of light. Eq. (6) can be written as the set of equations

$$\dot{\rho}_{11} = -i\Omega(t)(\rho_{10} - \rho_{01}) \tag{8}$$

$$\dot{\rho}_{01} = i\frac{\varepsilon_1 - \varepsilon_0}{\hbar}\rho_{01} - i\Omega(t)(\rho_{00} - \rho_{11}) \tag{9}$$

where $\Omega(t)$ is Rabi frequency given by $\Omega(t) = \mu_{01}E(t)/\hbar$. The remaining terms ($\rho_{00}$ and ρ_{10}) are derived from relations $\rho_{00} = 1 - \rho_{11}$ and $\rho_{10} = \rho_{01}^*$, respectively.

2.2 Rabi oscillation

Rabi oscillation, which involves generating population inversion in a two-level system highlights the power of coherent control and can be directly obtained from the OBEs. In this subsection, we discuss several aspects of Rabi oscillations, which are relevant for our studies.

2.2.1 Rotating-wave approximation

First we consider a steady electric field component of light $E(t) = E_0 \cos \omega t = \frac{E_0}{2}(e^{-i\omega t} + e^{i\omega t})$. Later on we'll extend this treatment to pulsed lasers, which we used in our work. Next, we transform the coherence terms of the density matrix to the rotating frame

$$\widetilde{\rho}_{01}(t) = e^{-i\omega t}\rho_{01}(t) \tag{10}$$

$$\widetilde{\rho}_{10}(t) = e^{+i\omega t}\rho_{10}(t). \tag{11}$$

We keep only slowly varying terms for $E(t)$, then the set of Eqs. (8) and (9) become

$$\dot{\rho}_{11} = -i\frac{\widetilde{\Omega}(t)}{2}(\widetilde{\rho}_{10} - \widetilde{\rho}_{01}) \tag{12}$$

$$\dot{\widetilde{\rho}}_{01} = i\Delta\widetilde{\rho}_{01} - i\frac{\widetilde{\Omega}(t)}{2}(\rho_{00} - \rho_{11}), \tag{13}$$

where $\widetilde{\Omega}(t) = \Omega_0 = \mu_{01}E_0/\hbar$ is the Rabi frequency in the rotating frame and Δ is the detuning between laser frequency and transition frequency of the two-level system given by

$$\Delta = \frac{\varepsilon_1 - \varepsilon_0}{\hbar} - \omega. \tag{14}$$

It is more intuitive to see the dynamics in terms of the Bloch representation. The Bloch vector in the rotating frame, $\tilde{\boldsymbol{r}} \equiv \left(2\mathrm{Re}\left(\tilde{\rho}_{01}\right) \; 2\mathrm{Im}\left(\tilde{\rho}_{01}\right) \; \rho_{11} - \rho_{00}\right)$, is related to $\boldsymbol{r}$ as,

$$\tilde{\boldsymbol{r}} = \begin{pmatrix} \tilde{u} \\ \tilde{v} \\ \tilde{w} \end{pmatrix} = \begin{pmatrix} \cos\omega t & \sin\omega t & 0 \\ -\sin\omega t & \cos\omega t & 0 \\ 0 & 0 & 1 \end{pmatrix} \boldsymbol{r}. \tag{15}$$

The OBEs are written as

$$\frac{d}{dt}\tilde{\boldsymbol{r}}(t) = \begin{pmatrix} 0 & -\Delta & 0 \\ \Delta & 0 & \Omega_0 \\ 0 & -\Omega_0 & 0 \end{pmatrix} \begin{pmatrix} \tilde{u} \\ \tilde{v} \\ \tilde{w} \end{pmatrix} = \tilde{\boldsymbol{\Omega}}(t) \times \tilde{\boldsymbol{r}}(t), \tag{16}$$

where $\tilde{\boldsymbol{\Omega}}(t)$ is the torque vector defined as

$$\tilde{\boldsymbol{\Omega}}(t) = \begin{pmatrix} -\Omega_0 \\ 0 \\ \Delta \end{pmatrix}. \tag{17}$$

The dynamics of two-level system corresponds to the rotation of the Bloch vector around the torque vector in the Bloch sphere.

2.2.2 Resonant interaction

We now apply the above framework to illustrate the population dynamics due to coherent interaction of light with a two-level system. We assume that the system is in the ground state before interaction of light, i.e., $\rho_{00}(0) = 1$, $\rho_{11}(0) = \rho_{01} = 0$. For resonant interactions we set $\Delta = 0$ and solve Eqs. (12) and (13) to obtain solutions

$$\rho_{11}(t) = \frac{1}{2}(1 - \cos\Omega_0 t) \tag{18}$$

$$\rho_{01}(t) = i\sin\Omega_0 t. \tag{19}$$

The solid red line in Fig. 1B shows the evolution of population for resonant interaction. The solid red line in Fig. 1A shows the evolution of the Bloch vector as a rotation around the $\tilde{u}$ axis. As the interaction with the continuous-wave (CW) light source increases, the population from the ground state is transferred to the excited state and back. We note that this enables perfect population inversion, i.e., $\rho_{00}(t)-\rho_{11}(t)=-1$, which is a consequence of the coherent interaction. For a system without any decay mechanisms, as we have assumed so far, these oscillations will go on indefinitely. In this case the final state of the system can be calculated from the product $\Omega_0 t$, which is known as the pulse area. Thus, a longer interaction with low-intensity light can be the same as a shorter interaction with a higher-intensity light. We use this property later on to extend the aforementioned CW-interaction treatment to interaction with high-intensity laser pulses.

2.2.3 Nonresonant interaction

For nonresonant interaction ($\Delta \neq 0$), the solution is

$$\begin{pmatrix} \tilde{u}(t) \\ \tilde{v}(t) \\ \tilde{w}(t) \end{pmatrix} = \begin{pmatrix} \dfrac{\Omega_0 \Delta}{\Omega_0^2+\Delta^2}(1 - \cos\Omega(\Delta)t) \\ -\dfrac{\Omega_0}{\sqrt{\Omega_0^2+\Delta^2}}\sin\Omega(\Delta)t \\ -\dfrac{\Omega_0^2\cos\Omega(\Delta)t+\Delta^2}{\Omega_0^2+\Delta^2} \end{pmatrix} \tag{20}$$

where $\Omega(\Delta)$ is the generalized Rabi frequency, defined by

$$\Omega(\Delta) = \sqrt{\Omega_0^2 + \Delta^2}. \tag{21}$$

Fig. 1B shows the population evolution for nonresonant interaction. The magenta-dashed, cyan-dotted, and blue dash-dotted lines represent in the case of $\Delta/\Omega_0=0$, 0.5, 1, and 2, respectively. The important effects of detuning are (i) the amplitude of population inversion is less than 1, i.e., perfect population inversion is not achievable and (ii) the oscillation becomes faster with increasing detuning.

2.2.4 Relaxation effects

We have ignored relaxation processes so far for the sake of simplicity. This situation, however, is unrealistic. The excited state in any system has a finite lifetime. Furthermore, the coherence between the upper and lower states can be destroyed due to the interactions with the environment. Consequently, the amplitude of Bloch vector $|\boldsymbol{r}|$ is less than unity during the system's evolution. We introduce phenomenological relaxation terms Γ_{11} and γ_{01} to rewrite Eqs. (12) and (13) as

$$\dot{\rho}_{11} = -\Gamma_{11}\rho_{11} - i\frac{\widetilde{\Omega}(t)}{2}\left(\widetilde{\rho}_{10} - \widetilde{\rho}_{01}\right) \tag{22}$$

$$\dot{\widetilde{\rho}}_{01} = -(\gamma_{01} - i\Delta)\widetilde{\rho}_{01} - i\frac{\widetilde{\Omega}(t)}{2}\left(\rho_{00} - \rho_{11}\right). \tag{23}$$

Γ_{11} is the longitudinal relaxation or population decay rate; it is related with the lifetime in the upper state T_{11} by

$$\Gamma_{11} = \frac{1}{T_{11}}. \tag{24}$$

γ_{01} is the transverse relaxation (dephasing) rate which represents the coherence decay time between the upper and lower states. γ_{01} is related to the population decay rate Γ_{11} as

$$\gamma_{01} = \frac{1}{2}\Gamma_{11} + \gamma_{01}^{*}, \tag{25}$$

where γ_{01}^{*} is called the pure-dephasing rate since it is independent of the population decay. We modify the OBEs in the Bloch representation to include the decay terms

$$\dot{\widetilde{u}} = -\gamma_{01}\widetilde{u} - \Delta\widetilde{v} \tag{26}$$

$$\dot{\widetilde{v}} = \Delta\widetilde{u} - \gamma_{01}\widetilde{v} + \Omega_0\widetilde{w} \tag{27}$$

$$\dot{\widetilde{w}} = -\Omega_0\widetilde{v} - \Gamma_{11}(\widetilde{w} + 1). \tag{28}$$

A sufficiently long interaction with light results in a steady state, which can be solved by setting the time derivative to zero. The equilibrium population inversion is

$$\widetilde{w}_{eq} = \frac{1}{1 + \frac{\Gamma\left(\gamma_{01}^2 + \Delta^2\right)}{\gamma_{01}\Omega_0^2}} - 1 \tag{29}$$

Next we discuss the effect of the relaxation terms on Rabi oscillation. We do not include pure-dephasing processes ($\gamma_{01}^{*} = 0$). The population evolution in the presence of decay terms cannot be obtained analytically and are calculated numerically. Fig. 2 shows the calculated population inversion including relaxation for (a) resonant ($\Delta = 0$) and (b) nonresonant ($\Delta \neq 0$) interaction. In both cases, the oscillation is smeared with increase in time. Furthermore, the equilibrium population inversion $\widetilde{w}_{eq}$ reduces as Γ_{11} increases. Comparing between the resonant and nonresonant cases, the

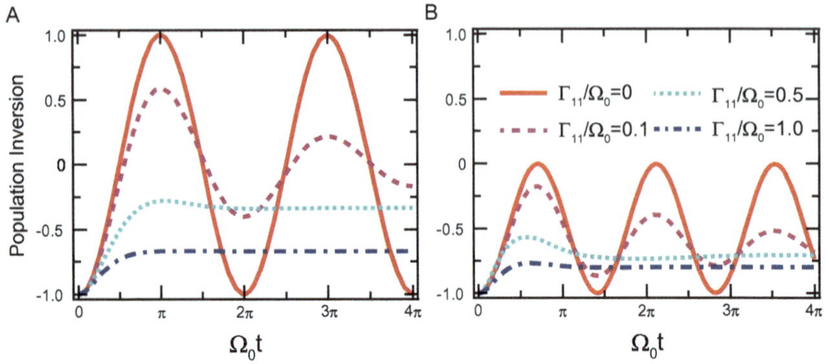

Fig. 2 Numerical solutions for population inversion for (A) resonant ($\Delta=0$) and (B) nonresonant ($\Delta\neq 0$) interactions. The red-solid, magenta-dashed, cyan-dotted, and blue dashed-dotted lines correspond to $\Gamma_{11}/\Omega_0=0$, 0.1, 0.5, and 1, respectively.

nonresonant case with no relaxation seems qualitatively similar to the resonant case with relaxation.

2.3 Coherent control in quantum dots—Previous studies

Significant developments in the growth and processing techniques for semiconductor nanostructures has enabled "atom-like" system realized in the solid-state structure, called semiconductor quantum dots(QDs). Three-dimensional confinement of charge carriers results in atom-like discrete energy levels in semiconductor QDs leading to the significant reduction of population decay rates and dephasing rates (Borri et al., 2001). These features have enabled coherent control in semiconductor QDs.

Evolution of a semiconductor QD due to coherent interaction with excitation light is generally studied using single-particle technique motivated by the interest in using them for quantum information. Single QDs are realized by using electron beam lithography and liftoff to open an aperture in this opaque aluminum or gold films deposited on the sample surface (Gammon et al., 1996). The earliest work was performed using resonant excitation for the coherent preparation and incoherent photoluminescence was detected (Kamada et al., 2001; Stievater et al., 2001). Advances in sample-preparation methods have led to the development of QDs in a diode structure, where the photocurrent is measured, which resulted in improved measurement accuracy (Ramsay et al., 2010a,b).

While experiments on single QDs are important to demonstrate the usefulness of QDs, they are impractical from the perspective of developing useful applications where isolation and control of single QDs might not be practical. Unfortunately reports on coherent control of semiconductor QD ensembles have been few and inconclusive. The difficulty arises due to large inhomogeneous broadening in the transition energy due to fluctuations in size, shape, and composition. The earliest work on a semiconductor QD ensemble was performed by using the conventional pump-probe measurement focusing on specific spectrum group of ensemble QDs (Borri et al., 2002). Later, the ensemble averaging was figured out to significantly smear out signals of Rabi oscillations in this approach (Kujiraoka et al., 2010).

Coherent control in QD ensembles, if implemented reliably, can have potential applications in diverse areas such as QD lasers, QD semiconductor amplifiers, and quantum memories beyond quantum information. The challenge of clear observation of coherent evolutions in semiconductor QD ensemble relies on unambiguously measuring homogeneous properties despite the inhomogeneous broadening. The conventional method, such as linear spectroscopy or time-resolved four-wave mixing cannot distinguish coherent evolutions in each frequency group. The advent of new types of nonlinear spectroscopy, namely *two-dimensional coherent spectroscopy (2DCS)*, have enabled us to distinguish between the homogeneous and inhomogeneous broadenings, and study many physical properties in semiconductor QD ensembles. We discuss the salient features of 2DCS in the next section.

3. Two-dimensional coherent spectroscopy

Two-dimensional coherent spectroscopy (2DCS) is a relatively new technique, inspired by the two-dimensional nucleus magnetic resonance (2D-NMR) spectroscopy (Ernst et al., 1987), extending the excitation and detection range from radio frequency to optical frequency. In an optical 2DCS experiment the transient four-wave mixing (TFWM) signal field is detected as a function of precisely-controlled delay between the excitation pulses. A 2D spectrum is obtained by taking a numerical Fourier transform of the TFWM signal with respect to the scanned delays. The main advantages of 2DCS are separation of inhomogeneous and homogeneous line widths, and identification of coupling between multiple resonances. As we will demonstrate later, these benefits are particularly useful to study semiconductor nanostructures. We begin with a brief discussion of the experimental and theoretical aspects of the TFWM technique and then proceed to describing 2DCS.

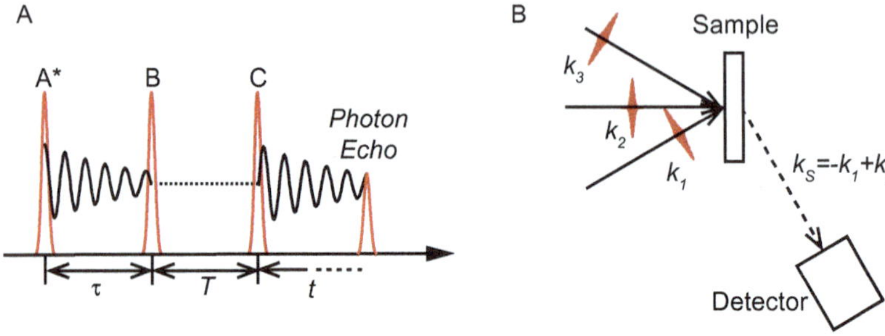

Fig. 3 (A) Time ordering of the excitation pulses and radiated signal for the rephasing transient four-wave mixing (TFWM) experiment. Delays between pulse A*, B, C and radiated signal are denoted as τ, T, and t, respectively. (B) Schematics of the experimental setup for measuring the TFWM signal using noncollinear geometry.

3.1 Transient four-wave mixing

TFWM experiments use either two or three pulses to excite the sample. The induced third-order nonlinear-signal intensity is measured as a function of a delay between pulses. Here, we restrict our discussion to the three-pulse TFWM experiment because 2DCS is a direct extension of this technique. Fig. 3A shows the pulse sequence for the rephasing (photon-echo) time ordering in TFWM measurements, where the first pulse is conjugate to the emitted FWM signal. Delays between the first, second, and third excitation pulses, and the emitted FWM signal are denoted as τ, T, and t, respectively. The FWM signal is usually spatially isolated from the excitation pulses and other nonlinear contributions through the noncollinear geometry shown in Fig. 3B; the excitation pulses propagate along direction $\boldsymbol{k}_1$, $\boldsymbol{k}_2$, and $\boldsymbol{k}_3$ and the TFWM signal is detected in the phase-matched direction $\boldsymbol{k}_s = -\boldsymbol{k}_1 + \boldsymbol{k}_2 + \boldsymbol{k}_3$. In the presence of inhomogeneous broadening, the FWM signal is emitted as a photon-echo pulse at $t = \tau$ due to the rephasing effect. The decay of the FWM signal as a functions of τ gives the homogeneous decay rate corresponding to the dephasing rate.

The three excitation pulses can be treated as perturbations on the system, which results in the third-order nonlinear-signal proportional to the product of the excitation pulses' electric fields. The theoretical treatment is further simplified if we assume Dirac-delta excitation pulses, which is reasonable when the probed frequency range is narrower than the excitation pulse spectrum. The TFWM signal for a single two-level system, following the aforementioned approximations, can be shown to be

$$S(\tau, T, t) = 2|E|^3 \mu_{01}^4 e^{-i\omega_{01}(\tau - t)} e^{-(\gamma_{01}\tau + \Gamma_{11}T + \gamma_{01}t)} e^{-\frac{1}{2}\sigma_{01}^2(\tau - t)^2} \Theta(\tau)\Theta(T)\Theta(t), \tag{30}$$

where $|E|$ is the electric-field strength of each excitation pulse, μ_{01} is the transition dipole moment and ω_{01} is the optical transition energy, γ_{01} is the dephasing rate, Γ_{11} is the population decay rate, and σ_{01} is the inhomogeneous linewidth for the transition. Θ is the Heaviside step function to ensure the correct time ordering of the excitation pulses. As is evident in Eq. (30), TFWM signal contains rich information. Since the TFWM uses the higher-order perturbation process, it is also sensitive to many-body effects when it is applied to a multilevel system. Especially, in a semiconductor, TFWM revealed numerous many-body phenomena such as the presence of biexcitons, exciton-exciton correlations, and exciton-phonon interactions (Chemla and Shah, 2001).

3.2 Perturbative solution of the OBE using double-sided Feynman diagram

The most precise calculations of TFWM signal in semiconductor nanostructures are based on microscopic theory (Haug and Koch, 2009). However, important physical insights are often gained through simpler, perturbative solution of the OBEs discussed above. Here we present how the TFWM signal can be calculated for a system using double-sided Feynman diagrams. Since the nth order polarization $P^{(n)}$ is related to density matrix $P^{(n)} = \langle \mu \rho^{(n)} \rangle$, our goal here is to perturbatively calculate the density matrix. Instead of rigorously solving the OBEs, we minimized mathematical descriptions by limiting ourselves to special cases, in which we treat with a single two-level system and approximate the excitation pulses as Dirac-delta-function in time. A comprehensive description can be found in, e.g., Mukamel (1995).

Here we discuss a two-level system composed of the upper and lower states. By adding the population decay rate and the dephasing rate to Eqs. (8) and (9), the dynamics of density matrix is determined by the coupled differential equations

$$\dot{\rho}_{11} = -\Gamma_{11}\rho_{11} - i\Omega(t)(\rho_{10} - \rho_{01}) \tag{31}$$

$$\dot{\rho}_{01} = -\gamma_{01}\rho_{01} + i\frac{\varepsilon_1 - \varepsilon_0}{\hbar}\rho_{01} - i\Omega(t)(\rho_{00} - \rho_{11}). \tag{32}$$

We need to solve Eqs. (31) and (32) perturbatively to obtain the desired dynamics. It should be mentioned here that Rabi solution we discussed in the last section is valid for *any* strength of the applied field while the perturbative solution is only appropriate in the weak-field limit. The advantages of using the perturbative calculation is that the OBEs with decay terms can be analytically calculated in the limit of Dirac-delta-function excitation pulses.

We write the density matrix element by expanding in terms of Rabi frequency by substituting $\Omega(t) \rightarrow \alpha\Omega(t)$ to tag the number of order in the perturbation, as

$$\rho_{ij} = \rho_{ij}^{(0)} + \alpha\rho_{ij}^{(1)} + \alpha^2\rho_{ij}^{(2)} + \cdots \tag{33}$$

By inserting Eq. (33) in Eqs. (31) and (32), and applying rotating approximation, we can derive the relationships in the nth order as

$$\dot{\widetilde{\rho}}_{11}^{(n)} = -\Gamma_{11}\widetilde{\rho}_{11}^{(n)} - i\left(\widetilde{\Omega}^*(t)\rho_{10}^{(n-1)} - \widetilde{\Omega}(t)\rho_{01}^{(n-1)}\right) \tag{34}$$

$$\dot{\widetilde{\rho}}_{01}^{(n)} = -\left(\gamma_{01} - i\frac{\varepsilon_1 - \varepsilon_0}{\hbar}\right)\widetilde{\rho}_{01}^{(n)} - i\widetilde{\Omega}^*(t)\left(\rho_{00}^{(n-1)} - \rho_{11}^{(n-1)}\right). \tag{35}$$

Here, we have expressed Rabi frequency as $\Omega(t) = \widetilde{\Omega}(t)e^{+i\omega t} + \widetilde{\Omega}^*(t)e^{-i\omega t}$, and kept only slowly varying terms. Assuming that the duration of pulse arriving at τ is much shorter than the dynamics of density matrix, Rabi frequency can be approximately written as a Dirac-delta function and Eqs. (34) and (35) can be analytically solved to obtain

$$\begin{aligned}\widetilde{\rho}_{11}^{(n)}(t) = \widetilde{\rho}_{11}^{(n)}(-\infty) - \int_{-\infty}^{t} \Gamma_{11}\widetilde{\rho}_{11}^{(n)}(t')dt' \\ - i\left(\widetilde{\Omega}_0^*\rho_{10}^{(n-1)}(\tau) - \widetilde{\Omega}_0\rho_{01}^{(n-1)}(\tau)\right)\end{aligned} \tag{36}$$

$$\begin{aligned}\widetilde{\rho}_{01}^{(n)}(t) = \widetilde{\rho}_{01}^{(n)}(-\infty) - \int_{-\infty}^{t} \left(\gamma_{01} - i\frac{\varepsilon_1 - \varepsilon_0}{\hbar}\right)\widetilde{\rho}_{01}^{(n)}(t')dt' \\ - i\widetilde{\Omega}_0^*\left(\rho_{00}^{(n-1)}(\tau) - \rho_{11}^{(n-1)}(\tau)\right).\end{aligned} \tag{37}$$

From these expressions, we notice that a single light-matter interaction, which increases the "order" of interaction by one, converts a population to coherence and vice versa. We can generalize this to mean that from the $(n-1)$th order to the nth order results in taking the system from $(i,j) \rightarrow (i \pm 1, j)$ or $(i, j \pm 1)$, where (i,j) represents $\rho_{i,j}$. Furthermore, the complex conjugation ($\widetilde{\Omega}(t)$ or $\widetilde{\Omega}^*(t)$) and overall sign of the interaction are related to the direction of the perturbative change, e.g., in Eq. (36) the transition from $\rho_{10}^{(n-1)}$ to $\widetilde{\rho}_{11}^{(n)}$ is associated by the factor $-i\widetilde{\Omega}_0^*$ while that from $\rho_{01}^{(n-1)}$ includes a factor of $+ i\widetilde{\Omega}_0$ component.

Double-sided Feynman diagrams are convenient to visualize such a perturbative evolution and can be used as a powerful and intuitive tool to calculate the TFWM signal. Fig. 4 shows four basic elements used in these

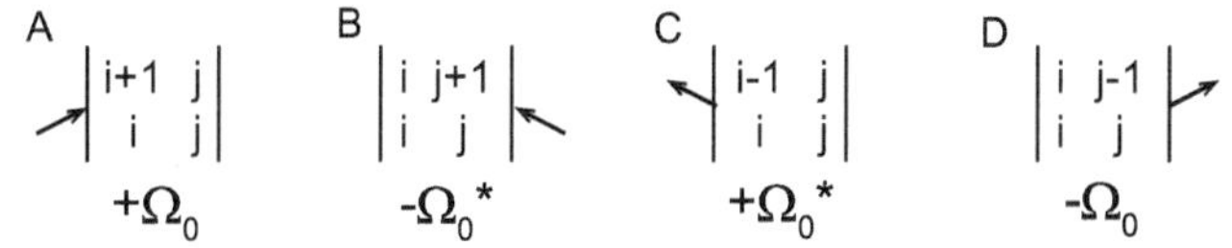

Fig. 4 (A)–(D) The four basic elements for the double-sided Feynman diagram. The corresponding factor gained by each interaction is shown below to each diagram.

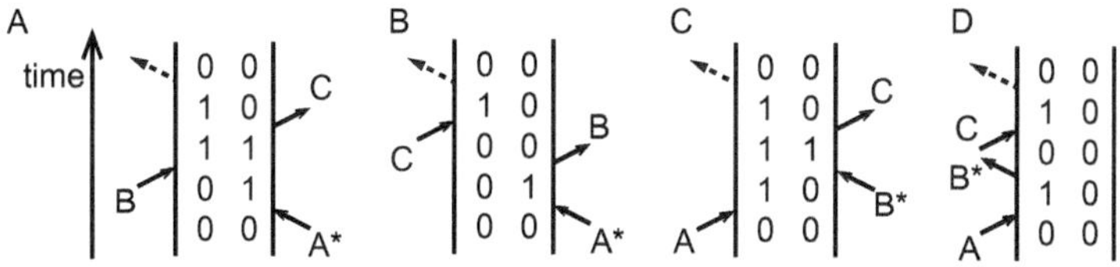

Fig. 5 The four possible double-sided Feynman diagrams representing the interactions leading to rephrasing (A and B) and nonrephasing (C and D) FWM signal for a two-level system.

diagrams, which represent all the transitions discussed above. The factor obtained with each interaction with an excitation pulse is shown below each diagram. A few basic rules follow a few basic principles. (i) Indices of density matrix are indicated by the indices $|i\,j|$, and excitation pulses are shown as arrows. (ii) The arrows are shown as incoming/outgoing when the acting field is absorbed/emitted. (iii) Arrow pointing from left to right (Fig. 4A and D) indicate interaction with the nonconjugate component of the excitation pulse ($\widetilde{\Omega}_0$), while arrow pointing from right to left represent interaction with the conjugate part of the excitation pulse ($\widetilde{\Omega}_0^*$) (Fig. 4B and C). (iv) Each interaction with the arrow on the right side of the diagram introduces a factor of -1.

The TFWM process is depicted by Feynman diagrams comprising four light-matter interactions—three with excitation pulses and one for emission of the TFWM signal—can be generated based on the fundamental components in Fig. 4. The evolution of the system can follow multiple quantum-mechanical pathways; all possible pathways need to be included to calculate the total TFWM signal. Fig. 5 shows the four possible Feynman diagrams that contribute to the FWM signal from a two-level system. Time (perturbative order) increases from the bottom to the top, as shown in Fig. 5A. The first excitation pulse is conjugate to the FWM signal in Fig. 5A and B, and they contribute to the rephrasing signal. The FWM signal in Eq. (30) was obtained by adding the contributions of these diagrams; it is useful to revisit that equation to highlight some key features. (i) The factor of 2 in Eq. (30) comes from the constructive interference between these pathways. (ii) The signal phase evolves with frequencies $-\omega_{01}$ and ω_{01} during delays τ and t, respectively, and manifests as the factor $e^{-i\omega_{01}(\tau-t)}$. The opposite signs indicate the relative conjugation between the first excitation pulse and the radiated signal. The amplitude of these frequencies can also be different in case of multilevel systems; these unique frequencies can be directly read from the corresponding Feynman diagrams. The Feynman diagrams in Fig. 5C and D show the second excitation pulse is conjugated to the signal and, thus, represent the nonrephasing signal. The same procedure to calculate the FWM signal, in principle, can be applied to more complicated multilevel systems; care must be taken to include all possible Feynman diagrams.

3.3 Principle of 2DCS

An important application of the FWM experiment is to measure the homogeneous dephasing rate even in the presence of inhomogeneity as discussed in Section 3.1. However, these measurements assume that the inhomogeneous linewidth is significantly larger than the homogeneous linewidth and that the homogeneous decay rate is constant across the inhomogeneous spectrum.

2DCS can be used to overcome these limitations, which is enabled by a couple of modifications. First, we measure the TFWM signal field as a function of multiple time axes by heterodyning the signal with a known reference pulse, as shown schematically in Fig. 6A, instead of the time-integrated FWM signal. Second, the TFWM signal is measured as function of two time delays with precise, subwavelength accuracy. These two advances enable us to take a Fourier transform of the FWM signal with respect to the two time delays that are scanned, which unfolds the one-dimensional FWM signal into two frequency axes as shown by simulation for an inhomogeneously broadened system in Fig. 6B. The vertical and horizontal axes in Fig. 6B are frequencies ω_τ and ω_t corresponding to Fourier transform with respect to τ and t, respectively. Since the first pulse is conjugate to the emitted FWM signal, the vertical frequency axis shows negative values. In the inhomogeneous limit ($\sigma_{01} \gg \gamma_{01}$), the homogeneous and inhomogeneous broadening are perfectly separated along the diagonal and cross-diagonal axes shown in Fig. 6B. In case of intermediate inhomogeneity, the homogeneous and inhomogeneous linewidths are obtained through lineshape analysis of 2DCS spectra (Siemens et al., 2010).

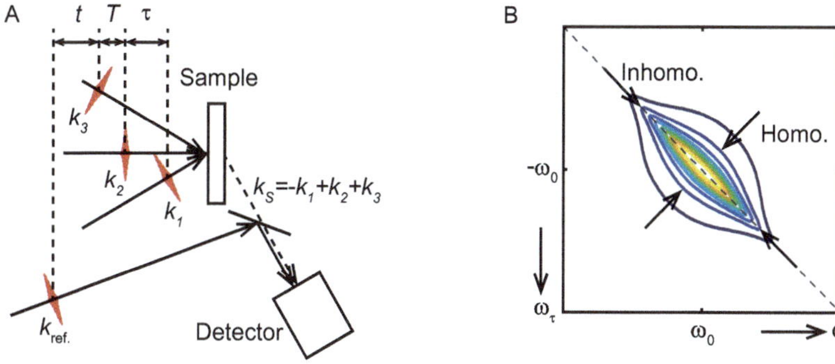

Fig. 6 (A) Schematics of an experimental setup for 2DCS. (B) Simulated rephasing 2D amplitude spectrum.

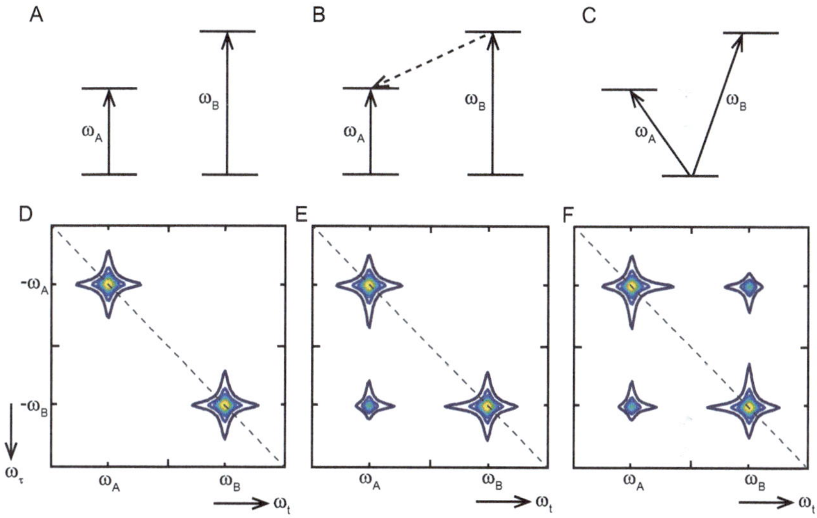

Fig. 7 Two two-level systems with (A) no, (B) incoherent, and (C) coherent coupling between the excited states. (D)–(F) Simulated rephasing 2D spectra clearly showing the coupling, or lack thereof, in examples (A)–(C).

Another advantage of 2DCS is the ability to unambiguously reveal coupling between multiple transitions through off-diagonal peaks. Fig. 7A–C shows three scenarios of coupling between two two-level systems with different transition energies ($\omega_A < \omega_B$) and corresponding 2DCS rephasing spectra (Fig. 7D–F). When the transitions are independent (Fig. 7A), only diagonal peaks, at the corresponding transition energies appear as shown in Fig. 7D. If the population transfer occurs from the higher- to lower-energy excited state through an incoherent relaxation process, a single additional off-diagonal peak appears at $(\omega_A, -\omega_B)$ as shown in Fig. 7E. Furthermore, if the two transitions are quantum-mechanically coupled, where two excited states share the common ground state shown in Fig. 7C, it manifests as two off-diagonal peaks shown in Fig. 7F. It is worth mentioning that the off-diagonal peaks cannot be identified by linear spectroscopy such as PL or absorption because they would overlap with strong diagonal peaks along one-dimensional axis (Jonas, 2003). In TFWM measurements, the coupling appears as a quantum beat, which can be difficult to differentiate from beats due to polarization interference (Koch et al., 1992).

These benefits of 2DCS have been widely leveraged to study physical phenomena in diverse materials such as atomic vapors (Dai et al., 2012; Li et al., 2013; Tian et al., 2003; Vaughan et al., 2007), colloidal nanocrystals (Wong and Scholes, 2011), nitrogen vacancy center in diamond (Huxter et al., 2013), light-harvesting complexes (Brixner et al., 2005), and proteins and other molecular systems (Fayer, 2009).

3.4 Experimental implementation of 2DCS

The aforementioned advantages of 2DCS vis-à-vis TFWM experiment is enabled by the detection of phase-sensitive FWM signal field with a precise control of time delays. Achieving these levels of control is nontrivial for optical frequencies in the range of 100s of THz. Numerous experimental techniques depending on the optical geometry, feedback technique, and analysis method have been developed in order to achieve the aforementioned precise control. A discussion of the nuances of the different techniques is beyond the scope of this chapter. We restrict ourselves to briefly distinguish between two broad categories of experiments based on the geometry of the excitation pulses, which can be collinear and noncollinear, as Fig. 8 (Nardin et al., 2015).

Background-free noncollinear detection of the FWM signal is the most commonly used geometry. Fig. 8A shows the noncollinear so-called "box" geometry for a one-quantum rephasing scan in which excitation pulses A, B, and C are incident on the sample from directions $\boldsymbol{k_A}$, $\boldsymbol{k_B}$, and $\boldsymbol{k_C}$, respectively, and the signal is emitted in the phase-matched direction, $\boldsymbol{k_s} = -\boldsymbol{k_A} + \boldsymbol{k_B} + \boldsymbol{k_C}$. The FWM signal field is measured using heterodyne interferometry with a fourth reference pulse whose amplitude and phase are known. The time delays are generated through three of nested Michelson interferometers with passive (Selig et al., 2008) or active phase-stabilization (Bristow et al., 2009). Alternatively, diffractive optics and spectral pulse

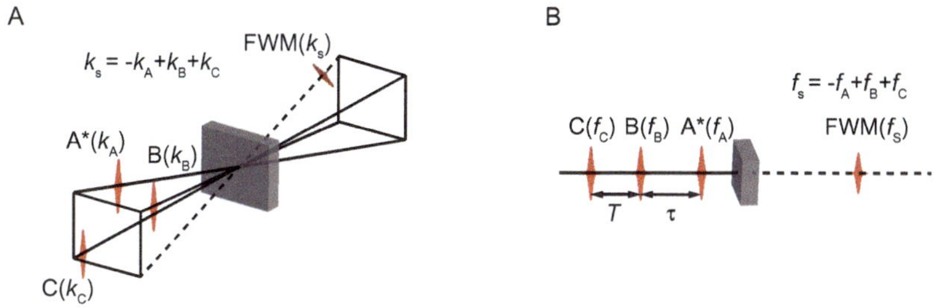

Fig. 8 Schematics of commonly used geometry for 2DCS experiments. (A) Noncollinear box geometry in which three pulses with wavevector k_A, k_B, and k_C are incident on the sample. FWM signal is detected in the phase-matched direction $k_s = -k_A + k_B + k_C$. (B) Collinear geometry in which three pulses with phase-modulated with frequencies f_A, f_B, and f_C are incident on the sample. FWM signal is detected at the frequency of $f_s = -f_A + f_B + f_C$.

shapers can be used to achieve a noncollinear pulse sequence (Grumstrup et al., 2007; Tian et al., 2003; Vaughan et al., 2007).

The phase matching condition essential for the detection of the FWM signal in noncollinear geometry depends on the far-field constructive interference between the FWM radiated from a collection of oscillators. While this leads to a straightforward spatial filtering of the nonlinear signal of interest, this technique cannot be applied to study a single nanoemitter such as a QD. This limitation is circumvented through collinear 2DCS experiments in which the excitation pulses are "tagged" with unique modulation frequencies and the FWM signal is filtered in the frequency domain. The collinear approach also allows us to focus the excitation pulses to a small, overlapping spot using high-numerical-aperture-microscope objective, which is commonly used in single QD spectroscopy. Fig. 8B shows a typical 2DCS experiment with collinear geometry. The center frequencies of excitation pulses A, B, and C are shifted by frequencies f_A, f_B, and f_C, respectively, using acousto-optic modulators (AOMs). The frequency shifts are imparted onto the FWM signal as a frequency shift f_s, which is used to filter out the FWM signal of interest. For instance, in the example of Fig. 8B, the rephrasing signal can be isolated by detecting the signal with a shift $f_s = -f_A + f_B + f_C$. The FWM signal is detected by mixing it with a reference beam using an additional AOM (Langbein and Patton, 2006). Another option is to use a reference pulse to generate a fourth-order population, which is detected as fluorescence (Wagner et al., 2005) or photocurrent (Nardin et al., 2013).

3.5 2DCS studies of semiconductor nanostructures

Owing to the interest in the development of optoelectronic devices as well as fundamental physics, semiconductor nanostructures have been a subject of intensive research for more than half a century. Along with the rapid development of the lasers, better understanding and numerous applications have been achieved at an incredible pace (Shah, 1999).

Coherent spectroscopy techniques are widely used to study light-matter interactions in semiconductor nanostructures (Cundiff, 2008). In recent years 2DCS has emerged as an important, state-of-the-art tool in these studies. Next, we will briefly discuss some of these results, which will also highlight the advantages of 2DCS from the perspective of studying semiconductor nanostructures.

3.5.1 Semiconductor quantum wells

Many-body effects manifest themselves in exciton dynamics in semiconductors, and significantly modify their electronic and optical properties. While difficult to infer from other techniques, it is relatively straightforward to qualitatively reveal these effects using 2DCS. The earliest work performed in semiconductor excitons are on GaAs/AlGaAs quantum wells (QWs). The researchers have successfully revealed the origin of many-body physics through 2DCS experiments in conjunction with theoretical calculations (Karaiskaj et al., 2010; Li et al., 2006; Zhang et al., 2007). Furthermore, 2DCS was successfully applied to study higher-order correlations by extending up to the seventh order nonlinearity (Turner and Nelson, 2010).

An interesting example to highlight the advantages of 2DCS is the study of coherent exciton dynamics in a double-QW structure comprising two QWs with different widths. The double QWs provide an attractive playground to explore fundamental quantum-mechanical phenomena and have potential for optoelectronic applications owing to the feasibility to change the barrier and well widths, which have been studied extensively for more than 20 years (Faist et al., 1994; Leo et al., 1991). Fig. 9A shows the electron and hole wavefunctions in the InGaAs/GaAs coupled QWs. The QWs have thicknesses of 8 and 10 nm with a barrier of 10 nm between them. Calculations show some overlap of the electron wavefunction between the two QWs, which leads to coupling between them. However, the nature of this coupling—incoherent (Fig. 7B) or coherent (Fig. 7C) had been a matter of longstanding debate. A recent 2DCS study unambiguously revealed the presence of coherent coupling between excitations in the two

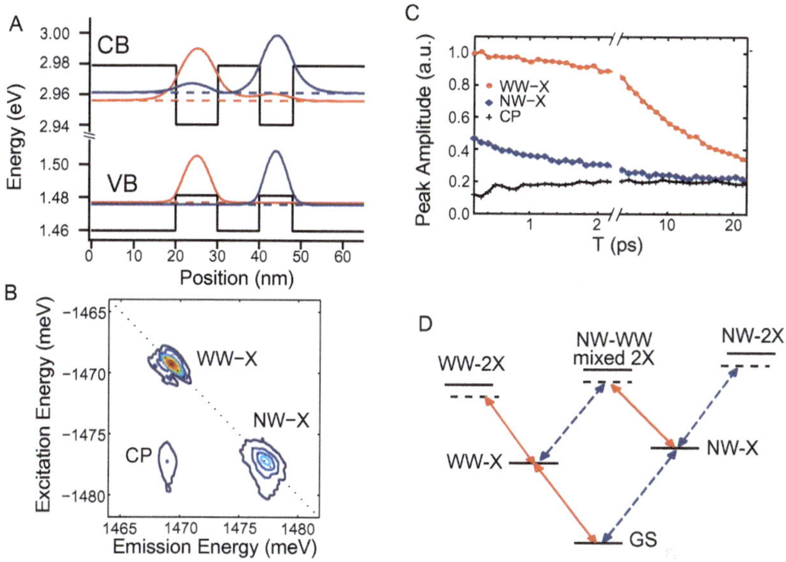

Fig. 9 (A) Double QWs with thicknesses of 10 and 8 nm with a barrier of 10 nm between them. The calculated wavefunctions and ground state energies for the electrons and holes in the two QWs. (B) A rephasing 2DCS spectrum with the cocircularly polarized pulses. (C) Time-dependent amplitude for each peak in the 2D spectrum. The values are normalized by WW-X. Clear oscillation for the CP peak is noticed until ~1 ps. (D) Energy level structure of the coupled double quantum well revealed by this work (Nardin et al., 2014).

QWs (Nardin et al., 2014). Fig. 9B shows the rephasing 2DCS spectrum. In addition to the diagonal peaks for the narrow-well (NW-X) and wide-well (WW-X) excitons, off-diagonal peak CP (cross-peak) was observed appearing at the excitation energy of the NW-X and emission energy of the WW-X. Although the CP looks similar to the 2D spectra in Fig. 7E, presence of coherent coupling cannot be ruled out. The origin of this coupling was extracted from the CP amplitude as a function of delay *T*, which is shown in Fig. 9C as black cross-marks. The amplitude increase indicates population transfer while coherent coupling is evidenced from the measured oscillations, which reveal a combination of the two possible coupling mechanisms at play. The interactions were phenomenologically modeled through many-body interactions, which are included in the few-level system shown in Fig. 9D. Similar effects were also confirmed by two-color 2DCS with the added advantage of measuring coupling in the absence of stronger diagonal peaks (Tollerud et al., 2014).

Although many-body interactions are qualitatively apparent in 2D spectra, quantifying them had been a challenge. Recent studies have extended the previous qualitative understanding of many-body effects to quantify these interactions by treating excitons as bosons (Singh et al., 2016). Well-studied phenomenon such as spectral diffusion have also been revisited to clearly demonstrate and quantify deviations from previously assumed assumptions of uniform "diffusion" of excitons to higher and lower energies (Singh et al., 2017; Richter et al., 2018). Thus, 2DCS continues to bring new insights to even well-studied systems such as QWs.

3.5.2 Semiconductor quantum dots

Fluctuations in size, shape, composition, and interface with surrounding matrix make the optical properties of semiconductor QDs highly inhomogeneous. As a result, the energy spectrum of a semiconductor QD ensemble typically exhibits much broader profile (inhomogeneous broadening) than the intrinsic linewidth (homogeneous broadening). The ability of 2DCS to separate between the homogeneous and inhomogeneous broadenings (Siemens et al., 2010) was critical to study excitons in interfacial semiconductor QDs (Moody et al., 2011a,b). Fig. 10Aand B shows the rephasing 2D spectra measured for an interfacial GaAs QD ensemble at low and high temperature, respectively (Moody et al., 2011a). The homogeneous linewidth increases with increasing temperature due to elastic exciton-phonon scattering. Furthermore, homogeneous linewidths was measured as a function of absorption energy. This measurement gives us important insights into the effect of morphology on electron-phonon coupling in semiconductor QDs. Fig. 10C clearly shows that the homogeneous linewidth monotonically increases with the increase of absorption energy with a greater change at elevated temperatures. This finding revealed greater exciton-phonon coupling for smaller QDs. Furthermore, high-resolution 2DCS revealed the couplings between QDs in this system (Kasprzak et al., 2011; Mermillod et al., 2016) with a recent demonstration of optical control of the coupling (Martin and Cundiff, 2018).

Another commonly studied class of QDs are the InAs self-assembled QDs. Contrary to the interfacial QDs, no clear size-dependence of physical properties that includes biexciton binding energy (Moody et al., 2013a), fine structure splitting (Moody et al., 2013b), and linewidths for the ground- (*s*-shell) and excited-state (*p*-shell) excitons (Suzuki et al., 2018b) is observed. These trends were attributed to the annealing effects in InAs

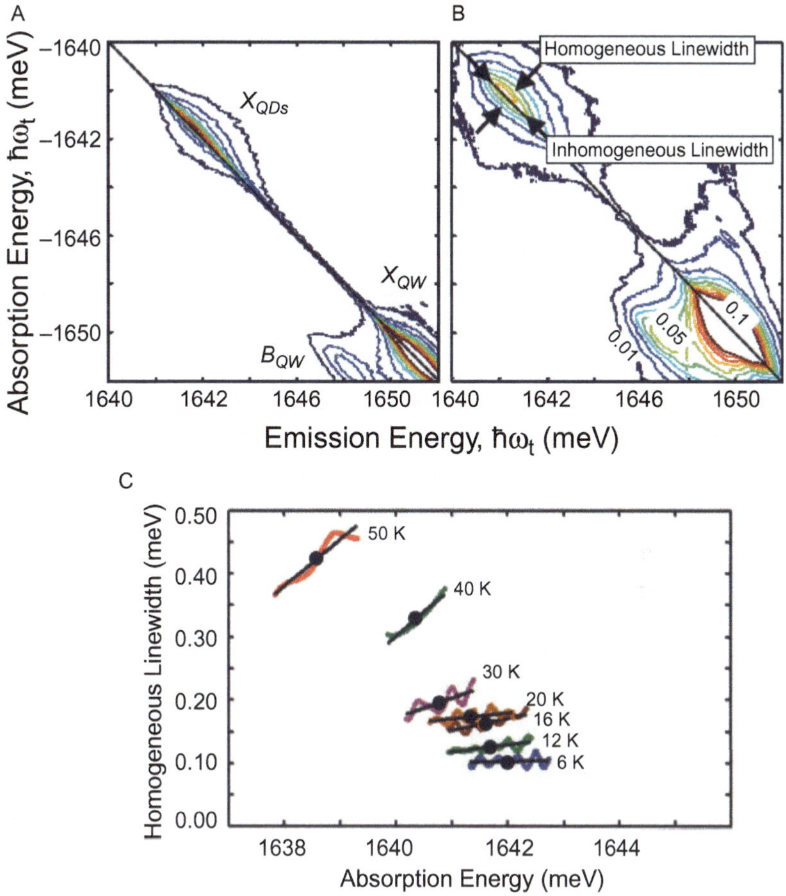

Fig. 10 Rephasing one-quantum spectra on interfacial GaAs QDs at the lattice temperature of (A) 6 K and (B) 50 K. (C) Homogeneous linewidths as a function of absorption energy at several lattice temperatures (Moody et al., 2011a).

QDs, which relieve the strain by reducing the QD/barrier lattice mismatch and minimize built-in piezoelectric fields that can influence the optical properties.

Most of the 2DCS studies are restricted to relatively lower intensities of the excitation pulses so that the nonlinear signal is restricted to the third order. Higher-intensity pulses are required to clearly demonstrate coherent control through Rabi oscillations, as we discussed earlier. With this motivation initial experiments revealed signatures of the fifth-order nonlinearity for an InAs QDs as a new peak in 2D spectrum (Moody et al., 2013c), which would otherwise overlap with strong exciton and trion signal in TFWM.

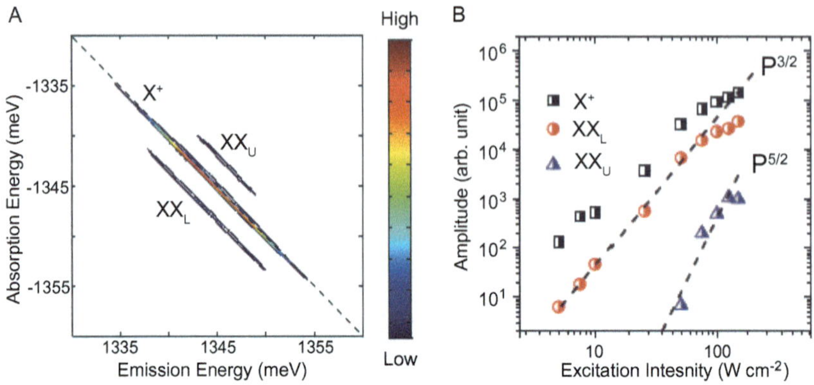

Fig. 11 (A) Rephasing amplitude 2D spectrum for an InAs self-assembled QDs ensemble. The polarization is set to cross-diagonal (*HVVH*). (B) Amplitude for each peak as a function of excitation intensity (Moody et al., 2013c).

Fig. 11A shows rephasing amplitude spectra measured in an ensemble of InAs QDs for the cross-linear polarization (*HVVH*), which was used to emphasize the biexciton contribution along with suppressing contribution from excitons. The peaks arising from the neutral exciton/biexciton system denoted as XX_L and XX_U are well separated from the strong trion peak (X^+) appearing along the diagonal. From the excitation-intensity-dependent measurement shown in Fig. 11B, the XX_U peak shows the fifth-order nonlinear interaction with the excitation pulses. While the appearance of the new peak is demonstrated by the increase of excitation intensity, the available power of the excitation source was still in the perturbative regime, and was insufficient for achieving population inversion. This limitation lead us to develop a novel prepulse experiment that brings us to "beyond perturbative regime", namely, "coherent control" of a semiconductor QD ensemble.

4. Coherent control of a semiconductor QD ensemble

As we discussed earlier, most studies of coherent control of QDs used single QDs to circumvent the problem of inhomogeneity although coherent control of QD ensembles is desired from a practical standpoint. Clear observation of coherent population evolution in an ensemble system has been lacking. In this section, we describe how we overcome this difficulty by implementing 2DCS in the pump-probe geometry (Suzuki et al., 2016, 2018a), which was used to reveal frequency-resolved coherent evolution of an inhomogeneously broadened QD ensemble. By exciting both the exciton and biexciton transitions, we reveal novel interplay between populations in these states, which is generally absent in a two-level system. Furthermore, our studies clearly revealed the effects of detuning and excitation-induced dephasing on the coherent population dynamics.

4.1 Experimental

We overcome the issue of low laser intensity mentioned at the end of the last section by implementing 2DCS in the pump-probe geometry to unambiguously reveal the coherent population manipulation in QD ensemble. We refer to this technique as *prepulse 2DCS*, where the prepulse is used as a pump and 2DCS is used as a probe. Fig. 12A shows the schematics of the experiment. 2DCS is performed in the rephasing time ordering with noncollinear "box" geometry, which was discussed earlier. In addition to the rephasing one-quantum measurement for 2DCS, where Fourier transformations are carried out with respect to τ and t, the prepulse is used and the delay between the prepulse pump and the 2DCS probe (pulse A*) is denoted as Δt. The prepulse and each pulse used in 2DCS have the same spectral profile and are generated by a mode-locked Ti:sapphire laser at a repetition rate of 76 MHz. The pulses have a band width of 14.8 meV at the full width at the half maximum. The spatial profile of the prepulse is twice as broad as the 2D pulses in order to reduce spatial inhomogeneity in excitation.

The sample is an InAs self-assembled QD ensemble with GaAs barriers, consisting of 10 quantum-mechanically isolated, epitaxially grown layers. It is thermally annealed postgrowth at 900 °C for 30 s, which results in a 100 meV in-plane confinement. The sample was unavoidably doped during

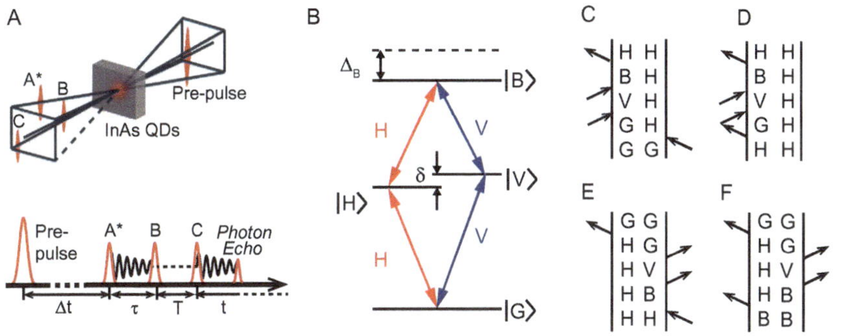

Fig. 12 (A) Illustration of experimental geometry (top) and time-sequence (bottom) for prepulse 2DCS. (B) Energy-level diagram for InAs QD excitonic system. (C)–(F) Coherent paths for the cross-linear polarization, *HVVH*.

growth resulting in approximately half of the QDs being charged with a hole, which forms a trion with a photoexcited electron and hole pair (Moody et al., 2013c).

Fig. 12B shows the energy level diagram and selection rules for the InAs QD excitonic system. The two non-degenerate horizontal ($|H\rangle$) and vertical ($|V\rangle$) exciton states have orthogonal linearly polarized transitions from the ground state ($|G\rangle$) due to the electron-hole exchange interaction with broken cylindrical symmetry of the QD shape. The energy of the biexciton state $|B\rangle$ is lower than the sum of the two exciton energies by $\Delta_B = 3.3\,\text{meV}$, which is much larger than the fine structure splitting energy $\delta = 19\,\mu\text{eV}$ between $|H\rangle$ and $|V\rangle$. A cross-linearly polarized excitation and detection scheme is used, i.e., *HVVH* for A*, B, C and the signal, respectively, which isolates the biexciton signal from the trion's (Moody et al., 2013a). The quantum paths for cross-linearly polarization in 2DCS are shown in Fig. 12C–F, where they start from population in the ground (Fig. 12C), the *H* exciton (Fig. 12D and E), and the biexciton states (Fig. 12F), respectively.

4.2 Four-level exciton-biexciton system

We will first discuss coherent control of the exciton-biexciton four-level system focusing on the QDs with exciton transition resonant to the center of the excitation light (Suzuki et al., 2016). Fig. 13A shows a 2D rephasing amplitude spectrum without the prepulse. The spectrum shows two peaks that are inhomogeneously broadened along the diagonal direction, indicated by dashed lines, due to QD size dispersion. The peak labeled Tr arises from the trion nonlinear response, whereas the peak labeled LP arises from the

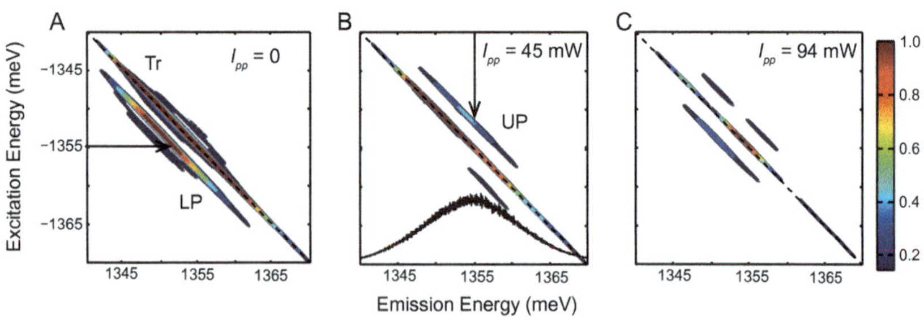

Fig. 13 Rephasing one-quantum 2D spectra for the prepulse power of (A) 0 mW, (B) 45 mW, and (C) 94 mW. Tr, LP, and UP represent the trion, lower, and upper peaks, respectively.

biexciton nonlinear response and is redshifted from the diagonal along the emission energy axis by the biexciton binding energy. The lower peak (LP) in Fig. 13A corresponds to the quantum path starting from the ground state shown in Fig. 12C.

When the prepulse is incident on the sample, the 2D spectrum drastically changes as shown in Fig. 13B and C for the prepulse power of 45 and 94 mW, respectively. The delay time Δt is 15 ps. As the I_{PP} increases, LP disappears and new peak labeled as UP appears (Fig. 13B). However, as I_{PP} increases further, the LP reappear and the UP disappears again (Fig. 13C). The polarization for the prepulse is set along the H direction, which transfers population from the ground to the H exciton, and from the H exciton to the biexciton states. As a result of the population transfer, the quantum paths starting from the H exciton (Fig. 12D and E) and the biexciton sates (Fig. 12F) also contribute to the 2D signal. Considering the rules of Feynman diagrams, the quantum path shown in Fig. 12D contributes to the signal at the position of LP in the 2D spectrum with the opposite sign compared to the signal from Fig. 12C. Similarly, signals from the quantum paths shown in Fig. 12E and F appear at the position labeled as the upper peak (UP) in Fig. 13B, which is redshifted in the excitation energy by the biexciton binding energy, with opposite signs with respect to each other. As a result, the LP represents the difference between ground state population ρ_{G} and the H exciton state population ρ_{H}, whereas the UP represents the difference between the H exciton population ρ_{H} and the biexciton population ρ_{B}.

After qualitatively noting the origin of the various peaks, we discuss the intensity dependence of these peaks in more detail. Fig. 14A shows peak amplitudes for the LP (red circles) and UP (blue circles) at energy positions resonant with the prepulse shown as the arrows in Fig. 13A and B, respectively. The signals are plotted as a function of the square root of the prepulse power, $\sqrt{I_{PP}}$, which is proportional to the pulse area. The power-dependent measurements are repeated for $\Delta t = 50$ and 100 ps, as shown in Fig. 14A. As Δt increases, the population dynamics after coherent interaction with the prepulse affects the signals significantly due to the relaxation from the biexciton to exciton, and the exciton to ground states.

In order to gain quantitative insight, we performed calculations using the optical Bloch equations (OBE) for the four-level system. Here, we extend the calculation for a single two-level system as discussed in Section 2 to the four-level system shown in Fig. 12B. We also use a Gaussian-shaped pulse for an electric field instead of a steady form. Fig. 14B shows the simulation results for the population of each component as a function of the pulse area at

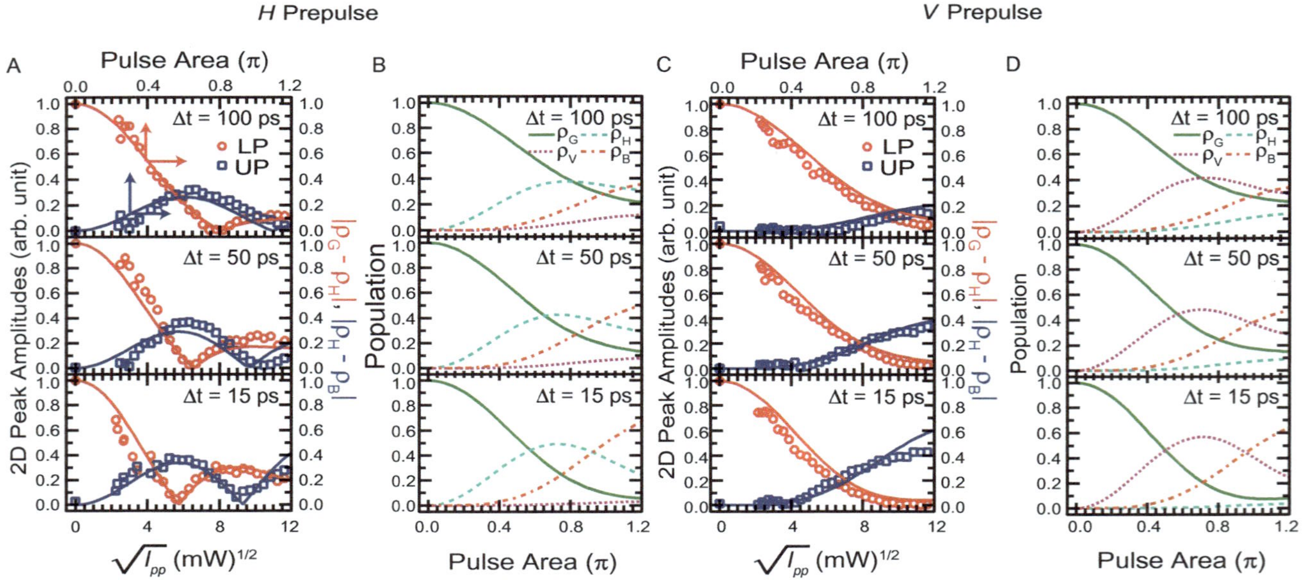

Fig. 14 Peak amplitudes of (A) *H*- and (C) *V*-polarized prepulse 2D rephasing spectra for the LP (red circles) and UP (blue squares). Solid lines are simulation results of $|\rho_G - \rho_H|$ (red) and $|\rho_H - \rho_B|$ (blue). (B), (D) Simulation results for population in each state, ρ_G(solid lines), ρ_H(dashed lines), ρ_V(dotted lines), and ρ_B(dot-dashed lines).

each delay after H polarized prepulse. The population differences $|\rho_G - \rho_H|$ and $|\rho_H - \rho_B|$ are also plotted as red and blue solid lines, respectively, in Fig. 14A.

To confirm the validity of these results, we perform experiments and simulations with a V-polarized prepulse, in which case the H exciton state cannot be populated, and the quantum paths starting with population in the H exciton state (Fig. 12D and E) do not contribute to the signal. Fig. 14C shows the 2D peak amplitudes of the LP (red circles) and UP (blue squares) at the resonant energies as a function of $\sqrt{I_{PP}}$ at indicated Δt with the simulation results as solid lines, while the simulation results for the population of each component is shown in Fig. 14D. Different behaviors of the population dynamics from the case of H polarized prepulse, which are consistent with the quantum pathways, are clearly observed.

4.3 Detuning

Having demonstrated the coherent control of a semiconductor QD ensemble for the resonant energy group, we now turn our attention to coherent control of population in QDs that are detuned from the excitation pulses (Suzuki et al., 2018a). We can qualitatively see the frequency-dependence by the fringes visible in Fig. 13B and C, which we will discuss in more detail.

We first revisit the prepulse 2D rephasing spectrum shown in Fig. 13C, which is shown on a different color scale in Fig. 15A to enhance the visibility. Because the center energy of the prepulse spectrum is 1355 meV, the signals for the resonant energy group can be tracked at the position where the black dotted line labeled as 0 intersects each peak as sown in Fig. 15A. Similarly, the detuning effect can be elucidated by focusing on different positions in each peak. In Fig. 15A, amplitude of each peak at intersections with the blue/red dotted lines labeled as +3 and −3 correspond to signals belonging to $\Delta = +3$ meV and $\Delta = -3$ meV, respectively, where Δ is the detuning. Further detuning effect for $\Delta = +5$ meV and $\Delta = -5$ meV can be observed along the green/orange-dotted lines labeled as +5 and −5.

Fig. 15B and D shows peak amplitudes for the LP and UP for $\Delta = \pm 3$ meV, respectively. The behavior for $\Delta = +3$ meV clearly differs from that for $\Delta = -3$ meV. This asymmetric behavior with respect to the sign of detuning is not seen in a two-level system and is a direct consequence of coherent excitation of the four-level exciton-biexciton system. The detuning energy, Δ, is defined as the energy difference between the center of the prepulse (1355 meV) and the energy separation between $|G\rangle$ and $|H\rangle$ states.

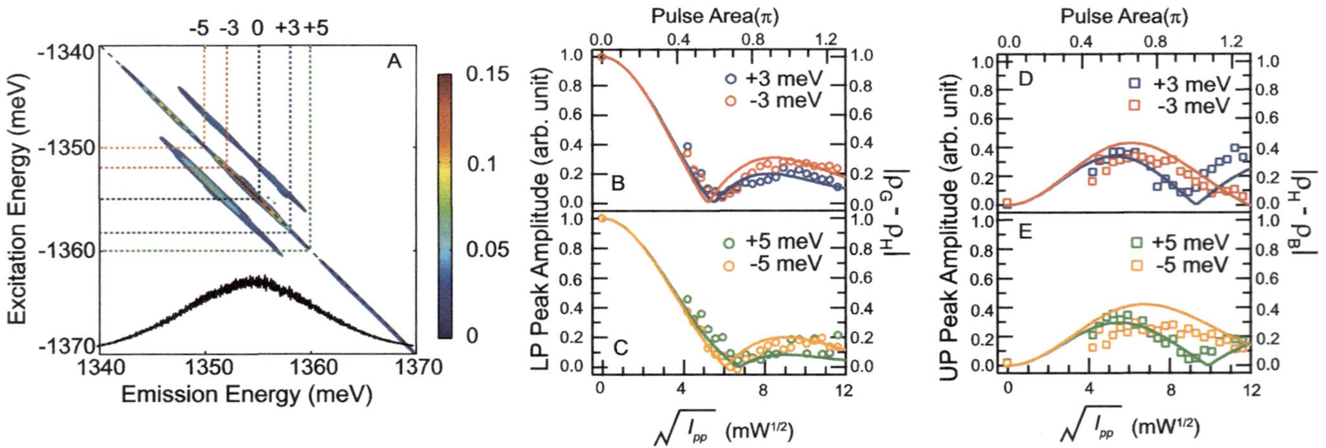

Fig. 15 (A) Rephasing 2DCS spectrum using prepulse power of 94 mW. (B)–(E) Peak amplitude of 2D rephasing spectra for the LP (B and C) and UP (D and E) belonging to detuned frequency group ($\Delta = \pm 3$, ± 5 meV). The solid lines are simulation results for $|\rho_G - \rho_H|$ in (B) and (C) and $|\rho_H - \rho_B|$ in (D) and (E).

The biexciton binding energy Δ_B lowers the transition energy between $|H\rangle$ and $|B\rangle$ by 3.3 meV compared to the $|G\rangle$ to $|H\rangle$ transition. Consequently, energy groups of $\Delta = +3$ and -3 meV no longer have a symmetric relationship with respect to the $|H\rangle$ to $|B\rangle$ transition, although they remain symmetric with respect to the $|G\rangle$ to $|H\rangle$ transition. This asymmetric relationship is confirmed for the higher detuned energy group of $\Delta = +5$ and -5 meV shown in Fig. 15C and E.

For quantitative insight, we perform calculations using the OBEs for the four-level system with considering detuning effects. We also include excitation-induced effects (EID), which is discussed later. The solid lines in Fig. 15B–E show calculation results as the population difference, $|\rho_G - \rho_H|$ in Fig. 15B and C and $|\rho_H - \rho_B|$ in Fig. 15D and E, respectively.

4.4 Two-level trion system

Next we focus on the trion signal appearing along the diagonal lines (Suzuki et al., 2018a). The energy structure and selection rules for InAs QDs change significantly depending on the existence of a hole in a QD. Fig. 16A shows the energy-level diagram of the trion system. From a hole in the spin-up (spin-down) state $|+\rangle$ ($|-\rangle$), a spin-up (spin-down) trion state $|T^+\rangle$ ($|T^-\rangle$) is created by right (left) circularly polarized light σ_+ (σ_-). Since

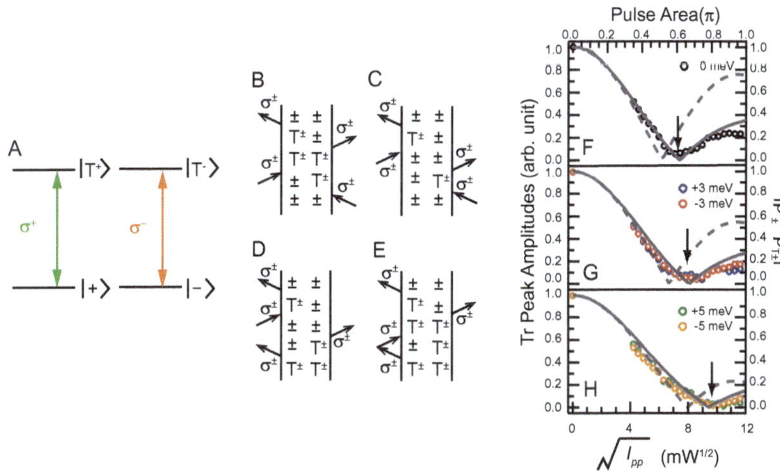

Fig. 16 (A) The energy-level diagram representing the trion transition $[|+\rangle(|-\rangle) \leftrightarrow |T^+\rangle(|T^-\rangle)]$ by right (left) circularly polarized light $[\sigma_+ (\sigma_-)]$. (B)–(E) The quantum paths for the trion two-level system. (F)–(H) Peak amplitude of the rephasing one-quantum 2DCS spectrum for the Tr peak belonging to the resonant (F) and detuned frequency groups (G and H).

spin-up and spin-down states are degenerate without an external field, a charged QD offers a doubly degenerate two-level system. Fig. 16B–E shows the quantum paths contributing to the cross-linearly polarized excitation and detection scheme (*HVVH*).

As a result, the trion two-level system exhibits qualitatively different behavior to the coherent control from the exciton-biexciton four-level system. The trion peak (Tr) in Fig. 13A corresponds to the quantum path ways starting from the ground state shown in Fig. 16B and C. With the prepulse, other quantum pathways starting with the population in the trion state (Fig. 16D and E) also contribute to the 2D signal at the position of the Tr in the 2D spectrum with the opposite sign compared to the signal from Fig. 16B and C. As a result, amplitude of the Tr peak represent the population difference between the ground and trion states.

Fig. 16F–H shows the peak amplitudes of Tr belonging to the resonant energy group ($\Delta = 0\,\text{meV}$) and detuned energy groups ($\Delta = \pm 3\,\text{meV}$, $\pm 5\,\text{meV}$) as a function of $\sqrt{I_{PP}}$. The detuning dependence is symmetric with respect to the absolute value of detuning for both red- and blue-detuned components, which is expected for a two-level system and in a sharp contrast to the exciton-biexciton four-level system. Additionally, as $|\Delta|$ increases, the I_{PP} for the peak minimum increases as indicated by the black arrows.

To confirm these conclusions, we perform the calculation using the OBEs for the two-level system. The dashed lines in Fig. 16F–H show the simulation results for the population differences $|\rho_{\pm} - \rho_{T^{\pm}}|$ without additional dephasing. Although they have good qualitative agreement, the experimental data deviate from simulation at higher powers. This phenomena is caused by EID and will be discussed in the next section; we have successfully reproduced the experimental data including the EID term as shown by the solid lines in Fig. 16F–H.

4.5 Excitation induced dephasing mechanisms in the femtosecond regime

Here we will discuss the EID effects, which are necessary to satisfactorily explain the experimental results, as shown in Fig. 16F–H, and discuss the underlying mechanisms (Suzuki et al., 2018a). Among EID mechanisms, the longitudinal acoustic (LA) phonons have been reported to play a central role for damping of Rabi oscillation in semiconductor QDs (Ramsay et al., 2010a,b). In this model, the confined carriers interact with the LA phonons of the crystal via the deformation potential. Since the occupation number of

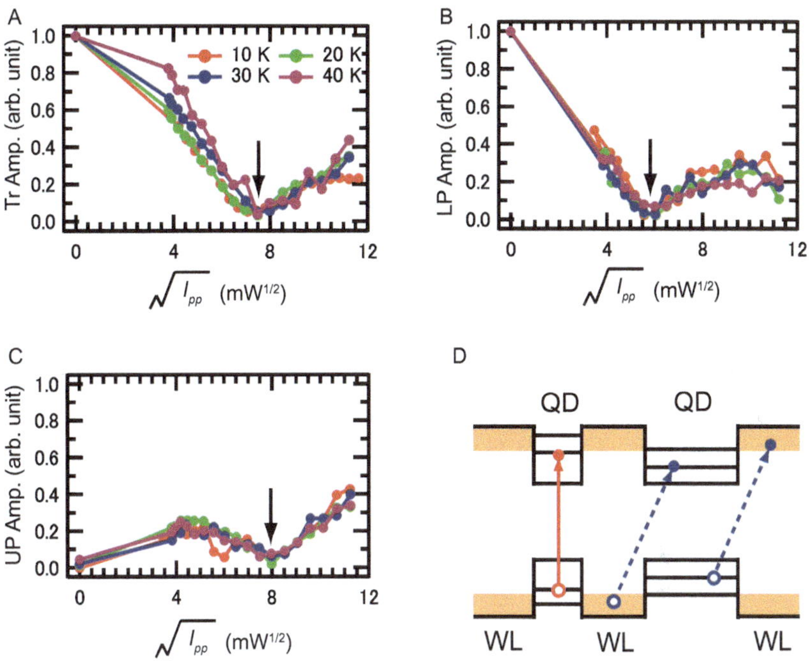

Fig. 17 (A)–(C) Temperature-dependent Tr-, LP-, and UP-peak amplitudes belonging to the resonant frequency group. The black arrows show zero crossing points. (D) Schematics of the wetting-layer-EID model. The red-solid arrow represents the intra-dot transition while the blue-dashed arrows indicate the crossed transitions explained in the text.

LA phonons is significantly controlled with the lattice temperature, the Rabi frequencies are reported to be renormalized depending on the lattice temperature (Ramsay et al., 2010b).

To check the role of phonons to the dephasing mechanism, we repeated the above measurements for several temperatures. Fig. 17A–C shows Tr, LP, and UP amplitudes belonging to the resonant energy group with varying lattice temperature from 10 to 40 K. As shown with the black arrows, the minimum points of the peaks with respect to the pump power do not change with temperature. This invariance is clearly different from reports explaining temperature-dependent behavior in the phonon-mediated model, where the oscillation period sensitively changes as a function of the lattice temperature (Ramsay et al., 2010b).

In order to rationalize this difference, we propose the wetting-layer model in which carriers created in the wetting layers by the laser pulses contribute to the additional dephasing as shown in Fig. 17D (Wang et al., 2005).

Due to the coupling between bound exciton states in QDs and the continuum states of the wetting layers, the existence of crossed transition involving bound electron (hole) states and hole (electron) continuum states has previously been confirmed as a broad resonances seen in single QD photoluminescence excitation (PLE) spectra (Htoon et al., 2001; Toda et al., 1999; Vasanelli et al., 2002). The excitation laser we have used is spectrally broad (14.8 meV) and can cover these crossed transitions. Moreover, the transition energy between the lowest bound hole and the electron states for a strongly confined QD (the left QD in Fig. 17D) corresponds to the crossed transition energy for a loosely confined QD (the right QD in Fig. 17D) in our ensemble measurements. Consequently, it is more likely to generate wetting-layer carries in our studies compared to the spectrally narrow (0.1–1 meV) laser used in single QD studies. Our studies thus reveal the unique EID mechanism in an ensemble of QDs due to excitation with ultrashort pulses, which would need to be accounted for while developing applications of coherent control in QD ensembles.

5. Summary and outlook

We have reviewed our recent studies on coherent control of a semiconductor QD ensemble along with the background necessary to understand our results. We have demonstrated coherent manipulation of exciton/biexciton/trion populations using a prepulse and measured these populations using 2DCS as a probe. Calculations based on OBEs were used to explain and understand the observations. These measurements also revealed the contribution and origin of EID, which contributes to the fast decay of Rabi oscillations.

For semiconductor QDs, recent advances of sample growth and processing techniques have created many playgrounds that include colloidal QDs (Pietryga et al., 2016), droplet epitaxially grown QDs (Gurioli et al., 2019), and electrically-switchable Si-based QDs (Watson et al., 2018). Each of these new-type of QDs has alternative and advantageous features in optical properties compared to conventional QDs grown by the Stranski–Krastanov method (Bimberg et al., 1999), which we have used. Colloidal QDs with impressive single-dot emission properties have been developed (Park et al., 2019), which seem promising from the perspective of coherent control. The recently discovered perovskite nanocrystals (Protesescu et al., 2015) have shown impressive coherent optical properties (Rainò et al., 2018; Utzat et al., 2019), which warrant more detailed and direct studies. These features enable us to coherently control them in more sophisticated ways, and thus lead semiconductor QDs to be used as multiple types of functional nodes in quantum engineering devices.

In order to investigate variable profiles in semiconductor QDs such as exciton fine structures or wide-gap materials, further advances for 2DCS as a spectroscopy technique is desired. Recently, 2DCS experiments have been developed for a wide range of frequencies from terahertz (Woerner et al., 2013) to ultraviolet frequencies (Tseng et al., 2009) with table-top experiments. With the recent developments in ultraviolet sources through high harmonic generation, 2DCS has also been demonstrated in extreme ultraviolet frequency with attosecond time resolution (Marroux et al., 2018), which gives access to studying diverse range of light-matter interactions. Furthermore unprecedented frequency resolution of frequency combs has also been leveraged to study atomic hyperfine states and their interactions (Lomsadze et al., 2018; Lomsadze and Cundiff, 2017).

We have successfully unraveled another profile of 2DCS that can be a sensitive tool for reading out the final state as a result of coherent control by independently tracking each QDs within a certain frequency group. Our work has so far directly served to motivate coherent controls of different ensemble systems (Poltavtsev et al., 2017; Weinzetl et al., 2019). In the light of the progress made in developing diverse set of materials and 2DCS techniques, we hope that this work can contribute to development of coherent control protocols in systems with significant inhomogeneous broadening.

References

Allen, L., Eberly, J.H., 1987. Optical Resonance and Two-Level Atoms. Dover Publications, Inc., New York.

Atatüre, M., Dreiser, J., Badolato, A., Högele, A., Karrai, K., Imamoglu, A., 2006. Quantum-dot spin-state preparation with near-unity fidelity. Science 312, 551–553.

Bimberg, D., Grundmann, M., Ledentsov, N.N., 1999. Quantum Dot Heterostructures. John Wiley and Sons, New York.

Borri, P., Langbein, W., Schneider, S., Woggon, U., Sellin, R.L., Ouyang, D., Bimberg, D., 2001. Ultralong dephasing time in InGaAs quantum dots. Phys. Rev. Lett. 87, 157401. 4 pages.

Borri, P., Langbein, W., Schneider, S., Woggon, U., Sellin, R.L., Ouyang, D., Bimberg, D., 2002. Rabi oscillations in the excitonic ground-state transition of InGaAs quantum dots. Phys. Rev. B 66, 081306. 4 pages.

Bristow, A.D., Karaiskaj, D., Dai, X., Zhang, T., Carlsson, C., Hagen, K.R., Jimenez, R., Cundiff, S.T., 2009. A versatile ultrastable platform for optical multidimensional Fourier-transform spectroscopy. Rev. Sci. Instrum. 80, 073108.

Brixner, T., Stenger, J., Vaswani, H.M., Cho, M., Blankenship, R.E., Fleming, G.R., 2005. Two-dimensional spectroscopy of electronic couplings in photosynthesis. Nature 434, 625–628.

Chemla, D.S., Shah, J., 2001. Many-body and correlation effects in semiconductors. Nature 411, 549–557.
Cundiff, S.T., 2008. Coherent spectroscopy of semiconductors. Opt. Express 16, 4639–4664.
Dai, X., Bristow, A.D., Karaiskaj, D., Cundiff, S.T., 2010. Two-dimensional Fourier-transform spectroscopy of potassium vapor. Phys. Rev. A 82, 052503 (6 pages).
Dai, X., Richter, M., Li, H., Bristow, A.D., Falvo, C., Mukamel, S., Cundiff, S.T., 2012. Two-dimensional double-quantum spectra reveal collective resonances in an atomic vapor. Phys. Rev. Lett. 108, 193201 (5 pages).
Ernst, R.R., Bodenhausen, G., Wokaun, A., 1987. Principles of Nuclear Magnetic Resonance in One and Two Dimensions. Oxford University Press, London.
Faist, J., Capasso, F., Sivco, D.L., Sirtori, C., Hutchinson, A.L., Cho, A.Y., 1994. Quantum cascade laser. Science 264, 553. 4 pages.
Fayer, M.D., 2009. Dynamics of liquids, molecules, and proteins measured with ultrafast 2D IR vibrational echo chemical exchange spectroscopy. Annu. Rev. Phys. Chem. 60, 21–38.
Gammon, D., Snow, E.S., Shanabrook, B.V., Katzer, D.S., Park, D., 1996. Fine structure splitting in the optical spectra of single GaAs quantum dots. Phys. Rev. Lett. 76, 3005–3008.
Gibbs, H.M., 1973. Incoherent resonance fluorescence from a Rb atomic beam excited by a short coherent optical pulse. Phys. Rev. A 8, 446–455.
Grumstrup, E.M., Shim, S.-H., Montgomery, M.A., Damrauer, N.H., Zanni, M.T., 2007. Facile collection of two-dimensional electronic spectra using femtosecond pulse-shaping technology. Opt. Express 15, 16681–16689.
Gurioli, M., Wang, Z., Rastelli, A., Kuroda, T., Sanguinetti, S., 2019. Droplet epitaxy of semiconductor nanostructures for quantum photonic devices. Nat. Mater. 18, 799–810.
Haug, H., Koch, S.W., 2009. Quantum Theory of the Optical and Electronic Properties of Semiconductors, fifth ed. World Scientific, Singapore.
Htoon, H., Kulik, D., Baklenov, O., Holmes, A.L., Takagahara, T., Shih, C.K., 2001. Carrier relaxation and quantum decoherence of excited states in self-assembled quantum dots. Phys. Rev. B 63, 241303. 4 pages.
Huxter, V.M., Oliver, T.A.A., Budker, D., Fleming, G.R., 2013. Vibrational and electronic dynamics of nitrogen-vacancy centres in diamond revealed by two-dimensional ultrafast spectroscopy. Nat. Phys. 9, 744–749.
Jonas, D.M., 2003. Optical analogs of 2D NMR. Science 300, 1515–1517.
Kamada, H., Gotoh, H., Temmyo, J., Takagahara, T., Ando, H., 2001. Exciton Rabi oscillation in a single quantum dot. Phys. Rev. Lett. 87, 246401. 4 pages.
Karaiskaj, D., Bristow, A.D., Yang, L., Dai, X., Mirin, R.P., Mukamel, S., Cundiff, S.T., 2010. Two-quantum many-body coherences in two-dimensional Fourier-transform spectra of exciton resonances in semiconductor quantum wells. Phys. Rev. Lett. 104, 117401. 4 pages.
Kasprzak, J., Patton, B., Savona, V., Langbein, W., 2011. Coherent coupling between distant excitons revealed by two-dimensional nonlinear hyperspectral imaging. Nat. Photonics 5, 57–63.
Koch, M., Feldmann, J., von Plessen, G., Göbel, E.O., Thomas, P., Köhler, K., 1992. Quantum beats versus polarization interference: an experimental distinction. Phys. Rev. Lett. 69, 3631–3634.
Kujiraoka, M., Ishi-Hayase, J., Akahane, K., Yamamoto, N., Ema, K., Sasaki, M., 2010. Optical Rabi oscillations in a quantum dot ensemble. Appl. Phys. Exp. 3, 092801. 3 pages.
Langbein, W., Patton, B., 2006. Heterodyne spectral interferometry for multidimensional nonlinear spectroscopy of individual quantum systems. Opt. Lett. 31, 1151–1153.

Leo, K., Shah, J., Göbel, E.O., Damen, T.C., Schmitt-Rink, S., Schäfer, W., Köhler, K., 1991. Coherent oscillations of a wave packet in a semiconductor double-quantum-well structure. Phys. Rev. Lett. 66, 201–204.

Li, X., Wu, Y., Steel, D., Gammon, D., Stievater, T.H., Katzer, D.S., Park, D., Piermarocchi, C., Sham, L.J., 2003. An all-optical quantum gate in a semiconductor quantum dot. Science 301, 809–811.

Li, X., Zhang, T., Borca, C.N., Cundiff, S.T., 2006. Many-body interactions in semiconductors probed by optical two-dimensional Fourier transform spectroscopy. Phys. Rev. Lett. 96, 057406. 4 pages.

Li, H., Bristow, A.D., Siemens, M.E., Moody, G., Cundiff, S.T., 2013. Unraveling quantum pathways using optical 3D Fourier-transform spectroscopy. Nat. Commun. 4, 1390. 9 pages.

Lomsadze, B., Cundiff, S.T., 2017. Frequency combs enable rapid and high-resolution multidimensional coherent spectroscopy. Science 357, 1389. 3 pages.

Lomsadze, B., Smith, B.C., Cundiff, S.T., 2018. Tri-comb spectroscopy. Nat. Photonics 12, 676–680.

Marroux, H.J.B., Fidler, A.P., Neumark, D.M., Leone, S.R., 2018. Multidimensional spectroscopy with attosecond extreme ultraviolet and shaped near-infrared pulses. Sci. Adv. 4, eaau3783. 7 pages.

Martin, E.W., Cundiff, S.T., 2018. Inducing coherent quantum dot interactions. Phys. Rev. B97, 081301. 5 pages.

Mermillod, Q., Wigger, D., Delmonte, V., Reiter, D.E., Schneider, C., Kamp, M., Höfling, S., Langbein, W., Kuhn, T., Nogues, G., Kasprzak, J., 2016. Dynamics of excitons in individual InAs quantum dots revealed in four-wave mixing spectroscopy. Optica 3, 377–384.

Moody, G., Siemens, M.E., Bristow, A.D., Dai, X., Karaiskaj, D., Bracker, A.S., Gammon, D., Cundiff, S.T., 2011a. Exciton-exciton and exciton-phonon interactions in an interfacial GaAs quantum dot ensemble. Phys. Rev. B 83, 115324. 7 pages.

Moody, G., Siemens, M.E., Bristow, A.D., Dai, X., Bracker, A.S., Gammon, D., Cundiff, S.T., 2011b. Exciton relaxation and coupling dynamics in a GaAs/ $Al_xGa_{1-x}As$ quantum well and quantum dot ensemble. Phys. Rev. B 83, 245316. 7 pages.

Moody, G., Singh, R., Li, H., Akimov, I.A., Bayer, M., Reuter, D., Wieck, A.D., Bracker, A.S., Gammon, D., Cundiff, S.T., 2013a. Influence of confinement on biexciton binding in semiconductor quantum dot ensembles measured with two-dimensional spectroscopy. Phys. Rev. B 87, 041304. 5 pages.

Moody, G., Singh, R., Li, H., Akimov, I.A., Bayer, M., Reuter, D., Wieck, A.D., Cundiff, S.T., 2013b. Correlation and dephasing effects on the non-radiative coherence between bright excitons in an InAs QD ensemble measured with 2D spectroscopy. Solid State Commun. 163, 65–69.

Moody, G., Singh, R., Li, H., Akimov, I.A., Bayer, M., Reuter, D., Wieck, A.D., Cundiff, S.T., 2013c. Fifth-order nonlinear optical response of excitonic states in an InAs quantum dot ensemble measured with two-dimensional spectroscopy. Phys. Rev. B 87, 045313. 5 pages.

Mukamel, S., 1995. Principles of Nonlinear Optics and Spectroscopy. Oxford University Press, New York.

Nardin, G., Autry, T.M., Silverman, K.L., Cundiff, S.T., 2013. Multidimensional coherent photocurrent spectroscopy of a semiconductor nanostructure. Opt. Express 21, 28617–28627.

Nardin, G., Moody, G., Singh, R., Autry, T.M., Li, H., Morier-Genoud, F., Cundiff, S.T., 2014. Coherent excitonic coupling in an asymmetric double InGaAs quantum well arises from many-body effects. Phys. Rev. Lett. 112, 046402. 5 pages.

Nardin, G., Autry, T.M., Moody, G., Singh, R., Li, H., Cundiff, S.T., 2015. Multidimensional coherent optical spectroscopy of semiconductor nanostructures: collinear and non-collinear approaches. J. Appl. Phys. 117, 112804. 6 pages.

Nilsson, M., Boström, F.V., Lehmann, S., Dick, K.A., Leijnse, M., Thelander, C., 2018. Tuning the two-electron hybridization and spin states in parallel-coupled InAs quantum dots. Phys. Rev. Lett. 121, 156802. 5 pages.

Park, Y.-S., Lim, J., Klimov, V.I., 2019. Asymmetrically strained quantum dots with nonfluctuating single-dot emission spectra and subthermal room-temperature linewidths. Nat. Mater. 18, 249–255.

Pietryga, J.M., Park, Y.-S., Lim, J., Fidler, A.F., Bae, W.K., Brovelli, S., Klimov, V.I., 2016. Spectroscopic and device aspects of nanocrystal quantum dots. Chem. Rev. 116, 10513–10622.

Poltavtsev, S.V., Reichelt, M., Akimov, I.A., Karczewski, G., Wiater, M., Wojtowicz, T., Yakovlev, D.R., Meier, T., Bayer, M., 2017. Damping of Rabi oscillations in intensity-dependent photon echoes from exciton complexes in a CdTe/(Cd,Mg)Te single quantum well. Phys. Rev. B96, 075306. 7 pages.

Protesescu, L., Yakunin, S., Bodnarchuk, M.I., Krieg, F., Caputo, R., Hendon, C.H., Yang, R.X., Walsh, A., Kovalenko, M.V., 2015. Nanocrystals of cesium lead halide perovskites ($CsPbX3$, X = cl, Br, and I): novel optoelectronic materials showing bright emission with wide color gamut. Nano Lett. 15, 3692–3696.

Rainò, G., Becker, M.A., Bodnarchuk, M.I., Mahrt, R.F., Kovalenko, M.V., Stöferle, T., 2018. Superfluorescence from lead halide perovskite quantum dot superlattices. Nature 563, 671–675.

Ramsay, A.J., Gopal, A.V., Gauger, E.M., Nazir, A., Lovett, B.W., Fox, A.M., Skolnick, M.S., 2010a. Damping of exciton Rabi rotations by acoustic phonons in optically excited InGaAs/GaAs quantum dots. Phys. Rev. Lett. 104, 017402. 4 pages.

Ramsay, A.J., Godden, T.M., Boyle, S.J., Gauger, E.M., Nazir, A., Lovett, B.W., Fox, A.M., Skolnick, M.S., 2010b. Phonon-induced Rabi-frequency renormalization of optically driven single InGaAs/GaAs quantum dots. Phys. Rev. Lett. 105, 177402. 4 pages.

Richter, M., Singh, R., Siemens, M., Cundiff, S.T., 2018. Deconvolution of optical multidimensional coherent spectra. Sci. Adv. 4, eaar7697. 7 pages.

Selig, U., Langhojer, F., Dimler, F., Löhrig, T., Schwarz, C., Gieseking, B., Brixner, T., 2008. Inherently phase-stable coherent two-dimensional spectroscopy using only conventional optics. Opt. Lett. 33, 2851–2853.

Shah, J., 1999. Ultrafast Spectroscopy of Semiconductors and Semiconductor Nanostructures. Springer, New York.

Siemens, M.E., Moody, G., Li, H., Bristow, A.D., Cundiff, S.T., 2010. Resonance lineshapes in two-dimensional Fourier transform spectroscopy. Opt. Express 18, 17699–17708.

Singh, R., Suzuki, T., Autry, T.M., Moody, G., Siemens, M.E., Cundiff, S.T., 2016. Polarization-dependent exciton linewidth in semiconductor quantum wells: a consequence of bosonic nature of excitons. Phys. Rev. B 94, 081304. 5 pages.

Singh, R., Richter, M., Moody, G., Siemens, M.E., Li, H., Cundiff, S.T., 2017. Localization dynamics of excitons in disordered semiconductor quantum wells. Phys. Rev. B 95, 235307. 14 pages.

Stievater, T.H., Li, X., Steel, D.G., Gammon, D., Katzer, D.S., Park, D., Piermarocchi, C., Sham, L.J., 2001. Rabi oscillations of excitons in single quantum dots. Phys. Rev. Lett. 87, 133603. 4 pages.

Stone, K.W., Gundogdu, K., Turner, D.B., Li, X., Cundiff, S.T., Nelson, K.A., 2009. Two-quantum 2D FT electronic spectroscopy of biexcitons in GaAs quantum wells. Science 324, 1169–1173.

Suzuki, T., Singh, R., Bayer, M., Ludwig, A., Wieck, A.D., Cundiff, S.T., 2016. Coherent control of the exciton-biexciton system in an InAs self-assembled quantum dot ensemble. Phys. Rev. Lett. 117, 157402. 5 pages.

Suzuki, T., Singh, R., Bayer, M., Ludwig, A., Wieck, A.D., Cundiff, S.T., 2018a. Detuning dependence of Rabi oscillations in an InAs self-assembled quantum dot ensemble. Phys. Rev. B 97, 161301. 5 pages.

Suzuki, T., Singh, R., Moody, G., Aßmann, M., Bayer, M., Ludwig, A., Wieck, A.D., Cundiff, S.T., 2018b. Dephasing of InAs quantum dot *p*-shell excitons studied using two-dimensional coherent spectroscopy. Phys. Rev. B 98, 195304. 9 pages.

Takagahara, T., 1999. Theory of exciton dephasing in semiconductor quantum dots. Phys. Rev. B 60, 2638–2652.

Tian, P., Keusters, D., Suzaki, Y., Warren, W.S., 2003. Femtosecond phase-coherent two-dimensional spectroscopy. Science 300, 1553–1555.

Toda, Y., Moriwaki, O., Nishioka, M., Arakawa, Y., 1999. Efficient carrier relaxation mechanism in InGaAs/GaAs self-assembled quantum dots based on the existence of continuum states. Phys. Rev. Lett. 82, 4114–4117.

Tollerud, J.O., Hall, C.R., Davis, J.A., 2014. Isolating quantum coherence using coherent multi-dimensional spectroscopy with spectrally shaped pulses. Opt. Express 22, 6719–6733.

Tseng, C., Matsika, S., Weinacht, T.C., 2009. Two-dimensional ultrafast Fourier transform spectroscopy in the deep ultraviolet. Opt. Express 17, 18788–18793.

Turner, D.B., Nelson, K.A., 2010. Coherent measurements of high-order electronic correlations in quantum wells. Nature 466, 1089–1092.

Utzat, H., Sun, W., Kaplan, A.E.K., Krieg, F., Ginterseder, M., Spokoyny, B., Klein, N.D., Shulenberger, K.E., Perkinson, C.F., Kovalenko, M.V., Bawendi, M.G., 2019. Coherent single-photon emission from colloidal lead halide perovskite quantum dots. Science 363, 1068. 5 pages.

Vasanelli, A., Ferreira, R., Bastard, G., 2002. Continuous absorption background and decoherence in quantum dots. Phys. Rev. Lett. 89, 216804. 4 pages.

Vaughan, J.C., Hornung, T., Stone, K.W., Nelson, K.A., 2007. Coherently controlled ultrafast four-wave mixing spectroscopy. J. Phys. Chem. A 111, 4873–4883.

Wagner, W., Li, C., Semmlow, J., Warren, W.S., 2005. Rapid phase-cycled two-dimensional optical spectroscopy in fluorescence and transmission mode. Opt. Express 13, 3697–3706.

Wang, Q.Q., Muller, A., Bianucci, P., Rossi, E., Xue, Q.K., Takagahara, T., Piermarocchi, C., MacDonald, A.H., Shih, C.K., 2005. Decoherence processes during optical manipulation of excitonic qubits in semiconductor quantum dots. Phys. Rev. B 72, 035306. 5 pages.

Watson, T.F., Philips, S.G.J., Kawakami, E., Ward, D.R., Scarlino, P., Veldhorst, M., Savage, D.E., Lagally, M.G., Friesen, M., Coppersmith, S.N., Eriksson, M.A., Vandersypen, L.M.K., 2018. A programmable two-qubit quantum processor in silicon. Nature 555, 633–637.

Weinzetl, C., Görlitz, J., Becker, J.N., Walmsley, I.A., Poem, E., Nunn, J., Becher, C., 2019. Coherent control and wave mixing in an ensemble of silicon-vacancy centers in diamond. Phys. Rev. Lett. 122, 063601. 6 pages.

Woerner, M., Kuehn, W., Bowlan, P., Reimann, K., Elsaesser, T., 2013. Ultrafast two-dimensional terahertz spectroscopy of elementary excitations in solids. New J. Phys. 15, 025039. 16 pages.

Wong, C.Y., Scholes, G.D., 2011. Biexcitonic fine structure of CdSe nanocrystals probed by polarization-dependent two-dimensional photon echo spectroscopy. J. Phys. Chem. A115, 3797–3806.

Zhang, T., Kuznetsova, I., Meier, T., Li, X., Mirin, R.P., Thomas, P., Cundiff, S.T., 2007. Polarization-dependent optical 2D Fourier transform spectroscopy of semiconductors. Proc. Natl. Acad. Sci. U. S. A. 104, 14227–14232.

CHAPTER NINE

Single-photon nonlinear optics with a semiconductor quantum dot

Shuo Sun[a,*] and Kai Müller[b]
[a]JILA and Department of Physics, University of Colorado, Boulder, CO, United States
[b]Walter Schottky Institut and Department of Electrical and Computer Engineering, Technische Universität München, Garching, Germany
*Corresponding author: e-mail address: shuosun@colorado.edu

Contents

1. Introduction

Nonlinear optics has enabled tremendous applications including frequency conversion (Fejer, 1994), soliton generation (Kivshar and Agrawal, 2003), Raman lasing (Pask, 2003), and nonlinear spectroscopy (Mukamel, 1995). Nonlinear optical phenomena typically occur at an intense optical field, but advances in optical materials and device engineering have resulted in progressively more efficient processes and enabled the observation of nonlinear processes at lower and lower optical powers

Semiconductors and Semimetals, Volume 105
ISSN 0080-8784
https://doi.org/10.1016/bs.semsem.2020.10.005

(Byer, 1997; Eaton, 1991; Koenderink et al., 2015; Lapine et al., 2014). One may wonder if it is possible to observe nonlinear optical phenomena at the fundamental limit, where the propagation of light can be changed by a single photon. Besides its profound impact in optics and quantum physics (Chang et al., 2014), the realization of single-photon nonlinear optics will enable many new applications including low-energy electro-optics (Willner et al., 2014), nonclassical light generation (Faraon et al., 2008; Müller et al., 2015b), deterministic photonic quantum gates (Chuang and Yamamoto, 1995; Milburn, 1989), and nondestructive photon detection (Imoto et al., 1985).

Self-assembled quantum dots in semiconductors provide a powerful resource as medium of single-photon optical nonlinearities. Similar with atoms, quantum dots have optical transitions that can absorb light one photon at a time (Michler et al., 2000), which is the origin of the optical nonlinearity. In addition, quantum dots possess at least an order of magnitude larger oscillator strength than trapped atoms (Guest et al., 2002) and can straightforwardly couple with nanophotonic structures that confine light in the subwavelength scales (Lodahl et al., 2015). These properties lead to the largest light-matter coupling cooperativity (Najer et al., 2019) and bandwidth (Ota et al., 2018) observed so far in any optical cavity or waveguide quantum electrodynamics systems. Such strong light-matter interactions can therefore generate giant optical nonlinearities at the single-photon level.

In this Chapter, we review the theoretical and experimental progress in generating single-photon optical nonlinearities with single self-assembled quantum dots. We first consider the use of the quantum dot as a two-level system. When strongly coupled with a nanophotonic cavity, the quantum dot induces a giant anharmonicity in the cavity energy levels, leading to a frequency mismatch between the single- and multi-photon resonances of the coupled system. Therefore, depending on the frequency of the incident light, the system will tend to transmit either the single-photon state or multi-photon states, which are referred to as the effect of photon blockade and photon-induced tunneling, respectively. We will discuss the theory of these effects, their experimental demonstrations, various considerations in improving these effects, and the applications of these effects in nonclassical light generation in Section 2.

While two-level quantum dots provide a giant optical nonlinearity at the single-photon level, they are fundamentally limited by a time-bandwidth tradeoff that makes deterministic single-photon switching impossible, and thus cannot be directly applied in optical and quantum information processing. A three-level system that possesses an extra memory state can overcome this limit, enabling a single photon to deterministically switch a second photon. In Section 3, we will review the theories and the experimental progress towards the realization of a spin-photon quantum switch between a single photon and a quantum dot possessing a spin qubit, and discuss the experiments on demonstration of a single-photon switch and transistor mediated by a quantum dot spin.

Finally, in Section 4 we will discuss the limitations in current experimental platforms and possible ways to improve the performance of these devices. We will also highlight new applications that can be enabled by using quantum dots as the medium of single-photon optical nonlinearities.

2. Quantum dot as a two-level system: Photon blockade and photon-induced tunneling

2.1 Jaynes-Cummings model

In this section we review single-photon optical nonlinearities where the quantum dot is treated as a two-level system. While we focus our discussion on quantum dots, the same physics holds for a large number of systems such as trapped atoms (Brune et al., 1996; Schuster et al., 2008) and superconducting circuits (Fink et al., 2008). The energy structure of a two-level system coupled to a resonator is well described by the Jaynes-Cummings Hamiltonian:

$$\mathbf{H} = \omega_a \mathbf{a}^\dagger \mathbf{a} + (\omega_a + \Delta)\boldsymbol{\sigma}^\dagger \boldsymbol{\sigma} + g\left(\mathbf{a}^\dagger \boldsymbol{\sigma} + \boldsymbol{\sigma}^\dagger \mathbf{a}\right),$$

where ω_a denotes the frequency of the resonator, $\mathbf{a}$ is the resonator mode operator, $\boldsymbol{\sigma}$ is the lowering operator of the two-level system, Δ is the detuning between the two-level system and the resonator, and g is the coupling strength between the two-level system and the resonator. The dissipation of the system is described by the cavity energy decay rate κ and the two-level system decay rate γ. Taking these rates into account leads to a quantum Liouville equation with complex eigen-energies (Laussy et al., 2012):

$$E_{\pm}^{n} = n\omega_a + \frac{\Delta}{2} - i\frac{(2n-1)\kappa + \gamma}{4} \pm \sqrt{\left(\sqrt{n}g\right)^2 + \left(-\frac{\Delta}{2} - i\frac{\kappa - \gamma}{4}\right)^2}$$

where $E_{\pm}^{n}$ describes the energy of the n-th rung. The real parts of $E_{\pm}^{n}$ yield the energies of the states whereas the imaginary parts yield their linewidth. The eigenstates are hybridized light-matter states known as polaritons and the eigen-energies form the so-called Jaynes-Cummings ladder. The rung n denotes the number of photons present in the system and each rung consists of two eigenstates. The eigen-energies of the first two rungs are presented in Fig. 1A as a function of the detuning Δ. They consist of a series of anticrossing branches labeled LP_n and UP_n for lower polariton and upper

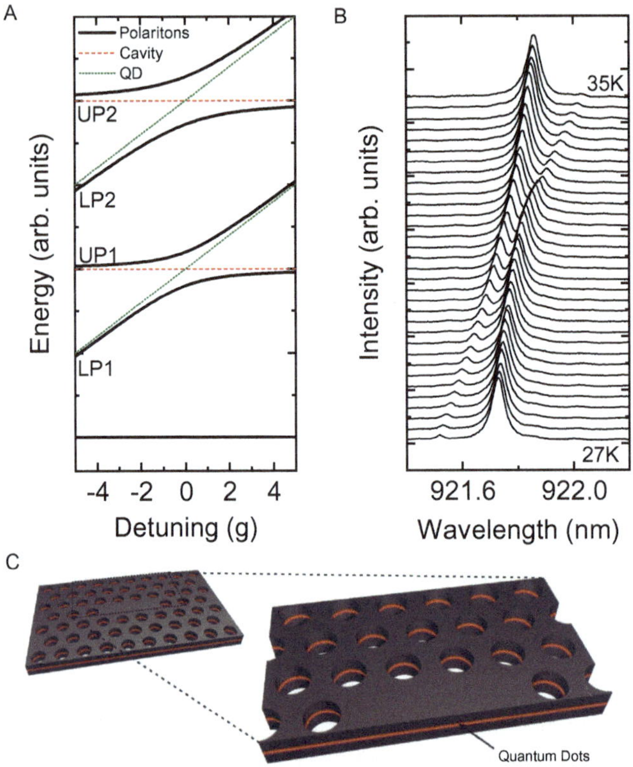

Fig. 1 (A) Schematic illustration of the energy structure of a strongly coupled system given by the Jaynes-Cummings ladder. (B) Cross-polarized reflectivity spectra of the coupled quantum-dot-cavity system obtained by temperature tuning the quantum dot through the cavity resonance. An anticrossing of the peaks clearly demonstrates the strong coherent coupling. (C) Schematic illustration of quantum dots embedded in a photonic crystal L3 cavity. *Panel (A and B): Adapted from Müller, K., Fischer, K.A., Rundquist, A., Dory, C., Lagoudakis, K.G., Sarmiento, T., Kelaita, Y.A., Borish, V., Vučković, J., 2015. Ultrafast polariton-phonon dynamics of strongly coupled quantum dot-nanocavity systems. Phys. Rev. X 5, 031006. Panel (C): Adapted from Müller, K., Rundquist, A., Fischer, K.A., Sarmiento, T., Lagoudakis, K.G., Kelaita, Y.A., Muñoz, C.S., Del Valle, E., Laussy, F.P., Vučković, J., 2015. Coherent generation of nonclassical light on chip via detuned photon blockade. Phys. Rev. Lett. 114, 233601 with permission. Copyright 2015, American Physical Society.*

polariton of the n-th rung. For large detuning, the polariton states adiabatically approach the uncoupled states. However, at resonance the polaritons anticross due to the coupling, and the splitting at resonance is given by $2g\sqrt{n}$. Observing this strong coupling requires that the linewidth of the polaritons are smaller than the splitting, i.e., $2g$ must be larger than κ and γ. Therefore, resonators with high quality factor and small mode volume are needed. For quantum dots, this has been achieved in micropillar resonators (Reithmaier et al., 2004), photonic crystal cavities (Yoshie et al., 2004) and tunable open cavities (Najer et al., 2019). While open cavities are inherently tunable by changing the cavity length, in micropillar resonators and photonic crystal cavities the detuning between the quantum dot and resonator can be controlled by tuning the resonator via inert-gas deposition (Mosor et al., 2005) or by tuning the quantum dot frequency via changing the sample temperature, applying electric field (Laucht et al., 2009), or applying strain (Sun et al., 2013).

Experimentally, the splitting of the lowest manifold can be observed either in photoluminescence or by measuring the frequency resolved transmission through the coupled system. For photonic crystal cavities with a linearly-polarized cavity mode, this transmission can be conveniently measured using cross-polarized reflectivity (Englund et al., 2007). Photons which are admitted to the cavity and subsequently re-emitted have a polarization that is rotated relative to the incident polarization. Aligning the cavity at 45 degree relative to the incident polarization results in a polarization rotation of 90 degree, i.e., the polarization of the signal is identical to the polarization of the detection channel. An example measurement of such a measurement is presented in Fig. 1B where a quantum dot is tuned through the cavity mode of a photonic crystal L3 cavity illustrated in Fig. 1C. A clear anticrossing is observed and a fit to the data reveals values of $g = 2\pi \times 12.3$ GHz and $\kappa = 2\pi \times 18.4$ GHz.

2.2 Photon blockade and photon-induced tunneling

In a strongly coupled quantum-dot-cavity system, non-classical light can be generated through optical nonlinearities at the single photon level which manifest in the anharmonicity of the Jaynes-Cummings ladder (Faraon et al., 2008; Reinhard et al., 2012). A laser tuned into resonance with the first rung of the ladder (blue arrow Fig. 1(A)) is not in resonance with subsequent higher rungs of the ladder. Therefore, the admission of a photon to the system reduces the probability for a second photon to be admitted, which is known as photon blockade. On the other hand, for a laser tuned in between the polariton branches, the admission of a photon to the system enhances the probability for additional photons to be admitted, which results in a preferential transmission of multi-photon pulses and known as photon-induced tunneling. To quantify the quantum character of light, the measured degree of second-order coherence (Glauber, 1963) is commonly used, which in a pulse-wise form is given by

$$g^{(2)}(0) = \langle \mathbf{n}(\mathbf{n}-1)\rangle / \langle \mathbf{n} \rangle^2,$$

where $\mathbf{n}=\mathbf{a}^{\dagger}\mathbf{a}$ is the photon number operator.

Due to the highly dissipative character of nanophotonic systems, the observed antibunching in a photon blockade is moderate when the quantum dot and cavity are in resonance (Faraon et al., 2008; Reinhard et al., 2012). Although a laser in resonance with the first rung is not in exact resonance with the transition from the first rung to the second rung, it still has significant overlap with the transition due to the linewidth of the states. Obtaining lower values of $g^{(2)}(0)$ and, thus, a better single-photon purity requires improvement of the g/κ ratio. To this end surface passivation of photonic crystal cavities has recently demonstrated Q factors of 160,000 in GaAs cavities (Kuruma et al., 2020).

An additional possibility to improve photon blockade is to increase the anharmonicity by exploiting a quantum-dot-cavity detuning (Müller et al., 2015b). This concept is visualized in Fig. 2(A), which presents the detuning dependent energy differences between the rungs for transitions leading to $n=1$, $n=2$, $n=3$ as solid, dashed, and dotted lines, respectively. With increasing quantum-dot–cavity detuning the energy difference between transitions to the first rung (solid lines) and subsequent transitions to higher rungs of the Jaynes-Cummings ladder (dashed and dotted lines) increases for exciting the more quantum-dot-like polariton. This is also the case for the energies of exciting the n-th rung of the Jaynes-Cummings ladder in an n-photon process (Fig. 2(B)). In addition, the pulse length of the excitation laser plays an important role (Müller et al., 2015a): longer excitation pulses increase the chance of re-excitation during the presence of the pulse while shorter pulses have a broader linewidth and thus increase the overlap with the different transitions. A measurement that demonstrates the influence of detuning and pulse length is presented in Fig. 2(C), which shows $g^{(2)}(0)$ as a function of the laser detuning for different pulse lengths and quantum-dot-cavity detunings.

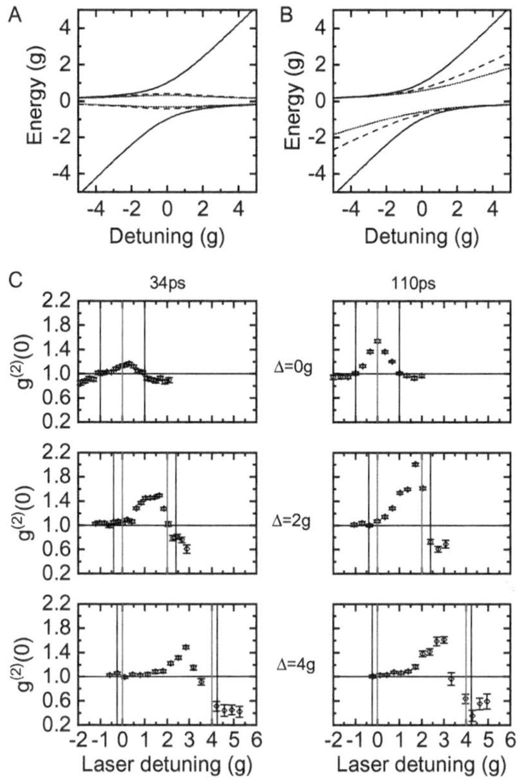

Fig. 2 (A) Transition energies for climbing the Jaynes-Cummings-ladder rung by rung. Transitions from the ground state to first rung are represented by solid lines and subsequent steps are presented as dashed and dotted lines. (B) Energies for exciting the *n*-th rung of the Jaynes-Cummings-ladder in an *n*-photon process. In (A) and (B) the energies are given relative to the cavity frequency. (C) Measured second-order coherence $g^{(2)}(0)$ as a function of the laser detuning for different quantum-dot-cavity detunings (top to bottom) and excitation pulse lengths (left to right). In each case, the red vertical line indicates the bare cavity frequency, the vertical green line indicates the bare quantum dot frequency, and the black vertical lines indicate the polariton frequencies. *Panel (A and B): Adapted from Müller, K., Rundquist, A., Fischer, K.A., Sarmiento, T., Lagoudakis, K.G., Kelaita, Y.A., Muñoz, C.S., Del Valle, E., Laussy, F.P., Vučković, J., 2015. Coherent generation of nonclassical light on chip via detuned photon blockade. Phys. Rev. Lett. 114, 233601. Panel (C): Adapted from Müller, K., Fischer, K.A., Rundquist, A., Dory, C., Lagoudakis, K.G., Sarmiento, T., Kelaita, Y.A., Borish, V., Vučković, J., 2015. Ultrafast polariton-phonon dynamics of strongly coupled quantum dot-nanocavity systems. Phys. Rev. X 5, 031006 with permission. Copyright 2015, American Physical Society.*

With increasing quantum-dot-cavity detuning, the value $g^{(2)}(0)$ obtained in the photon-blockade regime (driving the more quantum-dot-like polariton) decreases. Here the lowest value obtained is $g^{(2)}(0) = 0.34 \pm 0.07$.

2.3 Boosting photon blockade by homodyning

While optimizing the quantum-dot-cavity detuning and excitation pulse lengths can significantly reduce $g^{(2)}(0)$ as discussed above, the values obtainable from detuned Jaynes-Cummings systems with experimentally feasible g/κ ratios are worse than those from state-of-the-art single-photon sources (Trivedi et al., 2020). Therefore, in order to further improve $g^{(2)}(0)$, it is essential to understand where the limitation comes from. For a detuned Jaynes-Cummings system in the photon blockade regime, the signal consists of multiple components. Photons can be transmitted through the quantum-dot-like polariton where strong single-photon nonlinearities exist (Fig. 2A), but the excitation laser can also be coherently scattered off the detuned resonator mode. Although the laser-cavity detuning is large, the coherent scattering is in fact dominating the signal as the cavity mode has a Lorentzian spectral line shape and harmonic energy ladder.

This can be experimentally demonstrated as schematically illustrated in Fig. 3A. In addition to the emission at the laser frequency in resonance with the quantum-dot-like upper polariton, emission also occurs from the cavity-like lower polariton due to a phonon-mediated population transfer. This emission is detuned from the coherent scattering of the laser and samples only the population of the quantum-dot-like polariton. Consequently, when filtering on the phonon-mediated detuned emission it is possible to measure Rabi oscillations (Fig. 3B) and obtain a significantly better value of $g^{(2)}(0) = 0.03 \pm 0.01$ (Fig. 3C).

This low value confirms the generation of single photons with high single-photon purity. Moreover, the fact that the signal is detuned from the excitation laser is beneficial for some applications as it allows for simple frequency filtering to isolate the quantum signal from the pump laser. However, the single photons emitted in this way will have a poor indistinguishability as the phonon-mediated population transfer induces a timing jitter in the emission that is large compared to the short lifetime of the cavity-like polariton (Fischer et al., 2016a; Trivedi et al., 2020). Therefore, another technique is required to suppress the coherent scattering off the detuned cavity mode.

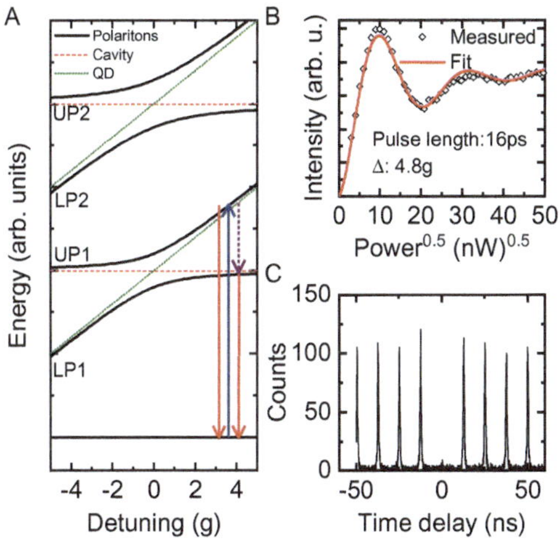

Fig. 3 (A) Schematic illustration of the excitation of UP1 in a detuned Jaynes-Cummings system with emission from UP1 and phonon-mediated emission from LP1. (B) Rabi oscillation observed upon exciting UP1 and detecting LP1. (C) Measurement of the second-order coherence for exciting UP1 with a π pulse and detecting LP1. *Adapted from Müller, K., Fischer, K.A., Rundquist, A., Dory, C., Lagoudakis, K.G., Sarmiento, T., Kelaita, Y.A., Borish, V., Vučković, J., 2015. Ultrafast polariton-phonon dynamics of strongly coupled quantum dot-nanocavity systems. Phys. Rev. X 5, 031006 with permission. Copyright 2015, American Physical Society.*

To this end, for a photonic crystal cavity system in cross-polarized geometry, a self-homodyne effect, which is intrinsic to photonic crystal cavities, can be used to interferometrically cancel the coherent scattering of the laser from the detuned cavity (Fischer et al., 2016b, 2018). Coherently scattered light from the cavity interferes with light scattered from the photonic crystal region outside of the cavity, which produces a Fano resonance. The Fano resonance can be tuned such that at the spectral position of the quantum-dot-like polariton, coherent scattering from the detuned cavity completely cancels out. This is visualized in Fig. 4A, which shows the simulated emission spectrum of a detuned Jaynes-Cummings system with a pulsed excitation laser in resonance with the quantum-dot-like polariton (UP1) without (blue) and with (black) self-homodyne suppression (Müller et al., 2016). The system parameters κ and g are taken from the measurement presented in Fig. 1B discussed above. Without self-homodyne suppression, the spectrum is dominated by a strong peak with the width of the excitation laser and detuned phonon-mediated emission from LP1. In contrast, with optimal suppression, the peak at the position of UP1 decreases in intensity and its

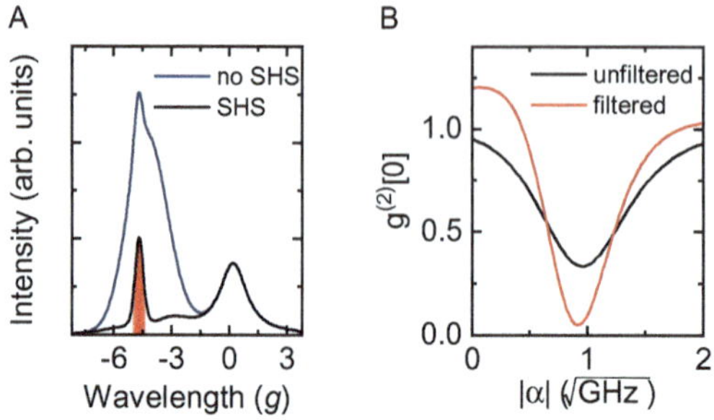

Fig. 4 (A) Simulated spectrum for resonant excitation of UP1 by a 16ps long π pulse at a quantum-dot-cavity detuning $\Delta = 4.5g$, with and without self-homodyne suppression. (B) $g^{(2)}(0)$ (second-order coherence) for the excitation conditions of (A) (resonant excitation of UP1) as a function of the self-homodyne suppression tuning parameter with and without spectral filtering on UP1 (red region in A). *Adapted from Müller, K., Fischer, K.A., Dory, C., Sarmiento, T., Lagoudakis, K.G., Rundquist, A., Kelaita, Y.A., Vučković, J., 2016. Self-homodyne-enabled generation of indistinguishable photons. Optica 3, 931–936 with permission. Copyright 2016, Optical Society of America.*

linewidth decreases from that of the laser to that of UP1, confirming the suppression of the coherent scattering. The resulting simulated value of $g^{(2)}(0)$ is presented in Fig. 4B as a function of a parameter that tunes the strength of the homodyne effect. Crucially, the simulation predicts that a very low value of $g^{(2)}(0)$ can be achieved for optimal suppression of the coherent scattering and with additional spectral filtering on UP1. This spectral filtering further suppresses the parts of the coherently scattered laser that were not perfectly suppressed in the homodyne suppression due to the large bandwidth of the short pulse length. An experimental test of this prediction confirmed the generation of single photons of high optical quality with $g^{(2)}(0) = 0.05 \pm 0.04$, an indistinguishability of $V = 0.96 \pm 0.05$ and an emission rate of $(55\ \text{ps})^{-1}$ (Müller et al., 2016).

2.4 Waveguide integrated self-homodyne architecture

The cross-polarized measurement technique discussed above is well-suited to study the physics of Jaynes-Cummings systems and their optical nonlinearities at the single-photon level. However, the photonic crystal platform also allows for on-chip integration by coupling the photonic crystal cavity to input and output waveguides (Faraon et al., 2007). Crucially, it was recently proposed that the interferometric cancellation technique discussed in Section 2.3 can also be adopted in an integrated geometry making use of a Partially Transmitting Element (PTE) (Fischer et al., 2017).

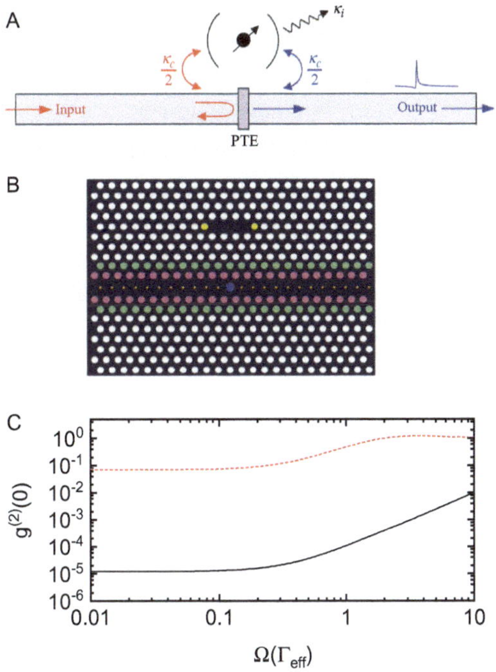

Fig. 5 (A) Schematic illustration of the integrated cancellation of coherent scattering using a partly transmitting element. A cavity is coupled to an input and output waveguide with rates $\kappa/2$ while a partly transmitting element (PTE) allows a part of the input light to directly couple to the output waveguide. (B) Proposed structure to realize the configuration presented in (A). Colored circles represent holes of the photonic crystal where the size and/or position is altered relative to the periodic crystal. (C) Simulated degree of second order coherence $g^{(2)}(0)$ of a detuned photon blockade ($\Delta = 6g$) as a function of the effective driving strength. With interferometric cancellation (black—PTE tuned to optimized suppression) the values of $g^{(2)}(0)$ are orders of magnitude lower compared to the case without interferometric cancellation (red—PTE fully blocking). *Adapted from Fischer, K.A., Kelaita, Y.A., Sapra, N.V., Dory, C., Lagoudakis, K.G., Müller, K., Vučković, J., 2017. On-chip architecture for self-homodyned nonclassical light. Phys. Rev. Appl. 7, 044002 with permission. Copyright 2016, American Physical Society.*

The concept is schematically illustrated in Fig. 5A. A cavity is coupled to input and output waveguides with rates $\kappa/2$ while a PTE allows a part of the input light to directly couple to the output waveguide. The direct emission from input to output waveguide can interfere with the coherent scattering through the cavity and the two components can be set to cancel out by tuning the transmission of the PTE. A design of an architecture to realize this concept is presented in Fig. 5B. The performance of this architecture can be estimated by quantum optical simulations based on fabricable Q-factor of the cavity of $\approx 51{,}000$ (in the absence of the waveguides) and coupling strength of $g = 2\pi \times 10$ GHz, lower than the best values recently demonstrated (Kuruma et al., 2020). The simulated power dependence of $g^{(2)}(0)$ is presented in Fig. 5C for the PTE fully blocking and tuned to optimal interferometric cancellation as dashed red and solid black lines, respectively. Here, an effective driving strength of 1 corresponds to a pulse area of π. While without interferometric cancellation (red dashed line) the purity of single-photon generation is moderate $g^{(2)}(0) \sim 0.1$, with optimal interferometric cancellation (solid black line) high-quality single photons with $g^{(2)}(0) < 10^{-4}$ are observed for effective driving strengths <1. While this architecture has not been experimentally realized yet, it gives an outlook to the potential of exploiting quantum dots for optical single-photon nonlinearities in future quantum technologies.

3. Quantum dot as a three-level system: A single-photon switch and transistor

3.1 Limit of two-photon interactions using a two-level system

As described in Section 2, a single two-level quantum dot strongly coupled to a nanophotonic cavity offers strong optical nonlinearity at the single-photon level, manifesting the effect of photon blockade and photon-induced tunneling. One fundamental question is whether we can employ this nonlinearity to realize a photon-photon logic gate, where the state of one photon can deterministically control the state of another photon. Successful realization of this device lies at the heart of both classical optical information processing and optical quantum computing. For example, it can significantly reduce the resource overhead in linear optical quantum computing by replacing the probabilistic measurement-based two-photon quantum gates (Brien, 2007; Kok et al., 2007).

Unfortunately, a fundamental limit exists in achieving ideal two-photon interactions with a two-level atom. To intuitively recognize this limit, we consider the generation of photon blockade with a specific device shown in Fig. 6A. The photon blockade is generated by a single two-level atom strongly coupled to an optical cavity. The cavity couples equally to two counter-propagating modes of a waveguide. The frequency detuning among the atom, the cavity and the incident photons can be arbitrary. This device resembles all the experimental demonstrations of photon blockade described in Section 2.

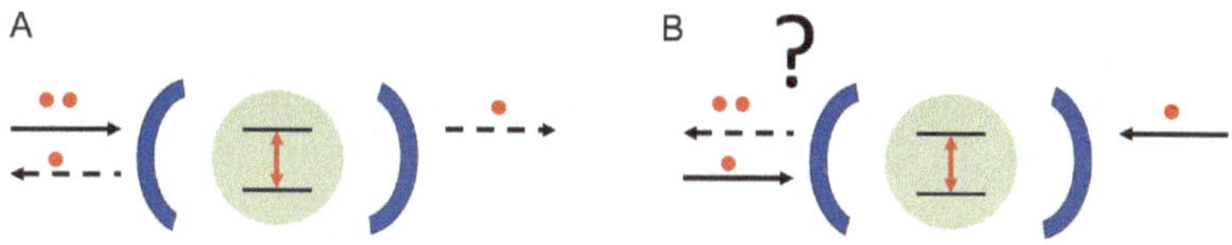

Fig. 6 (A) Ideal transformation of the photon blockade effect. (B) The transformation of the ideal photon blockade effect under time-reversal symmetry.

For two indistinguishable photons incident on the device from the left, an ideal photon blockade effect will transmit only one of the two photons and reflect the other one, as illustrated in Fig. 6A. According to time reversal symmetry, if two indistinguishable photons—one from the left and one from the right are incident on the device, the same device will transform both photons into the left-propagating mode of the waveguide, as illustrated in Fig. 6B. However, since the cavity couples equally to the left and right propagating mode of the waveguide, and the two photons are from both modes respectively, transforming both photons into the left-propagating mode violates the symmetry of this scattering problem. This suggests that the ideal photon blockade operation cannot be achieved with this device.

While the above analysis considers a specific device in which the two-level system couples to a two-sided optical cavity and a specific two-photon interaction functionality in the form of photon blockade, the conclusion applies generally to all photonic structures and all types of two-photon interactions. In fact, it has been shown that single-photon nonlinearity from a passive localized Kerr medium has a fundamental limit in achieving perfect two-photon quantum gates (Gea-Banacloche, 2010; Shapiro, 2006), where the state of one photon deterministically controls the state of another. To understand this limit, Xu et al. has studied the analytical properties of the two-photon scattering matrix of such localized quantum devices and demonstrated an inevitable frequency mixing and entanglement in the photon-photon interaction component of the two-photon scattering matrix (Xu et al., 2013). This means that the frequency of each single photon cannot be preserved after the interaction, leading to a fundamental error in the two-photon quantum gate. The same limit can also be understood from a time-bandwidth tradeoff argument (Brod and Combes, 2016; Nysteen et al., 2017; Rosenblum et al., 2011). Since the fundamental operating principle of the photon-photon interaction relies on the saturation of the two-level system, the incident photon pulse must be much shorter than the excited state lifetime so that the atom does not have enough time to decay back to the ground state before the scattering of the second photon. However, due to time-bandwidth uncertainty, a pulse that is significantly shorter than the atomic excited state lifetime must have a linewidth exceeding the interaction bandwidth of the atom. Thus, such a pulse will have some spectral components that cannot interact with the atom, leading to a non-negligible probability that the two photons do not interact with each other. The opposing requirements on the incident photon bandwidth suggests that perfect two-photon interactions are inherently prohibited when using a two-level system as the medium for nonlinearity.

3.2 Achieving perfect two-photon interactions with a three-level system

As we described above, the fundamental limit in the two-photon interactions set by the use of a two-level system arises from the tradeoff between the excited state saturation time and the atom-photon interaction bandwidth. One way to overcome this limit is to employ a three-level system with an additional memory state. By employing the additional memory state to register the first incident photon, we can extend the memory of the optical nonlinearity from the excited state lifetime to the memory state lifetime, while keeping the atom-photon interaction bandwidth to be the excited state linewidth. Thus, we are able to break the time-bandwidth tradeoff and surpass the fundamental limit in the two-photon interactions set by the use of a two-level system.

We now discuss a specific protocol for realizing a deterministic two-photon quantum gate, originally proposed by Duan and Kimble (2004). We consider a generic three-level system coupled to an optical cavity, as shown in Fig. 7A. We assume that the three-level system has a λ-type energy structure, with two ground states forming a stable spin qubit, denoted as

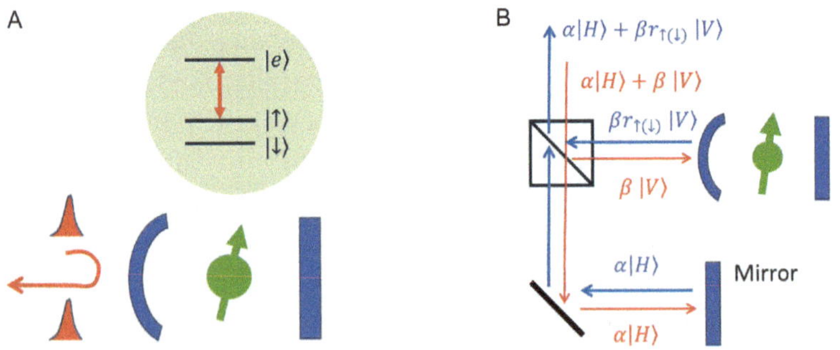

Fig. 7 (A) Schematic setup of a coherent spin-photon quantum interface. (B) Schematic setup realizing a quantum controlled phase flip gate between a single spin and a polarization-encoded photonic qubit.

$|\uparrow\rangle$ and $|\downarrow\rangle$, and one optically excited state denoted as $|e\rangle$. This energy structure exists in many qubit systems that are optically addressable, including cold atoms, trapped ions, charged quantum dots, and color centers. We assume that only one of the two possible transitions, say transition $|\uparrow\rangle \leftrightarrow |e\rangle$, is strongly and resonantly coupled with the cavity. The other possible transition $|\downarrow\rangle \leftrightarrow |e\rangle$ can be decoupled with the cavity for multiple reasons, such as the transition is not dipole allowed, the transition has an orthogonal polarization with the cavity, or the transition is far detuned from the cavity resonance.

The spin-dependent coupling between the three-level system and the cavity enables a coherent quantum interface between the spin qubit and an incident photon. To see this effect, we consider the cavity to be single-sided so that an incident photon is always reflected from the cavity. However, the phase of the reflected photon can be different depending on the state of the spin. If the spin is in the state $|\downarrow\rangle$, the cavity behaves as an empty cavity as it does not couple to any atomic transitions. For a single incident photon that is resonant with the cavity, the reflection coefficient is given by $r_\downarrow = -1$, indicating a π phase shift upon reflection. On the other hand, if the spin is initially in the state $|\uparrow\rangle$, the cavity behaves as a strongly coupled system with a reflection coefficient given by $r_\uparrow = (C-1)/(C+1)$, where $C = 2g^2/\kappa\gamma$ is the cooperativity, which is a function of the atom-cavity coupling strength g, the cavity energy decay rate κ, and the atomic transition $|\uparrow\rangle \leftrightarrow |e\rangle$ dipole decay rate γ. In the large cooperativity limit where $C >> 1$, we have $r_\uparrow = 1$. Thus, a single incident photon experiences no phase shift upon reflection from the cavity.

The coherent spin-photon quantum interface can be used to realize a Controlled Phase Flip (CPF) gate between the spin qubit and a single photon. Here we introduce a setup where the polarization degree of freedom is used to encode the photonic qubit, as shown in Fig. 7B. But the CPF gate can be applied on any other photonic qubit basis such as photon number, dual-rail, time-bin, and frequency-bin by slightly modifying the setup. In the polarization-based setup, a polarization beamsplitter is used to couple only one polarization component (the V polarization) of the incident photon to the cavity. The other polarization component (the H polarization) is directly reflected from a mirror. Therefore, the H polarization component will always reflect with no phase shift regardless of the spin state, whereas the V polarization component will pick up a spin-dependent π phase shift upon reflection. This realizes a spin-photon CPF operation.

The spin-photon CPF operation can be used to realize a deterministic single-photon switch, where the state of the first incident photon can control the state of the second photon. To see how this can be achieved, note that a single photon in the H/V polarization basis can flip the spin in the $|\pm\rangle$ basis, where $|\pm\rangle = (|\uparrow\rangle \pm |\downarrow\rangle)/\sqrt{2}$. Specifically, a single V-polarized photon will flip the spin from $|\pm\rangle$ to $|\mp\rangle$, whereas a single H-polarized photon will keep the spin state the same. Similarly, the spin-photon CPF gate allows a spin in the $|\uparrow(\downarrow)\rangle$ basis to switch a photon in the X and Y polarization basis, where $|X\rangle = (|H\rangle + |V\rangle)/\sqrt{2}$ and $|Y\rangle = (|H\rangle - |V\rangle)/\sqrt{2}$. Specifically, for an X-polarized incident photon, the spin will maintain its polarization if the spin is in the state $|\uparrow\rangle$ (since $|\uparrow\rangle|H\rangle \rightarrow |\uparrow\rangle|H\rangle$ and $|\uparrow\rangle|V\rangle \rightarrow |\uparrow\rangle|V\rangle$), but flip its polarization to Y if the spin is in the state $|\downarrow\rangle$ (since $|\downarrow\rangle|H\rangle \rightarrow |\downarrow\rangle|H\rangle$ and $|\downarrow\rangle|V\rangle \rightarrow -|\downarrow\rangle|V\rangle$). Sequentially applying both switching operations with a spin basis rotation in the middle will enable the first photon (in the H/V polarization basis) to switch the polarization of the second photon (in the X/Y polarization basis) via the mediation of the spin.

The spin-photon CPF gate can also be used to generate a two-photon quantum gate as described by Duan and Kimble (2004). Specifically, a CPF gate between two consecutive incident photons p_1 and p_2 (denoted as $CPF_{p_1,p2}$) acting on an arbitrary two-photon state $|\psi_{p_1,p2}\rangle$ can be decomposed as

$$\left(\mathbf{CPF}_{p_1,p_2}\otimes\mathbf{I}_s\right)\left|\psi_{p_1,p_2}\right\rangle|\phi_s\rangle = \mathbf{CPF}_{p_1,s}\mathbf{R}_s(-\pi/2)\mathbf{CPF}_{p_2,s}\mathbf{R}_s(\pi/2)\mathbf{CPF}_{p_1,s}\left|\psi_{p_1,p_2}\right\rangle|\phi_s\rangle,$$

where $|\phi_s\rangle = (|\uparrow\rangle + |\downarrow\rangle)/\sqrt{2}$ is the initial state of spin qubit, $\mathbf{I}_s$ is an identical operator acting on the spin qubit, $\mathbf{CPF}_{p_1,s}$ and $\mathbf{CPF}_{p_2,s}$ represents the CPF gates between the spin and the photons p_1 and p_2 respectively, $\mathbf{R}_s(\theta)$ is a single qubit gate that rotates the spin by an angle θ along the y-axis of the Bloch sphere. Intuitively speaking, the two-photon CPF gate is achieved via sequential scattering of the two photons from the device ($\mathbf{CPF}_{p_1s}$ and $\mathbf{CPF}_{p_2,s}$) to entangle both photons with the spin, followed by re-scattering of the first photon ($\mathbf{CPF}_{p_1,s}$ again) to disentangle the spin from the two photons.

3.3 Realization of a spin-photon quantum interface with a single quantum dot

As described in the previous section, to realize deterministic photon-photon interactions requires a coherent quantum interface between an optical photon and a three-level system. Kim et al. for the first time demonstrated such a quantum interface based on semiconductor quantum dots (Kim et al., 2013). To obtain the necessary three-level system from the quantum dot, they employed a neutral quantum dot and applied an external magnetic field along its growth direction (Faraday configuration). Fig. 8A shows the energy level structure of the quantum dot. The excited state was split into two, giving rise to a three-level system with a V-type energy level structure (Bayer et al., 2002). They strongly couple one of the optical transitions ($|g\rangle \leftrightarrow |+\rangle$) with a photonic crystal cavity (see Fig. 8B), and largely detune the other optical transition ($|g\rangle \leftrightarrow |-\rangle$) from the cavity. Under this configuration, they can controllably flip the polarization of an incident photon by controlling the state of the quantum dot between $|g\rangle$ and $|-\rangle$ (Fig. 8C). This realizes a Controlled-NOT (CNOT) gate between the quantum dot qubit and the reflected photon, as demonstrated by the transformation matrix probed experimentally (Fig. 8D).

A major limitation of this work is the short coherence time of the quantum dot qubit. The qubit they employed was formed by the ground state $|g\rangle$ and one of the excited state $|-\rangle$. The excited state $|-\rangle$ has a short lifetime of $\sim$500 ps due to the large spontaneous emission rate. This short lifetime limits the achievable functionality of this device. For example, this device cannot achieve a quantum phase switch where a single incident photon flips the quantum dot from the state $(|g\rangle + |-\rangle)/\sqrt{2}$ to the state $(|g\rangle - |-\rangle)/\sqrt{2}$, since the quantum dot qubit will quickly decay to the state $|g\rangle$ before one can measure its state upon reflection of a photon. For this reason, this device cannot be used to generate deterministic photon-photon interactions.

The spin of a charged quantum dot offers a much longer coherence time. Fig. 9A shows the energy level structure of a negatively charged quantum dot in the presence of a magnetic field applied perpendicular to its growth direction (Voigt configuration) (Bayer et al., 2002). The negatively charged quantum dot possesses an extra electron with two possible spin orientations ($|\uparrow\rangle$ and $|\downarrow\rangle$), leading to a stable spin qubit in the ground state manifold. Previous works have measured a coherence time of $T_2 \sim 1\,\mu s$ for this qubit (De Greve et al., 2011; Greilich et al., 2006; Press et al., 2010), which is much longer than the qubit based on a neutrally charged quantum dot. The quantum dot has two excited states due to the two possible hole spins, giving rise to four possible transitions. In the Voigt configuration, all the four possible transitions are optically allowed. The polarization of the two spin conserving transitions is orthogonal with the two spin non-conserving transitions. The optical dipole between the spin conserving and non-conserving transitions differ by a phase $\pi/2$.

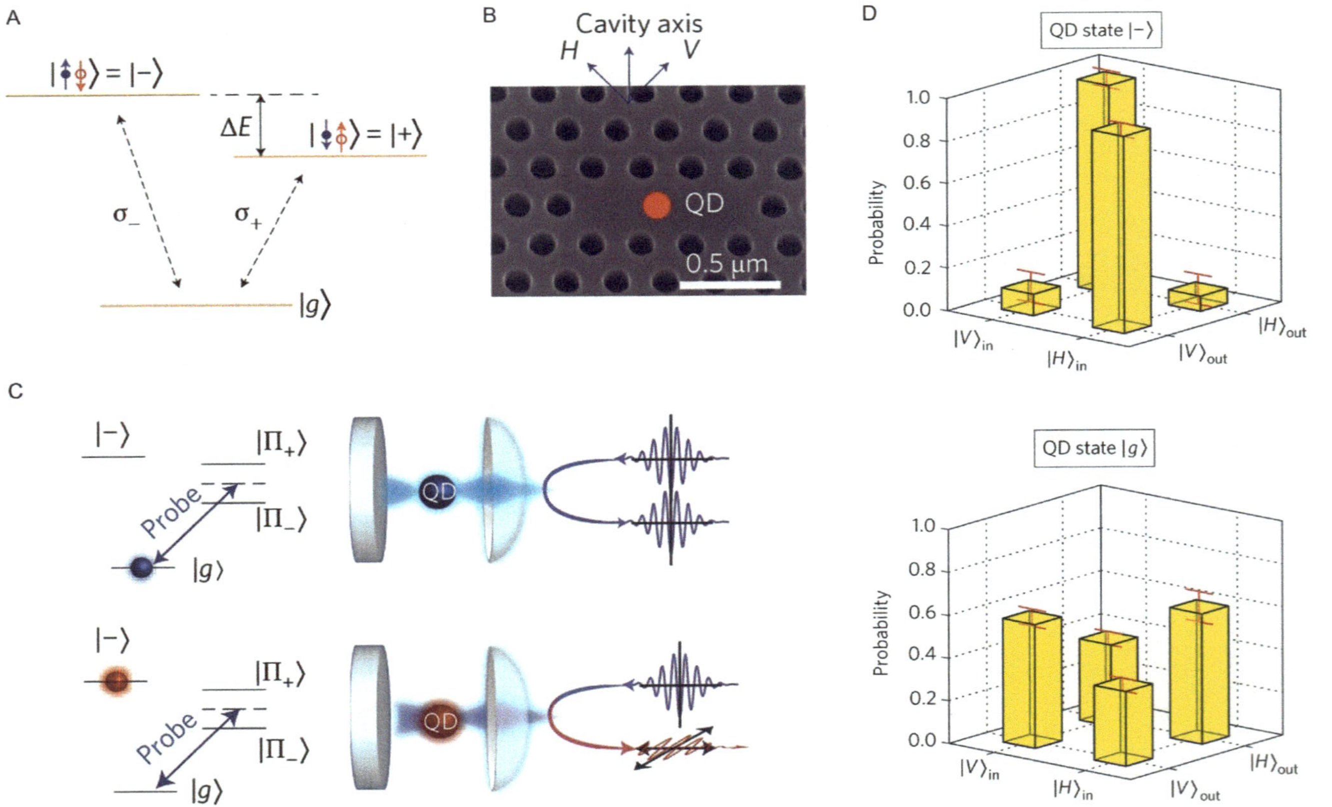

Fig. 8 (A) Energy level structure of a neutral quantum dot under a magnetic field. (B) SEM image of the fabricated photonic crystal cavity. (C) Illustration of the CNOT operation between the quantum dot qubit and a single photon. The polarization of an incident photon is preserved when the quantum dot is in state $|g\rangle$ (top), and is rotated when the quantum dot is in state $|-\rangle$ (bottom). The horizontal dashed lines indicate the degenerate energy level of the $|+\rangle$ quantum dot state and the cavity photon state, which are split into two polariton states Π_+ and Π_- in the strong coupling regime. *Adapted from Kim, H., Bose, R., Shen, T.C., Solomon, G.S., Waks, E., 2013. A quantum logic gate between a solid-state quantum bit and a photon. Nat. Photon. 7, 373 with permission. Copyright 2013, Springer Nature.*

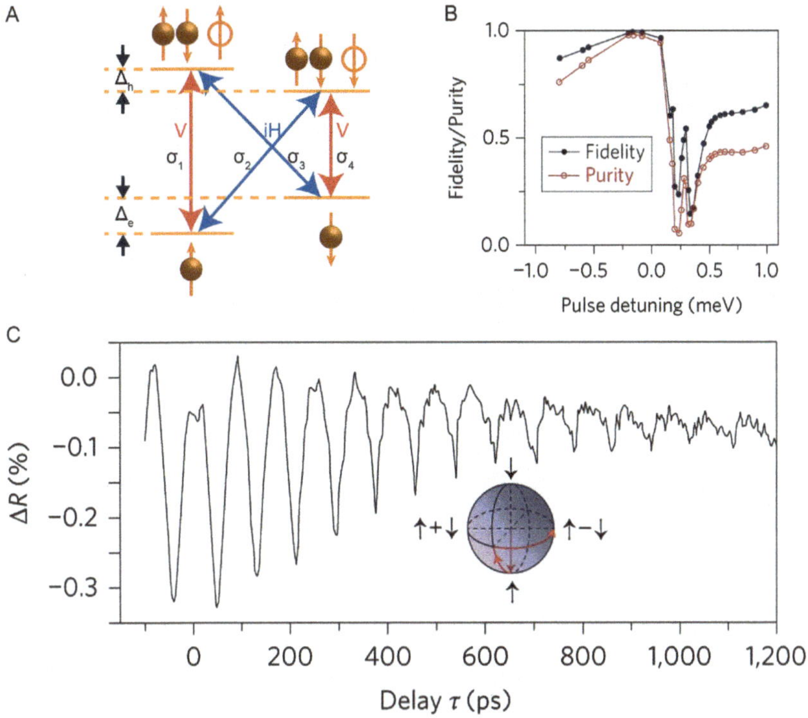

Fig. 9 (A) Energy level structure of a negatively charged quantum dot under a magnetic field applied perpendicular to its growth direction. The level structure is composed of two spin ground states and two excited states, with four allowed optical transitions labeled σ_1 to σ_4. (B) Theoretical model of the fidelity of the spin rotation and purity of the resulting state as a function of the detuning between the rotation pulse and the cavity resonance. (C) Measured change in cavity reflectivity induced by the quantum dot spin as a function of delay time τ between two $\pi/2$ spin rotation pulses. Ramsey interference fringes were observed, indicating coherent control of the spin in the spin Bloch sphere. *Panel (A): Adapted from Sun, S., Kim, H., Solomon, G.S., Waks, E., 2016. A quantum phase switch between a single solid-state spin and a photon. Nat. Nanotechnol. 11, 539–544 with permission. Copyright 2016, Springer Nature. Panel (B and C): Adapted from Carter, S.G., Sweeney, T.M., Kim, M., Kim, C.S., Solenov, D., Economou, S.E., Reinecke, T.L., Yang, L., Bracker, A.S., Gammon, D., 2013. Quantum control of a spin qubit coupled to a photonic crystal cavity. Nat. Photon. 7, 329–334 with permission. Copyright 2013, Springer Nature.*

To realize a spin-photon interface requires the capability of initialization, coherent control, and readout of a quantum dot spin coupled to an optical cavity. All these operations have been demonstrated for a bulk quantum dot using an all-optical technique (Press et al., 2008). In particular, coherent Rabi oscillations between the two spin states can be induced by using a picosecond optical pulse that drives a Raman transition through the two λ-systems (Press et al., 2008). The optical pulse must be circularly polarized to ensure that the population transfer through the two possible λ-systems interferes constructively. However, applying the same technique to a cavity-coupled quantum dot is not straightforward since the cavity may distort the polarization or the spectral shape of the optical drive. Carter et al. investigated this issue experimentally by using a charged quantum dot weakly coupled to a photonic crystal cavity (Carter et al., 2013). They discovered that a proper detuning between the quantum dot and the cavity can enable perfect spin control fidelity, even for a quantum dot inside the cavity (see Fig. 9B). This is because when the drive is detuned enough from the cavity, it can interact with the quantum dot through the broadband and polarization insensitive leaky modes of the photonic crystals. Using this technique, they coherently manipulated their quantum dot and observed Ramsey interference through the spin-modulated cavity reflectivity (see Fig. 9C). Note that the cavity reflectivity is probed through a cross-polarization setup. Thus, a change in the cavity reflectivity suggests a rotation in the polarization of the reflected photon. However, because the quantum dot is weakly coupled to the cavity, they can only change the cavity reflectivity by less than 0.5% with the quantum dot spin. This poor contrast limits the application of this device in optical switches or quantum logic gates.

In another work, Sun et al. demonstrated the first quantum switch between a solid-state spin and an optical photon by employing a strongly coupled system (Sun et al., 2016). In this experiment, they resonantly and strongly couple transition σ_1 of the quantum dot (labeled in Fig. 9A) with a photonic crystal cavity, and detune all other transitions from the cavity. Thus, the cavity behaves like a strongly coupled system when the quantum dot is in the state $|\uparrow\rangle$, but an empty cavity when the quantum dot is in the state $|\downarrow\rangle$. By initializing the spin to $|\uparrow\rangle$ and $|\downarrow\rangle$ respectively through optical pumping, they observed spin-dependent cavity spectrum as shown in Fig. 10A and B, measured using a similar cross-polarization setup. Remarkably, they were able to observe $\sim$60% change in cavity reflectivity at the cavity resonance, corresponding to two orders of magnitude improvement over the previous work.

In the same work, Sun et al. also demonstrated modulation of the cavity reflectivity (equivalent to rotation of photon polarization) through coherent manipulation of the spin. Here they used a pulse sequence shown in Fig. 11A. A continuous wave laser was used to constantly initialize the spin

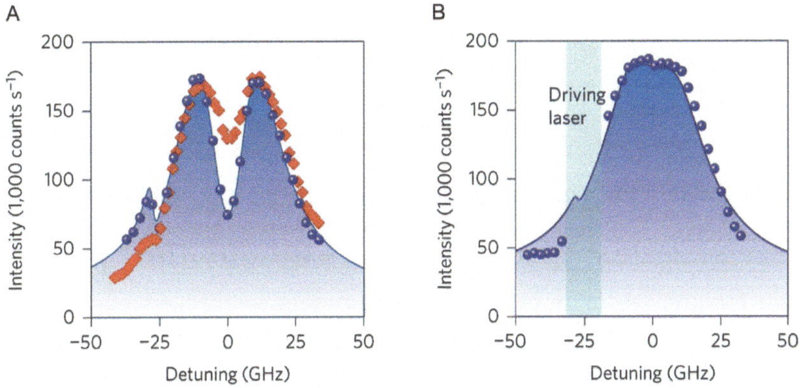

Fig. 10 (A) Cavity reflection spectrum with (blue circles) and without (red diamonds) an optical pumping laser. The optical pumping laser was resonant with transition σ_4 to initialize the quantum dot to the state $|\uparrow\rangle$. With the pumping laser, a strong suppression of the cavity reflectivity at the σ_1 resonance were observed, due to the strong coupling between the transition σ_1 and the cavity. (B) Cavity reflection spectrum when the pump laser was resonant with transition σ_2 to initialize the quantum dot to the state $|\downarrow\rangle$. The cavity exhibited a spectrum that closely resembled a bare cavity. *Adapted from Sun, S., Kim, H., Solomon, G.S., Waks, E., 2016. A quantum phase switch between a single solid-state spin and a photon. Nat. Nanotechnol. 11, 539–544 with permission. Copyright 2016, Springer Nature.*

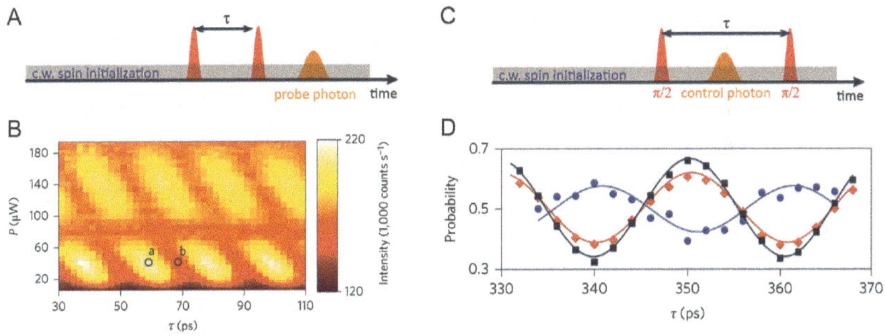

Fig. 11 (A) Pulse timing diagram for demonstration of cavity reflectivity modulation through coherent spin manipulation. (B) Reflected probe intensity as a function of rotation pulse power P and the delay time τ. (C) Pulse timing diagram for switching of the spin state by a single photon. (D) Occupation probability of the spin-down state as a function of the delay time τ, in the absence of control pulse (black squares), conditioned on detecting a reflected control photon polarized along the cavity axis (blue circles), and in the presence of the control field but not conditioning on the detection of a control photon (red diamonds). *Adapted from Sun, S., Kim, H., Solomon, G.S., Waks, E., 2016. A quantum phase switch between a single solid-state spin and a photon. Nat. Nanotechnol. 11, 539–544 with permission. Copyright 2016, Springer Nature.*

to the $|\uparrow\rangle$ state through optical pumping. The power of the laser is strong enough such that the spin is well-initialized at the end of each repetition cycle, but weak enough such that the optical pumping effect is negligible during the rotation and probe pulses. They used a pair of 6-ps pulses that were red detuned from the cavity by ~15 cavity linewidths to coherently control the spin through the leaky modes of the photonic crystals. By varying the power and the delay time between the two pulses, the spin can be coherently prepared into any superposition between $|\uparrow\rangle$ and $|\downarrow\rangle$, manifesting a two-fold oscillation in the spin population. Right after the second rotation pulse, a 63-ps probe pulse was used to measure the cavity reflectivity. As shown in Fig. 11B, the cavity reflectivity shows the featured two-fold oscillation as they varied the power and the delay time. This experiment completes their demonstration on switching the polarization of a photon with a single spin.

A quantum switch implementing a CPF gate would also exhibit the complementary effect, where the scattering of a single photon flips the spin in the $|\pm\rangle$ basis. To demonstrate this effect, Sun et al. used the pulse sequence shown in Fig. 11C. They again performed a Ramsey interference measurement by using two spin rotation pulses with a controllable delay time τ. But this time they injected a 63-ps weak laser pulse that serves as a control field before the second rotation pulse. The blue curve in Fig. 11D show the occupation probability of the spin-down state conditioned on the detection of a control photon, as a function of delay between the two rotation pulses. The black curve shows the same occupation probability when they blocked the control field. The Ramsey interference fringe conditioned on detecting a single control photon is shifted in phase by $\sim\pi$ radians relative to the case where there is no control photon, demonstrating that a single control photon flips the spin in the $|\pm\rangle$ basis.

The experimental works we mentioned above all utilized a quantum dot spin coupled with a photonic crystal cavity. Considerable efforts have also been put forward to realize a spin-photon interface using other photonic structures, including micropillar cavities (Androvitsaneas et al., 2016, 2019; Arnold et al., 2015; Wells et al., 2019) and nanobeam waveguides (Coles et al., 2016; Javadi et al., 2018). All of these works have reported spin-dependent optical transmittivity or reflectivity by initializing the spin to the state $|\uparrow\rangle$ and $|\downarrow\rangle$ respectively. However, none of them have demonstrated the full functionality of a spin-photon CPF operation as reported in (Sun et al., 2016). Specifically, to demonstrate the quantum nature of the spin-photon CPF gate requires the capability to switch the spin in the

$|\pm\rangle$ basis with a single photon. It is not clear whether these devices can achieve sufficient cooperativity for achieving a π-phase flip while simultaneously allowing optical coherent control of the quantum dot spin.

Before we move on, we would like to mention that the spin-photon quantum interface itself has many important applications in quantum information. Many of these applications have been explored theoretically or experimentally in the setting of quantum dot spins (Sun and Waks, 2017), including deterministic generation of spin-photon entanglement (Sun and Waks, 2014), cavity-enhanced spin readout (Sun et al., 2018a; Sun and Waks, 2016), and photon-mediated spin entanglement (Hu et al., 2008). We recommend the readers to go through these relevant references for details of these applications. In this chapter, we will only focus on using the spin-photon quantum interface for creating optical nonlinearities at the single-photon level.

3.4 Realization of a single-photon switch and transistor mediated by a quantum dot spin

Built upon the demonstrated spin-photon quantum switch, Sun et al. demonstrated the first single-photon switch and transistor mediated by a solid-state spin (Sun et al., 2018b). Fig. 12A shows working principle of the single-photon switch. A gate pulse first sets the internal quantum memory of the switch. If the gate pulse contains zero photons, the quantum dot spin stays in the state $|\downarrow\rangle$. But if the gate pulse contains one photon, it sets the spin to the state $|\uparrow\rangle$. Subsequently, the spin state controls the cavity reflection coefficient, thereby changing the polarization of reflected signal photons. Fig. 12B shows the pulse sequence to implement these two steps. They first prepared the quantum dot in the state $|+\rangle$ using an initialization pulse to optically pump the spin to the state $|\downarrow\rangle$ followed by a $\pi/2$ optical rotation pulse. The spin then freely evolves for a time τ, followed by a second identical rotation pulse. They injected the gate pulse between these two spin rotation pulses. If the free evolution time was set to be an integer number plus one-half of spin precession period, then in the absence of a gate photon the spin will evolve to the state $|-\rangle$ and the second rotation pulse will rotate it back to the state $|\downarrow\rangle$. But if a single gate photon reflects from the cavity, it flips the spin from the state $|-\rangle$ to $|+\rangle$. In this case the second rotation pulse rotates the spin to the state $|\uparrow\rangle$. A circularly polarized signal field then reflects off the cavity and undergoes a spin-dependent polarization rotation, as demonstrated in the spin-photon quantum switch (Sun et al., 2016). They were able to use a single gate photon to control the transmittivity of the signal photon by an absolute value of 0.24. This value differs from the ideal

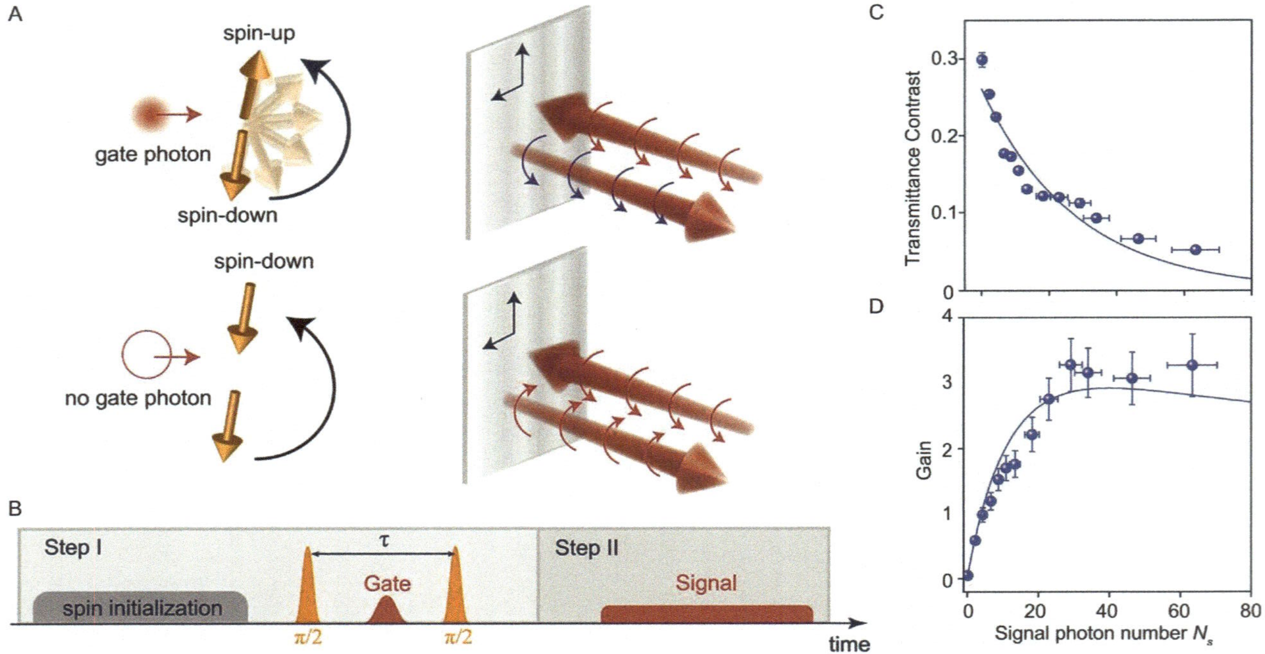

Fig. 12 (A) Schematic of the working principle of the single-photon switch and transistor. In the first step, a gate photon controls the state of the spin. In the second step, the spin determines the polarization of the signal field. (B) Pulse timing diagram to implement the single-photon switch and transistor. (C) Switching contrast as a function of average signal photon number. (D) Transistor gain as a function of average signal photon number. *Adapted from Sun, S., Kim, H., Luo, Z., Solomon, G.S., Waks, E., 2018. A single-photon switch and transistor enabled by a solid-state quantum memory. Science 361, 57–60 with permission. Copyright 2018, American Association for the Advancement of Science.*

contrast of unity due to both imperfect spin preparation fidelity of $F \approx 0.78$ and a finite cooperativity of $C \approx 2$.

A remarkable feature of the spin-mediated single-photon switch is that the device can switch many signal photons before the spin state resets. This is because the quantum dot spin in the device has a lifetime that is appreciably longer than the bandwidth of the switch, determined by the modified lifetime of the quantum dot excited state. This property enables a single-photon transistor where a single gate photon can control a signal composed of many photons, an important distinction from switches lacking a quantum memory (e.g., those based on passive two-level systems). To demonstrate the operation of a single-photon transistor, they measured their switching contrast while increasing the average number of photons in the signal pulse (see blue circles in Fig. 12C). The switching contrast remained nonvanishing with up to 60 signal photons, although it did get degraded as they increase the number of signal photons. This is because each signal photon can apply a weak back-action on the spin through inelastic Raman scattering, inducing an undesired spin-flip that resets the internal quantum memory. This weak back-action limits the number of signal photons that can reflect from the cavity before a spin-flip event resets the spin-state. From the numerical fit to a theoretical model (the solid line in Fig. 12C), they determined the spin can switch a signal field containing up to an average of 27.7 photons before its state was reset.

An important feature of transistors is that they exhibit a gain above unity. For a single-photon transistor, the gain can be defined as the change in the average number of transmitted signal photons induced by a single gate photon. As shown in Fig. 12D, the measured gain increased initially with the signal photon number but saturated at strong signal fields due to an increased probability of spin-flip from inelastic scattering. It tapered down as the signal photon number further increased due to the deviation from the weak excitation regime (the device can only scatter at most one photon per modified quantum dot excited state lifetime). They reported a maximum gain of 3.3 ± 0.4 with an incident photon number of 29.2 ± 3.2.

4. Summary and outlook

In summary, we have reviewed the theoretical and experimental progress in generating single-photon optical nonlinearities with single self-assembled quantum dots. When used as a two-level system, the quantum dot generates strong anharmonicity in the energy levels of a nano-cavity, which results in preferred transmission of single-photon or two-photon states depending on the incident light frequency. When used as a three-level system, the quantum dot—cavity system can realize an efficient quantum interface between a single photon and a single quantum dot spin. This capability enables spin-modulated cavity response, spin-photon quantum switching, and a single-photon switch and transistor mediated by a single spin.

Despite the rapid progress in quantum nonlinear optics with semiconductor quantum dots, the field is still at its infancy. Experiments demonstrated so far are all proof-of-concept, with limited switching contrast, input and output efficiency, and device yield. Numerous efforts have been invested to improve the performance of this platform and to make them practically useful. Employing new cavity designs with smaller mode volumes (Hu and Weiss, 2016) and improved fabrication technique (Kuruma et al., 2020) could potentially enable a much larger device cooperativity and thus improve the performance of these devices. Active charge stabilization could further help to improve the device performance by reducing the charge noise of the quantum dot (Luo et al., 2019). Currently, most of the experiments use the out-of-plane direction to excite and collect photons, leading to low coupling and collection efficiency. A scalable device suitable for quantum information processing will require a much higher efficiency because photon loss constitutes a dominant error mechanism for photonic qubits. Recent progress has been made in improving coupling efficiency of nanophotonic devices, including using adiabatic tapered structures to directly couple to fiber (Daveau et al., 2017), adopting designs that have better spatial mode-matching with a fiber (Giesz et al., 2016), or by directly coupling a nano-cavity to on-chip waveguides (Faraon et al., 2007). On-chip tuning (Grim et al., 2019) and hybrid integration (Kim et al., 2020) could further enable integration of multiple qubits and cascaded devices, paving the way towards a complex quantum photonic circuit on-a-chip.

References

Androvitsaneas, P., Young, A.B., Schneider, C., Maier, S., Kamp, M., Höfling, S., Knauer, S., Harbord, E., Hu, C.-Y., Rarity, J.G., 2016. Charged quantum dot micropillar system for deterministic light-matter interactions. Phys. Rev. B 93, 241409.

Androvitsaneas, P., Young, A., Lennon, J., Schneider, C., Maier, S., Hinchliff, J., Atkinson, G., Harbord, E., Kamp, M., Hofling, S., 2019. Efficient quantum photonic phase shift in a low Q-factor regime. ACS Photon. 6, 429–435.

Arnold, C., Demory, J., Loo, V., Lemaître, A., Sagnes, I., Glazov, M., Krebs, O., Voisin, P., Senellart, P., Lanco, L., 2015. Macroscopic rotation of photon polarization induced by a single spin. Nat. Commun. 6, 6236.

Bayer, M., Ortner, G., Stern, O., Kuther, A., Gorbunov, A., Forchel, A., Hawrylak, P., Fafard, S., Hinzer, K., Reinecke, T., 2002. Fine structure of neutral and charged excitons in self-assembled In (Ga) As/(Al) GaAs quantum dots. Phys. Rev. B 65, 195315.

Brien, J.L., 2007. Optical quantum computing. Science 318, 1567.

Brod, D.J., Combes, J., 2016. Passive CPHASE gate via cross-Kerr nonlinearities. Phys. Rev. Lett. 117, 080502.

Brune, M., Schmidt-Kaler, F., Maali, A., Dreyer, J., Hagley, E., Raimond, J., Haroche, S., 1996. Quantum Rabi oscillation: a direct test of field quantization in a cavity. Phys. Rev. Lett. 76, 1800.

Byer, R.L., 1997. Quasi-phase matched nonlinear interactions and devices. J. Nonlinear Opt. Phys. Mater. 6, 549–592.

Carter, S.G., Sweeney, T.M., Kim, M., Kim, C.S., Solenov, D., Economou, S.E., Reinecke, T.L., Yang, L., Bracker, A.S., Gammon, D., 2013. Quantum control of a spin qubit coupled to a photonic crystal cavity. Nat. Photon. 7, 329–334.

Chang, D.E., VuletiĆ, V., Lukin, M.D., 2014. Quantum nonlinear optics—photon by photon. Nat. Photon. 8, 685.

Chuang, I.L., Yamamoto, Y., 1995. Simple quantum computer. Phys. Rev. A 52, 3489.

Coles, R.J., Price, D.M., Dixon, J.E., Royall, B., Clarke, E., Kok, P., Skolnick, M.S., Fox, A.M., Makhonin, M.N., 2016. Chirality of nanophotonic waveguide with embedded quantum emitter for unidirectional spin transfer. Nat. Commun. 7, 11183.

Daveau, R.S., Balram, K.C., Pregnolato, T., Liu, J., Lee, E.H., Song, J.D., Verma, V., Mirin, R., Nam, S.W., Midolo, L., 2017. Efficient fiber-coupled single-photon source based on quantum dots in a photonic-crystal waveguide. Optica 4, 178–184.

De Greve, K., McMahon, P.L., Press, D., Ladd, T.D., Bisping, D., Schneider, C., Kamp, M., Worschech, L., Höfling, S., Forchel, A., Yamamoto, Y., 2011. Ultrafast coherent control and suppressed nuclear feedback of a single quantum dot hole qubit. Nat. Phys. 7, 872–878.

Duan, L.M., Kimble, H.J., 2004. Scalable photonic quantum computation through cavity-assisted interactions. Phys. Rev. Lett. 92, 127902.

Eaton, D.F., 1991. Nonlinear optical materials. Science 253, 281–287.

Englund, D., Faraon, A., Fushman, I., Stoltz, N., Petroff, P., Vučković, J., 2007. Controlling cavity reflectivity with a single quantum dot. Nature 450, 857–861.

Faraon, A., Waks, E., Englund, D., Fushman, I., Vučković, J., 2007. Efficient photonic crystal cavity-waveguide couplers. Appl. Phys. Lett. 90, 073102.

Faraon, A., Fushman, I., Englund, D., Stoltz, N., Petroff, P., Vučković, J., 2008. Coherent generation of non-classical light on a chip via photon-induced tunnelling and blockade. Nat. Phys. 4, 859–863.

Fejer, M.M., 1994. Nonlinear optical frequency conversion. Phys. Today 47, 25–33.

Fink, J., Göppl, M., Baur, M., Bianchetti, R., Leek, P., Blais, A., Wallraff, A., 2008. Climbing the Jaynes–Cummings ladder and observing its nonlinearity in a cavity QED system. Nature 454, 315–318.

Fischer, K.A., Müller, K., Lagoudakis, K.G., Vučković, J., 2016a. Dynamical modeling of pulsed two-photon interference. New J. Phys. 18, 113053.

Fischer, K.A., Müller, K., Rundquist, A., Sarmiento, T., Piggott, A.Y., Kelaita, Y., Dory, C., Lagoudakis, K.G., Vučković, J., 2016b. Self-homodyne measurement of a dynamic Mollow triplet in the solid state. Nat. Photon. 10, 163–166.

Fischer, K.A., Kelaita, Y.A., Sapra, N.V., Dory, C., Lagoudakis, K.G., Müller, K., Vučković, J., 2017. On-chip architecture for self-homodyned nonclassical light. Phys. Rev. Appl. 7, 044002.

Fischer, K., Sun, S., Lukin, D., Kelaita, Y., Trivedi, R., Vučković, J., 2018. Pulsed coherent drive in the Jaynes-Cummings model. Phys. Rev. A 98, 021802.

Gea-Banacloche, J., 2010. Impossibility of large phase shifts via the giant Kerr effect with single-photon wave packets. Phys. Rev. A 81, 043823.
Giesz, V.R., Somaschi, N., Hornecker, G., Grange, T., Reznychenko, B., De Santis, L., Demory, J., Gomez, C., Sagnes, I., Lemaitre, A., 2016. Coherent manipulation of a solid-state artificial atom with few photons. Nat. Commun. 7, 1–6.
Glauber, R.J., 1963. The quantum theory of optical coherence. Phys. Rev. 130, 2529–2539.
Greilich, A., Yakovlev, D., Shabaev, A., Efros, A.L., Yugova, I., Oulton, R., Stavarache, V., Reuter, D., Wieck, A., Bayer, M., 2006. Mode locking of electron spin coherences in singly charged quantum dots. Science 313, 341–345.
Grim, J.Q., Bracker, A.S., Zalalutdinov, M., Carter, S.G., Kozen, A.C., Kim, M., Kim, C.S., Mlack, J.T., Yakes, M., Lee, B., Gammon, D., 2019. Scalable in operando strain tuning in nanophotonic waveguides enabling three-quantum-dot superradiance. Nat. Mater. 18, 963–969.
Guest, J., Stievater, T., Li, X., Cheng, J., Steel, D., Gammon, D., Katzer, D., Park, D., Ell, C., Thränhardt, A., 2002. Measurement of optical absorption by a single quantum dot exciton. Phys. Rev. B 65, 241310.
Hu, S., Weiss, S.M., 2016. Design of photonic crystal cavities for extreme light concentration. ACS Photon. 3, 1647–1653.
Hu, C.Y., Young, A., O'brien, J.L., Munro, W.J., Rarity, J.G., 2008. Giant optical Faraday rotation induced by a single-electron spin in a quantum dot: applications to entangling remote spins via a single photon. Phys. Rev. B 78, 085307.
Imoto, N., Haus, H., Yamamoto, Y., 1985. Quantum nondemolition measurement of the photon number via the optical Kerr effect. Phys. Rev. A 32, 2287.
Javadi, A., Ding, D., Appel, M.H., Mahmoodian, S., Löbl, M.C., Söllner, I., Schott, R., Papon, C., Pregnolato, T., Stobbe, S., Midolo, L., Schröder, T., Wieck, A.D., Ludwig, A., Warburton, R.J., Lodahl, P., 2018. Spin–photon interface and spin-controlled photon switching in a nanobeam waveguide. Nat. Nanotechnol. 13, 398–403.
Kim, H., Bose, R., Shen, T.C., Solomon, G.S., Waks, E., 2013. A quantum logic gate between a solid-state quantum bit and a photon. Nat. Photon. 7, 373.
Kim, J.-H., Aghaeimeibodi, S., Carolan, J., Englund, D., Waks, E., 2020. Hybrid integration methods for on-chip quantum photonics. Optica 7, 291–308.
Kivshar, Y.S., Agrawal, G.P., 2003. Optical Solitons: From Fibers to Photonic Crystals. Academic Press.
Koenderink, A.F., Alu, A., Polman, A., 2015. Nanophotonics: shrinking light-based technology. Science 348, 516–521.
Kok, P., Munro, W.J., Nemoto, K., Ralph, T.C., Dowling, J.P., Milburn, G.J., 2007. Linear optical quantum computing with photonic qubits. Rev. Mod. Phys. 79, 135–174.
Kuruma, K., Ota, Y., Kakuda, M., Iwamoto, S., Arakawa, Y., 2020. Surface-passivated high-Q GaAs photonic crystal nanocavity with quantum dots. APL Photon. 5, 046106.
Lapine, M., Shadrivov, I.V., Kivshar, Y.S., 2014. Colloquium: nonlinear metamaterials. Rev. Mod. Phys. 86, 1093.
Laucht, A., Hofbauer, F., Hauke, N., Angele, J., Stobbe, S., Kaniber, M., Böhm, G., Lodahl, P., Amann, M., Finley, J., 2009. Electrical control of spontaneous emission and strong coupling for a single quantum dot. New J. Phys. 11, 023034.
Laussy, F.P., Del Valle, E., Schrapp, M., Laucht, A., Finley, J.J., 2012. Climbing the Jaynes-Cummings ladder by photon counting. J. Nanophoton. 6, 061803.
Lodahl, P., Mahmoodian, S., Stobbe, S., 2015. Interfacing single photons and single quantum dots with photonic nanostructures. Rev. Mod. Phys. 87, 347.
Luo, Z., Sun, S., Karasahin, A., Bracker, A.S., Carter, S.G., Yakes, M.K., Gammon, D., Waks, E., 2019. A spin–photon interface using charge-tunable quantum dots strongly coupled to a cavity. Nano Lett. 19, 7072–7077.

Michler, P., Kiraz, A., Becher, C., Schoenfeld, W., Petroff, P., Zhang, L., Hu, E., Imamoglu, A., 2000. A quantum dot single-photon turnstile device. Science 290, 2282–2285.
Milburn, G.J., 1989. Quantum optical Fredkin gate. Phys. Rev. Lett. 62, 2124.
Mosor, S., Hendrickson, J., Richards, B., Sweet, J., Khitrova, G., Gibbs, H., Yoshie, T., Scherer, A., Shchekin, O., Deppe, D., 2005. Scanning a photonic crystal slab nanocavity by condensation of xenon. Appl. Phys. Lett. 87, 141105.
Mukamel, S., 1995. Principles of Nonlinear Optical Spectroscopy. Oxford University Press, New York.
Müller, K., Fischer, K.A., Rundquist, A., Dory, C., Lagoudakis, K.G., Sarmiento, T., Kelaita, Y.A., Borish, V., Vučković, J., 2015a. Ultrafast polariton-phonon dynamics of strongly coupled quantum dot-nanocavity systems. Phys. Rev. X 5, 031006.
Müller, K., Rundquist, A., Fischer, K.A., Sarmiento, T., Lagoudakis, K.G., Kelaita, Y.A., Muñoz, C.S., Del Valle, E., Laussy, F.P., Vučković, J., 2015b. Coherent generation of nonclassical light on chip via detuned photon blockade. Phys. Rev. Lett. 114, 233601.
Müller, K., Fischer, K.A., Dory, C., Sarmiento, T., Lagoudakis, K.G., Rundquist, A., Kelaita, Y.A., Vučković, J., 2016. Self-homodyne-enabled generation of indistinguishable photons. Optica 3, 931–936.
Najer, D., Söllner, I., Sekatski, P., Dolique, V., Löbl, M.C., Riedel, D., Schott, R., Starosielec, S., Valentin, S.R., Wieck, A.D., Sangouard, N., Ludwig, A., Warburton, R.J., 2019. A gated quantum dot strongly coupled to an optical microcavity. Nature 575, 622–627.
Nysteen, A., McCutcheon, D.P.S., Heuck, M., Mørk, J., Englund, D.R., 2017. Limitations of two-level emitters as nonlinearities in two-photon controlled-phase gates. Phys. Rev. A 95, 062304.
Ota, Y., Takamiya, D., Ohta, R., Takagi, H., Kumagai, N., Iwamoto, S., Arakawa, Y., 2018. Large vacuum Rabi splitting between a single quantum dot and an H0 photonic crystal nanocavity. Appl. Phys. Lett. 112, 093101.
Pask, H.M., 2003. The design and operation of solid-state Raman lasers. Prog. Quant. Electron. 27, 3–56.
Press, D., Ladd, T.D., Zhang, B., Yamamoto, Y., 2008. Complete quantum control of a single quantum dot spin using ultrafast optical pulses. Nature 456, 218–221.
Press, D., De Greve, K., McMahon, P.L., Ladd, T.D., Friess, B., Schneider, C., Kamp, M., Höfling, S., Forchel, A., Yamamoto, Y., 2010. Ultrafast optical spin echo in a single quantum dot. Nat. Photon. 4, 367–370.
Reinhard, A., Volz, T., Winger, M., Badolato, A., Hennessy, K.J., Hu, E.L., ImamoĞlu, A., 2012. Strongly correlated photons on a chip. Nat. Photon. 6, 93–96.
Reithmaier, J.P., Sęk, G., Löffler, A., Hofmann, C., Kuhn, S., Reitzenstein, S., Keldysh, L., Kulakovskii, V., Reinecke, T., Forchel, A., 2004. Strong coupling in a single quantum dot–semiconductor microcavity system. Nature 432, 197–200.
Rosenblum, S., Parkins, S., Dayan, B., 2011. Photon routing in cavity QED: beyond the fundamental limit of photon blockade. Phys. Rev. A 84, 033854.
Schuster, I., Kubanek, A., Fuhrmanek, A., Puppe, T., Pinkse, P.W., Murr, K., Rempe, G., 2008. Nonlinear spectroscopy of photons bound to one atom. Nat. Phys. 4, 382–385.
Shapiro, J.H., 2006. Single-photon Kerr nonlinearities do not help quantum computation. Phys. Rev. A 73, 062305.
Sun, S., Waks, E., 2014. Deterministic generation of entanglement between a quantum-dot spin and a photon. Phys. Rev. A 90, 042322.
Sun, S., Waks, E., 2016. Single-shot optical readout of a quantum bit using cavity quantum electrodynamics. Phys. Rev. A 94, 012307.
Sun, S., Waks, E., 2017. Interfacing single quantum dot spins with photons using a nanophotonic cavity. In: Michler, P. (Ed.), Quantum Dots for Quantum Information Technologies. Springer.

Sun, S., Kim, H., Solomon, G.S., Waks, E., 2013. Strain tuning of a quantum dot strongly coupled to a photonic crystal cavity. Appl. Phys. Lett. 103, 151102.
Sun, S., Kim, H., Solomon, G.S., Waks, E., 2016. A quantum phase switch between a single solid-state spin and a photon. Nat. Nanotechnol. 11, 539–544.
Sun, S., Kim, H., Solomon, G.S., Waks, E., 2018a. Cavity-enhanced optical readout of a single solid-state spin. Phys. Rev. Appl. 9, 054013.
Sun, S., Kim, H., Luo, Z., Solomon, G.S., Waks, E., 2018b. A single-photon switch and transistor enabled by a solid-state quantum memory. Science 361, 57–60.
Trivedi, R., Fischer, K.A., Vučković, J., Müller, K., 2020. Generation of non-classical light using semiconductor quantum dots. Adv. Quant. Technol. 3, 1900007.
Wells, L.M., Kalliakos, S., Villa, B., Ellis, D.J.P., Stevenson, R.M., Bennett, A.J., Farrer, I., Ritchie, D.A., Shields, A.J., 2019. Photon phase shift at the few-photon level and optical switching by a quantum dot in a microcavity. Phys. Rev. Appl. 11, 061001.
Willner, A.E., Khaleghi, S., Chitgarha, M.R., Yilmaz, O.F., 2014. All-optical signal processing. J. Lightwave Technol. 32, 660–680.
Xu, S., Rephaeli, E., Fan, S., 2013. Analytic properties of two-photon scattering matrix in integrated quantum systems determined by the cluster decomposition principle. Phys. Rev. Lett. 111, 223602.
Yoshie, T., Scherer, A., Hendrickson, J., Khitrova, G., Gibbs, H.M., Rupper, G., Ell, C., Shchekin, O.B., Deppe, D.G., 2004. Vacuum Rabi splitting with a single quantum dot in a photonic crystal nanocavity. Nature 432, 200–203.

CHAPTER TEN

Quantum-light shaping and quantum spectroscopy in semiconductors

Mackillo Kira[a,b,*], Georgios Roumpos[c,†], and Steven T. Cundiff[a,b,c]
[a]Department of Electrical Engineering and Computer Science, University of Michigan, Ann Arbor, MI, United States
[b]Department of Physics, University of Michigan, Ann Arbor, MI, United States
[c]JILA, University of Colorado and National Institute of Standards and Technology, Boulder, CO, United States
*Corresponding author: e-mail address: mackkira@umich.edu

Contents

† Current address: Google Research, 1600 Amphitheatre Pkwy, Mountain View, CA 94043, United States.

Semiconductors and Semimetals, Volume 105
ISSN 0080-8784
https://doi.org/10.1016/bs.semsem.2020.10.006

1. Introduction to quantum spectroscopy

Laser spectroscopy already provides the most accurate measurement tool, as demonstrated by its broad use in high-precision measurements ranging from metrology (Cundiff and Ye, 2003) and gravitational wave detectors (Abbott et al., 2016) to exploring macroscopic quantum properties such as Bose–Einstein condensation (Anderson et al., 1995; Bradley et al., 1995; Davis et al., 1995) through laser-cooling techniques. Currently, its ultrafast forms push the boundaries to detect quantum phenomena down to zepto-second timescales (Hernández-García et al., 2013; Köhler et al., 2011; Mourou and Tajima, 2011), separate disorder from intrinsic features in quantum materials by applying novel techniques such as multidimensional coherent spectroscopy (Cundiff and Mukamel, 2013; Smallwood and Cundiff, 2018), and launch coherent electron motion within a solid to realize lightwave electronics (Goulielmakis et al., 2007; Langer et al., 2018), to mention a few exciting directions.

All these investigations can be understood through classical aspects of light, defining how precisely and broadly intensity, energy, and/or temporal profiles of light pulses can be adjusted. Indeed, tailoring these three aspects as well as the timing of repetitive pulses (Jones et al., 2000) down to a fraction of an optical cycle has given ultrafast laser spectroscopy (Fennel et al., 2010; Kirilyuk et al., 2010; Ulbricht et al., 2011) unprecedented ability to observe and control quantum processes at their natural time scales. Also its spatial resolution has been extended to subwave length scales by nanooptics (Cocker et al., 2016; Ropers et al., 2007) matching scales of small molecules. In this context, pulse-shaping techniques (Weiner, 2000), which modify the phase and amplitude of a pulse to any arbitrary shape, enable the most versatile toolbox to optimize and control (Král et al., 2007; Krausz and Ivanov, 2009; Winterfeldt et al., 2008) nontrivial quantum processes at their natural time and spatial scales.

The astonishing achievements of laser spectroscopy are all based on its ability to *directly* match, detect, and control single- to few-body quantum aspects of diverse quantum systems. However, the *direct detection and manipulation of multiparticle quantum states of matter* with light remains challenging if not impossible as long as only classical aspects of light pulses are considered. In the same way as for interacting many-particle systems, quantum aspects of any light source can be rigorously classified in terms of correlations that systematically assign connected particle clusters (Kira, 2015; Kira and Koch, 2012) within many-body systems. A cluster-expansion representation (Kira and Koch, 2008) yields one possible and convenient connection between quantum optics and many-body systems. Specifically, quantum-light sources exhibiting clear quantum signatures such as squeezing, antibunching, or entanglement are uniquely described by distinct photon correlations. Exploiting such quantum-optical correlations in ultrafast spectroscopy has been suggested (Kira and Koch, 2006b) to directly excite and detect complex many-body states in extended semiconductors ranging from bulk to two-dimensional (2D) and one-dimensional (1D) quantum systems. We focus here on discussing direct band gap semiconductors that yield strong light–matter interactions as well as pronounced Coulombic many-body effects among electronic excitations the light excites and couples with.

The key ideas of the resulting full-blown quantum (optical) spectroscopy and its *direct* access to intriguing many-body states is illustrated in Fig. 1; the wave symbols denote the light properties while the closed and open symbols assign electron and hole (electronic vacancies) excitations within an extended semiconductor (shaded area). The shaded loops indicate entanglement/corrections among photons and electron–hole pairs. An elementary light–matter interaction in direct-gap semiconductors involves photon absorption accompanied by creation of an electron–hole pair as well as its reverse, emission of a photon through electron–hole recombination, c.f. Kira and Koch (2012) for a textbook discussion.

The most accurate lasers output a coherent state $|\alpha\rangle$ characterized by a single property, a coherent amplitude α that implies first-order coherence and no higher-order correlations. Therefore, all coherent states are

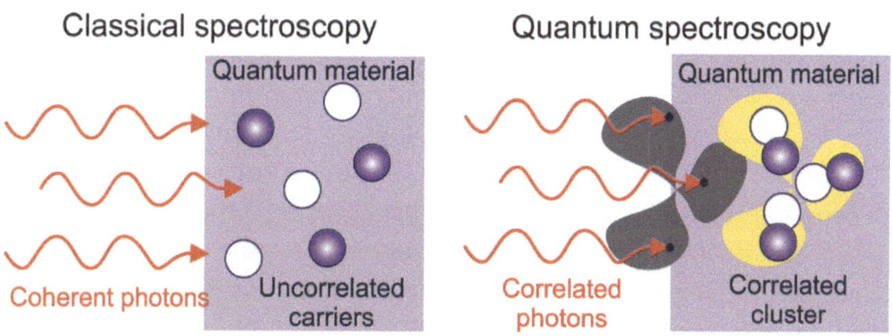

Fig. 1 Concept of classical vs quantum-optical spectroscopy. In classical spectroscopy (*left*), the photons (*wave symbols*) are uncorrelated and they create unbound pairs of electrons (*spheres*) and holes (*open circles*). In quantum spectroscopy (*right*), the photons are correlated (*yellow ellipse*) such that they directly excite a correlated electron-hole cluster (*yellow circle*).

uncorrelated, single-body states from the many-body perspective (Kira and Koch, 2006a, 2008). The careful analysis in Kira and Koch (2006b) showed that ultrafast excitations with coherent state light pulses can only *directly* excite uncorrelated electrons and holes in semiconductors as shown in Fig. 1A. This assignment also includes coherent electron–hole transition amplitude that can exhibit excitonic resonances due to Coulomb-bound electron–hole pair states, excitons (Kira and Koch, 2012). In other words, the single-body aspects of light are *directly* imprinted to matter upon photo-absorption excitations. Only the subsequent many-body dynamics can form higher-order electron–hole clusters that can affect the monitored single-body aspects. Consequently, classical spectroscopy can only indirectly excite and detect complex many-body states, which severely limits the capabilities of classical spectroscopies to access and especially control many-body states.

Since direct access to many-body states, especially to entanglement-related correlations, is critical for quantum-information science (QIS) applications such as quantum computing and quantum sensing, ultrafast spectroscopy on extended semiconductors does not yet constitute a realistic framework for QIS. However, (Kira and Koch, 2006b) shows that a quantum-light source, having given photon correlations, creates exactly identical particle correlations among electron–hole pairs when the photons are absorbed by light, as illustrated in Fig. 1B. This fundamental, quantum spectroscopic connection not only directly controls many-body excitations, but it works in the opposite direction—a specific many-body state imprints its correlations as such to the light emission, which can be detected using high-order photon-correlation measurements. These observations led to conceptualization (Kira and Koch, 2006b) of *quantum spectroscopy that extends ultrafast spectroscopy with quantum-light sources and detection in order to controllably interchange quantum-optical and many-body correlations in complex systems.* The direct capability to excite, control, and detect complex quantum many-body states would extend QIS for complex many-body systems, such as semiconductors.

Most current QIS efforts involve simple quantum systems such as photons, quantum dots, or trapped ions and superconducting qubits at cryogenic temperatures. Despite recent quantum-computing (Arute et al., 2019) and quantum-communication (Yin et al., 2020) demonstrations, it is still challenging to make such elements interact losslessly to create large-scale QIS systems, integrate them on a chip, or realize room-temperature applications. At the same time, the interaction, integration, and room-temperature aspects are already solved for extended semiconductors as the key components for existing information technology, although their QIS access to many-body states is still missing. Quantum spectroscopy provides a unifying gateway to introduce QIS directly on existing semiconductor platforms, including the vast benefits laser and semiconductor technologies offer. This chapter overviews the evolution of semiconductor quantum spectroscopy and introduces new methodology to realize it fully experimentally based on existing ultrafast spectroscopies.

1.1 Demonstration of semiconductor quantum spectroscopy

So far, semiconductor quantum spectroscopy has "only" been realized by algorithmically *projecting* a complete set of classical *measurements* into a response to a desired quantum-light pulse using an experimentally robust scheme (Kira et al., 2011). The first glimpses of quantum spectroscopy's advantages have already been seen in several semiconductor measurements that were projected to detect distinct electron–hole excitations. Unlike classical sources, the projected quantum source with a 120 fs duration and more than a million photons was shown to excite sufficiently narrow quantum-well absorption resonances such that one could characterize the excitation-induced broadening (Mootz et al., 2014) of a biexciton (an excitonic molecule with two electrons and holes). By projecting three-photon correlations of a Schrödinger's cat state to a similar source, a new quasiparticle—the dropleton—was discovered (Almand-Hunter et al., 2014) and shown to be a quantized liquid-like electron–hole droplet consisting of 6–14 particles.

Fig. 2 illustrates the differences of classically measured input and projected quantum spectroscopy via the dropleton study (Almand-Hunter et al., 2014). Specifically, Fig. 2A shows *classically* measured probe-absorption spectra as function of photon number N in a 120 fs-long pump pulse exciting ten 10 nm GaAs quantum wells at 4 K temperature. Such systems are well known to have a 1s-exciton state 10 meV below the band gap. The probe absorption is measured 16 ps after the pump, which is sufficiently long to form new many-body states even after a classical excitation. The pump and probe have an opposite circular polarization. Fig. 2A shows that the low-pump fields produce a clear resonance 2 meV below the 1s state, matching the energy of a biexciton state. We define the energy difference to the 1s exciton as binding energy. Elevating N of the pump starts to broaden this resonance and shift it to higher binding energies. Such behavior is unexpected for biexcitons because it should exhibit reduced binding as

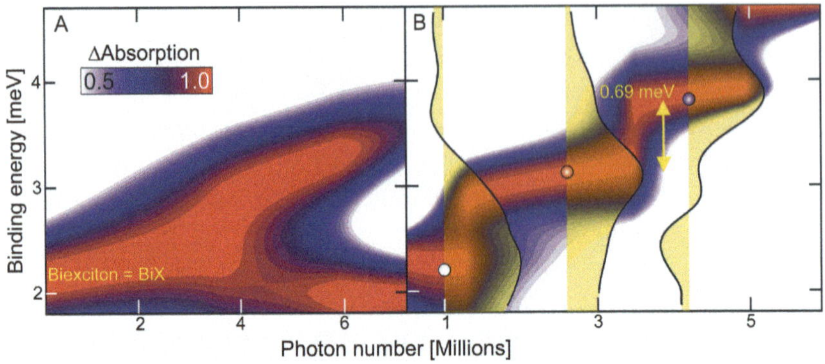

Fig. 2 Demonstration of quantum-optical spectroscopy comparing classical spectroscopy in **a** to quantum spectroscopy realized through projection techniques in **b**. *Adopted from Ref. Almand-Hunter, A.E., Li, H., Cundiff, S.T., Mootz, M., Kira, M., Koch, S.W., 2014. Quantum droplets of electrons and holes. Nature 506 (7489), 471–475. https://doi.org/10.1038/nature12994.*

fermionic blocking starts to reduce the Coulomb attraction at elevated excitation levels. However, the spectrum become so broad that the classical spectroscopy used here provides grounds only for speculations whether the broad features contain additional many-body states.

By using a special Schrödinger cat state in the projection (Almand-Hunter et al., 2014), the effect of three-photon correlations in the pump were deduced as function of N. Quantum spectroscopically, this corresponds to directly exciting three correlated electron–hole pairs, compared to uncorrelated states classical spectroscopy excites. Since the three electron–hole pair states are correlated beyond biexcitons (containing two pairs), the quantum spectroscopic projection should resolve features underneath the blurred biexciton resonance. Indeed, these studies revealed a crisp-clear, new resonance structure shown in Fig. 2B. In particular, the resonances appeared as well-resolved steps, which a correlation functional theory (Mootz et al., 2013) identified as being signatures of new liquid-like electron–hole arrangements—dropletons. In other words, this experiment–theory study demonstrated that quantum spectroscopy can directly access classically unforeseen many-body states in semiconductors. This work also verified the prediction (Kira and Koch, 2006a) that quantum spectroscopy provides a direct access to detect and control many-body states in complex systems.

1.2 Requirements of quantum-light shaping (QLS)

Since its formulation, quantum spectroscopy has provided concepts for how photon statistics can be applied to detect exciton correlations in coupled nanostructures (Carmele et al., 2009; Heindel et al., 2017), resolve quantum-Rabi splitting in a semiconductor quantum-dot microcavity (Schneebeli et al., 2008) and its experimental demonstration (Reinhard et al., 2012), as well as to resolve quantum-memory effects in quantum-dot microcavity emission, excite polaritons with quantum light (Carreño and Laussy, 2016; López Carreño et al., 2015), and detect a single polariton (Cuevas et al., 2018). On a quantum-source development and characterization side, Josephson junctions have been used for nonclassical photon-pair production (Leppäkangas et al., 2013), light bunching above semiconductor lasing threshold (Schulze et al., 2014) has been observed, plasmonic nanocavities have been applied to control quantum statistics (Zhao et al., 2015), methodology for intense thermal states (Smith et al., 2020) has been developed, and quantum fluctuations of terahertz fields have been detected (Riek et al., 2017). Besides motivating new directions for solids, quantum spectroscopy has become an active topic also in quantum chemistry, especially, by using entangled photons as a spectroscopic tool, as reviewed in Dorfman et al. (2016).

These examples illustrate that quantum spectroscopy delivers many-body state control and characterization required to extend QIS investigations for complex systems such as extended semiconductors. To become experimentally feasible extension of ultrafast spectroscopy, one should be able to design the quantum fluctuations of light, i.e., realize quantum-light shaping (QLS) at the same level of sophistication as classical light shaping is possible already. From the examples above, it is obvious QLS can dramatically enhance the control of interacting many-body states because pulses with tailored quantum fluctuations can excite specific many-body processes through correlated photon absorption events. Quantum spectroscopy of many-body systems poses three stringent requirements for QLS to become useful for many-body investigations:

1. The pulse must be intense in order to modify the many-body dynamics; the projection demonstrations (Almand-Hunter et al., 2014; Berger et al., 2014; Kira et al., 2011; Mootz et al., 2014) needed millions of photons per pulse.
2. The pulse must be ultrafast to match with the system dynamics, which often occur on femtosecond time scales.
3. The generation of quantum light should not modify pulse's intensity or temporal shape, not even at subcycle level to preserve the benefits of classical pulse shaping.

Because of these requirements, ultrafast and intense quantum sources have not yet been experimentally achieved. Therefore, the projective approach added on top of classical ultrafast spectroscopy still remains the most flexible way to realize quantum spectroscopy for complex systems. However, the projection approach is very costly because one first has to collect a large set of measurement data with high-precision classical laser spectroscopy. Then, this input data are processed with a mathematically involved projection algorithm based on the cluster-expansion transformation (Kira and Koch, 2008) to determine the response to a user-defined quantum source.

However, we can learn much from the existing quantum sources. If one relaxes the requirement for a specific ultrafast pulse shape, one can, e.g., postselect amplitude-squeezed sources through feedback filtering a laser's intensity by monitoring the electronic noise (Machida et al., 1987) or by utilizing parametric processes (Heidmann et al., 1987). If one settles for low-intensity sources, postselection yields more quantum-statistical flexibility (Bimbard et al., 2010) as shown by the realization of Fock (Lvovsky et al., 2001), NOON (Afek et al., 2010), and Schrödinger cat states (Chrzanowski et al., 2011; Ourjoumtsev et al., 2007), entanglement effects (Lee et al., 2011), as well as single-photon subtraction (Neergaard-Nielsen et al., 2006; Wenger et al., 2004), and/or addition (Parigi et al., 2007; Zavatta et al., 2004). Despite the differences in setups, these pioneering contributions are built around a common idea—accessing source's quantum fluctuations via postselection. In the reminder of this chapter, we develop an experimental algorithm for QLS that reveals specified quantum fluctuations of an ultrafast laser pulse train without modifying the duration of the pulses. The algorithm satisfies the requirements for realizing quantum spectroscopy by combining a set of simple experimental ideas and operations—postselection, averaging, and differences of the measured outcomes.

2. Theoretical background for quantum-light shaping (QLS)

To formulate QLS precisely, we start from the general quantum-optics formulation of ultrafast light pulses, their quantum fluctuations, and detection based on the Born's law. The light quantization is comprehensively presented in several quantum-optics textbooks, (cf. Cohen-Tannoudji et al., 1989; Gerry and Knight, 2005; Grynberg et al., 2010; Kira and Koch, 2012; Leonhardt, 1997; Loudon, 2000; Meystre and Sargent III, 2007; Scully and Zubairy, 2002; Vogel and Welsch, 2006; Walls and Milburn, 1994), using different approaches. To specify our method and to highlight those aspects important for the QLS, we briefly overview the needed core elements; the notation is adopted from Kira and Koch (2012).

Generally, the quantum aspects of a high-quality laser pulse can be described by a single quantized mode of the electromagnetic field

$$\hat{\mathbf{E}}(\mathbf{r},t) = i\mathcal{E}\left[\hat{B}\,\mathbf{u}(\mathbf{r},t) - \hat{B}^{\dagger}\,\mathbf{u}^{\star}(\mathbf{r},t)\right], \qquad \left[\hat{B},\,\hat{B}^{\dagger}\right]_{-} = 1, \quad (1)$$

where $\hat{B}$ is a Boson operator and $\mathbf{u}(\mathbf{r},\,t)$ is the spatial or temporal mode studied. In general, $\mathbf{u}(\mathbf{r},\,t)$ follows either from the Helmholtz equation, yielding a spatial mode, or from the Maxwell's wave equation, yielding a temporal mode. Therefore, $\mathbf{u}(\mathbf{r},\,t)$ is still defined by the classical electromagnetism while quantization exclusively describes the inherent quantum fluctuations within the amplitude (whose quantized form is $\hat{B}$). The quantization procedure settles the scale of the amplitude via the vacuum-field amplitude $\mathcal{E}$. As the main consequence of quantization, a precise value for light's amplitude is replaced by a distribution that follows from the density matrix $\hat{\rho}_{\text{light}}$.

Both $\hat{B}$ and $\hat{B}^{\dagger}$ can be expressed (Kira and Koch, 2012; Walls and Milburn, 1994) in terms of canonical operators

$$\begin{cases} \hat{x}_{\theta} \equiv \frac{1}{2}\left(\hat{B}\,e^{-i\theta} + \hat{B}^{\dagger}\,e^{i\theta}\right) \\ \hat{y}_{\theta} \equiv \frac{1}{2i}\left(\hat{B}\,e^{-i\theta} - \hat{B}^{\dagger}\,e^{i\theta}\right) \end{cases} \Rightarrow \begin{cases} \hat{B} = \hat{x}_0 + i\hat{y}_0 \\ \hat{B}^{\dagger} = \hat{x}_0 - i\hat{y}_0 \end{cases}, \quad [\hat{x}_{\theta},\,\hat{y}_{\theta}]_{-} = \frac{i}{2}, \quad (2)$$

where we have added a phase θ between the photon operators. The identified $\hat{x}_{\theta}$ and $\hat{y}_{\theta}$ form a canonical quadrature pair because they satisfy the scaled canonical commutation relation for any choice of θ. As we insert the definitions (2) into Eq. (1), the electric field becomes

$$\hat{\mathbf{E}}(\mathbf{r},t) = -2\mathcal{E}\,\hat{y}_{\theta}\,|\mathbf{u}(\mathbf{r},t)| = -2\mathcal{E}\,\hat{x}_{\theta+\frac{\pi}{2}}\,|\mathbf{u}(\mathbf{r},t)|, \quad (3)$$

after we define $\mathbf{u}(\mathbf{r},\,t) = |\mathbf{u}(\mathbf{r},\,t)|\,e^{-i\theta}$ at a fixed time and position. The quantized field operator is Hermitian defined by the quadrature $\hat{x}_{\theta+\frac{\pi}{2}} = \hat{y}_{\theta}$.

The Boson operators in Eq. (1) are directly connected with the elementary problem of the harmonic oscillator which defines both the quantized-light Hamiltonian $\hat{H}_{\text{light}} = \hbar\omega(\hat{B}\hat{B}^{\dagger} + \frac{1}{2})$. In this connection, the Fock states $|n\rangle$ are orthonormal eigenstates of $\hat{H}_{\text{light}}$ where n has integer values. State $|0\rangle$ corresponds to the vacuum and $|n>0\rangle$ is an n-photon state that is a true quantum-light state.

The most classical field is defined by a coherent state $|\alpha\rangle$ that is an eigenstate of $\hat{B}$, i.e.,

$$\hat{B}|\alpha\rangle = \alpha|\alpha\rangle \qquad \langle\alpha|\hat{B}^{\dagger} = \langle\alpha|\,\alpha^{\star}. \tag{4}$$

As a result, one can replace $\hat{B}$ by a complex valued amplitude α, i.e., $\hat{B}\to\alpha$, when normally ordered expectation values $\langle\alpha|[\hat{B}]^J\hat{B}^K|\alpha\rangle = [\alpha^{\star}]^J\alpha^K$ are computed with coherent states. This is an experimentally relevant reduction because light detectors measure normally ordered quantities (Walls and Milburn, 1994). This reduces Eq. (1) to its classical form such that coherent states are as classical as quantized state can be.

Generally, the quantum statistic refers to any representation that describes *all* aspects of the quantum fluctuations. Using coherent states to the represent the quantum statistics always yields a diagonal density matrix, $\hat{\rho}_{\text{light}}$ (Glauber, 1963; Sudarshan, 1963), Husimi function, $Q(z)$ (Kira et al., 2011; Walls and Milburn, 1994), and Wigner function, $W(z)$ (Kira et al., 2011; Walls and Milburn, 1994),

$$\begin{aligned}\hat{\rho}_{\text{light}} &= \int d^2\alpha\,|\alpha\rangle P(\alpha)\langle\alpha| \;\Leftrightarrow\; Q(z) = \frac{1}{\pi}\int d^2\alpha\, e^{-|z-\alpha|^2}\,P(\alpha)\\ &\Leftrightarrow\; W(z) = \frac{2}{\pi}\int d^2\alpha\, e^{-2|z-\alpha|^2}\,P(\alpha)\end{aligned} \tag{5}$$

respectively. The integration weight, $P(\alpha)$, is the Glauber–Sudarshan function and $\alpha = x + iy$ is a phase-space coordinate. Both $P(\alpha)$ and $Q(z)$ are phase-space distributions that uniquely define the quantum statistics. The diagonality of the expression (5) follows entirely from the overcompleteness, compare relation (A.6). Especially, the nonorthogonality of the coherent states leads to the feature that $P(\alpha)$ cannot be interpreted as a probability distribution. In fact, $P(\alpha)$ can become partially negative valued for some values of the coherent amplitude. This possible negativity discriminates between classical and quantum-light sources: The negativity of $P(\alpha)$ implies pure quantum sources because one cannot then construct $\hat{\rho}_{\text{light}}$ as a probabilistic average over classical $|\alpha\rangle\langle\alpha|$ states. When $P(\alpha)$ is positive definite, the source is semiclassical. For example, the displaced thermal state,

$$\begin{aligned}\hat{\rho}_{|\beta\rangle,\,\Delta n} &= \int d^2\alpha\,|\alpha\rangle\,P_{|\beta\rangle,\,\Delta n}(\alpha)\,\langle\alpha| \;\Leftrightarrow\; P_{|\beta\rangle,\,\Delta n}(\alpha) = \frac{1}{\pi\Delta n}\,e^{-\frac{|\alpha-\beta|^2}{\Delta n}}\\ &\Leftrightarrow\; Q_{|\beta\rangle,\,\Delta n}(z) = \frac{1}{\pi(1+\Delta n)}\,e^{-\frac{|z-\beta|^2}{1+\Delta n}}\end{aligned} \tag{6}$$

is a semiclassical field.

In contrast to P, the Husimi function can be directly measured because it defines a positive definite probability distribution in a phase-space measurement that simultaneously records the orthogonal x and y quadratures. If the measurement has a nonideal quantum efficiency, η, the η-deteriorated detection yields (Leonhardt and Paul, 1993)

$$Q_\eta(z) = \frac{1}{\pi}\int d^2\alpha\, e^{-|z-\sqrt{\eta}\,\alpha|^2}\, P(\alpha) \Leftrightarrow P_\eta(\alpha) = \frac{1}{\eta}\, P\left(\frac{\alpha}{\sqrt{\eta}}\right)$$
$$\Leftrightarrow\; W_\eta(z) = \frac{2}{\pi}\int d^2\alpha\, e^{-2|z-\sqrt{\eta}\,\alpha|^2}\, P(\alpha), \tag{7}$$

where we have also included the connection with the Wigner function.

2.1 Quantum theory vs measurements: Born's law

Actual measurements are directly related to Hermitian operators $\hat{R}$; in our discussion, $\hat{R}$ stands for the system response to laser-pulse excitation or an operator needed to characterize quantum statistics of light. Once $\hat{R}$ is known, the theoretical description produces all the eigenstates, $\hat{R}\,|\phi_R\rangle = R\,|\phi_R\rangle$, and the state of the system can be described by the quantum-mechanical probability distribution $p(R)$ defining the probability of detecting outcome R. Born's law clearly formulates how a measurement of the property $\hat{R}$ is connected with the quantum theory. An individual measurement will always produce one of the possible R values. However, the selection process of one particular outcome follows randomly from the $p(R)$ distribution such that it is insufficient to deduce the actual state of the system from a single measurement. For that, one must perform $\mathcal{N}$ repetitive measurements where one collects an ensemble of measurement outcomes.

As soon as $\mathcal{N}$ becomes large enough, the theoretically predicted expectation value $\langle\hat{R}\rangle$ and the average of the measurements yield the same result

$$\langle\hat{R}\rangle = \lim_{\mathcal{N}\to\infty}\frac{1}{\mathcal{N}}\sum_{j=1}^{\mathcal{N}} R_j = \lim_{\mathcal{N}\to\infty}\sum_R \frac{n(R)}{\mathcal{N}}\, R = \sum_R p(R)\, R = \mathrm{Tr}\left[\hat{R}\,\hat{\rho}\right],$$
$$p(R) \equiv \lim_{\mathcal{N}\to\infty}\frac{n(R)}{\mathcal{N}}, \tag{8}$$

where $\hat{\rho}$ is the density matrix of the system and the jth measurement produces outcome R_j. Alternatively, R labels all possible outcomes of the experiment and $n(R)$ specifies how many times the event R occurs in the measurements that are performed $\mathcal{N}$ times. Obviously, $\sum_R \frac{n(R)}{\mathcal{N}} R$ is just an average of all measurements events, written in a convenient mathematical format. If the outcome of each measurement is stored separately, one can also histogram the quantum-mechanical probability distribution $p(R)$ from an experimental estimate $p_{\text{exp}}(R) \equiv \frac{n(R)}{\mathcal{N}}$; see the last expression in Eq. (8). In other words, repetitive measurements construct also the quantum-mechanical probability distribution governing the microscopic physics, which can be directly applied to measure, e.g., the Husimi function.

To generalize the measurements for QLS, we also consider an observable $\hat{R}\,\hat{\Pi}$ where $\hat{\Pi}$ produces a phase-space outcome $\Pi(z_{c,\,j})$ that conditions the $\hat{R} \to R_j$ measurement. As this combination is inserted to Eq. (8), we get

$$\langle \hat{R}\,\hat{\Pi} \rangle = \lim_{\mathcal{N}\to\infty} \frac{1}{\mathcal{N}} \sum_{j=1}^{\mathcal{N}} R_j \Pi(z_{c,j}), \tag{9}$$

which defines how the conditional measurements are averaged. We show in Section 3 that $\langle \hat{R}\,\hat{\Pi} \rangle$ shapes the quantum-optical response when conditioning is applied to any measurement.

2.2 Detection in quantum-light shaping

Fig. 3 presents the schematics of our quantum-light shaper (QLS). To measure conditional information, we consider a QLS that splits the incoming source $\hat{B}_{\text{in}}$ into $\hat{B}_s$ in the shaping arm and $\hat{B}_c$ in the conditioning arm. Specifically, the beam splitter (blue plate) transforms the phase-space counts z_{in} and z_{vac} in the input and in the incoming vacuum, respectively, to phase-space counts z_s and z_c in the shaping and conditioning arms, respectively. To account for the pure QLS aspects, we analyze the transfer properties without a sample (black box). We also assume that the beam splitter is lossless such that it reflects fraction f_c^2 while it transmits fraction $f_s^2 = 1 - f_c^2$ of the incoming light. We introduce a convention where both f_s and f_c are real valued, and assign the phase shifts such that transmission into the horizontal direction produce $+f_s$ while transmission into the vertical direction yields $-f_s$; reflection f_c is chosen to be positive in all directions.

Using the transmission–reflection assignment above, the beam splitter transforms the phase-space counts via

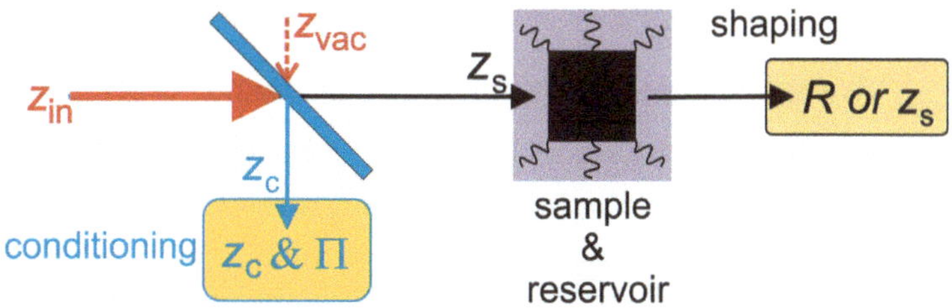

Fig. 3 Schematic representation of the full QLS setup with a sample (*black box*). The detection in the conditioning arm measures z_c while shaping arm detects z_s or sample response R. Formally, the incoming source (vacuum) produces a count z_{in} (z_{vac}), despite not being detected. The user can implement any z_c-dependent pattern function $\Pi(z_z)$ to weight the z_c outcomes. The *shaped light* is propagated through the sample which may interact with a reservoir (*springs*). The sample's response is recorded with some typical spectroscopic detection (*box symbol*), producing the outcome R. In case R stands for a phase-space measurement, we have $R = z_s$.

$$\begin{cases} z_s = f_s\, z_{\text{in}} + f_c\, z_{\text{vac}} \\ z_c = f_c\, z_{\text{in}} - f_s\, z_{\text{vac}} \end{cases} \Leftrightarrow \begin{cases} z_{\text{in}} = f_s\, z_s + f_c\, z_c \\ z_{\text{vac}} = f_c\, z_s - f_s\, z_c \end{cases} \tag{10}$$

In this situation, the quantum statistics of the incoming field is described a two-mode density matrix $\hat{\rho} = \hat{\rho}_{\text{light}} \otimes \hat{\rho}_{\text{vac}}$ where $\hat{\rho}_{\text{light}}$ ($\hat{\rho}_{\text{vac}}$) is the single-mode density matrix of incoming source (vacuum) field.

Since the source and vacuum parts factorize, the corresponding two-mode Husimi function reduces to $Q^{(2)}(z_{\text{in}}, z_{\text{vac}}) = Q_{\text{light}}(z_{\text{in}})\, \frac{1}{\pi} e^{-|z_{\text{vac}}|^2}$ where the Husimi function of the vacuum follows from Eq. (6) by setting β and Δn to zero. This can be expressed in terms of z_s and z_c counts by inserting transformation (10) into $Q^{(2)}(z_{\text{in}}, z_{\text{vac}})$ yielding $Q^{(2)}(z_s, z_c) = Q^{(2)}(f_s z_s + f_c z_c, f_c z_s - f_s z_c)$ that defines the probability distribution to detect counts z_s and z_c counts simultaneously. It is straightforward to include quantum efficiency η in the z_c detection with help of Eq. (7), producing a probability distribution

$$Q^{(2)}(z_s, z_c) = \frac{1}{\pi} e^{-|F_c z_s - F_s z_c|^2} Q_\zeta(F_s z_s + F_c z_c), \tag{11}$$

that contains the η-deteriorated Husimi function of the incoming source after we identify the effective division fractions and the effective quantum efficiency

$$F_s = \sqrt{\frac{f_s^2}{f_s^2 + \eta f_c^2}}, \qquad F_c = \sqrt{\frac{\eta f_c^2}{f_s^2 + \eta f_c^2}}, \qquad \zeta = f_s^2 + \eta f_c^2, \tag{12}$$

respectively.

3. Theory of the quantum-light shaping

To derive the fundamental properties of the QLS, we consider the schematic setup shown in Fig. 3. We assume that the sample to be characterized (black box) is sufficiently localized such that it is only excited by the z_s beam. Physically, the sample may be some complicated matter system, such as extended semiconductors an organic molecules, that is generally embedded within a reservoir (shaded box). In quantum spectroscopy (Kira and Koch, 2006b), one excites distinct many-body states of the matter via the quantum fluctuations of light.

To derive the general QLS scheme, we do not have to specify the exact details of the sample. In particular, it is enough to know that only z_s and the reservoir interact with the sample and that z_c is directed such that it interacts neither with the sample nor with the reservoir. These conditions are very reasonable and realizable in ultrafast spectroscopy. With these assumptions, we can write the system Hamiltonian in the form

$$\hat{H}_{\text{sys}} = \hat{H}_{\text{mat,r},s} + \hat{H}_c, \tag{13}$$

where $\hat{H}_{\text{mat,r},s}$ contains the full many-body interactions of the matter, the reservoir, and the shaped light source z_s. The remaining $\hat{H}_c$ contains the Hamiltonian involving only the light in the conditioning arm. Since z_c is decoupled from the matter, reservoir, and z_c degrees of freedom, $\hat{H}_c$ must commute with $\hat{H}_{\text{mat,r},s}$, such that $\left[\hat{H}_{\text{mat,r},s},\ \hat{H}_c\right]_- = 0$. This relation does not imply that z_c and z_s are not correlated because the joint quantum-statistical distribution $Q^{(2)}(z_s,\ z_c)$ can be correlated even for two systems described via mutually commuting operators, see Section 2.2.

We next consider a generic situation where the quantum-optical response of the sample to the source z_s is determined by measuring the dynamics of the sample property $\hat{R}$. In practice, $\hat{R}$ can be, e.g., the intensity, the spectrum, or the measured Husimi function of the transmitted light. However, we do not need to specify the explicit form of the measurement in more detail to determine the generic properties of the QLS. We only need to assume that the light modes z_s and z_c are initially decoupled from the sample and the reservoir, implying $\hat{\rho}_{\text{sys}}(t=0) = \hat{\rho}_{\text{mat,r}}(0)\otimes\hat{\rho}_{s,c}(0)$ for the initial density matrix. In the Heisenberg picture, the quantum dynamics of the characterized property $\hat{R}$ follows from the linear evolution

$$\begin{aligned}\langle\hat{R}(t)\rangle &= \mathrm{Tr}\left[e^{\frac{i}{\hbar}\hat{H}_{\mathrm{sys}}t}\hat{R}(0)e^{-\frac{i}{\hbar}\hat{H}_{\mathrm{sys}}t}\hat{\rho}_{\mathrm{sys}}(0)\right] \\ &= \mathrm{Tr}\left[e^{\frac{i}{\hbar}\hat{H}_{\mathrm{mat,r},s}t}e^{\frac{i}{\hbar}\hat{H}_{c}t}\hat{R}(0)e^{-\frac{i}{\hbar}\hat{H}_{c}t}e^{-\frac{i}{\hbar}\hat{H}_{\mathrm{mat,r},s}t}\hat{\rho}_{\mathrm{sys}}(0)\right],\end{aligned} \tag{14}$$

where we have used the decomposition (13) and $\left[\hat{H}_{\mathrm{mat,r},s},\ \hat{H}_c\right]_- = 0$ that allows us to separate the conditioning part from the rest. Since $\hat{R}(0)$ involves only the z_s-induced matter properties and $\hat{H}_c$ does not contain them, also these operators commute. As a consequence, we have

$$e^{\frac{i}{\hbar}\hat{H}_c t}\hat{R}(0)e^{-\frac{i}{\hbar}\hat{H}_c t} = \hat{R}(0)e^{\frac{i}{\hbar}\hat{H}_c t}e^{-\frac{i}{\hbar}\hat{H}_c t} = \hat{R}(0) \ \Rightarrow\ \hat{R}(t) = e^{\frac{i}{\hbar}\hat{H}_{\mathrm{mat,r},s}t}\hat{R}(0)e^{-\frac{i}{\hbar}\hat{H}_{\mathrm{mat,r},s}t}. \tag{15}$$

Hence, we see that $\hat{R}(t)$ does not depend on the conditioning degree of freedom. Altogether, these results convert the system response (14) into

$$\begin{aligned}\langle\hat{R}(t)\rangle &= \mathrm{Tr}\left[\hat{R}(t)\ \hat{\rho}_{\mathrm{sys}}(0)\right] \\ &= \mathrm{Tr}\left[e^{\frac{i}{\hbar}\hat{H}_{\mathrm{mat,r},s}t}\hat{R}(0)e^{-\frac{i}{\hbar}\hat{H}_{\mathrm{mat,r},s}t}\hat{\rho}_{\mathrm{mat,\ r}}(0)\otimes\hat{\rho}_{s,c}(0)\right]\end{aligned} \tag{16}$$

after having applied $\hat{\rho}_{\mathrm{sys}}(0) = \hat{\rho}_{\mathrm{mat,\ r}}(0)\otimes\hat{\rho}_{s,c}(0)$. We can compute $\langle\hat{R}(t)\rangle$ by evaluating the trace separately over the z_c and the matter–reservoir–z_s degrees of freedom,

$$\begin{aligned}\langle\hat{R}(t)\rangle &= \mathrm{Tr}_{\mathrm{mat,r},s}\left[\mathrm{Tr}_c\left[\hat{R}(t)\ \hat{\rho}_{\mathrm{mat,\ r}}(0)\otimes\hat{\rho}_{s,c}(0)\right]\right] \\ &= \mathrm{Tr}_{\mathrm{mat,r},s}\left[\hat{R}(t)\ \hat{\rho}_{\mathrm{mat,\ r}}(0)\otimes\mathrm{Tr}_c\left[\hat{\rho}_{s,c}(0)\right]\right]\end{aligned} \tag{17}$$

because only the $\hat{\rho}_{s,c}$ part depends on z_c. We may now rewrite the system response in the form

$$\langle\hat{R}(t)\rangle = \mathrm{Tr}\left[\hat{R}(t)\ \hat{\rho}_{\mathrm{mat,\ r}}(0)\otimes\hat{\rho}_s(0)\right],\quad \hat{\rho}_s(0)\equiv\mathrm{Tr}_c\left[\hat{\rho}_{s,c}(0)\right], \tag{18}$$

where we identified a reduced density matrix $\hat{\rho}_s(0)$ that contains only the z_s degree of freedom. At this point, we have not specified the quantum statistics of $\hat{\rho}_s(0)$. In a case where the incoming field is a genuine quantum source, $\langle\hat{R}(t)\rangle$ becomes the quantum-optical response to $\hat{\rho}_s(0)$. When $\hat{\rho}_s(0)$ is a semiclassical state, $\langle\hat{R}(t)\rangle$ is the response measured via classical laser spectroscopy.

3.1 Fundamental postselection aspects

The QLS allows us to simultaneously measure the system response $\hat{R}(t)$ and the conditioning at the initial time $t = 0$ with a pattern function $\hat{\Pi}$ that depends only on z_c. Such measurements yield an average

$$\langle \hat{R}(t)\, \hat{\Pi}(0) \rangle = \mathrm{Tr}\left[\hat{R}(t)\, \hat{\Pi}(0)\, \hat{\rho}_{\mathrm{sys}}(0)\right]. \tag{19}$$

As in the derivation of Eq. (17), we implement partial traces,

$$\begin{aligned} \langle \hat{R}(t)\, \hat{\Pi}(0) \rangle &= \mathrm{Tr}_{\mathrm{mat,r},s}\left[\mathrm{Tr}_c\left[\hat{R}(t)\, \hat{\Pi}(0)\, \hat{\rho}_{\mathrm{mat,r}}(0) \otimes \hat{\rho}_{s,c}(0)\right]\right] \\ &= \mathrm{Tr}_{\mathrm{mat,r},s}\left[\hat{R}(t)\, \hat{\rho}_{\mathrm{mat,r}}(0) \otimes \mathrm{Tr}_c\left[\hat{\Pi}(0)\, \hat{\rho}_{s,c}(0)\right]\right] \end{aligned} \tag{20}$$

because $\hat{R}(t)\, \hat{\rho}_{\mathrm{mat,r}}(0)$ does not depend on z_c, based on Eq. (15), which allows for the regrouping in the last step. The trace over the conditioning degrees of freedom produces an effective density matrix

$$\hat{\rho}_{\Pi} \equiv \mathrm{Tr}_c\left[\hat{\Pi}(0)\, \hat{\rho}_{s,c}(0)\right], \qquad \mathrm{Tr}[\hat{\rho}_{\Pi}] = 1 \tag{21}$$

that contains only the z_s degrees of freedom. Since the choice of the pattern function is completely free, we focus on cases where $\mathrm{Tr}\left[\hat{\Pi}(0)\, \hat{\rho}_{s,c}(0)\right]$ is nonvanishing such that we can renormalize $\hat{\Pi}$ to produce the usual norm (21). Therefore, we can assume that $\mathrm{Tr}[\hat{\rho}_{\Pi}] = 1$ is satisfied from here on.

With these considerations, the pattern function—which conditions the sample response (20)—can be cast into the form

$$\langle \hat{R}(t)\, \hat{\Pi}(0) \rangle = \mathrm{Tr}\left[\hat{R}(t)\, \hat{\rho}_{\mathrm{mat,r}}(0) \otimes \hat{\rho}_{\Pi}(0)\right] \tag{22}$$

which agrees exactly with the quantum-optical response (18). Consequently, if $\hat{\rho}_{\Pi}(0)$ corresponds to a quantum field, the conditioning $\hat{\Pi}$ shapes the quantum-optical response of matter even if the original excitation source is semiclassical. This intriguing property follows directly from the linearity of quantum mechanics, irrespective of the potentially complicated many-body and reservoir dynamics of the sample excitations. This feature constitutes the principle behind our QLS and we utilize this profound consequence in our QLS measurements. More specifically, Eq. (22) proves the conditional shaping of source $\hat{\rho}_{\Pi}$ when average (1) [or (9)] is used in experiments to shape the quantum response $R_Q \equiv \langle \hat{R}(t)\, \hat{\Pi}(0) \rangle$. In the following, we determine how the pattern function must be chosen to shape a quantum source from semiclassical lasers.

3.2 Defining the pattern function

For the realization of QLS, we need a pattern function that generates $\hat{\rho}_{\Pi}$ from a semiclassical source entering the experimental setup, shown in Fig. 3. Here, we make use of the feature that the density-matrix $\hat{\rho}_{s,c}$ and the Husimi function $Q^{(2)}(z_s, z_s)$ are equivalent representations of the quantum statistics, compare Section 2. Since $Q^{(2)}(z_s, z_c)$ is a phase-space probability, the operator averages follow from phase-space averages when the pattern function $\hat{\Pi}$ is expressed by its matching phase-space form $\Pi(z_c)$. In other words, we simply have to replace $\hat{\Pi}$ by $\Pi(z_c)$ and trace via $\int d^2 z_c$ to convert the density-matrix expressions into the matching Husimi-function forms. Therefore, the conditional quantum statistics $\hat{\rho}_{\Pi}$ in Eqs. (21) and (22) become

$$Q_{\Pi}(z_s) = \int d^2 z_c \, Q^{(2)}(z_s, z_c) \, \Pi(z_c) \tag{23}$$

that holds for any incoming source.

To show that a pattern function can shape a quantum $Q_{\Pi}(z_s)$ from a semiclassical source, we assume that the incoming source is a Gaussian displaced by $\beta \equiv \alpha_{\text{in}}$ in Eq. (6). The excess noise of this source is defined by Δn_{in}. We assume that the conditioning is performed in a realistic experiment with the quantum efficiency η. Since we want to define which quantum statistics are truly shaped, we assume the state is perfectly detected without quantum-statistical deterioration. The corresponding $Q^{(2)}(z_s, z_c)$ is given by the result (11). Using thermal state (6), $Q^{(2)}(z_s, z_c)$ becomes a product of Gaussians allowing us to rearrange the exponents into many equivalent forms. One possibility is

$$\begin{aligned} Q^{(2)}(z_s + \alpha_s, z_c + \alpha_c) &= \frac{\kappa^2}{\pi} \, e^{-|\kappa z_s - z_c/\xi|^2} \, Q_c(z_c + \alpha_c), \\ \alpha_s = f_s \, \alpha_{\text{in}}, \quad \alpha_c &= f_c \sqrt{\eta} \, \alpha_{\text{in}}. \end{aligned} \tag{24}$$

According to our QLS configuration, one detects the displacements α_s and α_c in the shaping and in the conditioning arms, respectively, when the incoming source has displacement α_{in}. The η-deteriorated Husimi function in the conditioning arm is

$$Q_c(z_c) = \frac{1}{\pi(1 + \eta f_c^2 \Delta n_{\text{in}})} \, e^{-\frac{|z_c - \alpha_c|^2}{1 + f_c^2 \eta \Delta n_{\text{in}}}}, \tag{25}$$

based on Eq. (6). Eq. (24) also contains a κ scaling and a ξ magnification factor

$$\kappa^2 \equiv 1 - \epsilon_s^2\,, \quad \xi^2 \equiv \frac{1}{\epsilon_s^2\,\epsilon_c^2}, \quad \text{with } \epsilon_s^2 \equiv \frac{f_s^2\,\Delta n_{\rm in}}{1 + [f_s^2 + \eta f_c^2]\,\Delta n_{\rm in}}, \quad \epsilon_c^2 \equiv \frac{\eta f_c^2\,\Delta n_{\rm in}}{1 + \eta f_c^2\,\Delta n_{\rm in}}, \tag{26}$$

whose physical meaning is clarified below.

With this information, the conditional quantum statistics (23) becomes

$$Q_\Pi(z_s + \alpha_s) = \int d^2 z_c\, \frac{\kappa^2}{\pi}\, e^{-|\kappa z_s - z_c/\xi|^2}\, Q_c(z_c + \alpha_c)\, \Pi(z_c + \alpha_c). \tag{27}$$

We now introduce the pattern function of QLS via

$$\Pi(z_c + \alpha_c) \equiv \frac{P(z_c/\xi + \alpha_s)}{\xi^2\; Q_c(z_c + \alpha_c)}, \qquad \Delta n_{\rm in} > 0. \tag{28}$$

Here, we require $\Delta n_{\rm in} > 0$ to produce a nonvanishing result because $1/\xi$ is proportional to $\Delta n_{\rm in}$, see Eq. (26). This follows from the fact that the correlation between z_s and z_c vanishes for a perfect coherent state laser because ideally classical fields are beam split without creating correlations tagging the (z_s, z_c) events, compare Fig. 4B. The needed $\Delta n_{\rm in}$ fluctuations can be generated in experiments, e.g., by modulating the amplitude and phase of the laser, as shown in Section 5.

Inserting the pattern function (28) into Eq. (27) produces

$$\begin{aligned} Q_\Pi(z_s + \alpha_s) &= \int d^2 z_c\, \frac{\kappa^2}{\pi \xi^2}\, e^{-|\kappa z_s - z_c/\xi|^2}\, P(z_c/\xi + \alpha_s) \\ &= \int d^2 z\, \frac{\kappa^2}{\pi}\, P(z)\, e^{-|(\kappa z_s + \alpha_s) - z|^2} \end{aligned} \tag{29}$$

This result has identically the same form as the transformation from the Glauber–Sudarshan function to the Husimi function, given by Eq. (7). Therefore, the shaping produces the quantum statistics

$$Q_\Pi(z_s + \alpha_s) = \kappa^2 Q_P(\kappa\, z_s + \alpha_s) \;\rightarrow\; Q_\Pi(z_s) = Q_P(z_s), \text{ for } \kappa \rightarrow 1. \tag{30}$$

As the central result for our QLS, we note that Q_P is here the Husimi function defined by that Glauber–Sudarshan function that has been identified in the pattern function (28). Since the choice of the pattern function is completely within the user's discretion, the conditioning-average (1) shapes a quantum source when we use the appropriate $P(z_c)$ of a quantum source in the pattern function. Therefore, our QLS provides an extremely versatile approach to shape any desired quantum source, as demonstrated in Section 5.

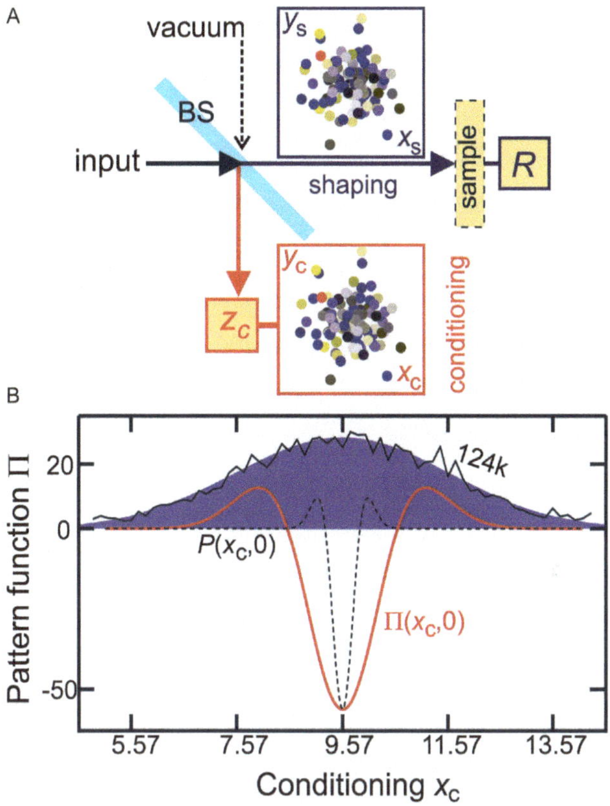

Fig. 4 Quantum-light shaping and conditioning. (A) The incoming semiclassical light pulse is an ensemble of amplitude-phase modulated laser pulses. These are divided by a beam splitter (BS) into conditioning and shaping arms. The shaping-arm pulse excites the sample whose response *R* is measured. The squares are phasor diagrams that illustrate simulated conditioning ($z_c = x_c + iy_c$) and shaping (x_s, y_s) counts detected simultaneously without the sample. The *circles* with equal colors mark an individual (z_c, z_s) outcome; one event singled out with a *red circle*. (B) Histogrammed experimental conditioning counts (*black line*, scaled, 124K counts) and a thermal-state distribution (*shaded area*, scaled) along the x_c axis are compared with Π (*red line*) and *P* function (*black dashed line*, scaled) corresponding to a regularized displaced Fock-state $|1\rangle$.

4. Quantum-light shaping algorithm

The QLS theory of Section 3 reveals an extraordinary possibility for quantum spectroscopy—our *QLS setup makes it possible to condition any quantum state to an ultrafast laser pulse train without modifying the duration or any other classical aspect of the pulses*. Most important, the approach satisfies the stringent requirements posed in Section 1.2 for realizing quantum spectroscopy in complex systems. We will next summarize the all-experimental algorithm and key ideas for QLS based on the detailed derivations in Section 3.

Fig. 4A illustrates how QLS is to be executed from an experimental point of view. As in the conceptual Fig. 3, a beam splitter divides a *classical* incoming laser pulse into conditioning and shaping arms, and only the shaping-arm light excites a sample whose generic observable R is recorded. The incoming laser light is assumed to have some fluctuations that can be realized either by active modulation (see Section 5) or by using moderately noisy laser; the role of the modulation noise is explained below. To condition the quantum-optical response $\langle \hat{R} \rangle_Q$ from R, our algorithm includes a pulse-to-pulse approach elaborated in Section 5 to records each R_j outcome for each incident laser pulse, indexed by j, as given in Eq. (9). Technically, we then simultaneously perform dual-channel balanced homodyne detection of each conditioning arm pulse and label the measured phasor outcome as $z_{c,j}$ where the c indicates conditioning. Based on the results in Section 3.2, collecting $(R_j, z_{c,j})$-outcome pairs $\mathcal{N}$ times allows us to construct the QLS average

$$\langle \hat{R} \rangle_Q \equiv \frac{1}{\mathcal{N}} \sum_{j=1}^{\mathcal{N}} R_j \, \Pi(z_{c,j}), \tag{31}$$

where the user-defined pattern function $\Pi(z)$ conditions $\langle \hat{R} \rangle_Q$ to be the response to *any* desired quantum source matching the laser's temporal form. Therefore, $\Pi(z)$ shapes only the light's quantum fluctuations, independent of the type of the sample or the observable $\hat{R}$.

Fig. 4A illustrates 200 simulated phasor measurements $z_c = x_c + iy_c$ and $z_s = x_s + iy_s$ in the conditioning and shaping arms, respectively; circles with matching colors indicate the simultaneously recorded events. The measurements yield a spread across a broad phase-space region due to the fluctuations within the modulated ensemble. However, the joint (z_c, z_s) counts are correlated because the detection of a given conditioning event $z_{c,j}$ most likely follows from a coherent state $|\alpha_c = z_{c,j}\rangle$, which suppresses the uncertainty in the fluctuations $z_{s,j}$ to those of a coherent state and not the extended fluctuations due to the modulation. The closeness of $z_{c,j}$ and $z_{s,j}$ counts demonstrate the existence of conditioning–shaping correlations that tag $z_{s,j}$ to be closer to $z_{c,j}$ than the overall spread. At the same time, these correlations vanish for a coherent state, as discussed in Section 3.2. Therefore, one needs to modulate the laser, both to observe these correlations and to sample an extended region of the phase space.

The general QLS setup of Fig. 4A simultaneously records $z_{c,j}$ together with the system response R_j. Histogramming the z_c events yields the probability distribution (Leonhardt, 1997) $Q_c(z_c)$ that uniquely defines the quantum fluctuations of the light in the conditioning arm. We have shown in Section 3.2 that the conditioning–shaping correlations yield a straightforward QLS average (31) that is identical to the sample's response to a source whose quantum fluctuations is characterized by the Glauber–Sudarshan distribution

$$P(z + \alpha_s) = \xi^2\ Q_c(\xi\ z + \alpha_c)\ \Pi(\xi\ z + \alpha_c), \tag{32}$$

regardless of how complicated the light-induced many-body dynamics are (Kira and Koch, 2006a). Especially, our QLS scheme provides flexibility for quantum-optical spectroscopy because the choice of the shaped quantum source is completely within the user's discretion through the particular choice of $P(z)$. The shaped source itself becomes truly nonclassical (i.e., quantum) only if the chosen P is partially negative (Kira and Koch, 2012) because it must then possess properties beyond a classical probability. While the sum (31) implements averaging, the negative parts of P include differential spectroscopy to the QLS algorithm. Technically, α_s (α_c) is the displacement center for shaped (conditioning) light, and the magnification factor $\xi > 1$ increases as more light is diverted to the conditioning arm, as defined by Eqs. (24)–(26).

4.1 Feasible pattern functions

Our QLS algorithm (31) and (32) becomes experimentally viable only if $\Pi(z_c)$ changes sufficiently slowly. To design the experiments of Section 5, we carefully analyze the viability of $\Pi(z_c)$ in those particular experiments. Fig. 4B shows the measured $Q_c(z_c)$ distribution (black line) resulting from a modulated laser (see Section 5) along the x_c axis and the $Q_c(z_c)$ distribution (shaded area) of a thermal-state; the modulated ensemble indeed mimics thermal fluctuations. Quantum sources often have a highly singular $P(z)$, which propagates to $\Pi(z_c)$. We implement a *physical regularization* where the quantum state's density matrix is expressed in terms of Fock states with a small amount of thermal fluctuations $\Delta\varepsilon \ll 1$ added. As explained in Appendix A.1, such states are physical and yet produce an analytical $P(z)$ that approaches the desired nonanalytic function for $\Delta\varepsilon \to 0$. The dashed line in Fig. 4B shows the $P(z_c)$ that corresponds to a displaced Fock-state $|1\rangle$ state (with $\Delta\varepsilon = 0.12$). We see that $P(z_c)$ changes rapidly on the scale of the

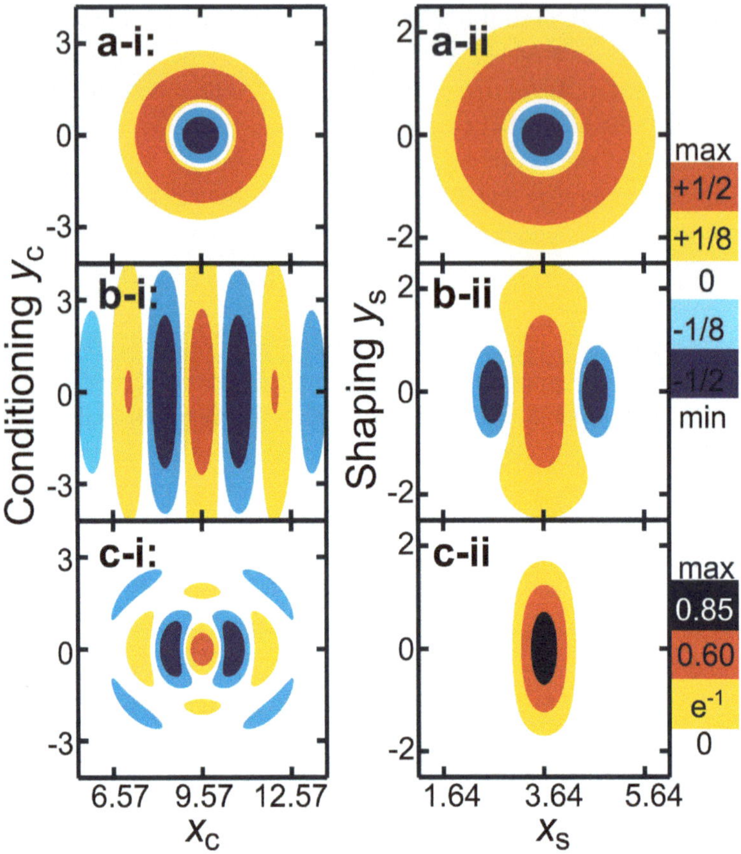

Fig. 5 Pattern function and predicted quantum distributions. The Π (i) and the target Wigner function (ii) are shown for a regularized displaced (A) Fock-state |1⟩, (B) cat state, and (C) squeezed-state shaping. The contours are defined with respect to the maximum and minimum of the functions.

recorded z_c. However, $P(z_c)$ appears in the pattern function (red line) through ξ magnification (32) such that all details of $\Pi(z_c)$ become well resolved.

Fig. 5A–C compares the $\Pi(z_c)$ (i) of three different quantum sources with the targeted Wigner function $W(z_s)$ (ii) for physically regularized and displaced Fock $|1\rangle$, Schrödinger cat, and squeezed state, respectively. We see that all pattern functions have negative regions (blue) without which the averaging (31) cannot shape nonclassical features beyond those already present in the input. The connection of the $\Pi(z_c)$ and shaped $W(z_s)$ is particularly clear for the Fock-state $|1\rangle$ shaping because they are topologically similar.

This connection is less obvious for the cat state while Π of a squeezed state is more complicated than the narrow Gaussian W it generates. The regularized Wigner functions reproduce the ideal sources with greater than 97% fidelity. These will be used as pattern functions for experiments analyzed in Section 5.

4.2 Fidelity, accuracy, and *P* magnification

We see that κ of Eq. (30) controls the fidelity at which the quantum source is shaped. Based on Eq. (26), the fidelity of QLS approaches 100% when more light is diverted to conditioning, i.e., when f_s becomes small. Figs. 5–9 explore QLS with $\Delta n_{\text{in}} = 8.24$, $f_c^2 = 0.874$, and, $f_s^2 = 0.126$. As we insert these numbers to Eq. (26), we find that our experiment of Section 5 produces the shaped sources with $\kappa = 0.942$, which explains why Fig. 9 contains all the quantum features of the intended quantum sources. One can improve the fidelity by dividing more light to conditioning, which naturally decreases the light intensity in the shaping arm. If we want to keep the intensity of the shaped quantum source constant while f_s is decreased, we must therefore increase the intensity of the incoming laser pulse. Current lasers can produce in excess of 10^9 photons per pulse such that one can sacrifice a large fraction of them to the conditioning and still have a high-intensity source in the shaping arm.

The situation without conditioning follows when setting $\Pi(z_c) = 1$. Compared with this situation, quantum-shaping protocol (1) weights detection events by $\Pi(z_c) \neq 1$ which typically is much larger than unity in some parts of the phase space (see Fig. 4B). When we compare conditioned and unconditioned measurements, the ensemble of $\mathcal{N}$ Husimi-function measurements approaches the actual quantum distribution with an accuracy proportional to

$$\text{QLS accuracy} \propto \frac{\text{Max}(|\Pi|)}{\sqrt{\mathcal{N}}}, \tag{33}$$

based on the central-limit theorem (Billingsley, 1995). Therefore, the QLS experiments must be repeated Max[$|\Pi|$] more times than regular measurements without conditioning. This means that one typically wants to minimize the magnitude of $|\Pi|$ to minimize the experimental effort when using the QLS. This goal can be reached by adjusting f_s and Δn_{in} appropriately. In our experiments, we have used a polarizing beam splitter to control f_s. We also control Δn_{in} through the modulation of incoming laser pulse's intensity,

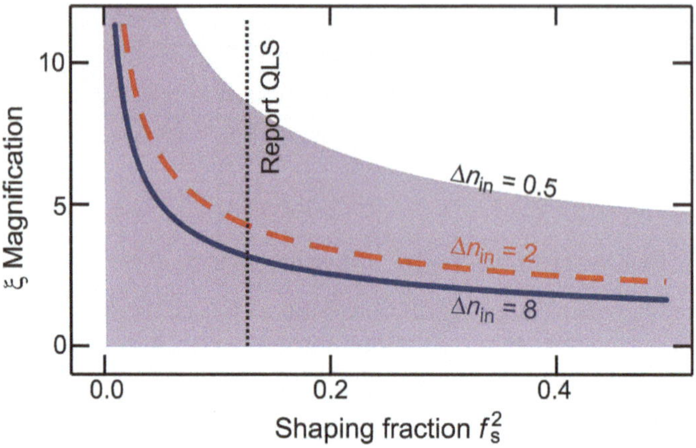

Fig. 6 Magnification of the Glauber–Sudarshan function in the QLS pattern function. The pattern function (28) stretches $P(z)$ by ξ that is shown as function of the fraction f_s^2 diverted into the shaping arm. The incoming source is assumed to have thermal fluctuations $\Delta n_{\text{in}} = 0.5$ (*shaded area*), $\Delta n_{\text{in}} = 2$ (*dashed line*), and $\Delta n_{\text{in}} = 8$ (*solid line*). The *vertical line* indicates $f_s^2 = 0.126$ used in Fig. 5 and our QLS experiments of Section 5.3.

see Section 5 for further details. Therefore, f_s and Δn_{in} can be adjusted within a relatively large parameter space to optimize the QLS operation.

As an additional experimental requirement, Π should have sufficiently broad features to be resolvable in experiments. Very often, the Glauber–Sudarshan function of true quantum sources produces very narrow features even after regularization, discussed in Appendix A.1, which must be managed in any experimentally viable QLS. Fortunately, the pattern function (28) offers a direct solution to this resolution problem because the Glauber–Sudarshan function appears in Π as via a ξ-magnified $P(z/\xi)$ because the phase space is stretched by ξ. Hence, narrow features are significantly broadened if ξ becomes sufficiently large. Inserting the experimental inputs (defined above) to Eq. (26), we find that the magnification is $\xi = 3.184$ in our shaping experiments. As shown in Fig. 5, this choice produces both a well-resolved and finite Π; its experimental feasibility is demonstrated in Fig. 9.

Like the fidelity, the magnification can be optimized by changing Δn_{in} and f_s^2. Fig. 6 analyzes the ξ magnification as function of the fraction f_s^2 for incoming lasers with $\Delta n_{\text{in}} = \frac{1}{2}$ (dashed line), $\Delta n_{\text{in}} = 2$ (solid line), and $\Delta n_{\text{in}} = 8$ (shaded area). For illustration, we assume the perfect-detection limit $\eta = 1$ but this does not essentially change the stretching capabilities of the pattern function. These results show that the magnification increases as Δn_{in} approaches zero while it saturates as Δn_{in} becomes larger. For a fixed Δn_{in}, the phase-space magnification increases considerably as f_s^2 becomes smaller, such that one can indeed broaden the pattern function to be well resolved in the experiment. At the same time, Eq. (28) shows that one should not be overstretch $P(z_c/\xi)$ to become wider than $Q_c(z_c)$. When too much stretching is applied, Π increases without bounds away from the displacement center, which increases the error in the ensemble averaged QLS based on Eq. (33). Therefore, one has two competing conditions when trying to minimize Eq. (33) while maximizing the magnification of $P(z_c)$ features through $\Pi(z_c)$. In other words, one tries to maximize ξ by decreasing Δn_{in} while making Π smaller by increasing Δn_{in}. Therefore, one typically needs a clearly nonzero Δn_{in} to produce an experimentally optimal pattern function.

We also may conclude that a nonperfect quantum efficiency, i.e., $\eta < 1$, in the conditioning measurement only changes the actual values of κ and ξ and the $Q_c(z_c)$ function detected. Especially, Eqs. (28) and (30) show that the QLS still shapes any quantum source despite the η deterioration in the conditioning arm. One simply has to include the deterioration when determining the pattern function. After that, η only influences the optimal f_s and Δn_{in} choice needed to realize an experimentally feasible QLS implementation.

5. All-experimental quantum-light shaping

We realize QLS experimentally with a three-channel homodyne detection setup, which is described in detail in Sections 5.1 and 5.2. We start with 120 fs laser pulses that have the quantum fluctuations of a coherent state $|\alpha_{\text{in}}\rangle$ corresponding to $|\alpha_{\text{in}}|^2 = 104.9$ photons on average. We then apply amplitude and phase modulation to generate an ensemble of pulses matching thermal Δn_{in}=8.24 photon fluctuations around the displacement center $\alpha_{\text{in}} = 10.24$. Since the ensemble is created by scanning $|\alpha\rangle$, the duration of the pulses is not changed. As compared to projective approaches, we do not measure $\langle R\rangle_{|\alpha\rangle}$ separately for each $|\alpha\rangle$ but use the entire ensemble of pulse-to-pulse measurements as input to the QLS algorithm (31), which significantly reduces the experimental collection time. To enable postselection, we use the setup shown in Fig. 4A and direct $f_c^2 = 87.4\%$ of the input light to the conditioning arm and $f_s^2 = 12.6\%$ to the shaping arm. By diverting more light to conditioning, one increases the amount of information for tailoring quantum fluctuations in the shaping arm. Unlike in many other postselection schemes, the input is a semiclassical pulse such that it can have any intensity and temporal shape adjustable by classical pulse shaping. We implement balanced homodyne detection (Smithey et al., 1993) with multiple channels to record z_s and z_c phase-space counts. The source and the local oscillators all originate from the same pulsed laser, as described below.

5.1 Single-channel setup

To present the all-experimental QLS setup in detail, we first discuss the implementation of an individual channel, and then describe the setup as a whole to experimentally realize QLS following the theoretical blueprints laid out in Sections 2.2 and 3. With a small modification, this setup can also measure photon correlation functions (McAlister and Raymer, 1997). A schematic of the homodyne detection technique (Bachor and Ralph, 2004) is shown in Fig. 7A. A strong local oscillator (LO) and a weak source of orthogonal linear polarization are combined in the polarizing beamsplitter PBS1. Balanced detection is performed by taking the difference of the pair of detectors. The half waveplate HWP1 is adjusted to balance the detectors in the absence of a source. When the LO is a strong coherent state and the source is a weak signal, the difference photocurrent is described by the quadrature-resembling operator

$$\hat{q} = \langle \hat{L} \rangle \left(e^{-i\theta_{LO}} \hat{B}^{\dagger} + e^{i\theta_{LO}} \hat{B} \right), \tag{34}$$

where $\hat{L}$ and $\hat{B}$ are the photon operators for the LO and the source, respectively, and θ_{LO} is the LO phase relative to the source. As a general property, $\hat{B}$ contains only that part of the source that overlaps temporally and spatially with the LO. We are working in the pulsed mode, which means that for each LO pulse we perform one measurement of the operator $\hat{q}$. The amplitude of the LO $\langle \hat{L} \rangle$ is extracted from measurements of $\hat{q}$ with the source blocked. We normalize (34) appropriately to obtain the $\hat{x}_\theta$ in Eq. (2).

In Fig. 7B, we show a simplified schematic diagram of the detection electronics. We use low noise and high quantum efficiency photodiodes. The difference photocurrent is amplified by the charge-sensitive preamplifier, and further by the shaping amplifier. Data acquisition is performed using a digitizer. A detailed characterization of an electronics setup very similar to ours can be found in Windpassinger et al. (2009). The raw data consist of a Gaussian pulse train where the height of each pulse is proportional to the number of photoelectrons in the initial pulse. Depending on the relative phase of the source and the local oscillator, the pulse heights can be positive

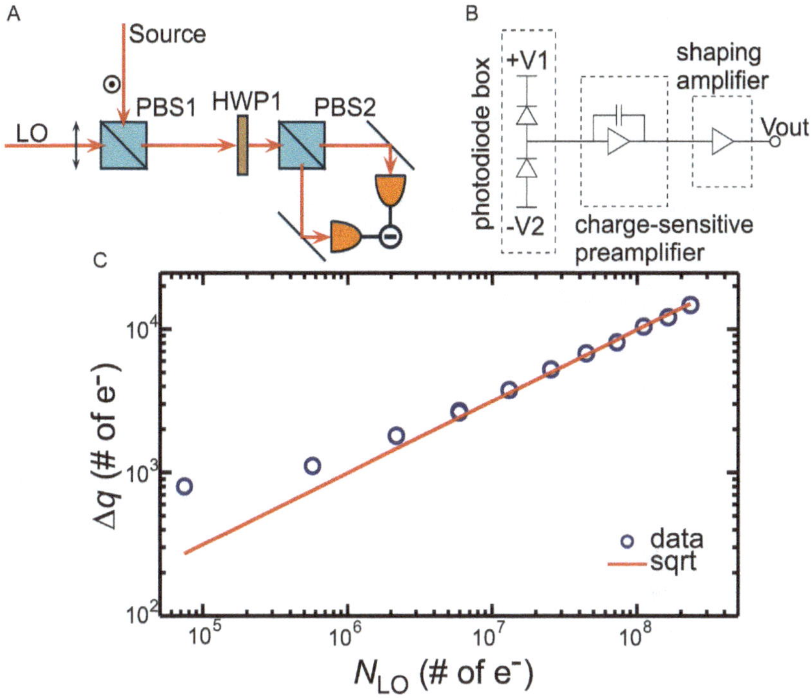

Fig. 7 (A) Experimental schematic the homodyne detection technique. (B) Schematic of the detector electronics. In every pulse, the photodiode box produces a current pulse whose number of electrons is a measurement the operator $\hat{q}$ in (34). The pulse is then integrated by the charge-sensitive preamplifier and then shaped to a Gaussian pulse of fixed width by the shaping amplifier. The height of the output pulse is proportional to the number of electrons in the input current pulse. (C) The standard deviation Δq is measured as a function of the number of photoelectrons N_{LO} per LO pulse.

or negative. The widths of the pulses are fixed by the choice of the shaping amplifier. For each pulse, we obtain the pulse height by multiplying the scope trace with an appropriate window function and integrating over one period. We calibrate the amplification using a function generator. For each pulse, we extract the difference number of photoelectrons from the two photodiodes by measuring the operator $\hat{q}$ of Eq. (34) in units of numbers of electrons.

To verify shot-noise performance, Fig. 7C presents the measured standard deviation Δq as a function of the number of LO photoelectrons per pulse, N_{LO}, when only the LO is present. The LO intensity is measured independently, and we use the expected quantum efficiency of $\eta = 80\%$ to extract the number of photoelectrons N_{LO}. We find a square root behavior of Δq as a function of N_{LO} above 7×10^6 photoelectrons per pulse, which confirms that we can measure the shot noise of the LO pulse. This measurement can alternatively be viewed to detect the vacuum noise of the empty-source port amplified by the LO.

We use a home-made cavity-dumped femtosecond laser with repetition rate 190 kHz. The center wavelength is tuned to 800 nm, and the bandwidth to 8 nm (full-width-half-maximum, FWHM); the corresponding pulse duration is 120 fs (FWHM). The same laser provided both the LO and the source as described below.

5.2 Multichannel setup

A schematic of the multichannel experimental QLS setup is shown in Fig. 8A. The horizontally polarized laser beam is attenuated by polarizer Pol1. Using the half waveplate HWP1 and the polarizing beamsplitter PBS1, we can control the relative intensities of the two local oscillators LO1 and LO2. Polarizer Pol2 ensures that LO1 is vertically polarized with a very good extinction ratio. The beam sampler BS1 splits roughly 4% of LO2 into the input source $\hat{B}_{\text{in}}$. The input source pulse is further attenuated by the combination of variable retarder VR2 and polarizer Pol3, and is combined with LO1 on the polarizing beamsplitter PBS2. After adjusting the angle of Pol3, PBS2 divides fraction $f_s^2 = 0.126$ of the input source to the channel Ch3 as the shaped source $\hat{B}_s$ while it directs fraction $f_c^2 = 0.874$ to the conditioning, $\hat{B}_c$, into channels Ch1 and Ch2. At the same time, we modulate the intensity of the input source pulse by changing the peak-to-peak voltage of the square pulse train applied to VR2.

The conditioning counts are recorded at the channels Ch1 and Ch2. More specifically, the 50/50 nonpolarizing beamsplitter NPBS1 divides $\hat{B}_c$ into Ch1 and Ch2; also the LO2 is combined in through the other input port of NPBS1. The Ch1 and Ch2 constitute a four-port homodyne setup while Ch3 is a two-port homodyne setup. Each homodyne detection channel is composed of a half waveplate, a polarizing beamsplitter cube, two photodiodes, and the appropriate electronics. Only shows the schematics of Ch1 are shown because the remaining two channels Ch2 and Ch3 are identical to Ch1. The quarter waveplate QWP1 is aligned with its fast and slow axes parallel to the polarizations of the source and LO2 such that it rotates the phase of LO2 by $\pi/2$ with respect to the source. Therefore, in the phasor diagram of $z_c \equiv q_x + iq_y$, channels Ch1 ($q \to q_x$) and Ch2 ($q \to q_y$) measure the orthogonal q_x and q_y, respectively.

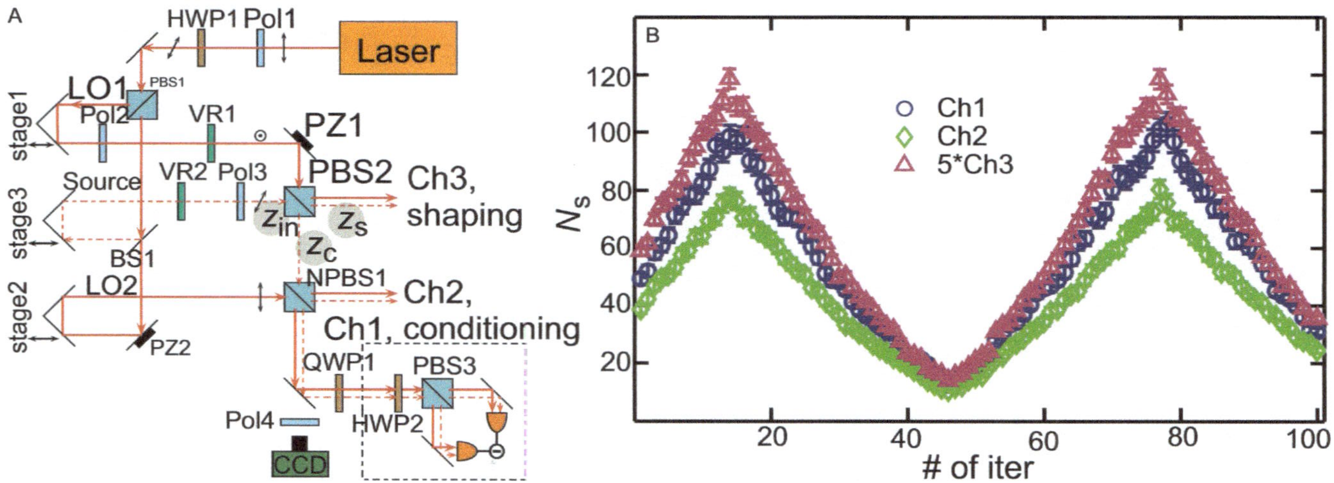

Fig. 8 (A) Schematic of the experimental setup (see text). For clarity, the LO beams are shown with a continuous line, and the source beam is shown with a dashed line separated from the LO beams. In reality, beams copropagate. Channels Ch1 and Ch2 perform the conditioning measurement, while Ch3 performs the measurement in the shaping arm (see Fig. 3). (B) Measured source intensities N_s at the three different channels for 100 consecutive iterations. Because N_s for Ch3 is small, we plot $5N_s$ for this channel. The error bars show the statistical uncertainties. We modulate the source intensity to simulate the intensity noise required by our postselection scheme.

The phases of LO1 and LO2 are scanned independently using PZ1 and PZ2 which are mirrors whose position is controlled by a piezoelectric actuator. The CCD camera and polarizer Pol4 are useful to verify the alignment of the three different beams by flipping the mirror in front of the CCD. We can then observe beam interference when the LO phases are scanned slowly. Nonzero retardance of the variable retarder VR1 allows us to deflect part of LO1 toward the CCD and verify the spatial overlap of LO1 with the shaped source $\hat{B}_s$. But during a measurement, VR1 needs to have zero retardance, so that LO1 is directed toward Ch3.

As discussed in Section 3.2, our all-experimental QLS algorithm (31) requires that the intensity and phase fluctuations of the input pulse z_{in} exceed that of a coherent state. We modulate the intensity of z_{in} using the variable retarder VR2 in Fig. 8A. The retardance of VR2 is determined by the peak-to-peak voltage of the applied 2 k Hz square wave. By applying slow AM modulation to this square wave, we can modulate the source intensity, while keeping all LO intensities fixed. Fig. 8B shows the measured source intensities at the three different channels for 100 consecutive iterations. Each iteration includes roughly 1000 data points. During each iteration, we scan the phases of LO1 and LO2. This is necessary in the shaping arm (Ch3) to perform tomography on the shaped quantum source. In the conditioning arms (Ch1 and Ch2), this means that in the phasor diagram we measure the projections of the source distribution along the axes at angles θ_{LO} and $\theta_{LO} + \pi/2$. We then obtain the projections along the x- and y-axes, q_x and q_y by an appropriate linear transformation. Because the source intensities are not exactly the same for Ch1 and Ch2, we need to scale the raw data to the average intensity of Ch1 and Ch2. To simulate the phase noise required by our postselection scheme, we multiply the vector (q_x, q_y) by the appropriate rotation matrix and add the same phase to θ_{LO1} in the data for Ch3. In the current experiment, the conditioning arm measures roughly 10 times more photons per pulse than a typical balanced homodyne detection experiment (see Fig. 8B).

It is important to notice that our QLS approach has no fundamental upper limit for increasing the intensity of the quantum source, and it can be applied to condition quantum features to any ultrafast pulse simply by adjusting the LO. Therefore, *our all-experimental QLS approach satisfies the stringent QLS conditions laid out in Section 1.2.* The main challenges are technical ones such as increasing the photon number for the LO, conditioning, and shaping arms while avoiding saturation of detectors as intensities increase.

5.3 All-experimental, ultrafast, and intense quantum sources

To demonstrate a proof-of-principle viability for all-experimental QLS, we condition and detect the three, very different, quantum-light sources analyzed in Fig. 5. These quantum-source targets have 14.25 photons on average, which is roughly an order of magnitude higher than many of the earlier quantum-source demonstrations (Afek et al., 2010; Chrzanowski et al., 2011; Lvovsky et al., 2001; Neergaard-Nielsen et al., 2006; Ourjoumtsev et al., 2007; Parigi et al., 2007; Wenger et al., 2004; Zavatta et al., 2004) that applied a different postselection strategy than we do. On top of shaping quantum statistics, our QLS approach can also freely and independently adjust temporal and intensity aspects of ultrafast pulse by applying well-known pulse-shaping techniques (Weiner, 2000) and technical improvements to realize high-sensitivity detection of high-intensity fields.

To demonstrate all-experimental QLS, we perform experiments without an explicit sample in the shaping arm. Therefore, our $\hat{R}$ observable is then $\hat{x}_s$ quadrature of the shaping-arm light because we use a balanced homodyne measurement (see Section 5.1) to detect $x^{\theta}_{s,j} \equiv \cos\theta \, \mathrm{Re}[z_{s,j}] + \sin\theta \, \mathrm{Im}[z_{s,j}]$ counts along. The quadrature direction θ is scanned to reconstruct the Wigner function $W(z_s)$ (Leonhardt, 1997) using our QLS algorithm (31) to post-select three very different quantum-light sources of Fig. 5.

Fig. 9A shows how the constructed $W(z_s)$ develops as more measurements are collected when no shaping is applied, i.e., the pattern function is flat $\Pi = 1$. We see that the pulse ensemble's fluctuations have a Gaussian shape exceeding those of a coherent state (half width of red contour would be 0.589 for a coherent state). Such excess fluctuations are necessary for QLS as explained in Section 3.2 in order to postselect these semiclassical quantum fluctuations into any quantum form by applying the pattern functions of Fig. 5 in the QLS conditioning (31). Shaping of a displaced Fock-state $|1\rangle$, a displaced Schrödinger cat state, and a squeezed state is shown in Fig. 9B–D, respectively. The QLS measurement accurately reproduces the analytic Wigner functions presented in Fig. 5. In all cases, the Wigner function of conditioned measurement approaches the shaping target with fidelity greater than 90%; the precise fidelity is indicated within each frame. These results demonstrate that our postselection, QLS algorithm:

1. delivers all-experimental shaping of very diverse quantum sources,
2. conserves the temporal shape of the incoming field,
3. has appreciable intensity, which in principle is not fundamentally limited.

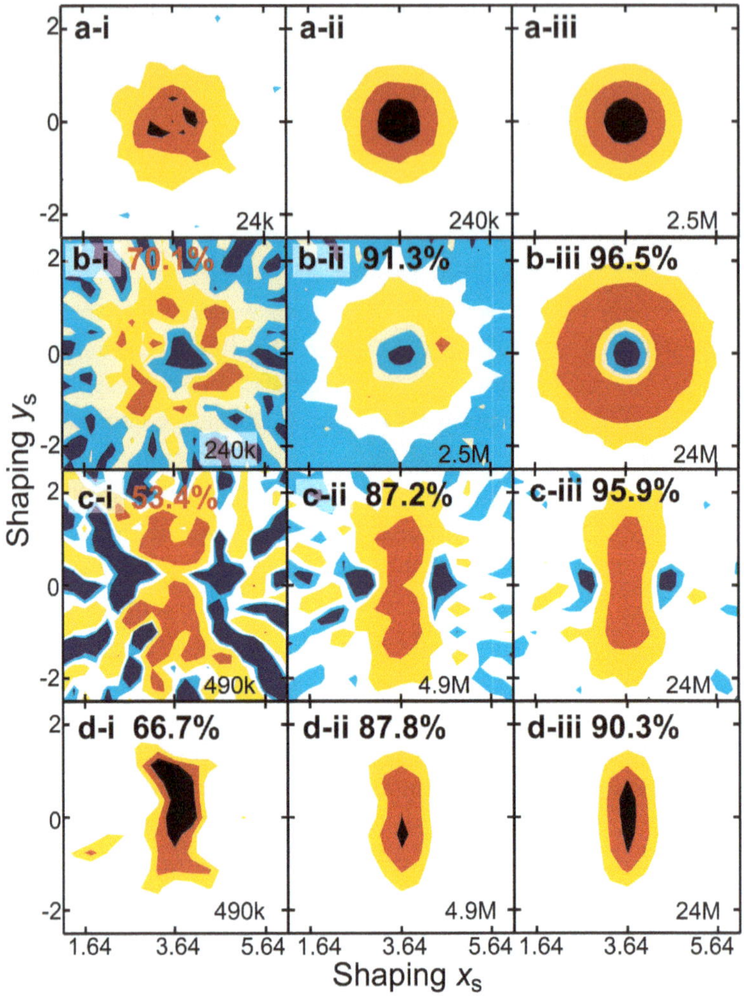

Fig. 9 Experimental QLS. Reconstructed Wigner function from $\mathcal{N}$ measurements. (A) No shaping applied, i.e., $\Pi = 1$. (B) Fock-state $|1\rangle$ conditioning by Π in Fig. 5A. (C) Schrödinger cat-state conditioning by Π in Fig. 5B. (D) Squeezed state Π in Fig. 5C. Contour colors chosen as in Fig. 5, $\mathcal{N}$ given in k(10^3) or M(10^6) counts. *Percentages* indicate the fidelity to reproduce the targeted quantum source.

In particular, our result *satisfy all there points listed as requirements for quantum-light shaping in Section 1.2, demonstrating a new, all-experimental direction to realize quantum spectroscopy of complex systems.* To the best of our knowledge, this capability has not been established before.

In more detail, the second point in the list above follows because the temporal LO mode is fixed, and the Wigner function measurement produces major distortions (Vogel and Grabow, 1992) if the signal—here the shaped source—does not match the local oscillator given here by the incoming 120 fs pulse. Since the QLS produces the intended $W(z_s)$, it *exclusively shapes the quantum fluctuations* of an ultrafast input light pulse without modifying its classical aspects. Therefore, one can use QLS in conjunction with classical pulse shaping to realize complete design freedom of laser pulses. It is noteworthy, that QLS can also compensate low quantum efficiency in conditioning by modifying the pattern function accordingly, as discussed in Section 3. Concerning the third point of the list, the demonstrated photon number of 14.25 is not yet quantum spectroscopically large, although it is that for most quantum-information studies operating with combinations of zero and one photons. Quantum spectroscopically relevant range is roughly from tens to millions of photons. In our experiments, the major limiting factor is the saturation of photodiodes mostly due to LO photons. Since we present here a proof-of-principle demonstration for all-experimental QLS, we did not attempt to optimize the conditions or improve photodetection of high-intensity field. This technical limitation can be addressed straightforwardly, which should improve the dynamic range of high-intensity quantum sources further.

6. Discussion and outlook

Conceptually, our all-experimental QLS algorithm (31) combines the ideas of projective approaches with conditional measurements. In projective approaches, one applies ordinary laser spectroscopy to deduce the average response $\langle \hat{R} \rangle_{|\alpha\rangle}$ to a classical, i.e., coherent state $|\alpha\rangle$, excitation. A set of $\langle \hat{R} \rangle_{|\alpha\rangle}$ results can be collected by scanning the amplitude $\alpha = x + iy$ throughout the (x, y) phase space. Mathematically, $\langle \hat{R} \rangle_Q = \int d^2\alpha \; P(\alpha) \; \langle \hat{R} \rangle_{|\alpha\rangle}$ *projects* (Sudarshan, 1963) a set of classical responses to the response $\langle \hat{R} \rangle_Q$ to quantum light, i.e., light with a nonclassical Glauber-Sudarshan $P(\alpha)$ function (Glauber, 1963; Sudarshan, 1963). Unfortunately, quantum-light often has a pathological $P(\alpha)$, which makes the evaluation of $\langle \hat{R} \rangle_Q$ challenging because experimental measurements always contain some noise. This was solved in Lobino et al. (2008) by applying regularization of $P(\alpha)$ when characterizing changes in a maximum likelihood density matrix of a single-mode light. In many-body experiments, one cannot determine the relevant density matrix due to its high dimensionality, which compromises the $\langle \hat{R} \rangle_Q$

projection even for a regularized $P(\alpha)$. This problem is solved in Kira et al. (2011) by converting the measured $\langle \hat{R} \rangle_{|\alpha\rangle}$ into a precisely known phase-space function, which yields integrability of $\langle \hat{R} \rangle_{Q}$ for any $P(\alpha)$.

Projective approaches pose challenges to both the experiment and subsequent analysis. Since one must perform $\mathcal{N}$ repetitive measurements at $N_\alpha \gg 1$ phase-space points, the total number of measurements is $N_\alpha \mathcal{N}$, where $\mathcal{N}$ is the number of measurements needed to characterize the response to a single coherent state. The resulting large dataset makes the data processing computationally intensive. The projective approach also suffers from exponential overhead that scales like $[N_\alpha]^J$ when projection is applied to J input sources. These challenges mean that projective approaches are best suited for few-source data mining to identify potentially interesting quantum light and the many-body states to which they couple.

Clearly, it would be preferable to simply perform experiments directly with light that has the desired quantum $P(\alpha)$. If pulses with the desired $P(\alpha)$ did exist, each generated pulse would have identical quantum fluctuations. However, the quantum fluctuations may be determined only from an ensemble of measurements because a single detection yields only one outcome, e.g., a phasor count. At the same time, the nonclassical fluctuations can be described via the projective $P(\alpha)$ integral over the coherent states $|\alpha\rangle$. The QLS algorithm (31) shapes fluctuations by exchanging the order of the averaging and the $P(\alpha)$ integral by performing pulse-by-pulse measurements with an ensemble of coherent states and then summing over *all* the outcomes weighted by a pattern function Π. This step combines an experimental Monte-Carlo sampling of the projective integral with the generalization of the postselection approach (Aharonov et al., 1964). As compared to projective approaches, the QLS algorithm has the significant advantage that it only requires $\mathcal{N}$ measurements, i.e., a factor of $[N_\alpha]^J$ fewer, removing the related exponential overhead.

Since all-experimental QLS is demonstrated in Section 5.3 for a single mode, its extension to multi mode quantum sources are the natural next step. This will allow generating multiple quantum-light pulses whose mutual timing is as accurate as the best ultrafast spectroscopies have. This means that a suboptical-cycle precisions is not out of reach because we have demonstrated that QLS approach does not alter the temporal aspects of input pulses. Consequently, one can synchronize the arrival of, e.g., multiple single-photon sources to a sample, removing failed co-incidences in quantum-optical studies. However, there is price to pay—each quantum source requires a separate conditioning unit, which can become technically cumbersome. But yet again, the perfect co-incidences and inherent Monte-Carlo sampling eliminate both failed attempts and exponential overhead, which many other approaches suffer from.

The multimode QLS would also allow to generate arbitrary entangled sources among multiple sources. Such investigations could be started by combining two of our QLS units, and using a two-dimensional Glauber–Sudershan function as the basis of pattern function to be used in conditioning. This could be the method of choice to explore correlated multiphoton transitions in complex systems, generalize multidimensional coherent spectroscopy to the quantum-optics domain, or perform quantum spectroscopy with entangled sources, few exiting directions to mention.

In conclusion, we have presented a brief summary of quantum spectroscopy and its requirements, especially with respect to creating intense, ultrafast quantum-light sources. We have both proposed and demonstrated a viable all-experimental QLS, and expect these QLS capabilities to create completely new possibilities for quantum spectroscopy and broader quantum-information science.

Appendix

A.1 Quantum-light sources

Many of the standard quantum-optical states have nonanalytic Glauber–Sudarshan functions. To make the QLS experimentally viable, one obviously needs to regularize nonanalytic $P(\alpha)$ without distorting the main quantum-optical properties. We will next present the regularization details needed to produce viable pattern functions for QLS.

A.1.1 Physical regularization of quantum sources

In our QLS, we apply a physical regularization of quantum sources that can be expressed via

$$\hat{\rho}_{\text{light}} \equiv \frac{\hat{Z}^{\dagger}|0\rangle\langle 0|\hat{Z}}{\mathcal{N}}, \qquad \hat{Z}^{\dagger} = \sum_{n=0} c_n \left[\hat{B}^{\dagger}\right]^n, \tag{A.1}$$

where $\mathcal{N}$ is the norm and $\hat{Z}^{\dagger}$ is a creation-operator expansion c_n coefficients define the state. We see now that any pure state can be expressed with Eq. (A.1) because $\hat{Z}^{\dagger}$ induces any expansion in terms of Fock states.

This series can now be regularized by replacing the vacuum state of Eq. (A.1) by a thermal state (6):

$$\hat{\rho}_{\Delta\varepsilon,\mathrm{reg}} \equiv \frac{\hat{Z}^{\dagger}\hat{\rho}_{\Delta\varepsilon}\hat{Z}}{\mathcal{N}(\Delta\varepsilon)}, \quad \hat{\rho}_{\Delta\varepsilon} = \sum_{n=0}^{\infty} |n\rangle p_{\Delta\varepsilon}(n)\langle n|, \quad p_{\Delta\varepsilon}(n) = \frac{\Delta\varepsilon^{n}}{(1+\Delta\varepsilon)^{n+1}} \tag{A.2}$$

where $\mathcal{N}(\Delta\varepsilon)$ is the norm and $\hat{\rho}_{\Delta\varepsilon}$ is the density matrix of the thermal state having thermal fluctuations of size $\Delta\varepsilon$. Clearly, $\hat{\rho}_{\Delta\varepsilon,\mathrm{reg}}$ approaches the original state at the limit $\Delta\varepsilon \to 0$.

It is also straightforward to show that $\hat{\rho}_{\Delta\varepsilon,\mathrm{reg}}$ is always a physical density matrix. Obviously, it is Hermitian and normalized to unity, as any density matrix must be. Therefore, we only need to show its positivity to prove that it is a physical state. Since $\hat{\rho}_{\Delta\varepsilon,\mathrm{reg}}$ is Hermitian, we can always find a complete set of orthonormal eigenstates $\hat{\rho}_{\Delta\varepsilon,\mathrm{reg}}|\phi_{\lambda}\rangle = \varrho_{\lambda}|\phi_{\lambda}\rangle$. We can then formally compute any eigenvalue using

$$\begin{aligned}\varrho_{\lambda} = \langle\phi_{\lambda}\big|\hat{\rho}_{\Delta\varepsilon,\mathrm{reg}}\big|\phi_{\lambda}\rangle &= \sum_{n=0}^{\infty}\frac{\langle\phi_{\lambda}\big|\hat{Z}^{\dagger}\big|n\rangle\, p_{\Delta\varepsilon}(n)\,\langle n\big|\hat{Z}\big|\phi_{\lambda}\rangle}{\mathcal{N}(\Delta\varepsilon)} \\ &= \sum_{n=0}^{\infty}\frac{|\langle\phi_{\lambda}|\hat{Z}^{\dagger}|n\rangle|^{2}}{\mathcal{N}(\Delta\varepsilon)}\, p_{\Delta\varepsilon}(n) \geq 0\end{aligned} \tag{A.3}$$

that must be positive definite because all sum arguments are that. Therefore, $\hat{\rho}_{\Delta\varepsilon,\mathrm{reg}}$ has only positive eigenvalues, which allows for an association of a positive definite probability with each of its eigenstates. Hence, $\hat{\rho}_{|\alpha,\gamma\rangle,\Delta n}$ is a real physical quantum state because it is Hermitian, normalized to unity, and positive definite.

The Glauber–Sudarshan function of the physically regularized source becomes

$$\begin{aligned}P_{\Delta\varepsilon,\mathrm{reg}}(\alpha) = \frac{1}{\pi\Delta\varepsilon\mathcal{N}(\Delta\varepsilon)}\, e^{-\frac{|\alpha|^{2}}{\Delta\varepsilon}} &\sum_{n=0}^{\infty}|c_{n}|^{2}\sum_{k=0}^{n}\frac{n!\; n!\; (-s)^{k}|s\,\alpha|^{2(n-k)}}{k!\;(n-k)!(n-k)!} \\ &+ \mathrm{Re}\left[\sum_{n=1}^{\infty}\sum_{n'=0}^{n-1} c_{n}^{\star}\, c_{n'}\sum_{k=0}^{n}\frac{n!\; n'!\; (-s)^{k}(s\,\alpha^{\star})^{n'-k}(s\,\alpha)^{n-k}}{k!\;(n-k)!(n'-k)!}\right] \\ s \equiv \frac{1+\Delta\varepsilon}{\Delta\varepsilon}, \quad \mathcal{N}(\Delta\varepsilon) &= \sum_{n=0}^{\infty} n!\; |c_{n}|^{2}\;(1+\Delta\varepsilon)^{n}.\end{aligned} \tag{A.4}$$

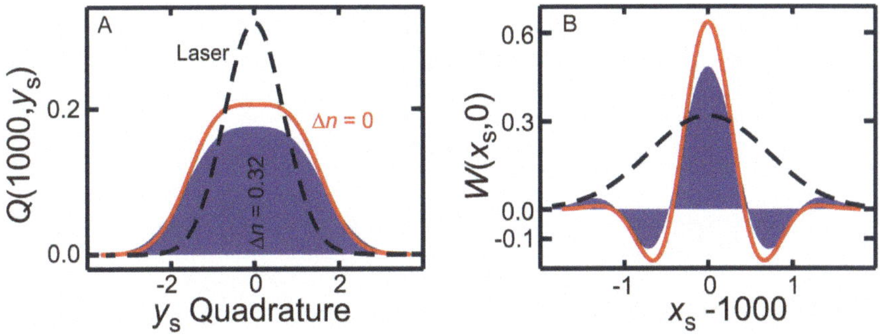

Fig. A.1 Quantum statistics of a Schrödinger cat state $n|\alpha = 1000, \gamma = i\rangle$. Cross-section of **a** the Husimi function and **a** the Wigner function. The *shaded area* corresponds to physical regularization with $\Delta\varepsilon = 0.32$ while the *solid line* has $\Delta\varepsilon = 0$. The *dashed line* indicates the extension of the semiclassical state $|\alpha = 1000, \gamma = 0\rangle$ using $\Delta\varepsilon = 0.32$ regularization.

In many physically relevant cases, we only need a finite number of Fock states to describe the quantum source such that c_n vanishes above certain limiting value. In that case, the n sum of Eq. (A.4) is limited such that $P_{\Delta\varepsilon,\mathrm{reg}}$ becomes a Gaussian times a polynomial. This function is analytic for $\Delta\varepsilon > 0$, as it should be for a regularized distribution.

Several $P_{\Delta\varepsilon,\mathrm{reg}}(\alpha)$ are presented in Appendix A.1.2. To determine the quantum-statistical changes the regularization introduces, we construct the Wigner and the Husimi function of Schrödinger's cat state by inserting $P_{\Delta\varepsilon,\mathrm{reg}}(\alpha)$ into Eq. (7). Fig. A.1A shows a y cross section of the Q function while Fig. A.1B presets an x cross section of the W function for the regularized $\Delta\varepsilon = 0.32$ (shaded area) and the unregularized $\Delta\varepsilon = 0$ (solid line) source. The dashed line shows the phase-space distribution of a coherent state whose center coincides with the cat state.

We see that the regularized cat state shows an interference pattern in x direction. In particular, W dips to negative values even when the state is regularized. In fact, the regularization by $\Delta\varepsilon = 0.32$ changes the quantum statistics only a little so that the physical regularization maintains the main features of the intended quantum source. In our experimental QLS of Section 5.3, we have regularized the Fock-state $|1\rangle$ by $\Delta\varepsilon = 0.12$, the cat state by $\Delta\varepsilon = 0.44$, and the squeezed state by $\Delta\varepsilon = 0.178$. Only the squeezed state has a truncated c_n, up to Fock-state $|4\rangle$. Fig. 5 presents examples for displaced Fock-state $|1\rangle$, Schrödinger's cat state, and squeezed state and the Wigner function that follows from regularization. In all cases, the features of the unregularized quantum sources are reproduced with high fidelity.

A.1.2 Regularized Glauber–Sudarshan functions

Fock-state $|n\rangle$ can be expressed in terms of Eq. (A.1) when we choose $c_j = \delta_{j,n}/\sqrt{n!}$ in the $\hat{Z}$-operator. As this result is inserted to Eq. (A.4) we find

$$P_{\Delta\varepsilon,|n\rangle}(\alpha) = \frac{1}{\pi\Delta\varepsilon} e^{-\frac{|\alpha|^2}{\Delta\varepsilon^2}} \sum_{k=0}^{n} \frac{n!\,(-s)^k |s\,\alpha|^{2(n-k)}}{k!\,(n-k)!(n-k)!}, \qquad s = \frac{1+\Delta\varepsilon}{\Delta\varepsilon}. \tag{A.5}$$

as the $\Delta\varepsilon$-regularized P function for $|n\rangle$. For example, $P_{\Delta\varepsilon,|1\rangle}(0) = -\frac{s}{\pi\Delta\varepsilon}$ is negative valued, which show that Fock-state $|1\rangle$ is a quantum source. A similar quantum property, i.e., negativity, is found for all $|n > 0\rangle$.

Coherent states are typically defined by introducing a displacement operator $D(\alpha) \equiv e^{\alpha\hat{B}^\dagger - \alpha^\star\hat{B}}$ that acts upon a vacuum state yielding $|\alpha\rangle \equiv D(\alpha)|0\rangle$ and $|\alpha\rangle = \sum_{n=0}^{\infty} \frac{\alpha^n}{\sqrt{n!}} e^{-\frac{1}{2}|\alpha|^2} |n\rangle$, c.f. Kira and Koch (2012). All $|\alpha\rangle$ form a complete set of states, but they are not orthogonal

$$\langle\alpha|\alpha'\rangle = e^{-\frac{1}{2}|\alpha-\alpha'|^2 + i\mathrm{Im}[\alpha'\alpha^\star]}, \qquad \frac{1}{\pi}\int d^2\alpha\, |\alpha\rangle\langle\alpha| = \mathbb{I} \tag{A.6}$$

such that they form an overcomplete set of basis states.

A coherent state $|\alpha\rangle$ can be expressed in terms of Eq. (A.1) when we choose $c_j = \frac{\alpha^n}{n!} e^{-|\alpha|^2/2}$ in the $\hat{Z}$-operator. A straightforward evaluation of Eq. (A.4) produces $\Delta\varepsilon$-regularized P function

$$P_{\Delta\varepsilon,|\alpha\rangle}(z) = \frac{1}{\pi\Delta\varepsilon} e^{-\frac{|z-(1+\Delta\varepsilon)\,\alpha)|^2}{\Delta\varepsilon}}, \tag{A.7}$$

where the phase-space coordinate is expressed in terms of z to avoid confusion with the displacement α. We see that any coherent state has a positive definite $P_{\Delta\varepsilon,|\alpha\rangle}(z)$ for all values of $\Delta\varepsilon$ as it should for a classical field. Alternatively, we may choose $\hat{Z}^\dagger = D(\alpha)$ which contains both $\hat{B}$ and $\hat{B}^\dagger$ operators in $\hat{Z}^\dagger$ expansion. This choice also regularizes $P_{|\alpha\rangle}$; only the exponent of Eq. (A.7) is replaced by $|z-\alpha|^2/\Delta\varepsilon$. This explicit choice does not alter the $\Delta\varepsilon \to 0$ limit such that we may regularize $P_{\Delta\varepsilon,|\alpha\rangle}(z)$ in physically different ways.

Schrödinger's cat state is a linear superposition of two different coherent states:

$$|\alpha,\gamma\rangle \equiv N\left(e^{-i\mathrm{Im}[\alpha\gamma^\star]}|\alpha+\gamma\rangle + e^{+i\mathrm{Im}[\alpha\gamma^\star]}|\alpha-\gamma\rangle\right), \; N = \frac{1}{\sqrt{2+2e^{-2|\gamma|^2}}} \tag{A.8}$$

where the norm N follows from the property (A.6) of the coherent states. This is compatible with Eq. (A.1) when we make an identification

$$\hat{Z}^{\dagger} \equiv N\left(e^{-i\mathrm{Im}[\alpha\gamma^{\star}]}D(\alpha+\gamma)+e^{+i\mathrm{Im}[\alpha\gamma^{\star}]}D(\alpha-\gamma)\right), \quad |\alpha,\gamma\rangle = N\,\hat{Z}^{\dagger}|0\rangle. \tag{A.9}$$

This produces the physically regularized Glauber–Sudarshan function

$$\begin{aligned} P_{|\alpha,\gamma\rangle,\Delta\varepsilon}(z+\alpha) = \frac{N}{\pi\Delta\varepsilon}\Bigg[& e^{-\frac{|z-\gamma|^2}{\Delta\varepsilon}} + e^{-\frac{|z+\gamma|^2}{\Delta\varepsilon}} \\ & + 2e^{-\frac{|z|^2}{\Delta\varepsilon}} e^{+|\gamma|^2(2+\frac{1}{\Delta\varepsilon})} \cos(4+\frac{2}{\Delta\varepsilon})\mathrm{Im}[\gamma^{\star}z]\Bigg], \end{aligned} \tag{A.10}$$

following a similar derivation as that for a coherent state. This is clearly analytic while the cosine part produces negative regions in between the $\pm\,\gamma$ displacement centers. Therefore, $|\alpha,\,\gamma\rangle$ is true quantum sources for $\gamma \neq 0$.

A squeezed-vacuum state is defined by a Taylor expansion,

$$|0,r\rangle \equiv \frac{1}{\sqrt{\cosh r}}\sum_{n=0}^{\infty}\frac{\sqrt{(2n)!}}{n!}\left(\frac{1}{2}\tanh r\right)^{n}|2n\rangle \tag{A.11}$$

that contains only the even Fock states (Walls and Milburn, 1994). We see that the Fock-state coefficients decay faster than exponentially for sufficiently large n. Therefore, one can always express the squeezed states using only J first terms of the sum (A.11). Especially, this provides a physical state that approaches the squeezed vacuum via physical wave functions.

We also see that the squeezed state (A.11) can be expressed in terms of Eq. (A.1) when we choose $c_{2j+1}=0$ and $c_{2j}=\sqrt{(2j)!}(\frac{1}{2}\tanh r)^{j}/[j!\sqrt{\cosh r}]$. Eq. (A.4) produces then the regularized P function of the squeezed state:

$$\begin{aligned} P_{|0,r\rangle,\Delta\varepsilon}(\alpha) = & \frac{e^{-\frac{|\alpha|^2}{\Delta\varepsilon}}}{\pi\Delta\varepsilon\mathcal{N}(\Delta\varepsilon)}\left(\sum_{j=0}^{J}\left(\frac{(2j)!}{j!}\right)^{2}\left(\frac{\tanh r}{2}\right)^{2j}\sum_{k=0}^{n}\frac{(-s)^{k}|s\,\alpha|^{2(2j-k)}}{k!\,(2j-k)!(2j-k)!}\right. \\ & \left. +\mathrm{Re}\left[\sum_{j=1}^{J}\sum_{j'=0}^{j-1}\frac{(2j)!\,(2j')!}{j!j'!}\left(\frac{\tanh r}{2}\right)^{j+j'}\sum_{k=0}^{n}\frac{(-s)^{k}(s\,\alpha^{\star})^{2j'-k}(s\,\alpha)^{2j-k}}{k!\,(2j-k)!(2j'-k)!}\right]\right) \\ s \equiv & \frac{1+\Delta\varepsilon}{\Delta\varepsilon}, \quad \mathcal{N}(\Delta\varepsilon)=\sum_{j=0}^{J}\frac{(2j)!}{j!j!}(1+\Delta\varepsilon)^{n}\left(\frac{1}{2}\tanh r\right)^{2j}. \end{aligned} \tag{A.12}$$

This converges for all of r for any finite J. In practice, one wants to minimize J to the lowest value that describes the squeezed vacuum $|0, r\rangle$ accurately in sum (A.11). Once this condition is satisfied, one can directly apply Eq. (A.12) to generate the Glauber–Sudarshan function of a general squeezed state, $|\alpha_0, r\rangle$. One simply displaces $P_{|0,r\rangle,\Delta\varepsilon}(\alpha)$ by α_0, i.e., $P_{|\alpha_0,r\rangle,\Delta\varepsilon}(\alpha) = P_{|0,r\rangle,\Delta\varepsilon}(\alpha - \alpha_0)$.

References

Abbott, B.P., et al., 2016. Observation of gravitational waves from a binary black hole merger. Phys. Rev. Lett. 116 (6), 061102. https://doi.org/10.1103/PhysRevLett.116.061102. LIGO Scientific Collaboration and Virgo Collaboration.

Afek, I., Ambar, O., Silberberg, Y., 2010. High-NOON states by mixing quantum and classical light. Science 328, 879–881.

Aharonov, Y., Bergmann, P.G., Lebowitz, J.L., 1964. Time symmetry in the quantum process of measurement. Phys. Rev. 134, B1410–B1416.

Almand-Hunter, A.E., Li, H., Cundiff, S.T., Mootz, M., Kira, M., Koch, S.W., 2014. Quantum droplets of electrons and holes. Nature 506 (7489), 471–475. https://doi.org/10.1038/nature12994.

Anderson, M.H., Ensher, J.R., Matthews, M.R., Weiman, C.E., Cornell, E.A., 1995. Observation of Bose-Einstein condensation in a dilute atomic vapor. Science 269, 198–201.

Arute, F., et al., 2019. Quantum supremacy using a programmable superconducting processor. Nature 574, 505–510. https://doi.org/10.1038/s41586-019-1666-5.

Bachor, H.-A., Ralph, T.C., 2004. A Guide to Experiments in Quantum Optics, second ed. Wiley-VCH.

Berger, C., Huttner, U., Mootz, M., Kira, M., Koch, S.W., Tempel, J.S., Aßmann, M., Bayer, M., Mintairov, A.M., Merz, J.L., 2014. Quantum-memory effects in the emission of quantum-dot microcavities. Phys. Rev. Lett. 113 (9), 093902. https://doi.org/10.1103/PhysRevLett.113.093902.

Billingsley, P., 1995. Probability and Measure, third ed. John Wiley & Sons, Cambridge, UK.

Bimbard, E., Jain, N., MacRae, A., Lvovsky, A.I., 2010. Quantum-optical state engineering up to the two-photon level. Nat. Photon. 4, 243–247.

Bradley, C.C., Sackett, C.A., Tollett, J.J., Hulet, R.G., 1995. Evidence of Bose-Einstein condensation in an atomic gas with attractive interactions. Phys. Rev. Lett. 75 (9), 1687–1690. https://doi.org/10.1103/PhysRevLett.75.1687.

Carmele, A., Knorr, A., Richter, M., 2009. Photon statistics as a probe for exciton correlations in coupled nanostructures. Phys. Rev. B 79 (3), 035316. https://doi.org/10.1103/PhysRevB.79.035316.

Carreño, J.C.L., Laussy, F.P., 2016. Excitation with quantum light. I. Exciting a harmonic oscillator. Phys. Rev. A 94 (6), 063825. https://doi.org/10.1103/PhysRevA.94.063825.

Chrzanowski, H.M., Bernu, J., Sparkes, B.M., Hage, B., Lund, A.P., Ralph, T.C., Lam, P.K., Symul, T., 2011. Photon-number discrimination without a photon counter and its application to reconstructing non-gaussian states. Phys. Rev. A 84 (5), 050302. https://doi.org/10.1103/PhysRevA.84.050302.

Cocker, T.L., Peller, D., Yu, P., Repp, J., Huber, R., 2016. Tracking the ultrafast motion of a single molecule by femtosecond orbital imaging. Nature 539, 263–267. https://doi.org/10.1038/nature19816.

Cohen-Tannoudji, C., Dupont-Roc, J., Grynberg, G., 1989. Photons & Atoms, third ed. Wiley, New York.

Cuevas, Á., López Carreño, J.C., Silva, B., De Giorgi, M., Suárez-Forero, D.G., Sánchez Muñoz, C., Fieramosca, A., Cardano, F., Marrucci, L., Tasco, V., Biasiol, G., del Valle, E., Dominici, L., Ballarini, D., Gigli, G., Mataloni, P., Laussy, F.P., Sciarrino, F., Sanvitto, D., 2018. First observation of the quantized exciton-polariton field and effect of interactions on a single polariton. Sci. Adv. 4 (4). https://doi.org/10.1126/sciadv.aao6814.

Cundiff, S.T., Mukamel, S., 2013. Optical multidimensional coherent spectroscopy. Phys. Today, 7, 44 (2013) 66 (7), 44–49. https://doi.org/10.1063/PT.3.2047.

Cundiff, S.T., Ye, J., 2003. Colloquium: femtosecond optical frequency combs. Rev. Mod. Phys. 75 (1), 325–342. https://doi.org/10.1103/RevModPhys.75.325.

Davis, K.B., Mewes, M.O., Andrews, M.R., van Druten, N.J., Durfee, D.S., Kurn, D.M., Ketterle, W., 1995. Bose-Einstein condensation in a gas of sodium atoms. Phys. Rev. Lett. 75, 3969–3973.

Dorfman, K.E., Schlawin, F., Mukamel, S., 2016. Nonlinear optical signals and spectroscopy with quantum light. Rev. Mod. Phys. 88 (4), 045008. https://doi.org/10.1103/RevModPhys.88.045008.

Fennel, T., Meiwes-Broer, K.H., Tiggesbäumker, J., Reinhard, P.G., Dinh, P.M., Suraud, E., 2010. Laser-driven nonlinear cluster dynamics. Rev. Mod. Phys. 82 (2), 1793–1842. https://doi.org/10.1103/RevModPhys.82.1793.

Gerry, C.C., Knight, P.L., 2005. Introductory Quantum Optics, first ed. Cambridge University Press, Cambridge, UK.

Glauber, R.J., 1963. Coherent and incoherent states of the radiation field. Phys. Rev. 131, 2766–2788.

Goulielmakis, E., Yakovlev, V.S., Cavalieri, A.L., Uiberacker, M., Pervak, V., Apolonski, A., Kienberger, R., Kleineberg, U., Krausz, F., 2007. Attosecond control and measurement: lightwave electronics. Science 317 (5839), 769–775. https://doi.org/10.1126/science.1142855.

Grynberg, G., Aspect, A., Fabre, C., 2010. Introduction to Quantum Optics: From the Semi-Classical Approach to Quantized Light, first ed. Cambridge University Press, Cambridge, UK.

Heidmann, A., Horowicz, R.J., Reynaud, S., Giacobino, E., Fabre, C., Camy, G., 1987. Observation of quantum noise reduction on twin laser beams. Phys. Rev. Lett 59, 2555–2558.

Heindel, T., Thoma, A., von Helversen, M., Schmidt, M., Schlehahn, A., Gschrey, M., Schnauber, P., Schulze, J.H., Strittmatter, A., Beyer, J., Rodt, S., Carmele, A., Knorr, A., Reitzenstein, S., 2017. Photon statistics as a probe for exciton correlations in coupled nanostructures. Nat. Commun. 8, 14870. https://doi.org/10.1038/ncomms14870.

Hernández-García, C., Pérez-Hernández, J.A., Popmintchev, T., Murnane, M.M., Kapteyn, H.C., Jaron-Becker, A., Becker, A., Plaja, L., 2013. Zeptosecond high harmonic kev x-ray waveforms driven by midinfrared laser pulses. Phys. Rev. Lett. 111 (3), 033002. https://doi.org/10.1103/PhysRevLett.111.033002.

Jones, D.J., Diddams, S.A., Ranka, J.K., Stentz, A., Windeler, R.S., Hall, J.L., Cundiff, S.T., 2000. Carrier-envelope phase control of femtosecond mode-locked lasers and direct optical frequency synthesis. Science 288 (5466), 635–639. https://doi.org/10.1126/science.288.5466.635.

Kira, M., 2015. Hyperbolic bloch equations: atom-cluster kinetics of an interacting bose gas. Ann. Phys. 356, 185–243. https://doi.org/10.1016/j.aop.2015.02.030.

Kira, M., Koch, S.W., 2006a. Many-body correlations and excitonic effects in semiconductor spectroscopy. Prog. Quantum Electron. 30, 155–296. https://doi.org/10.1016/j.pquantelec.2006.12.002.

Kira, M., Koch, S.W., 2006b. Quantum-optical spectroscopy of semiconductors. Phys. Rev. A 73 (1), 013813. https://doi.org/10.1103/PhysRevA.73.013813.

Kira, M., Koch, S.W., 2008. Cluster-expansion representation in quantum optics. Phys. Rev. A 78 (2), 022102. https://doi.org/10.1103/PhysRevA.78.022102.

Kira, M., Koch, S.W., 2012. Semiconductor Quantum Optics, first ed. Cambridge University Press, Cambridge, UK.

Kira, M., Koch, S.W., Smith, R.P., Hunter, A.E., Cundiff, S.T., 2011. Quantum spectroscopy with Schrödinger-cat states. Nat. Phys. 7, 799–804.

Kirilyuk, A., Kimel, A.V., Rasing, T., 2010. Ultrafast optical manipulation of magnetic order. Rev. Mod. Phys. 82 (3), 2731–2784. https://doi.org/10.1103/RevModPhys.82.2731.

Köhler, J., Wollenhaupt, M., Bayer, T., Sarpe, C., Baumert, T., 2011. Zeptosecond precision pulse shaping. Opt. Express 19 (12), 11638–11653. https://doi.org/10.1364/OE.19.011638.

Král, P., Thanopulos, I., Shapiro, M., 2007. Colloquium: coherently controlled adiabatic passage. Rev. Mod. Phys. 79 (1), 53–77. https://doi.org/10.1103/RevModPhys.79.53.

Krausz, F., Ivanov, M., 2009. Attosecond physics. Rev. Mod. Phys. 81 (1), 163–234. https://doi.org/10.1103/RevModPhys.81.163.

Langer, F., Schmid, C.P., Schlauderer, S., Gmitra, M., Fabian, J., Nagler, P., Schüller, C., Korn, T., Hawkins, P.G., Steiner, J.T., Huttner, U., Koch, S.W., Kira, M., Huber, R., 2018. Lightwave valleytronics in a monolayer of tungsten diselenide. Nature 557, 76–80. https://doi.org/10.1038/s41586-018-0013-6.

Lee, N., Benichi, H., Takeno, Y., Takeda, S., Webb, J., Huntington, E., Furusawa, A., 2011. Teleportation of nonclassical wave packets of light. Science 332, 330–333.

Leonhardt, U., 1997. Measuring the Quantum State of Light, first. Cambridge University Press, Cambridge, UK.

Leonhardt, U., Paul, H., 1993. Realistic optical homodyne measurements and quasiprobability distributions. Phys. Rev. A 48, 4598–4604.

Leppäkangas, J., Johansson, G., Marthaler, M., Fogelström, M., 2013. Nonclassical photon pair production in a voltage-biased Josephson junction. Phys. Rev. Lett. 110 (26), 267004. https://doi.org/10.1103/PhysRevLett.110.267004.

Lobino, M., Korystov, D., Kupchak, C., Figueroa, E., Sanders, B.C., Lvovsky, A.I., 2008. Complete characterization of quantum-optical processes. Science 322, 563–566.

López Carreño, J.C., Sánchez Muñoz, C., Sanvitto, D., del Valle, E., Laussy, F.P., 2015. Exciting polaritons with quantum light. Phys. Rev. Lett. 115 (19), 196402. https://doi.org/10.1103/PhysRevLett.115.196402.

Loudon, R., 2000. The Theory of Quantum Light, third ed. Oxford University Press, Oxford, UK.

Lvovsky, A., Hansen, H., Aichele, T., Benson, O., Mlynek, J., Schiller, S., 2001. Quantum state reconstruction of the single-photon fock state. Phys. Rev. Lett. 87, 050402.

Machida, S., Yamamoto, Y., Itaya, Y., 1987. Observation of amplitude squeezing in a constant-current-driven semiconductor laser. Phys. Rev. Lett. 58, 1000–1003.

McAlister, D.F., Raymer, M.G., 1997. Ultrafast photon-number correlations from dual-pulse, phase-averaged homodyne detection. Phys. Rev. A 55 (3), R1609–R1612. https://doi.org/10.1103/PhysRevA.55.R1609.

Meystre, P., Sargent III., M., 2007. Elements of Quantum Optics, fourth ed. Springer, Berlin.

Mootz, M., Kira, M., Koch, S.W., 2013. Pair-excitation energetics of highly correlated many-body states. New J. Phys. 15 (9), 093040. https://doi.org/10.1088/1367-2630/15/9/093040.

Mootz, M., Kira, M., Koch, S.W., Almand-Hunter, A.E., Cundiff, S.T., 2014. Characterizing biexciton coherences with quantum spectroscopy. Phys. Rev. B 89 (15), 155301. https://doi.org/10.1103/PhysRevB.89.155301.

Mourou, G., Tajima, T., 2011. More intense, shorter pulses. Science 331 (6013), 41–42. https://doi.org/10.1126/science.1200292.

Neergaard-Nielsen, J.S., Nielsen, B.M., Hettich, C., Mølmer, K., Polzik, E.S., 2006. Generation of a superposition of odd photon number states for quantum information networks. Phys. Rev. Lett. 97, 083604.

Ourjoumtsev, A., Jeong, H., Tualle-Brouri, R., Grangier, P., 2007. Generation of optical 'Schrödinger cats' from photon number states. Nature 448, 784–786.

Parigi, V., Zavatta, A., Kim, M., Bellini, M., 2007. Probing quantum commutation rules by addition and subtraction of single photons to/from a light field. Science 317, 1890–1893.

Reinhard, A., Volz, T., Winger, M., Badolato, A., Hennessy, K.J., Hu, E.L., Imamoğlu, A., 2012. Strongly correlated photons on a chip. Nat. Photon. 6, 93–96. https://doi.org/10.1038/nphoton.2011.321.

Riek, C., Sulzer, P., Seeger, M., Moskalenko, A.S., Burkard, G., Seletskiy, D.V., Leitenstorfer, A., 2017. Subcycle quantum electrodynamics. Nature 541, 376–379. https://doi.org/10.1038/nature21024.

Ropers, C., Neacsu, C.C., Elsaesser, T., Albrecht, M., Raschke, M.B., Lienau, C., 2007. Grating-coupling of surface plasmons onto metallic tips: a nanoconfined light source. Nano Lett. 7 (9), 2784–2788. https://doi.org/10.1021/nl071340m.

Schneebeli, L., Kira, M., Koch, S.W., 2008. Characterization of strong light-matter coupling in semiconductor quantum-dot microcavities via photon-statistics spectroscopy. Phys. Rev. Lett. 101 (9), 097401. https://doi.org/10.1103/PhysRevLett.101.097401.

Schulze, F., Lingnau, B., Hein, S.M., Carmele, A., Schöll, E., Lüdge, K., Knorr, A., 2014. Feedback-induced steady-state light bunching above the lasing threshold. Phys. Rev. A 89 (4), 041801. https://doi.org/10.1103/PhysRevA.89.041801.

Scully, M.O., Zubairy, M.S., 2002. Quantum Optics, third ed. Cambridge University Press, Cambridge, UK.

Smallwood, C.L., Cundiff, S.T., 2018. Multidimensional coherent spectroscopy of semiconductors. Laser Photon. Rev. 12 (12), 1800171. https://doi.org/10.1002/lpor.201800171.

Smith, R.P., Martin, E.W., Kira, M., Cundiff, S.T., 2020. Toward direct optical excitation of excitonic many-body effects using intense thermal states. OSA Continuum 3 (5), 1283–1301. https://doi.org/10.1364/OSAC.392972.

Smithey, D.T., Beck, M., Raymer, M.G., Faridani, A., 1993. Measurement of the Wigner distribution and the density matrix of a light mode using optical homodyne tomography: application to squeezed states and the vacuum. Phys. Rev. Lett. 70, 1244–1247.

Sudarshan, E.C.G., 1963. Equivalence of semiclassical and quantum mechanical descriptions of statistical light beams. Phys. Rev. Lett. 10, 277–279.

Ulbricht, R., Hendry, E., Shan, J., Heinz, T.F., Bonn, M., 2011. Carrier dynamics in semiconductors studied with time-resolved terahertz spectroscopy. Rev. Mod. Phys. 83 (2), 543–586. https://doi.org/10.1103/RevModPhys.83.543.

Vogel, W., Grabow, J., 1992. Statistics of difference events in homodyne detection. Phys. Rev. A 47, 4227–4235.

Vogel, W., Welsch, D.G., 2006. Quantum Optics, third ed. Wiley, Darmstadt.

Walls, D.F., Milburn, G.J., 1994. Quantum Optics, first ed. Springer-Verlag, New York.

Weiner, A.M., 2000. Femtosecond pulse shaping using spatial light modulators. Rev. Sci. Instr. 71, 1929–1960.

Wenger, J., Tualle-Brouri, R., Grangier, P., 2004. Non-gaussian statistics from individual pulses of squeezed light. Phys. Rev. Lett. 92, 153601.

Windpassinger, P.J., Kubasik, M., Koschorreck, M., Boisen, A., Kjærgaard, N., Polzik, E.S., Müller, J.H., 2009. Ultra low-noise differential AC-coupled photodetector for sensitive pulse detection applications. Meas. Sci. Technol. 20 (5), 055301. https://doi.org/10.1088/0957-0233/20/5/055301.

Winterfeldt, C., Spielmann, C., Gerber, G., 2008. Colloquium: optimal control of high-harmonic generation. Rev. Mod. Phys. 80 (1), 117–140. https://doi.org/10.1103/RevModPhys.80.117.

Yin, J., et al., 2020. Entanglement-based secure quantum cryptography over 1,120 kilometres. Nature 582, 501–505. https://doi.org/10.1038/s41586-020-2401-y.

Zavatta, A., Viciani, S., Bellini, M., 2004. Quantum-to-classical transition with single-photon-added coherent states of light. Science 306, 660–663.

Zhao, D., Gu, Y., Chen, H., Ren, J., Zhang, T., Gong, Q., 2015. Quantum statistics control with a plasmonic nanocavity: multimode-enhanced interferences. Phys. Rev. A 92 (3), 033836. https://doi.org/10.1103/PhysRevA.92.033836.

Index

Note: Page numbers followed by "*f*" indicate figures and "*t*" indicate tables.

N

O

P

R

S

Kolibri 360 GmbH
Ferdinand-Jühlke-Straße 7
99095 Erfurt, Deutschland
produktsicherheit@kolibri360.de